Transnational Dynamics in Southeast Asia

The **Institute of Southeast Asian Studies (ISEAS)** was established as an autonomous organization in 1968. It is a regional centre dedicated to the study of socio-political, security and economic trends and developments in Southeast Asia and its wider geostrategic and economic environment. The Institute's research programmes are the Regional Economic Studies (RES, including ASEAN and APEC), Regional Strategic and Political Studies (RSPS), and Regional Social and Cultural Studies (RSCS).

ISEAS Publishing, an established academic press, has issued more than 2,000 books and journals. It is the largest scholarly publisher of research about Southeast Asia from within the region. ISEAS Publishing works with many other academic and trade publishers and distributors to disseminate important research and analyses from and about Southeast Asia to the rest of the world.

Transnational Dynamics in Southeast Asia

The Greater Mekong Subregion and Malacca Straits Economic Corridors

EDITED BY

Nathalie Fau • Sirivanh Khonthapane • Christian Taillard

INSTITUTE OF SOUTHEAST ASIAN STUDIES
Singapore

First published in Singapore in 2014 by
ISEAS Publishing
Institute of Southeast Asian Studies
30 Heng Mui Keng Terrace, Pasir Panjang
Singapore 119614

E-mail: publish@iseas.edu.sg • Website: bookshop.iseas.edu.sg

ISEAS Library Cataloguing-in-Publication Data

Transnational dynamics in Southeast Asia : the Greater Mekong subregion and Malacca Straits economic corridors / edited by Nathalie Fau, Sirivanh Khonthapane and Christian Taillard.

1. Southeast Asia— Economic integration.
2. Greater Mekong River Subregion—Economic integration.
3. Malacca, Strait of—Economic integration.
4. Southeast Asia—Economic conditions.
5. Greater Mekong River Subregion—Economic conditions.
6. Growth triangles—Southeast Asia.

I. Fau, Nathalie.
II. Sirivanh Khonthapane.
III. Taillard, Christian, 1942-

HC441 T775 2014

ISBN 978-981-4517-89-8 (soft cover)
ISBN 978-981-4519-70-4 (E-book PDF)

Cover Photo: Reproduced with kind permission of Marie Mellac.

Credit of maps: All the maps in this book have been created by the authors themselves. “Source” of maps refers only to the origin of the data.

Typeset by International Typesetters Pte Ltd
Printed in Singapore by Markono Print Media Pte Ltd

CONTENTS

PART II: NATIONAL POLICIES RELATED TO REGIONAL INTEGRATION

PART III: NEW NODES OF ECONOMIC CORRIDORS: URBAN PAIRS AND TWIN BORDER CITIES

LIST OF FIGURES

LIST OF MAPS

LIST OF TABLES

PREFACE

TRANSITER SOUTHEAST ASIA, A FRANCO-ASIAN SCIENTIFIC COOPERATION PROGRAMME

The main goal of the Transiter Programme "Transnational Dynamics and Territorial Redefinitions in Southeast Asia: The Greater Mekong Subregion and Malacca Straits economic corridors", is to better understand the role played by the economic corridors in the ongoing transition between the development of cross-border trade, towards a broader transnational integration process. This objective will be reached basing the research work on concrete examples, studied according to a multi-disciplinary approach of monitoring evaluation during a three-year period, in order to value and estimate the potentialities of these new dynamics, as well as anticipate on their potential negative impacts on the local populations and natural environments to be transected by the corridors. Although the primary focus of research is on Southeast Asia, the programme also analyses the conditions when possible to compare the transnational dynamics taking place along the Straits of Malacca's maritime corridors with the Greater Mekong Subregion (GMS) continental corridors. It also details and reviews the multi-disciplinary research available on this region and, in particular, on Lao PDR, which is a "key country" within the subregion and the main counterpart within this scientific cooperation programme.

VIEWS OF LAO COUNTERPART

Following the policy of the government in transforming Lao PDR from land-locked to land-linked status, there have been substantial investments in transport infrastructure. Such initiative is driven by development partners, mainly the Asian Development Bank (ADB), through the Greater Mekong

Subregion (GMS) framework and bilateral cooperation between Laos and its neighbours. This gives birth to the economic corridor development concept, focusing on developing the transport infrastructure to connect Lao PDR with her giant neighbours, including the construction of the East-West Economic Corridor (EWEC) and the North-South Economic Corridor (NSEC), and the promotion of economic opportunities along the corridors. Along with the improvement in the hardware, the government of Laos has gradually improved the trade and investment climate in the country with the objective of making it easier for the private sector to maximize the benefits from this increased connectivity and integration. Despite the massive investment in hardware, together with gradual reforms in trade and investment regulations, the expansion of trade and investment in the subregion and in Lao PDR in particular remains below expectations and the transport cost has not reflected this development.

In light of the importance of the ongoing integration of the Lao economy at the subregional, regional and global level, the National Economic Research Institute (NERI) within the Ministry of Planning and Investment and the Centre Asie du Sud-Est (CASE) — from the French Centre National de la Recherche Scientifique (CNRS) and the Ecole des Hautes Etudes en Sciences Sociales (EHESS) — have developed a Lao-French Scientific Cooperation Project to research on the impacts of the regional economic corridors in the GMS with specific focus on the transition from cross-border trade to transnational integration. The project aims to build upon strong complementarities between the two institutes, in disciplinary terms, thematic areas and geographies of interest, allowing a further impulse in multi-disciplinary approach to research on transnational integration. These exchanges have resulted in the outline of a bilateral cooperation project, based on the existing synergies between the NERI and the CASE, with the milestones of jointly organizing the mid-term workshop in July 2009 under the title "Researching the Impacts in Laos of the Economic Corridors in the Greater Mekong Subregion: From Cross-Border Trade to Transnational Integration" supported by French National Research Agency (ANR), French Agency for Development (AFD) and the Lao Government, and a final international research symposium in Vientiane in December 2010.

The project has had multiple dimensions of long-lasting benefits to Lao PDR as well as the counterpart agencies. At the researcher level, this has been achieved through the participation in the fieldwork with the French counterparts. By doing so, the skills and competence of our researchers have been strengthened. At the institutional level, NERI had the opportunity to cooperate and share experiences with the reputable French counterpart raising the status of external cooperation to a more advanced level. The successful organization of the final international symposium helped NERI to network with researchers, not only from GMS and Malacca Straits in Southeast Asia but also from Central America, creating the new frontier for NERI to further expand its cooperation.

Finally, at the country level, the findings from the researchers have created a wealth of knowledge on the roles of economic corridors in the ongoing transition between the developments of cross-border trade towards a broader transnational integration process of Lao PDR with understandings on the comparable cases of the Straits of Malacca. Young researchers from NERI have had the opportunity to learn from Malacca's maritime corridors. Several experiences can be drawn such as the development of dry ports in Malaysia (ie. Nazery Khalid's contribution) which could be initiated and developed in Lao PDR at the head of the international bridges crossing the Mekong, or the establishment of pair cities which command areas linking interior productive areas and external nodes, or the corridor leading importance to combine national integration and territorial planning (aménagement du territoire) which could also be promoted. The different types of classification of stages of corridor development, starting from transport corridors, multimodal, logistic corridors, economic corridors (ie. Ruth Banomyong's contribution) provides us with the basis for the assessment of the level of development of the corridors crossing Lao PDR. In addition, the results of the studies will be used by planners and researchers to review related policies on subregional, regional and global integration including to developing national integration strategy as well as to feed into the ongoing preparation of national development strategy up to the 2020s.

VIEWS OF FRENCH COUNTERPART

The Southeast Asia component of the ANR Transiter programme has been developed as scientific cooperation composed of three phases, each concluded with a scientific seminar. The first phase of the research programme led to the Franco-Lao workshop "Researching the Impacts in Laos of the Economic Corridors in the Greater Mekong Subregion: From Cross-Border Trade to Transnational Integration", held in Vientiane in July 2009. This seminar gave the opportunity to present and discuss the first results of Lao and French scholars' fieldwork research from NERI and CASE on the transition process from border trading practices to transnational integration. This first stage has allowed us to compare issues, data sources and fieldwork performed by researchers from both institutes, mainly economists from the NERI and pluridisciplinary researchers from the CASE.

The second phase of this research programme was concluded by a Southeast Asian symposium held in Vientiane in December 2010, "Transnational Dynamics and Territorial Redefinitions in Southeast Asia: The Greater Mekong Subregion and Malacca Straits economic corridors", prepared by the same institutions, NERI and CASE, and supported by the French National Research Agency (ANR), the Agency of the Research Institute for Development (AIRD) and the Lao Government. This conference followed a year after the Central American Regional symposium on transnational dynamics and territories, held in San José, Costa Rica in December 2009. These regional symposiums hence encompass the two cultural areas included in the Transiter Programme.

These regional seminars have been the subject of separate publishing projects centred on the comparison of the transnational integration processes at work in each of the two case studies chosen in Southeast Asia: the Greater Mekong Subregion and the Malacca Straits. These seminars have also made it possible to initiate a comparison between these two transnational integration dynamics and those of the Central American isthmus, the second study area of the Transiter programme. At this second level, the first basic element in the comparison of these two emerging integrations concerns their location at the crossroads between major north-south and east-west international and subregional trade routes. They also share spatial profiles with a North-South structure resulting from the respective positions of sea coasts, rivers and highlands. These two north-south profiles, combined with the above mentioned connections between world and subregional traffic, lead to a dynamic of spatial recompositions where emerging transnational North-South corridors play an essential role.

The third and last phase of this programme was concluded by an international conference, held in Paris in February 2012, "Transnational Dynamics and Territorial redefinitions: Cross Perspectives from Southeast Asia and Central America", has compared the transnational dynamics generated in the Central American isthmus and in Southeast Asia. This conference has, first of all, enabled a thorough comparison to be made of the transnational dynamics studied in both parts of the Transiter programme: Southeast Asia and the Central American isthmus. The comparative approach was then broadened to include the research results of other French programmes dealing with similar issues for other subregions on all continents. A final synthesis will deal with the impact of the different transnational integrations taken into account on ongoing regionalization in the globalization process.

The last two phases of the programme have closely associated French researchers and their partners from the subregions studied who have studied similar scientific issues. The field of scientific cooperation founded on different cultural approaches and interpretations, initiated in the first place with Lao researchers, has been extended to all the Asian partners. There were two objectives to this dialogue which took into account the different outlooks of the participants involved: firstly, to submit the results of the French research teams to the scrutiny of their Asian partners, and secondly, to ask them to present their own results in such a way as to lead to a constructive debate, the fruit of balanced scientific cooperation, which is taken into account in this work, in both the diversity of the contributions and the two comparative conclusions.

Sirivanh Khonthapane (NERI) and Christian Taillard (CASE)

ACKNOWLEDGEMENTS

Three mixed laboratories of the National Centre for Scientific Research (CNRS) — the SEDET (Paris 7 University), the Southeast Asia Centre (CASE, CNRS/EHESS) and the CEMCA (CNRS/MAE, Mexico) issued a call for projects concerning southern countries, initiated in France by the National Research Agency (ANR) in 2007. The project, entitled "Transnational Dynamics and Territorial Redefinitions: A Comparative Approach to Central America and Southeast Asia" (Transiter), was approved in 2008 and funded by the French National Research Agency (ANR) and the French Development Agency (AFD)

This project, steered by Laurent Faret, a professor at Paris 7 University and a researcher at the SEDET, has made it possible not only to finance many research areas for experienced researchers, but also for young doctoral students. Thanks to frequent meetings, organized and supported by both the SEDET and the CASE, it has been possible to set up comparative grids enabling us to study the regional integration processes at work along the continental and maritime corridors of Southeast Asia. This book is the result of a collective multi-disciplinary effort over a period of four years, punctuated by numerous seminars, roundtables and symposia. We would like to thank all those taking part in these sessions analysing different types of regional cooperation in Asia and the rest of the world, and the discussants in our seminars and symposia whose constructive criticisms and suggestions have enabled our research to progress.

Special thanks are due to our Asian partners, too numerous to mention by name, who greatly contributed to the progress of our research. We are also grateful to the National Economic Research Institute (NERI) of the Ministry for Planning and Investment of the Lao PDR, which co-steered this research programme and brought together all the contributors from Vientiane to discuss the results. We also benefited from a fruitful cooperation with researchers from the Maritime Institute of Malaysia (MIMA) in Kuala Lumpur and their knowledge of the Malacca Straits.

This book could not have existed without all the help we received to publish it. We thank Caroline Charras-Wheeler who quickly and efficiently revised the English in the book as a whole, and also translated several of the papers. Our sincere thanks also go to Muriel Charras who patiently and competently managed the Asian part of this programme for the CASE, and was in charge of formatting the manuscript. Thanks also to Vatthana Pholsena who provided a precious link between our programme and ISEAS. Finally, we thank Mrs Triena Ong, the Managing Editor of ISEAS, who took interest in the comparative approach we developed, and Sheryl Sin for her excellent editorial skills.

Nathalie Fau, Sirivanh Khonthapane and Christian Taillard

LIST OF CONTRIBUTORS

Abdul Rahim Anuar, Associate Professor, College of Law, Government & International Studies, Universiti Utara Malaysia.

Abdul Rahim was trained as an economist. He is currently an Associate Professor at the School of International Studies, Universiti Utara Malaysia. He specializes in trade and development of Asia Pacific countries. His publications include: Abdul Rahim Anuar and Mohammad Haji Alias, "Structural Transformation of the Malaysian Economy", in *Political Economy of Structural Transformation: Comparative Islamic Perspective*, edited by Choudhury M.A. and Mohammad Haji Alias (2003). His current research focus on border trade among Malaysia, Thailand and Indonesia.

Bounthavy Sisouphanthong, Vice-Minister, Ministry of Planning and Investment, Lao PDR.

Bounthavy Sisouphanthong is the Vice-Minister of Planning and Investment of the Lao PDR since 2007. He obtained his Ph.D. in Economics from the University of Paris Nord in 2008. He is the co-author (with Christian Taillard) of *Atlas of Laos: The Spatial Structures of the Economic and Social Development of the Lao People's Democratic Republic* (2000); and the third *Lao PDR National Human Development Report* (2006).

Bouté, Vanina, Associate Professor, University of Picardie/CASE, France.

Vanina Bouté is an Associate Professor at the University of Picardie (France) and a Research Associate at the Southeast Asia Centre (CASE, Paris). She is currently a research fellow at the Research Institute on Contemporary Southeast Asia (IRASEC) in Bangkok. She completed her Ph.D. in Anthropology at the *École Pratique de Hautes Études* (Paris, France) in 2005.

She is the author of *Mirroring the Power: The Phounoy of Northern Laos — Ethnogenesis and Dynamics of Integration* (2011, in French), and has co-edited with Vatthana Pholsena, *Laos: Power and Society* (2012, in French). Her current research focus on migration and dynamics of change among ethnic groups in the border regions of northern Laos.

Colin, Sébastien, Associate Professor, INALCO/ASIES, France.

Sébastien Colin is a geographer and an Associate Professor at the National Institute of Oriental Languages and Civilizations (INALCO, Paris). His current research interests include Chinese border policy, cross-border dynamics and territorial changes in the border regions of China, as well as Chinese maritime policy. He is the author of a book entitled *China and Its Borders* (2011) and several article in French on the abovementioned topics. His papers in English include: "A border opening onto numerous geopolitical issues: The Yanbian Korean Autonomous Prefecture", China Perspectives (Online) 48, July–August 2003; "Swinging Borders: The Sino-Korean Border During the Sunshine Policy", in *De-Bordering Korea: Tangible and Intangible Legacies of the Sunshine Policy*, edited by Valérie Gelézeau, Koen de Ceuster and Alain Delissen (2013).

Charras, Muriel, Senior Researcher, CNRS/CASE, France.

Muriel Charras was trained as a geographer. She is a CNRS senior researcher at the Southeast Asia Centre (CASE) in Paris. She specializes in Outer Indonesia and has carried out extensive fieldwork, especially in Sulawesi, Bali and Sumatra. Her recent publications include "The Reshaping of the Indonesian Archipelago after 50 years of Regional Imbalance", in *Regional Autonomy in a Multi-Cultural Indonesia*, edited by M. Erb, C. Faucher and P. Sulistiyanto (2005). Her current research focus on settlement process in pre-colonial Southeast Asia from the case study of Sriwijaya/Palembang hinterland.

Farhat, Amel, Ph.D. candidate, National University of Singapore.

Amel Farhat is a Ph.D. candidate at the Department of Geography, National University of Singapore. She obtained a Master of Arts from Paris VII — Denis Diderot University in 2011, majoring in the study of the Geography of Emerging and Developing Countries. Her interests include the analysis of transport system, notably their transnational dynamics, and the regional geography of Southeast Asia. Her current Ph.D. research investigates the multiscalar and systemic approach of transport between Malaysia and Singapore.

Fau, Nathalie, Associate Professor, University Paris VII/SEDET, France.

Nathalie Fau is Associate Professor of Geography at the University Paris Diderot-Paris 7, Senior Researcher at SEDET and at Research Institute on Contemporary Southeast Asia (IRASEC) and Research Associate at the Maritime Institute of Malaysia (MIMA) in Kuala Lumpur. She obtained her Ph.D. in Geography from University of Paris in 2003. Since the late 1990s, her work has focused on regional integration processes in Southeast Asia, with a focus on maritime regions and infrastructures. She specializes in the Straits of Malacca and has researched on policies of joint administration put into place by the bordering States (in areas such as the fight against pollution and piracy and measures to guarantee navigational safety); transverse flows which unite the two shores of the Straits (namely the reorganization of the port system, and the flow of migrants) as well as the cultural links (namely Malay identity). She has presented papers and spoken at various seminars and conferences and published articles and chapters in books on various topics.

Franck, Manuelle, Professor, INALCO/HSTM, France.

Manuelle Franck is Professor of Geography at the National Institute of Oriental Languages and Cultures (INALCO), Paris, France, where she teaches human geography of Southeast Asia. Her research works focus on urbanisation processes in Southeast Asian secondary cities, regional geography of Southeast Asian countries and on regional integration in the Malacca Straits. Her publications address issues related to territorial dynamics of Southeast Asian countries, especially Thailand and Indonesia, and to the metropolisation process of secondary cities in Indonesia.

Grillot, Caroline, Postdoctoral Fellow, Max Planck Institute, Halle, Germany.

Caroline Grillot is a social anthropologist (Paris X-Nanterre University) and a sinologist (INALCO) and currently a Postdoctoral Fellow at the Max Planck Institute, Halle, Germany. She spent more than ten years in China, notably in Shandong University (1994–95) and Sichuan University (1998–2000). She also assisted UNESCO (Beijing 2003) in the Human and

Social Sciences sector, translated several Chinese novels into French (Bleu de Chine) and participated in the musical alternative scene in China. Her researches have focused on social marginalities in China and Southeast Asia. She has recently achieved her Ph.D. in social anthropology under the joint supervision of Lisa Wynn (Macquarie University of Sydney, Australia, 2012) and Pál Nyíri (Vrije Universiteit Amsterdam, The Netherlands, 2013) on the topic of cross-border marriages in the Sino-Vietnamese borderlands. In 2010, she also published a book on this topic entitled *Volées, Envolées, Convolées... Vendues, en fuite ou re-socialisées: les fiancées vietnamiennes en Chine* (Connaissances et Savoirs). Her current research interest focuses on the issue of non-existence of immigrants in China, cross-border traders of the Sino-Vietnamese border, and the notion of chaos in China.

Lainé, Elsa-Xuan, Ph.D. candidate and lecturer, INALCO, France.

Elsa Lainé graduated from Development Studies at Institut d'Etudes Politiques de Paris and Urban Environmental Management at Asian Institute of Technology in Bangkok. She is currently a Ph.D. candidate in Geography at the Institut National des Langues et Civilisations Orientales (INALCO) in Paris. Her field of research deals with the transformations of border towns under regionalization in Lao PDR and Thailand, more specifically under the Greater Mekong Subregion framework. She participated in international workshops in Vientiane with contributions focusing on twin cities along the Mekong border.

Mellac, Marie, Associate Professor, University Bordeaux III/ADES, France.

Marie Mellac is an Associate Professor in Geography and a research associate at the ADES research institute (University of Bordeaux-National Center for Scientific Research (CNRS)), France. She specializes in research on development and management of natural resources with special focus on land and forestry. Her recent work focuses on legal geography and explores land and the elaboration of environmental legislation in Southeast Asia and in Madagascar. She has recently published a chapter entitled "Land Reform and Changing Identities in Two Tai-Speaking Districts in Northern Vietnam", in *Moving Mountains: Ethnicity and Livehoods in Highland China, Vietnam and Laos*, edited by J. Michaud and T. Forsyth (2011).

Muszafarshah Mohd Mustafa, Senior Lecturer at Universiti Utara Malaysia.

Muszafarshah Mohd Mustafa is a Senior Lecturer in Economics at the School of Economics, Finance & Banking, College of Business (COB), Universiti Utara Malaysia (UUM) in Malaysia. He worked previously in the private sector and was attached to the Northern Corridor Implementation Authority (NCIA) in Penang, Malaysia for a year in 2012 as an economic analyst. His publications and consultancy works include "Regional Disparities, Income Inequality & Poverty: A Cumulative Causation from Malaysia's Experience", in *Poverty and Global Recession in Southeast Asia*, edited by Aris Ananta and Richard Barichello (2012) and a *Feasibility Study on the Establishment of Malaysia-Thailand Special Border Economic Zone for the Economic Planning Unit* (2009). His areas of interest are public economics and development economics.

Nazery Khalid, Senior fellow, Maritime Institute of Malaysia (MIMA).

Nazery Khalid joined MIMA in April 2004 as a Research Fellow at the then Center for Economic Studies and Ocean Industries. His research focus is on Maritime Economics, specifically shipping, ports, logistics, maritime support services, seaborne trade issues and offshore oil and gas industry. Nazery Khalid has published extensively in refereed journals, magazines and newspapers and has contributed chapters in books on a broad range of topics on maritime economics. His research works on ship financing and multimodal transport development in Malaysia has been published by MIMA. He has also edited two conference proceedings on the Straits of Malacca. Nazery Khalid was a Visiting Scholar at the prestigious Logistics and Operations Management Section at Cardiff Business School, Cardiff University in October 2010 and at University Paris Diderot in October–November 2011. He also taught Maritime Economics to MBA in Shipping Management students at a Malaysian university.

Pholsena, Vatthana, Research Fellow, CNRS, France.

Vatthana Pholsena is a Research Fellow at the French National Centre for Scientific Research (CNRS). Currently based in Singapore, she is also the representative for the Institute of Research on Contemporary Southeast Asia (IRASEC). Her current research interests include state formation and social change in frontier regions (Laos-Vietnam), social history of revolution and the Vietnam Wars, and dynamics of ethnicity in Laos. She is the author of

Post-War Laos: The Politics of Culture, History and Identity (2006) and *Laos: Un pays en mutation* (2011).

Ruth Banomyong, Associate Professor, Thammasat University, Thailand.

Ruth Banomyong is an Associate Professor in the Department of International Business, Logistics and Transport Management at the Faculty of Commerce and Accountancy, Thammasat University in Thailand. He received his Ph.D. in 2001 in the field of International Logistics within the Logistics and Operations Management Section (LOMS) at Cardiff Business School (UK). His papers have appeared in refereed journals such as *International Journal of Physical Distribution and Logistics Management, International Journal of Logistics: Research and Applications, Asia Pacific Journal of Marketing and Logistics, Maritime Policy and Management.* He has co-authored four books in the areas of logistics and supply chain management and international trade.

Sirivanh Khonthapane, Assistant Minister of Planning and Investment, Lao PDR.

Sirivanh Khonthapane is Assistant Minister at the Ministry of Planning and Investment, Lao PDR. She received her Master degree in economics in the former Soviet Union in 1981 at University of Practice Lumumba and a Master degree in Public Management in 2007 at Lee Kuan Yew School of Public Policy, National University of Singapore. She was the Director General of National Economic Research Institute of Lao PDR for ten years. Her main publications focus on labour migration and gender, labour market in Lao PDR, economic intergation in the region, SMEs, and community production knowledge (One District One Product). She contributed to the First–Seventh Five-Year Socio-economic Development Plans of Lao PDR, and worked as National Advisor for National Indicative Plan Development for Lao National Mekong Secretariat in 2012–13.

Taillard, Christian, Senior Researcher, CNRS/CASE, France.

Christian Taillard is a geographer, specialized in mainland Southeast Asia, and CNRS Emeritus Senior Researcher at the Southeast Asia Centre (CASE) in Paris.

His publications in English on Laos cover a range of themes including village irrigation systems ("Village-based economic systems, the development of an irrigated area: Tha Ngon in Laos", in *Two Decades of L'espace Géographique: An Anthology* (1993), regional planning (*Atlas of Laos: Spatial Structures of the Economic and Social Development in the Lao PDR*, co-authored with Bounthavy Sisouphanthong (2000), and impacts of hydroelectric dams on villages downstream (research programme in progress).

A second set of publications deals with processes of urbanization and metropolization in Southeast Asia: *Territoires de l'urbain en Asie du Sud-Est: une métropolisation en mode mineur*, edited by M. Frank, Ch. Goldblum et Ch. Taillard (2012); *Vientiane, architectures d'une capitale: traces, formes, structures et projets*, edited by P. Clément, S. Clément-Charpentier, Ch. Goldblum, Bounleuam Sisoulath et Ch. Taillard (2011); and "Danang, a New Metropolis in Central Vietnam", in *Trends of Urbanization and Suburbanization in Southeast Asia*, edited by Tôn Nut Quynh Trân, Fanny Quertamp-Nguyên, and Claude de Miras (2012).

A third set relates to transnational integration between Northeast and Southeast Asia (*Intégrations régionale en Asie orientale*, edited by Ch. Taillard (2004)), which also involves land economic corridors in the Greater Mékong Subregion ("Un exemple réussi de régionalisation transnationale en Asie orientale: les corridors de la Région du Grand Mékong", in *L'Espace Géographique* (2009/1)).

Tan, Danielle, Lecturer, Sciences Po Lyon, France.

After completing two post-doctoral fellowships at ANU (Australian National University, Canberra) and at IIAS (International Institute for Asian Studies, Leiden, the Netherlands), Danielle Tan is currently teaching Political Science and Asian Studies at Sciences Po Lyon. She is also a Research Associate at the Research Institute on Contemporary Southeast Asia (IRASEC) in Bangkok and at the Institute of East Asian Studies in Lyons, France. Her main research interests include Chinese networks and migration in Southeast Asia, comparative politics and the political economy of the Greater Mekong Subregion, with a specific focus on ethnographic enquiries into the meaning and discourse of "the rise of China". She completed her Ph.D. in Political Science at Sciences Po/CERI (Paris, France) in December 2011. Her dissertation was entitled "From Communism to Neoliberalism: The Part Played by Chinese Networks in the Transformation of the State in Laos". She is currently co-editing a volume with Pál Nyíri (Vrije Universiteit Amsterdam, the Netherlands) gathering together the most recent research on the renewed Chinese presence in Southeast Asia.

Vignat, Ella, Ph.D. candidate, INALCO, France.

Ella Vignat is a Ph.D. candidate at the Institut National des Langues et Civilisations Orientales (INALCO) in Paris. She was a garment engineer and spent half her life in Asia with diverse working experiences in clothing industry, language teaching and humanitarian relief.

PART I

TRANSNATIONAL INTEGRATION PROCESSES IN SOUTHEAST ASIA

1

DEFINITIONS AND PROBLEMATICS OF TRANSNATIONAL DYNAMICS

Nathalie Fau

Three mixed laboratories of the National Centre for Scientific Research (CNRS) — the SEDET (CNRS/Paris 7 University), the South-East Asia Centre (CASE, CNRS/EHESS) and the CEMCA (Ministry of Foreign Affairs, Mexico) issued a call for projects concerning southern countries, opened in France by the National Research Agency (ANR) in 2007. The project, entitled "Transnational dynamics and territorial redefinitions, a comparative approach to Central America and Southeast Asia" (Transiter), was approved in 2008. This book sums up four years of research on the Southeast Asian theme, carried out thanks to scientific cooperation with the National Economic Research Institute (NERI) of the Ministry for Planning and Investment of the Lao Popular Democratic Republic. To study transnational dynamics in Southeast Asia, two areas were chosen: the Greater Mekong Subregion and the Malacca Straits.

1. TRANSITER: A COMPARISON OF TRANSNATIONAL INTEGRATION DYNAMICS IN MAINLAND AND INSULAR SOUTHEAST ASIA

The Greater Mekong Subregion, which came into being at the end of the 1980s, a decade characterized by the fashion for growth triangles or development squares strewn along the Asian Pacific seaboard, regroups the five countries of the Indochinese peninsula and two provinces of southern China (Yunnan was joined by Guangxi in 2004). The Asian Development Bank (ADB) took advantage of post-Cold War changes in national territorial strategies to promote this transnational integration programme from 1992 onwards, and has continued to support the most dynamic programme in East Asia. It proposed to base regional integration on the relaunch of trade flows that had been interrupted first by colonization, then by decades of war. The grouping of the five peninsular countries in ASEAN and the opening of borders with China enabled the 312 million inhabitants of the Greater Mekong Subregion to be brought together for the first time in this transnational programme. The restoration, over two decades, of the regional network via economic corridors, has now revived, in a new form, the old caravan route network which criss-crossed the peninsula in pre-colonial times. These corridors have since become so important that they have become the subject of rivalry not only between the peninsular powers, Thailand and Vietnam, but also between the regional powers of Japan and China (cf. Christian Taillard's chapter).

In contrast to the GMS, the Malacca Straits region is not really institutionally defined, but it was chosen because it is a coherent spatial unit, both by its dynamics and by the creation in the north and south of the straits of cross-border cooperation zones or "growth triangles". The Malacca Straits are situated between the east coast of Sumatra in Indonesia on the one hand, and the south of Thailand, the west coast of peninsular Malaysia and the city state of Singapore on the other. Their northern limit is an imaginary line stretching from Ujung Baka (in the far north of the island of Sumatra) to Laem Phra Chao (in the far south of the island of Phuket in Thailand), and their southern limit a line joining Than Datok (in the far southeast of peninsular Malaysia) and Tanjung Pergam (in the far north of the island of Bintan in Indonesia). In this book, the "Malacca Straits" area studied also includes the Singapore Straits which is involved in the same dynamics. The originality of the Malacca Straits is that it is both a trade zone and a major transit zone for international trade, of which the neighbouring countries have always been an integral part, and also a region in its own right, modelled, in spite of borders, by longstanding commercial and cultural relations between the two shores. Dynamics of competition and complementarity form the components of this region: Singapore sets the style for change by asserting its primacy in the services sector, Malaysia is trying to supplant the city state in the shipping and high-tech industry sectors, for example agro-industries processing products from the plantations that are increasingly developing in Sumatra. For both Indonesia and Thailand, the shores of the Malacca Straits appear more peripheral to their national centres of gravity (Java Sea and central plains), but these countries nevertheless take part in the transnational integration initiatives shaping each end of the straits (cf. Nathalie Fau's chapter).

Although of very different sizes and at different stages of the institutionalization process, these two transnational areas have numerous common features justifying their comparability. In the long term, they are both shaped by centuries-old maritime and land routes linking India and China, and more generally, South Asia and Northeast Asia. They are also the embodiment of the deliberate policies of States to be involved in the creation of economic spaces capable of taking part in the regionalization of the globalization process. They also provide a link between areas with considerable economic, political and social differences. Their dynamism still relies on the involvement of non-State actors, not only Asian development banks, but also private investors, whether multinational companies or networks of businessmen (Chinese, Minangkabau or family networks). Finally, their time-frame and the increasing complexity of their regional organizations are also very similar. The interest of this comparison also resides in the possibility of studying the similarities and specificities of transnational integration structures and processes in maritime and mainland contexts. Is the classic division between mainland and maritime Southeast Asia still significant for the analysis of transnational spaces?

The Transiter programme is therefore resolutely comparative. The objective is to study interactions between transnational dynamics and territorial redefintions, by making use of common theoretical tools and analysis grids. In order to facilitate exchanges between researchers in different scientific fields (geographers, historians, town planners and anthropologists), the vocabulary specific to regional integration and to cross-border areas was determined during research seminars. This chapter gives an account of these debates.

2. TWO CENTRAL CONCEPTS: REGION AND REGIONALIZATION

The Region: From "Infra-national" to "Supranational" Space

For French speakers, the term "region" designates an infra-national space, either an intermediate administrative unit between national and local levels (in France, between departments and the country as a whole), or a relevant geographical unit, i.e. coherent in its character or means of operation (natural or industrial region) (Taillard 2004). However, for English speakers, influenced by economists or political analysts, the term refers to a supranational scale, halfway between national and continental.

Although the use of the word "region" in the supranational sense is relatively recent for French speakers, the idea is in fact much older. Pierre Beckouche (2008) describes four stages in the forming of the supranational regions which were defined first of all in Europe before being applied to other continents. The beginnings of this concept of region date back to the nineteenth century and Monroe's doctrine of non-intervention by a third country in the affairs of the American continent, then to the creation of the International Union of American Republics. In Europe, the agreements concerning free navigation on the Rhine (1831) were the first stage of this concept. The second stage was linked to the Second World War: the forming of a multinational group was seen as a way of overcoming hostile rivalries between States and encouraging European agreement. Beyond Europe, Winston Churchill promoted regional agreements with the aim of restructuring world politics. Also, Article XXIV of the GATT authorized the setting up of regional free trade zones, even customs unions, as long as the common external tariff for third countries was not greater than that existing before their establishment. The third stage dates back to the 1960s and 1970s, with the trade agreements of the European Union or the Comecon, and the appearance of regional commissions in United Nations bodies. Finally, the fourth and most important stage dates from the 1990s, with the multiplication of Regional Trade Agreements.

Pierre Beckouche dates the rise in regionalization from the beginning of the 1990s: of the 186 active Regional Trade Agreements registered with the WTO, only 27 are prior to 1990. The United States, which for many years was hostile to regional agreements, preferring in the first place a global organization, was only converted fairly recently, when it signed agreements with Canada (1988) and Mexico (1993). By a domino effect, the setting up of regional agreements in Europe and then in North America led to the creation of new regional blocs: ASEAN, created in 1967, and developing towards the AFTA, but also Mercosur (Argentina, Brazil, Uruguay, Paraguay) and the Andes Pact (Ecuador, Colombia, Peru and Venezuela) at the beginning of the 1990s, ALENA and also the West African Economic and Monetary Union (UEMOA) in 1994. Many countries are thus engaged in the process of regional integration in forms ranging from sectorial cooperation to political union, with transfers of skills and sovereignty. These processes evolve in a context of globalization around the three main blocs of the Triad.

Free trade, which was supposed to promote global multilateralism, has mainly favoured economic regionalization: trade progresses more quickly within each region than with the rest of the world. Intra-regional trade has now reached 50 per cent of world goods imports compared with 40 per cent in 2000. In Europe, trade flows have developed more between European Community countries than between each of these countries and the rest of the world. However, the forming of these regions is envisaged by the WTO as the first stage of the general liberalization of trade and a way of including in globalization the economies of developing countries that would have been placed at a disadvantage by simple bilateral agreements between industrialized countries. This new regionalization is also seen as a tool for setting up the regulation of the effects of globalization at the level of new emerging regions, since it has proved easier to find agreements at regional than at international level, and these agreements are, above all, better suited to specific regional conditions and priorities. Finally, it reflects the emergence of a polycentric world, after the end of the bipolar world created by the Cold War, and is thus a way of counterbalancing the hegemony of the United States.

De jure versus *De facto* Regionalization

Regionalization therefore designates a process of economic or political organization at work in a supranational area, halfway between world and national level, and existing in varying forms. Two type of regional integration are usually distinguished: one, political and institutional (*de jure*), such as Europe, the other mostly economic (*de facto*), such as development corridors.

The very meaning of the term "regional integration" differs according to academic discipline. For political analysts the term is often understood as "a citizens' allegiance to a wider context than the original community", seen here in terms of Nation States (Badie and Smouts 1999, p. 66). Their research, like that of Sophie Boisseau de Rocher (1998), has shown that *de jure* regionalization, on account of the long gestation period of institutional structures, is not the driving element behind the regional integration process in Southeast Asia. For economists, regional integration is above all concerned with multinationals and company strategies; a region is integrated when actual and monetary relations are so intense that markets react simultaneously to the same crises. The spread of the 1997 Asian financial crisis from Thailand to Indonesia and then to Northeast Asia, has only served to reveal the dependence of the ten ASEAN member countries on their northern neighbours. Although economic integration is closely linked to the company strategy that governs international trade, regional integration also has important links with spatial organization, which brings us to the third, geographical meaning. Without good connections and efficient transport networks governed by infrastructures resulting from national investment strategies, the redeployment of firms would not have been possible at regional level. Integration therefore provides structure for a system when the subsets "making it up are conveniently linked to each other and to the centres of power, and when all parts of the territory are well supplied with services, goods, information and means of work, and their products are efficiently collected and redistributed" (Brunet 1997, p. 11).

In Southeast Asia, regional economic integration has preceded institutional integration. Philippe Hugon (1998) thus contrasts the Southeast Asian regional integration model with that of Africa:

> In most East Asian countries, we can note a convergence between rapid growth, inclusion in an international environment, a change in the sectorial structure of production and foreign trade, the central role of strong States and a relatively unplanned regionalization process. On the contrary, despite a high level of institutionalization, in Sub-Saharan Africa there is a concomitance between external marginalization, the maintaining of the sectorial structure of production and trade, the weakness of States despite a strong official line and the sparseness of regional relations, at least officially.

This author emphasizes the reverse order of regionalization in Africa and Asia.

3. A PROCESS: TRANSITION FROM CROSS-BORDER TO TRANSNATIONAL

It is in the context of globalization and the restructuring of the world into strongly connected regional subsets that cross-border and transnational constructions are emerging.

The evolution of the perception of the border gives a good idea of this change in the organization of world space since the 1970s. The term "border", initially taken from military vocabulary to designate a place where the army came face to face with the enemy or foreigners, has been defined since the Treaty of Westphalia in 1648 as an imaginary or symbolic line marking the territory over which a Nation State can legitimately exert its sovereignty. At this time, it was politically and legally discontinuous, making a break between the internal and external spheres. This concept, invented in Europe, was then widely disseminated, especially via colonization. With the globalization of the economy, the very idea of the border changes: it is no longer a break, but becomes a seam, and what counts is no longer its material existence but its obliteration. In the absence of conflict, borders become new places for maximizing exchanges, and hitherto marginal areas become central. This change of status of the border shows the emergence of new cross-border and transnational structures.

The term "cross-border" appeared fairly recently. It was used for the first time in 1969 at a European symposium on border regions, where the expression "cross-border hinterland" was used to designate an overlapping urban and socio-economic cross-border network (Hamez 2004). However, it was not until the 1980s that it came into general use, especially in the context of the building of Europe. A "cross-border space" is characterized by the specific linking of at least two contiguous border areas *across* an international limit. The prefix "across" is used to designate a new type of space that is not the result of the mere juxtaposition of two border areas, but of new territorial dynamics. The perimeter of these cross-border areas is a radius of a few kilometres, whose limits are quite vague. The term of "cross-border territory" is used when the area considered is the subject of a common institutionalized project. Cross-border dynamics thus depend on factors of proximity.

Although cross-border areas depend on factors of proximity, "transnational spaces" connect spaces of varying depth which are the whole or part of national territories, and lead to the emergence of new regional architecture. The term "transnational" nevertheless has two different meanings. The prefix "trans" can be used to designate either something that lies "beyond" (transalpine) or something that goes "across" (trans-Siberian). In the first sense, the term "transnational" is commonly used to define phenomena which are created in the global area beyond the national state context, and which are realized at least partially outside the control or the intermediate action of States (Badie and Smouts 1999, p. 66). Transnational companies ignore national boundaries, bypassing the state and adopting a global vision and field of action. The term "transnational" is different from the term "international", which implies an idea of cooperation between several nations, and that of "multinational" (concerning several nations). In this respect, the transnational fields include smuggling and illegal immigration, and activities like drug trafficking which escape state control: these flows cross borders without being noticed or accounted for. In another sense, the one used in this text, the term "transnational space" is synonymous with "subregional economic zone"; the latter is no longer centred on Nation States, but on the linking of spaces belonging to different countries, relying on relations of proximity but also on territorial differences leading to open cooperation, less institutionalized than supranational organizations such as common markets or regional unions (see Christian Taillard's chapter).

These cross-border integration processes are strongly dynamic today, on account of the hopes that are pinned on cooperation of this type, both by politicians and the private sector. They are increasing in numbers and have been taking off for the last decade in most existing cultural areas. Studies on Europe (Hamez 2004; Amilhat-Szary and Fourny 2006), especially on the European Union, have paid particular attention to these zones on account of the multiplication of European programmes (e.g. INTERREG) favouring the setting up of cross-border and transnational areas. But these new regional structures are also emerging in Latin America, Asia and Africa; the U.S.-Mexican border, Asian and Latin American development corridors and cross-border African national parks are all examples of these processes.

4. A DYNAMIC: FROM GROWTH TRIANGLES TO CROSS-BORDER ECONOMIC DEVELOPMENT ZONES

In Southeast Asia, these new regional structures take at least two forms, often with very similar mechanisms but different spatial logic: growth triangles, which are cross-border economic development zones, and development corridors.

Growth Triangles and Cross-border Economic Development Zones

During the 1980s and until the middle of the 1990s, many terms were used to designate cross-border economic cooperation zones in Asia: growth triangles or polygons, natural economic entity (Scalpino 1992), soft regionalism (Scalpino 1994), extended metropolitan region (McGee and Macleod 1993), cross-border industrial export zone. Taken separately, each of these names only designates one kind of the subregional economic zones growing up in Southeast Asia. Articles published on this subject usually concentrate on case studies, especially for the SIJORI (cross-border economic cooperation zone bringing together Singapore, Johor in Malaysia and Batam in Indonesia) or the Hong Kong-Guangdong zone; they then try to establish a model. The term "growth triangle" was the longest-lasting and best-known. It was used for the first time on 21 December 1989 by Goh Chok Tong, the then Deputy Prime Minister of Singapore, following a bilateral agreement signed with the Federated State of Johor (Joint Committee on Business Cooperation) on the one hand, and with the Indonesian government for the development of Batam island on the other. Theoretically, the aim of this cross-border cooperation was to promote the regional economic development of three territories (Singapore-Johor and Riau) with different advantages, in order to create a more extensive economic area with greater potential. The complementarity of the three actors was supposed to enable an overall synergy to be created, likely to valorize the region, make it attractive to investors, and increase the economic efficiency of the SIJORI region. This term was later used to designate the Hong Kong-Guangdong zone, since its development was so similar to that of the SIJORI; it was then generalized to designate cross-border and subregional spaces in East Asia.

From the middle of the 1990s onwards, the term "growth triangle" became generic, although it referred to at least two different types of economic cooperation zone in Southeast Asia (Chia 1993). The cross-border economic zones of Singapore-Johor in Malaysia-Riau in Indonesia (SIJORI) and Hong Kong-Guangdong (HKG) are one type. At the time they were the only zones actually created: they operated as cross-border conurbations whose metropolis (Singapore or Hong Kong) included a hinterland on the other side of the border (Batam in the Riau archipelago and Johor in one case and Guangdong province in the other). The operation of these areas, first created as an experiment, was essentially pragmatic and was not governed by any form of institutionalization.

Following the economic success of the SIJORI and Hong Kong-Guangdong "growth triangles", a second generation of cross-border cooperation sprang up on paper all over Asia. The IMT-GT (Indonesia-Malaysia-Thailand Growth Triangle) or Northern Malacca Straits Triangle associates four states in the north of the Malaysian peninsula centred on Penang, two provinces in northern Sumatra (Aceh and North Sumatra) centred on Medan and fourteen provinces in southern Thailand. The BIMP-EAGA or Great Eastern Zone includes Brunei Darussalam, the states of Kalimantan West and Sulawesi North in Indonesia, Sabah and Sarawak in Malaysia and Mindanao and Palawan in the Philippines. Geographical proximity does not exist within these regions which regroup much more widespread areas than those of the first generation. Although they are based on the economically successful models of SIJORI or HKG, these projects are nonetheless a criticism of them. Informal agreements like those that were used in SIJORI or HKG were seen as too uncertain to ensure the right distribution of economic repercussions. The resentment and unvoiced grudges that emerged were explained by the lack of institutionalization. Since the areas concerned by these cross-border cooperation projects were much larger, the fear of losing national sovereignty was even greater. Consequently, the precise definition of objectives, forms, fields and cooperation rules is aimed at limiting the risk of the emergence of centrifugal forces. The other major difference with the first generation is the absence of differing levels of economic development between the various components and growth centres at the origin of rising economic growth. The potential of areas situated on either side of the border is often the same and cooperation relies mainly on the joint development of resources and infrastructures in order to attract investors, without necessarily increasing exchanges inside each zone.

In East Asia, other subregional triangles or zones have been identified such as the Tumen zone associating territories in North Korea, China and Russia, or the Bohai-Yellow corridor zone involving South Korea, Japan and North China. This distribution of "growth triangles" and subregional zones can be explained by the role of the ADB which initially provided the frame for this development model before extending it to the whole of the Asia-Pacific region.

Corridors

At the end of the 1990s, growth triangles and zones were either replaced or made more complex by the addition of a new tool to the ADB's proposals — development corridors, also known as economic corridors (Rimmer 2004). There is an important body of literature concerning corridors (Debrie and Comtois 2010), written by geographers and economists but also by international or regional organizations (IMF, ADB, United Nations). To clarify this concept, we will deal successively with transport corridors, urban corridors and economic or development corridors.

Transport corridors, in geography, and especially in the geography of transport, are trade routes concentrating flows of passengers and freight. The transport corridor is a historical model of trade links between main cities favouring better connectivity between the different areas of a territory. These corridors are a real framework, enabling national poles and main urban nodes to be located, the hinterland disenclosed, and access to natural resources provided (Debrie and Comtois 2010). In nineteenth-century Europe, the building of a network between urban and industrial centres led to the emergence of linear routes that became more complex as technical innovations increased the variety of means of transport. A transport corridor is thus marked by its multimodality: a waterway may be doubled by a railway line and by a road link and connected to ports and airports. On this account, transport corridors are areas where riches accumulate (trade and economic activities) in the same space: since distribution costs are lower than elsewhere, economies of scale can be made, and economic and demographic concentration is also favoured.

These corridors are classified according to the main means of transport they develop: maritime, river, road, rail or air corridors. Corridors or sea routes are increasingly structural elements of world trade; with the development of containerization, world sea traffic is growing rapidly: it represented one billion tonnes in 1960, 3.7 billion in 1990, 5.8 billion in 2000 and more than 8 billion in 2009. It is estimated that more than three-quarters of world trade uses sea transport. These shipping routes are corridors a few kilometres wide whose general shape is designed to bypass the discontinuities imposed by continents and take account of unavoidable strategic passages such as canals, capes and straits. Rail corridors are declining on most continents where they have been replaced by road corridors, which are more flexible and make "door-to-door" service possible. However, the development of containerization has enabled the creation of railway "land bridges" providing a link between two sections of sea routes, resulting in a shorter and/or cheaper journey than a single sea trip. The success of freight transport in the United States is thus linked to the development of multimodal rail transport which enables distances to be considerably shortened by avoiding the Panama Canal or the Magellan Straits. A container from Singapore takes 36 days by sea but only 19 days using the Seattle-Chicago-New York land bridge, and transport services between the East Coast of the United States and the Asia-Pacific region take 6 to 14 days less according to destination. In the United States, these services were developed with the collaboration of the main international shipping lines, which were interested in the American land bridge for Far East-Europe relations: it has proved a more economical, faster and more reliable way of shipping containers than sending them entirely by sea. Multimodal transport equipment, whether for rail or road, is usually provided by the shipping companies which sometimes have their own terminals.

Even though one means of transport is often dominant, transport corridors are, above all, integrated logistics routes whose function is to ensure the fluidity and continuity of ever-increasing trade flows. The private sector, especially international haulage companies, directly favours this integration of routes:

> The development of logistics industry has enabled many freight forwarders to take control of larger segments of the supply chain. The level of functional integration of land distribution is increasing rapidly. Many distribution functions that used to be separated are now controlled by a single entity. In a conventional situation, the shift from one segment to the other was characterized by additional costs and delays, either administrative or physical. With an increasing level of functional integration, many intermediate steps in the transport chain have been removed. Mergers and acquisition have permitted the emergence of large logistics operators that control many segments of the supply chain (Notteboom and Rodrigue 2005).

Hubs, gateways and logistic nodes give structure to these corridors (Prentice 1996). Transhipment hubs are located at the junction of several sea routes, sometimes without an obvious link to the hinterland (Notteboom and Rodrigue 2005): rather than gateways to a hinterland, these hubs are just "pawns in a game of chess", i.e., they can be substituted in a global network (Debrie and Guerrero 2008). These transhipment hubs can be seen as logistic enclaves ensuring the smooth internal operation of maritime routes. The port is also often an essential component in the construction of land corridors (Notteboom and Rodrigue 2005). In addition to its function as a transit base and its industrial function with the creation of industrial-port complexes, it also has a major logistic function: the port is no longer just a point of contact between two spaces (sea and land) but between two transport systems; it has become an important stage in the logistic transport chain whose role is to provide a solution to the continuity of flows. Ports are "gateways" enabling the connection to the exterior of flows of goods using land routes structured by hubs which collect and distribute them along these corridors. Integrated logistic transport corridors have favoured the emergence of new nodes at intermodal interfaces, especially dry ports and inland container depots (Notteboom and Rodrigue 2005; ESCAP 2009).

Dry ports, strategically situated at the convergence of different means of transport, make it possible to lighten the congestion of sea ports and transmit their economic dynamism to the interior. Thus, in Germany, Duisberg, the biggest internal port and the biggest international logistics centre in Europe, integrates rail, road and river transport in one place, making it able to process and forward freight from over eighty European destinations.

Urban corridors are linear spaces structured by transports, which, by providing greater accessibility and connectivity, favour the development of a linear urban region. The first structured urban corridors appeared in North America: Jean Gottmann (1961) accurately describes the operation of the North American megalopolis breaking with the classic urbanization model of city centre and suburbs, and adopting a polycentric design. The urban region of the east coast of the United States stretches continuously for more than 1,000 km and includes seven metropolises of over a million inhabitants (Boston, Providence, New York, Baltimore, Washington and Richmond). The spatial integration in a megalopolis relies on a dense transport mesh with motorway and rail networks that make up the junction and the main route of the entire continent (Main Street) supporting intense relations and traffic flows: commuter shuttle services, economic exchanges or freight transport. According to Jean-Paul Rodrigue (2001), links between the cities in these corridors may be of three different types. In the first type, based on Christaller and Lösch's central place theory, they are hierarchical in nature, with the development of transport leading to ever-increasing competition and rivalry between centres that remain independent. In the second type, such as the American megalopolis, the creation of a transport corridor favours the specialization and interdependence of centres. In the third type, the most recent since it stems from the creation of integrated logistic routes, key cities on the corridors, multimodal logistics platforms, hubs and gateways all play the role of interfaces between regional, national and global systems.

Since Gottmann's studies were published, urban corridors have been identified at local, regional and even world level: Tokyo-Osaka-Kobe in Japan, the European spine from London to Milan in Europe or, in the Asia-Pacific region, the Tokyo-Jakarta corridor, sometimes known as a "mega-urban corridor" and described as "… an emerging international network of arterial air, surface and sea transportation corridors, telecommunication linkages and decision making pathways to integrate local and national development into an increasingly globalised system of production, commerce, finance and consumption" (Jones and Douglass 2008, p. 54). Rimmer (2004) also proposes a mega-urban corridor covering the whole of Southeast Asia and stretching from Bangkok in the north to the south of the Malaysian peninsula, including Kuala Lumpur and Singapore, and continuing along the east coast of Sumatra to Java and Bali. But, as Dean Forbes (1997) points out, "Yeung breaks this region into three, differentiating between an Indo-China economic zone, a Singapore centred region and another based on Jakarta" and "Douglass divides these regions into just two. These are a group comprising Bangkok, Kuala Lumpur and Singapore, and an overlapping Singapore, Jakarta, Surabaya cluster". On a larger scale, urban corridors are also often designated in Asia by the term "Extended Metropolitan Region" or "Desa-Kota", popularized by McGee (1993) and qualifying urban regions including several metropolises, and hinterlands developing both rural and non-rural activities, linked by transport routes.

Economic corridors were popularized by the United Nations and major international organizations. They took over the idea of the transport corridor and translated it into several different forms: development corridor, trade corridor, or growth corridor. In 2002, the United Nations launched the project entitled "Capacity-building in developing interregional land and land-cum-sea transport linkages" (ESCAP 2009), whose aim is to identify, in each world region, the interregional transport links which would contribute to better integration and favour economic development. According to United Nations planners (Demyck and Médina 2010), "transnational corridors are the new geographical space where urban development and competition strategies are deployed". It is not just a question of linking cities via more efficient communication routes, either by creating them or improving existing ones, but of developing a new type of multi-polar transnational space, connecting existing and emerging urban regions. In theory, this does not concern megalopolises, but should create new external conditions that many large and medium sized cities can take advantage of, especially in interior or border regions or in outlying pioneering areas. The function of these corridors is to favour the setting up of new productive activities thanks to improved accessibility, the development of energy infrastructures and the capacity for processing local products. Corridors should favour the development of outlying areas rather than the main structural nodes. This model of development via corridors has been implemented by the United Nations and Regional Banks on all continents.

These different forms of corridors as development strategies have several common features (Debrie and Comtois 2010). Corridors are seen first of all as *planning tools*, whose aim is to channel investment projects and strengthen economic growth processes. Their establishment allows for a reduction of the cost of imported goods, improved access to international markets, a better industrial network and improved interdependence of different economic sectors. In Latin America, East-West corridors (bi-oceanic corridors, IIRSA — Initiative for the Integration of Regional Infrastructure in South America) are aimed at linking the two ocean seaboards but also at linking areas of production, especially for agriculture, forests and energy, to those of consumption (Bender 2001), which are mainly located in coastal areas, and it is sometimes less expensive to import consumer goods by sea than to send them from inland regions. The aim of corridors is thus to link faraway agricultural regions to urban markets, whether internal or external, by decreasing transport costs and times. In this context, border zones and outlying areas should receive specific treatment.

Corridors are also designed as *regional integration tools*, no longer just national ones. They transcend physical, political, administrative, social and economic borders. In North America, Europe, Asia and West Africa, corridors are the medium used for economic integration projects. The objective of the nine, then ten pan-European corridors launched by the European Commission and the European Conference of Transport Ministers is to link the countries of Central and Eastern Europe to Western Europe; "First of all, a corridor is seen as a complement to the trans-European network assigned by the Treaty of Maastricht as an element of the cohesion of the European Union. It is considered to be a leading multimodal route connecting countries in the process of adhering to the European Union" (Debrie and Comtois 2010, p. 134). After the ALENA was established, eight so-called "trade" corridors were suggested to improve the capacity, continuity and fluidity of supply chains in North America. The Pacific Corridor is thus a multimodal transport route stretching from the Rockies to the Pacific coast, favouring trade between Western Canada, the American East Coast and Mexico. The objectives of these corridors are therefore political as much as economic: they are intended to improve cohesion and integration.

Corridors, therefore, are more than just transport routes. They are seen as *new tools of governance*. The public action of States has become more complex, with the necessity not only of increasing numbers of bilateral agreements, but also of taking into account the private sector which finances and exploits these corridors, and whose power has grown with decentralization policies. The management of transnational corridors necessitates a new framework of governance involving the establishment of collaboration with a multitude of actors. The ideal presented is that of decentralized international bilateral relations where the role of the State is of secondary importance, and replaced by initiatives of the "down-top" rather than the "top-down" type. In Latin America, it is explicitly stated that the creation of corridors should not be the subject of public centralized planning but should, on the contrary, be the result of the decentralized initiative of the private sector in partnership with national and international financial institutions such as the Inter-American Development Bank (IDB). The aim is to mobilize private capital to build these corridors.

5. PROBLEMATICS AND PLAN OF THE WORK

The increasing numbers of programmatic initiatives raise the question of the feasibility and reality of these transnational integration projects. Thus, in Latin America, the twelve integration routes have all been presented in planning programmes in strictly identical fashion, without any form of hierarchy, as if they all had the same potential. One of the directions of the Transiter programme's work has thus been to analyse the arguments of international and local organizations and national and local authorities to compare them with realities observed in the field, so as to identify the complementary features, but also the contradictions arising from these different arguments; the strategies of international bodies and financial partners are not always echoed by those found in national or local planning programmes.

Whereas the first part of this book presents the main outlines and comparisons of transnational dynamics in the Greater Mekong Subregion and Malacca Straits (see Map 1.1), the second part compares the arguments and objectives of international bodies with the national strategies of China (Collin), Vietnam (Mellac), Laos (Bounthavy), Myanmar (Vignat), Indonesia (Charras) and Malaysia (Khallid). The third part raises the question of the possible emergence of new urban nodes on corridors, and estimates their role in the revalorization of outlying border regions. The national policies of Southeast Asian countries have up till now mainly aimed at connecting outlying areas to the national capital via the construction of new infrastructures; however, the development of transnational regions changes the perspective of these regions by inciting them to play the role of a relay point in a mainly polycentric regional structure. The contributions in this part deal with the nodes emerging on the corridors, whether border cities (Colin), urban pairs (Franck and Farhat) or twin cities (Lainé, Grillot and Abdul Ramin Anuar, Muszafarshah Mohd Mustafa and Farhat). Another question, included in the final part, is the observed impact of these transnational projects on local socioeconomic development, especially in Laos (Pholsena, Bouté and Tan). Although corridors can integrate outlying areas, they can also merely cross them (tunnel effect)

Map 1.1
The Corridors Network of the GMS and the Malacca Straits Region

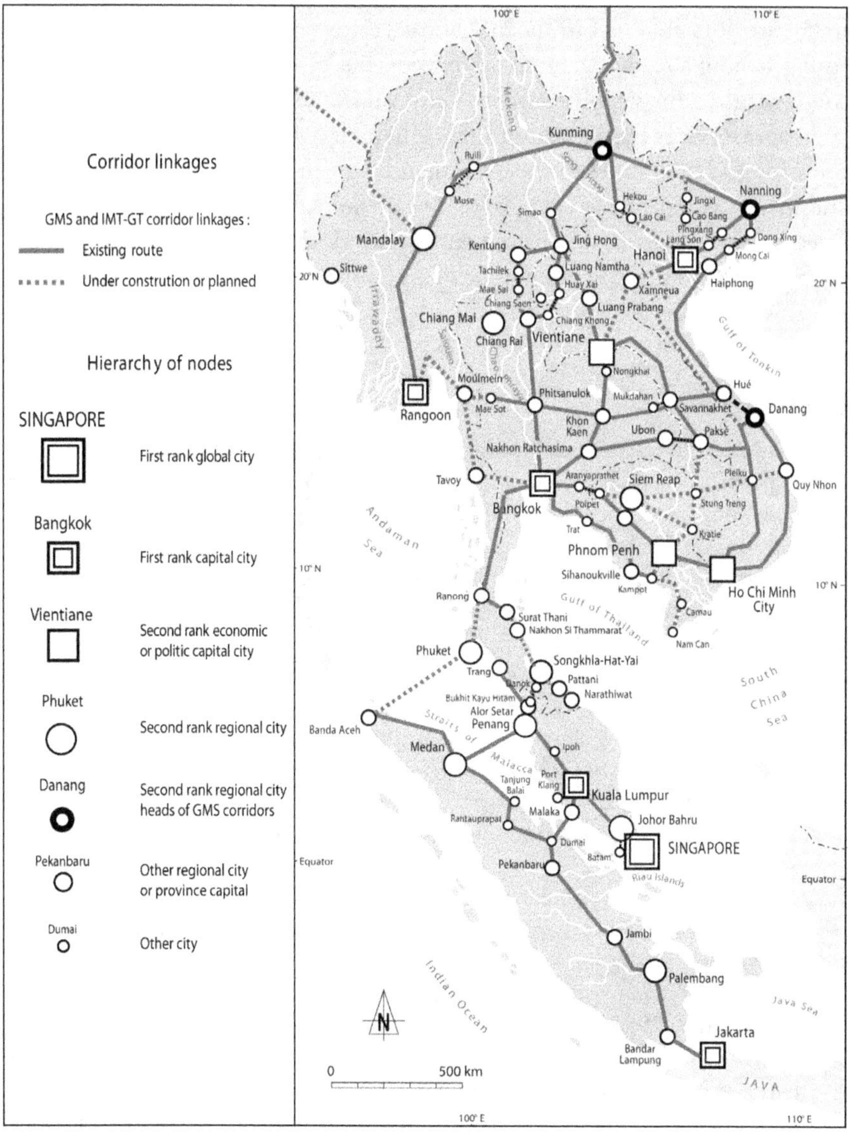

Map from M. Franck
Sources: C. Taillard (2009); N. Fau (2003); Asian Development Bank.

or drain their resources (pump effect). The Maputo transport corridor between South Africa and Mozambique has only created a corridor towards its capital, without any redistribution of growth to outlying areas. The first corridors in the European Union (Euro-corridors) mainly aimed at strengthening already dynamic urbanized areas of Europe in order to increase their weight and visibility on the international scene. Improvements in transport services tend to extend the sphere of influence of the regions with the strongest economies to the detriment of the most recently included zones and increase their role as existing economic poles. The Transiter programme's research aims to estimate, for the study period, the impact on local societies of regional planning, sectorial policies and political strategies that accompany improvements in transport services.

References

Amilhat-Szary Anne-Laure and Marie-Christine Fourny, eds. *Après les frontières, avec la frontière, nouvelles dynamiques transfrontalières en Europe* [After the border, with the border, new cross-border dynamics in Europe]. La Tour d'Aigues, Editions de l'Aube, 2006.

Badie, Bertrand and Marie-Claude Smouts. *Le retournement du monde, Sociologie de la scène internationale* [The upheaval of the world, Sociology of the international stage]. Presses de Science Po., 1999.

Beckouche, Pierre. *Les régions Nord-Sud* [The North-South regions]. Paris, Berlin, 2008.

Bender, S. "Trade Corridors: The Emerging Regional Development Planning Unit in Latin America". In *New Regional Development Paradigms*, vol. 2, *New Regions, Concepts, Issues and Practices*, edited by D. Edgington, A. Fernandez, and C. Hoshino. United Nations Centre for Regional Development/Greenwood Press, 2001.

Boisseau de Rocher, Sophie. *L'ASEAN et la construction regional en Asie du Sud-Est* [ASEAN and Regional Construction in Southeast Asia]. Harmattan, 1998.

Brunet, Roger. *Territoires de France et d'Europe* [Territories of France and Europe]. Paris, Berlin, 1997.

Chia, Siow Yue and Lee Tsao Yuan. "Subregional Economic Zones: A New Motive Force in Asia-Pacific Development". In *Pacific Dynamism and the International Economic System*, edited by Fred Bergsten and Marcus Naland. Washington, D.C.: Institute of International Economics, 1993.

Debrie, J. and Guerrero D. "(Re)spatialiser la question portuaire: pour une lecture géographique des arrière-pays européens" [(Re)spatializing the port question: a geographical explanation of European hinterlands]. *l'Espace géographique* 37, no. 1 (2008).

Debrie, Jean and Claude Comtois. "Une relecture du concept de corridors de transport: illustration comparée Europe/Amérique du Nord" [A review of the concept of transport corridors: an illustrated comparison between Europe and North America]. *Les Cahiers Scientifiques du Transport*, no. 58 (2010): 127–44.

Demyck, Noëlle and Lucine Médina. "Les corridors logistiques en Amérique centrale, nouvel avatar des modèles de développement?" [Logistics corridors in Central America: a new form of development model?]. Paper presented at the international conference "Transport et développement des territoires", Paris, 8–10 September 2010.

Economic and Social Commission for Asia and the Pacific (ESCAP). *Major Issues in Transport: Transport Infrastructure: Development and Operationalization of Dry Ports and Intermodal Transport Corridors*. ESCAP, 14–16 December 2009.

Forbes, Dean. "Metropolis and Megaurban Region in Pacific Asia". *Tijdschrift voor Economische en Sociale Geographic*, vol. 88, no. 5 (1997): 457–68.

Gottmann, Jean. *Megalopolis: The Urbanized Northeastern Seaboard of the United States*, 1961.

Hamez, G. *Du Transfrontalier au transnational: approche géographique, l'exemple de la frontière franco-belge* [From cross-border to transnational, geographical approach, the example of Franco-Belgian border]. Ph.D. dissertation, Department of Geography, Université Paris 1, 2004.

Hugon, Philippe, ed. "La régionalisation comparée en Afrique sub-saharienne et en Asie de l'est" [Regional integration: a comparison between South East Asia and Sub-Saharan Africa]. *Revue Tiers Monde* 39, no. 155 (1998).

Jones, Gavin and Mike Douglas, eds. *The Rise of Mega-Urban Regions in Pacific Asia: Urban Dynamics in a Global Era*. Singapore: National University of Singapore Press, 2008.

McGee, T.G. and Scott Macleod. "Emerging Extended Metropolitan Regions in the Asia-Pacific Urban System: A Case Study of the Singapore-Johor-Riau Growth Triangle". Paper presented at the forum on urban systems in Asia-Pacific, held at Hong Kong Chinese University, 11–13 February 1993.

Notteboom, Theo E. and Jean-Paul Rodrigue. "Port Regionalization: Towards a New Phase in Port Development". *Maritime Policy and Management*, vol. 32 no. 3 (2005): 305.

Notteboom, Theo E., Jean-Paul Rodrigue, and Economic and Social Commission for Asia and the Pacific (ESCAP). *Major Issues in Transport: Transport Infrastructure: Development and Operationalization of Dry Ports and Intermodal Transport Corridors*, 14–16 December 2009. <http://www.unescap.org/ttdw/FAMT/FAMT1/Documents/English/FAMT_SGO_2E.pdf>.

Prentice, B.E., ed. *Transport Gateway and Trade Corridors*. Saskatoon: University of Saskatchewan Printing Services, 1996.

Rimmer, Peter J. "Global Flows, Local Hubs, Platforms and Corridors: Regional and Economic Integration in Northeast Asia". *Journal of International Logistics and Trade*, vol. 1, no. 2 (2004): 599–624.

Rodrigue, Jean-Paul. "Freight, Gateways and Mega-urban Regions: The Logistical Integration of the Bostwash Corridor". *Tijdschrift voor Economische en Sociale Geographic*, vol. 95, no. 2 (2001): 147–61.

Scalpino, Robert A. "The United States and Asia: Future Prospects". *Foreign Affairs* (winter 1991–92): 19–40.

———. "Challenges to the Sovereignty of the Modern State". In *The Making of Security Community in the Asia-Pacific*. Kuala Lumpur: ISIS Malaysia, 1994.

Taillard, Ch., ed. "Introduction". In *Intégrations régionales en Asie orientale* [Regional integrations in East Asia]. Paris: Les Indes Savantes, 2004.

2

THE CONTINENTAL GRID OF ECONOMIC CORRIDORS IN THE GREATER MEKONG SUBREGION TOWARDS TRANSNATIONAL INTEGRATION

Christian Taillard

In the context of the regionalization process of globalization in Asia, the Asian Development Bank (ADB) has skilfully seized opportunities arising from a reversal in post-Cold War national territorial strategies. Between 1992 and 1994, it promoted the Greater Mekong Subregion (GMS) and has accompanied its development ever since, through the implementation of a regional integration initiative at the scale of continental Southeast Asia. It is the most dynamic ADB-assisted transnational integration programme in East Asia. Far from succumbing to the Asian crisis in 1997, regional integration has actually found its second wind.[1]

The term "transnational", as used to describe this form of integration, does not correspond to the first acceptation of the prefix *trans*: *beyond* (ie. trans-Alpine), nor to supranational configurations such as the Mekong Commission, for instance. The term is used in its second acceptation: *across* (ie. trans-Siberian), and refers to emerging recompositions within the Greater Mekong Subregion which integrate national constructions by blending two scales of the term "region": infra-national in French-speaking terminology and supranational for the Anglo-Saxons. These recompositions involve entire or partial networking of national territories entailing the emergence of new regional architectures, founded on converging interests between partner-nations. These configurations enable them to recover, on this new scale, a part of national power lost in the whirlwind of globalization.

This process was related to a redefinition of State and regional territorial strategies: regional integration through the expansion of trade in the peninsula severely disrupted, first by colonization, then by decades of war. Five of the six partner-nations of the Greater Mekong Subregion have undertaken a redefinition of their territorial policies. China, Lao PDR and Vietnam have operated a strategic reversal towards transnational integration, while Myanmar and Cambodia have reappeared on the regional stage. Only Thailand has maintained its position: peninsular centre at the main sea entrance portal of the Greater Mekong Subregion.

The end of the Cold War marks far-reaching regional and economic geopolitical transformations: Communist nations shifted from a centralized economy to a market economy. Thailand readapted its model of integration in the world economy in the aftermath of the 1997 Asian economic crisis. These transformations related to two new regional strategies: the regrouping of the five peninsular nations in the ASEAN and the opening of Chinese borders helped to unite, for the first time under the umbrella of this transnational programme, all the nations of the Greater Mekong Subregion.

1. THE MOST ADVANCED PROGRAMME FOR TRANSNATIONAL INTEGRATION IN EAST ASIA

Changes in the Scale of Integration: From River Basin to Peninsula

The Greater Mekong Subregion does not refer to a river basin project like the Mekong Commission but rather to a strategic vision for transnational integration on a continental level.

The basin represents barely 31 per cent of the total area of the Greater Mekong Subregion (2,570 million km^2), which is on an entirely different scale. The territorial importance of the diverse components is thus profoundly changed. Myanmar, marginal in the Mekong basin, ranks first (26 per cent of the area) and the Chinese share has increased, going from 17 per cent with Yunnan to 24.5 per cent with the addition of the Guangxi, enabling China to surpass Thailand (20 per cent) and Vietnam (13 per cent). Lao PDR and Cambodia declined in importance, 9 and 7 per cent respectively (see Table 2.1).

Table 2.1
Indicators showing the Relationships between the Nations of the GMS

Country or Chinese Province	Cambodia		Yunnan + Guangxi		Lao PDR		Myanmar		Thailand		Vietnam		Total GMS	
		%		%		%		%		%		%		%
Area (1,000 km²)	181.0	7.1	630.7	24.5	236.8	9.2	678.5	26.4	514.0	20.0	329.0	12.8	2,570.6	100
including Mekong basin	153.6	19	1,698	21	202.1	25	32.3	4	186.0	23	64.7	8	808.5	100
Incl. Yunnan			395.5	15.3										
Incl. Guangxi			235.2	9.2										
population (millions)	14.1	4.5	92.3	29.6	5.9	1.8	50.6	16.2	66.5	20.9	83.8	26.7	312.0	100
GDP (M$)	5,400	1.6	75,360	23.1	3,200	1.0	8,700	2.6	180,600	55.5	52,100	16.0	325,360	100
per capita income ($)	371		702		456		107		2,523		568		795	
growth rate		6.0		11.4		7.0		2.9		4.5		8.4		6.7
FDI (M$)	131		436		17		556		3,437		2,400			
tourists (millions)	1,421	6.9	2,231	10.9	1,095	5.4	0.660	3.2	11,500	56.0	3,468	17.0	20,376	100
export-import	6,406	2.1	7,978	2.6	1,268	0.4	5,428	1.8	215,169	71.0	66,742	22.0	302,990	100
intra GMS trade														
1992		20.1		1.2		63.3		23.4		2.8		4.7		5.7
1997		40.0		1.8		66.8		17.5		4.6		8.9		7.4
2002		24.5		1.9		67.8		35.6		8.9		13.9		12.6

Sources: Data from 2005, except Yunnan and Guangxi 2003–04, Exim Bank, Thailand.

The twin expansion, from the Mekong river basin to the Greater Mekong Subregion and from one to two Chinese provinces, does not only modify regional balance in demographic terms but also in economic terms. The Greater Mekong Subregion now contains 312 million people. The two Chinese provinces (30 per cent) lead the file, having surpassed other peninsular countries such as Vietnam (27 per cent), Thailand (20 per cent) and Myanmar (16 per cent). The importance of Cambodia and Lao PDR is far lower (4 per cent and 2 per cent respectively).

The regional GDP exceeds 325 billion dollars, thereby reflecting altogether quite different power balances between partner-countries. Thailand is the leader with 55 per cent of generated wealth, widening the gap with the two Chinese provinces and Vietnam (23 and 16 per cent) while Myanmar, Cambodia and Lao PDR have been sidelined (3 and 1 per cent).

Economic Corridors at the Centre of Territorial Strategy of the Greater Mekong Subregion

To develop the Greater Mekong Subregion, the ADB opted for a territorial strategy at the outset, aiming for peninsular linkage via five economic corridors, meridian and transversal (see Map 2.1), with priority in the first decade of the programme for reconstruction of road transport infrastructure and the re-establishment of trade affected by disruptive colonial compartmentalization. After the Asian crisis, the Bank made it imperative to link the reconstruction of transport infrastructure to the signing of specific free trade agreements, corridor by corridor, anticipating the global ASEAN Free Trade Agreement (AFTA) that will be applied to the new members in 2015.

Since 2002, integrated development of corridors has been the focus in the second decade of the Greater Mekong Subregion. The economic corridor is defined as a transnational integration strategy, concentrating activities on either side of the principal artery, in which transport infrastructure financing is closely bound to developmental opportunities in production, trade and multisector-based investments.

Transport investments have diversified in comparison to the previous decade: along with roads and bridges, local feeder roads and ports connect them to the main maritime routes. They also include railroads, airports and river navigation for industrial development and tourism. These investments also largely surpass the field of transport and extend to what may be termed interconnectivity, including interconnections of electrical and telecommunications networks, construction of gas and oil pipelines

Map 2.1
Two Generations of Economic Corridors in the GMS
(according GMS Transport Strategy)

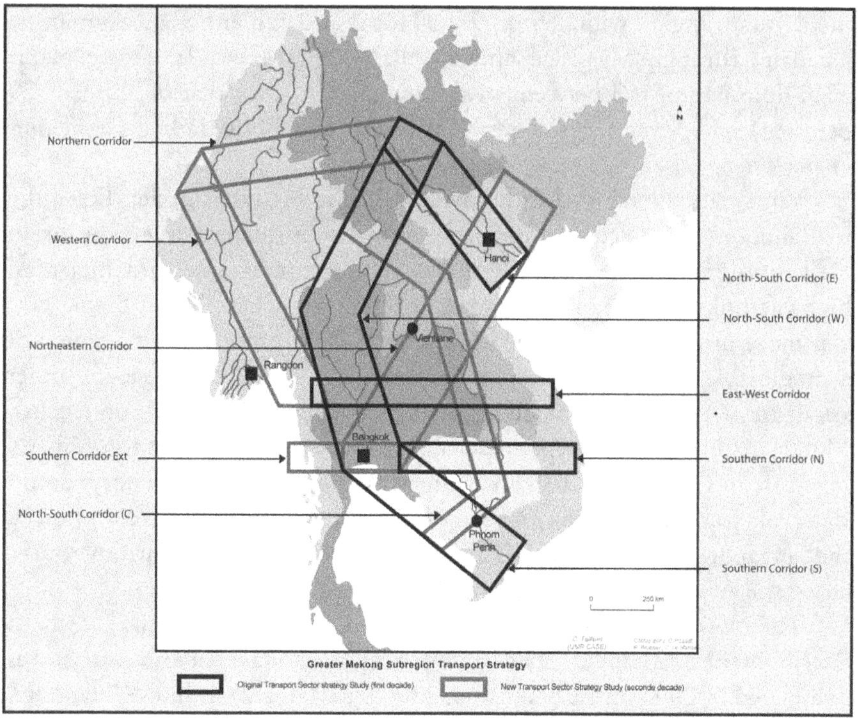

Source: Ch. Taillard and C. Lemasne, according GMS Transport Strategy.

and creation of cross-border free development zones. They also aim to make border crossing easier in order to develop trade and investment and encourage the private sector.

2. A CORRIDOR-BY-CORRIDOR APPROACH FOCUSING ON THE MEKONG AND THE CHAO PHRAYA DURING THE FIRST DECADE (1992–2002)

From one decade to the next, there was a visible change in scale of the networking of the Greater Mekong Subregion as the ADB map of regional strategy for transport shows (see Map 2.1). The first five corridors, meridian and transversal, are shown in dark grey. They have a pilot role during the first phase of the strategy for regional integration. In black is shown the extension in progress of the network during the second phase.

The North-South Corridor: Kunming-Bangkok

The North-South corridor, the most populated (30.4 million inhabitants) of all those planned at the regional level, traverses the "northern development quadrangle" which links Jing Hong and Chiang Rai, Yunnan to Thailand through Myanmar and Lao PDR (see Map 2.2). This explains Thai domination (53 per cent of the population) over that of Yunnan (29 per cent) and Myanmar (17 per cent), with Lao PDR trailing far behind (1 per cent).

With China having already invested in the highway linking Kunming to Rangoon, for quick and easy access to the Burmese market, the Asian Development Bank suggested the development of a second meridian highway linking Yunnan to Bangkok through the Chao Phraya basin. This was more promising in the medium term since it proposed the opening of a predictable market for a rapidly-expanding range of Chinese consumer goods from Yunnan and Sichuan (mainly motorcycles and pick-up trucks). Moreover, Sino-Thai entrepreneurs are recognized partners and top ASEAN investors in China, especially in Yunnan. As for the ADB, this north-south Bangkok-Kunming corridor is seen on a much broader scale as a section of the Singapore-Beijing trans-Asian highway, nearly doubling maritime traffic along the Asian eastern continental façade.

The North-South corridor is divided into two branches in the "northern development quadrangle". The western branch had taken a lead since it was completed in 2004 on Burmese financing (see Map 2.2). It links Chiang Rai to Jing Hong in Sipsong Panna, via Mae Sai and Kentung in Burma, with a two-lane 375 km-long asphalt road. The eastern branch through Huay Xai and Luang Namtha in Lao PDR is more interesting since it is shorter, only 245 km. The Laotian branch was based on tripartite financing, which is quite frequent nowadays in the Greater Mekong Subregion. Thailand, China and Lao PDR invested 30 million dollars each for this section, Laos' contribution being an ADB loan. The money was deployed to build a three-lane highway,[2] commissioned in 2007 for heavy vehicles, while the Burmese branch will be dedicated to light transport vehicles and tourism. China and Thailand too made a joint decision to finance the fourth bridge on the Mekong between Chiang Khong and Huay Xay (a joint investment of 45 million dollars), Laotian branch in late 2012. From late 2008 until now, the boundary between Myanmar and Yunnan on the Burmese branch of the corridor has been closed for political reasons. This will provide a new opportunity for the Lao branch when the bridge is opened.

Map 2.2
Spatial Structure Related to Economic Corridor Linkages in the Greater Mekong Region

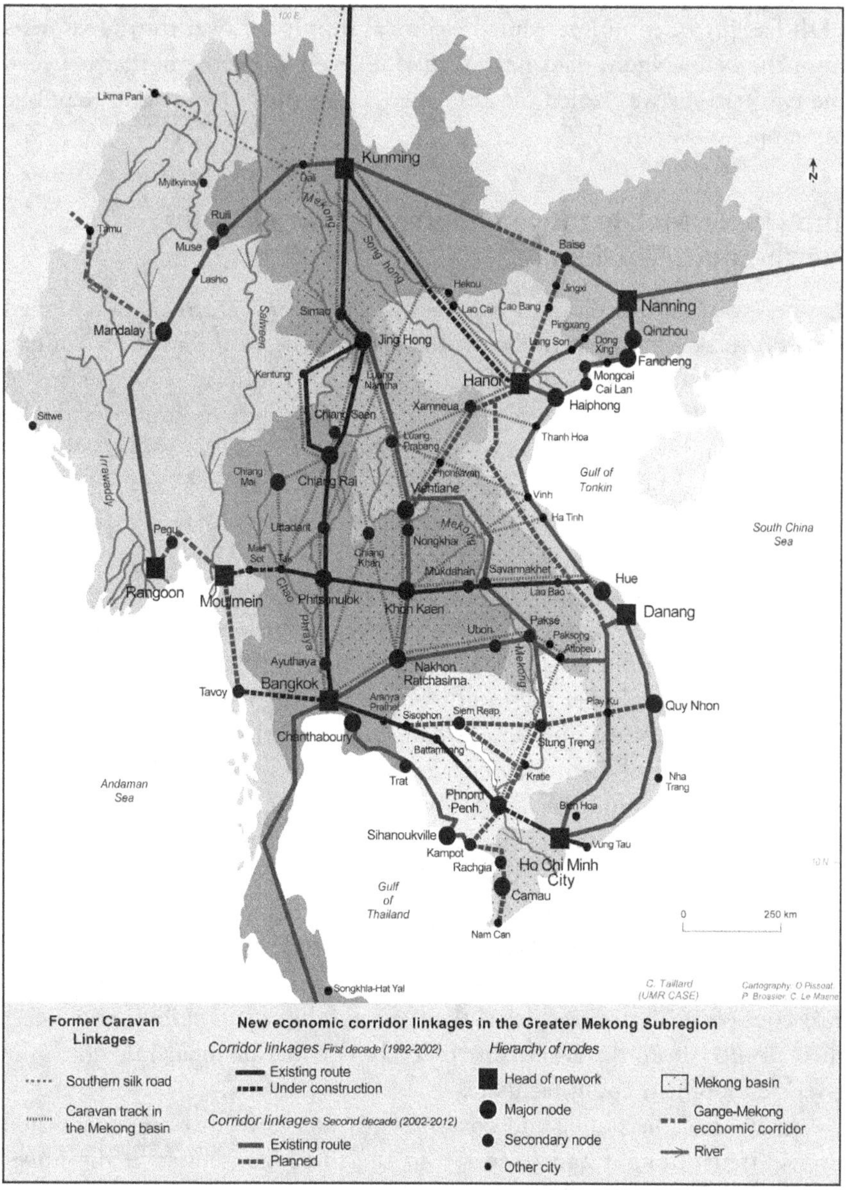

Source: Ch. Taillard, P. Brossier, G. Sillère, revised (2011).

Along the eastern branch a high-voltage electrical transport line will allow the Jing Hong dam (1,500 MW), the fourth constructed on the Mekong in Yunnan, to export electricity to Thailand from 2015. The Laotian branch is a safer option for this line than the Myanmar branch since Lao PDR has provided uninterrupted electrical supply for over thirty-five years from the Nam Ngum dam near Vientiane, even when the border between the two nations was sealed off after the Lao People's Democratic Republic came into power in 1975.

The Upper Mekong River Course to Complete the North-South Corridor

As the North-South corridor was taking shape, an agreement was signed in 2000 to develop river traffic on the Mekong between Simao in Yunnan and Luang Prabang in Lao PDR (see Map 2.2). It offered a new means of transport to China to export its manufactured goods to countries in the peninsula, and to Yunnan to export production to the West. It also linked north Thailand by boat to Luang Prabang, the former royal capital of Lane Xang, at present on the UNESCO heritage site list, nowadays a main tourist river route.

The upper stream part of the project was highly controversial and sparked off debate over the ecological impact of the desilting and dredging activities required to make the waterways navigable. It is often forgotten that colonial powers and later the International Mekong Committee undertook such work without causing serious damage to the riparian environment in the middle course of the river, which is now navigable. Oddly, Thai ecologists were supported by military forces who demanded, for security reasons, an agreement with Lao PDR to redefine river border limits. They contested the Franco-Thai treaty signed in 1893 which declares all Mekong islands to be sovereign property of Laos, regardless of their proximity to the Thai river bank. Dredging was likely to displace the talweg stretch that the Thai military wanted to be recognized as the international border between the two countries. This rationale masked yet another tacit motive: to protect their country from the eventuality of a Chinese marine incursion which the river development could facilitate.

In light of the public and government outcry downstream, aggravated by recurrent drought and retention in the three dams built in the upper course in China, the Mekong Commission entrusted an Australian academic specialist with the preparation of an impact study. The discouraging conclusions of this study led the Chinese government to stop dredging works in 2003 at the end of the first phase, allowing 150-tonne boats to navigate between Simao and Chiang Saen, the section which opens the Thai market to China. Against all expectations, China refused to commit itself to the other two phases which would have allowed the navigation of 300-tonne and later 500-tonne boats. The progress of the two road branches of the North-South corridor explains this unusual decision since China had to invest five million dollars toward dredging operations in the adjacent section of the river between Myanmar and Laos, a superfluous and less profitable investment compared to financing the North-South corridor road.

Thailand and China are expanding their partnership in Lao PDR with the joint financing of the Laotian branch of the North-South corridor, in Thai territory. Thailand has built in Chiang Saen a new riparian port downstream from the previous one to receive Chinese boats and has snapped up China's offer to establish a Sino-Thai economic zone to broaden bilateral business exchange. This offer was made in partnership with the National High-tech Industrial Development Zone of Kunming, including a free industrial zone, a logistics park, a business park and a cultural centre, modelled on those that Yunnan has constructed in Ruili at the Burmese border (Chia 2004). For China, it is a cross-border zone of a new kind since it has been relocated on the Thai-Lao border. The Thai planning authority, the National Economic and Social Development Board, has prepared a study identifying 112 projects for an investment potential of 476 million dollars at Chiang Saen, Mae Sai and Chiang Khong in the Thai part of the northern quadrangle. Lately this authority decided to relocate the Sino-Thai economic zone project in Chiang Khong close to the bridge on the Mekong River completing the Lao branch of the North-South corridor.

The East-West Corridor

The East-West corridor is a 1,450-km link between the two East and West maritime facades of the peninsula (see Map 2.2). Among the three routes under ADB scrutiny between Khon Kaen, the capital of northeast Thailand, and Vinh, Danang or Qui Nhon in Vietnam, finally the median route through the Lao Bao Pass, the lowest in the Annam mountain chain, was chosen. For Thailand, this option has the advantage of providing Khon Kaen with the much-needed radial road

Minh City), and the construction of a bridge on the Mekong between Savannakhet and Mukdahan (75 million dollars). The ADB, World Bank and Japan have jointly financed the reconstruction of Highway 9 linking Quang Tri to Savannakhet and the section of National Highway 1 linking Danang (460 million dollars). Japan is also studying the construction of a new coastal port along the western extremity of the corridor near Moulmein.

This corridor is the biggest challenge for the Greater Mekong Subregion since it does not cater to any of the peninsular capitals although it travels across provinces in which 15.3 million people live. It is dominated by Thailand (51 per cent) followed by Myanmar and Vietnam (28 per cent and 16 per cent, respectively), surpassing Lao PDR (5 per cent). However, it is the most advanced corridor and was finished at the end of 2009. Japan is the top donor, financing the eastern section of the East-West corridor (700 million dollars in a total of 1.5 billion dollars allocated to the Greater Mekong Subregion over the period 2004–07). It has financed the modernization of Danang port, the tunnel under the Hai Van pass (cloudy pass) between Danang and Hue (a project of national interest for Vietnam totalling 150 million dollars located on Highway 1 linking Hanoi to Ho Chi Minh City), and the construction of a bridge on the Mekong between Savannakhet and Mukdahan (75 million dollars). The ADB, World Bank and Japan have jointly financed the reconstruction of Highway 9 linking Quang Tri to Savannakhet and the section of National Highway 1 linking Danang (460 million dollars). Japan is also studying the construction of a new coastal port along the western extremity of the corridor near Moulmein.

Thailand, for its part, has invested 23 million baht to improve road linkage between Khon Kaen and Mukdahan. It will jointly finance with Lao PDR the transformation of Savannakhet airport into a shared international airport to serve the two twin cities on the Mekong River banks (1.2 billion baht), a first for the Greater Mekong Subregion. It is equally keen to develop the free development zones of Savannakhet and Seno, a former military base situated at the crossroads between the East-West corridor and National Highway 13. This free zone is likely to evolve in the future into a transnational zone including Mukdahan city on the Thai bank. Thailand is also investing in the western section of the corridor that it finances between Mae Sot and Moulmein. It is also developing a logistical hub at Mae Sot and working on a free development zone planned on the other side of the Burmese border.

Given the diversity of active financing sources and the number of countries concerned, four out of six within the Greater Mekong Subregion, the East-West corridor is the most international of all those planned. It is also the corridor which allocates the maximum financing to cross-border industrial zones (19 per cent), with 20 million dollars for the first phase from Savannakhet-Seno in Laos, 30 million for Mae Sot in Thailand and 25 million for the Lao Bao in Vietnam. A slightly lower proportion has been allocated to interconnection networks (18 per cent) while the transport infrastructure share (61 per cent) is the lowest compared to the other Greater Mekong Subregion corridors. The Thai body, the National Economic and Social Development Board, has undertaken a study to identify investments for the private sector with 74 projects totalling 364 million dollars.

The Emergence of a Second East-West Corridor: Ubon-Pakse-Danang

Another corridor features on Map 2.2 among projects for the first decade: linking Aranya Prathet-Siemreap-Stung Treng in Cambodia to Playku and Quy Nhon in Vietnam. It is the only corridor planned and not commissioned during the first decade of the Greater Mekong Subregion. However, a new corridor was constituted without the "regional" label. It takes the southern variant studied for the East-West corridor, linking Ubon in Thailand, via Pakse and Attopeu in Laos, to Danang in Vietnam via a section of the new Ho Chi Minh road linking the southern capital to the central capital through central highlands.

The new corridor was constituted after the construction of a new bridge on the Mekong at Pakse, with Japanese financing, and road investments to open out the Bolovens plateau, using bilateral aid. This corridor crosses the entire zone of border plateaus between Lao PDR and Vietnam. Apart from trade in these highlands (mainly cash crops such as coffee and tea), it provides access to the hydropower dam locations identified on the outskirts of the Bolovens plateau. These dams are of increasing interest to Vietnam, which is facing serious energy shortages for its rapid development, particularly in the central region.

The Southern Corridor: Bangkok-Phnom Penh-Ho Chi Minh City-Vung Tau

A second main transversal corridor links the three southern peninsular capitals: Bangkok, Phnom Penh and Ho Chi Minh City, with Japan and ADB as the principal financing parties (see Map 2.2). It is likely to be extended during the second stage, to the west up to the Andaman Sea along the gas pipeline which transports Burmese gas to Bangkok. A new road from Ratchaburi to Tavoy with Thai financing will serve to link it to the fourth capital, Rangoon. This corridor will connect the two maritime façades of the peninsula, by crossing four countries, much like the East-West corridor.

The corridor is highly populated, with 24.8 million people (excluding its extension to Myanmar) nearly equally distributed between Thailand and Vietnam (39 and 38 per cent) and has the maximum potential for trade revival, given the importance of the cities and deltas covered. At the interconnection of the power networks will be added the production of thermal energy and a proposed Bangkok-Phnom Penh gas pipeline. This explains why 48 per cent of the investments are in the energy sector. Like the East-West corridor, it is being tested for a single border customs inspection. Industrial zone projects are mushrooming on both sides of the Cambodian and Vietnamese border, while casinos have sprung up on Cambodian territory close to the Thai border.

3. THE EXTENSION OF ROAD CORRIDORS AS A REGIONAL LINKAGE IN THE SECOND DECADE (2002–12)

The second decade of the Greater Mekong Subregion witnessed a diversification of the road corridors with the creation of new ones to complete the meridian and transversal corridors of the first period, and an extension of the network since the introduction of a second Chinese province, Guangxi, in late 2004, thereby modifying the initial linkage (see Map 2.2).

The Addition of One Transversal and Three Meridian Corridors

To the corridor-by-corridor approach of the first period is added an integrated approach to regional linkage. The new sectorial transport plan includes corridors commissioned in the context of national programmes and bilateral aid. This plan integrates the Chinese-financed corridor linking Kunming to Rangoon, implemented in the 1980s before the Greater Mekong Subregion initiative, to give China an easy approach to the Burmese market and a secure access to the Indian Ocean. Strangely, this corridor is presented in two sections, southern and western, even though it forms a whole that has operated for a long time (see Map 2.1).

The plan also includes a corridor linking Kunming to Cambodia through Lao PDR. It branches out in the north on the North-South corridor and arrives at Luang Prabang and later Vientiane. The mountainous shape of this section and its two-lane structure prevent it from vying with the North-South corridor for heavy traffic. It then follows the Mekong valley to the Cambodian border, today a dead end. This Laotian section, along National Highway 13, has been completed with funding from various international bilateral and multilateral institutions. The new element is the missing infrastructure on Cambodian territory from the Laotian border to Kratie, to be financed by Chinese loans, and Japanese financing for the bridge crossing the Mekong and for reviving the Sihanouk City port.

To strengthen its central position on the Southern corridor and in the context of bilateral relations with Cambodia, Thailand has developed the duplication of this corridor with a new coastline link between the deltas of the Chao Praya and Mekong rivers. This new corridor also involves the prospective cooperation agreement signed between Cambodia and Thailand, to operate the offshore oil and gas fields located on both sides of their maritime boundaries in the Gulf of Thailand. A four-lane road already links Bangkok to Trat through Chanthaboury. The extension is underway in Cambodia, via Sihanouk City and Kampot, deploying highly diversified financing that includes Thai, Korean, World Bank and ADB sources. The extension has been planned in Vietnamese territory via Rach Gia up to Ca Mau with ADB financing.

An important omission in the sectorial transport plan is related to Vietnam, where two existing corridors are missing on Map 2.1. There is no sign of the coastline highway, known as Mandarin, to connect the two deltas and their metropolises, and offer access to Guangxi in the North and toward Cambodia in the South. The second omission concerns the Ho Chi Minh highway which partially overtakes the Mandarin road in the western highlands. This new road along the Cambodian and Laotian border is an alternative to prevent the main coastal road being cut off in the event of typhoons and give transversal access to these countries.

The Integration of Guangxi by Way of the New Northeastern Corridor

The most remarkable change in the new decade is undoubtedly the integration of Guangxi into the Greater Mekong Subregion. It is part of the Chinese strategy to turn the two southwestern provinces of Yunnan and Guangxi, bordering the Indochinese peninsula, into a crucial instrument in developing relations with ASEAN, in the light of the free trade treaty signed with this regional organization in 2002 which has been gradually implemented until 2012. It is striking to note that this schedule corresponds to the second decade of the Greater Mekong Subregion.

Initially, the Chinese government decided to make Kunming, the capital of Yunnan, a regional road, rail and air hub for the Greater Mekong Subregion (see Maps 2.3 and 2.4), to link the three main economic centres on the Chinese coastline (Beijing, Shanghai and Guangzhou-Hong Kong) with countries in the south (Myanmar, Thailand and Vietnam) as well as in the southwest (India). Furthermore, Beijing has equipped Kunming with a business and financial centre controlling a network of border free zones and cross-border zones (with twin cities: Ruili-Muse in Myanmar, Hekou-Lào Cai in Vietnam and Mohan-Boten in Laos) which are excellent tools for developing trade between Yunnan and other countries in the Greater Mekong Subregion. Beijing has also created a research centre within the university, dedicated to this transnational programme, and has constructed an international conference centre to make Kunming a metropolis at the regional level. A new airport has been planned 22 km northeast of the city for an investment of 1.6 billion dollars.

In the second stage, the Chinese government called for the integration of Guangxi in late 2004 into the Greater Mekong Subregion to make Nanning an exchange hub with ASEAN countries. An annual China-ASEAN fair has been created. The city has been equipped with a logistical platform connecting a dry port, 22 km south of the city, with the three seaports of the Beibu Gulf (Gulf of Tonkin): Fangcheng, Qinzhou and Beihai. Managed under a single authority, they handle 50 Mt of freight and 2.1 M containers annually. Nanning hence has facilities for maritime trade with ASEAN countries, despite stiff competition from the port network along the Pearl River delta of Hong Kong harbour. Nanning can also vaunt its position as a hub in the Kunming-Nanning corridor, implemented at the end of the first decade, which stretches up to Guangdong and Hong Kong (see Map 2.3). Two billion dollars invested in road transportation and additional financing dedicated to navigation on the Zhujiang between Nanning and the Pearl River delta demonstrate the potential of this new multimodal corridor.

Map 2.3
Road Corridors of the GMS and their Connections with China and India

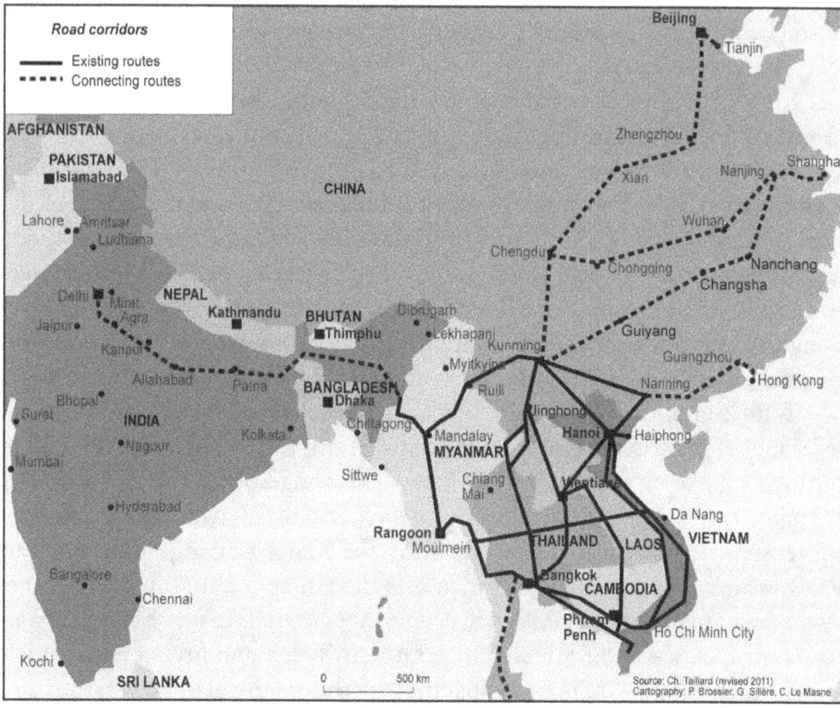

Source: Ch. Taillard, P. Brossier, G. Sillère, C. Lemasne, revised (2011).

The North-East Corridor, a New Opportunity for Vietnam and Lao PDR

The Greater Mekong Subregion offers a second advantage to Nanning: the Northeastern Nanning-Bangkok corridor via Hanoi and Vientiane (see Map 2.2) for an investment of 455 million dollars. This is strategic for China since it will connect Singapore to Guangzhou and Hong Kong by road and rail. On the cross-border scale, this Nanning-Hanoi corridor is duplicated into two other routes, to the east along the southern China Sea coast, and to the west via Cao Bang. Nanning will then have, like Kunming, three pairs of twin cities at the border: Pingsang-Lang Son on the main corridor, Dongxing-Mong Cai on the coastal highway and Jingxi-Cao Bang on the western highway, to develop business relations with Vietnam. The Chinese section of the North-East corridor is served by a tripartite financing agreement between the ADB, the central government and the two provinces (comparable in size to the largest countries in the Peninsula and with a similar investment capacity). The Vietnamese section is financed only by the government and the ADB. The absence of Chinese investment is worth noting, unlike the corridors traversing Myanmar and Lao PDR.

This North-East corridor has equal strategic importance for Lao PDR since it will replace Vientiane on an international highway. This corridor extends the Friendship Highway, a strategic road built by Thailand in the 1960s between Bangkok and Nongkhai to protect northeast Thailand from Communist threats and support the Royal Laotian government, which at the time only controlled the Mekong river plain. In 1994, the Friendship Bridge in Nongkhai, the first ever to be built on the Mekong in Lao PDR, thanks to Australian financing, provided an uninterrupted connection between Bangkok and Vientiane. The Laotian capital is, however, a dead end and Bangkok is now keen to trade overland with Hanoi via Lao PDR.

With economic reforms in Vietnam since 1986, Hanoi, the northern metropolis, is catching up with Ho Chi Minh City to become a credible business centre at the peninsular level. The congestion of Haiphong and Bangkok ports should also propel development of overland exchange. Moreover, the new corridor will open up the Xiang Khoang plateau in Lao PDR, which, like the Bolovens, is among the last agricultural pioneer fronts in the peninsula and has massive hydropower potential. Lastly, Xiang Khoang has large open-cast iron mines. These mines will demand investments which largely surpass the financing capacities of the peninsular powers but are feasible for China which is facing an acute ore shortage.

The missing Laotian section, between Vientiane and Xamneua, was included in the perspective of the Bangkok-Hanoi diagonal identified in the Atlas of the Lao PDR (Sisouphanthong and Taillard 2000). It is now integrated into the Bangkok-Nanning North-East corridor and included in the seventh National Plan of Laos. Now that the East-West corridor is complete, Japan, the top donor of the transversal corridors, can participate in the investment. The Japan Overseas Development Corporation (Ministry of Economy, Trade and Industry, METI) has also cited this corridor since 2004 among the integrating peninsular highways (Kunihiko 2004, p. 57).

Furthermore, this corridor provides a road linking the four preferred destinations of Japanese financing in continental Asia: Hong Kong, Guangzhou, Hanoi and Bangkok. This corridor may be also of interest to China since it offers open access to new and substantial mining resources at its doorstep. It may be foreseeable for China and Japan to jointly finance this corridor, each one's well-understood pragmatic interest could finally supersede their regional rivalry.

4. THE EMERGENCE OF A REGIONAL RAILWAY LINKAGE IN THE THIRD DECADE (2012–22)

Parallel to the gradual establishment of road corridors during the first two decades of the Greater Mekong Subregion, studies were undertaken to reconnect national rail networks on a regional scale. Since late 2010, the Chinese strategy of developing a national network of high-speed lines has changed the approach of the rail corridors in the Greater Mekong Region.

The Initial Approach: Connecting the National Rail Networks Running Along Coastlines

Initially, priority was given to the integration of existing rail networks by opening border crossings to traffic on long-closed railway lines inherited from the colonial period (especially between Hanoi and Kunming and Hanoi and Nanning) or World War II (between Bangkok and the Gulf of Martaban in Burma), and the completion of the missing links at the regional level.

The approach then focused on railway lines running along both coastlines of the Indochinese Peninsula (see Map 2.4). In the east, the first project initially aims at modernizing the railway which runs along the South China Sea in Vietnam, linking the two cities, and the Hanoi-Lang Son section which extends it further north (7 billion dollars); it also aims to restore train services between Nanning and Hanoi (a breaking of load now being imposed on border crossings). It also includes the construction of the missing section between Phnom Penh and Ho Chi Minh City for which research funding (estimated at 1.1 billion dollars for the missing section, and more than 7 billion dollars to upgrade the existing tracks) is well underway on the Vietnamese side, whereas the Cambodian section is funded by the China Railway Engineering Corporation Group. This first project constitutes an important component of the ASEAN-promoted meridian Singapore-Kunming railway, covering a distance of nearly 1,800 km in the Greater Mekong Subregion.

To the west, a second alternative or complementary project linking Bangkok to Kunming is much further advanced. Here, also, there is a question of the railway crossing the Three Pagodas Pass then running along the Gulf of Martaban up to Rangoon (see Map 2.3). This axis then follows the meridian spine of the Burmese rail network via Mandalay to reach Lashio. The missing section to be built here is much longer between the latter town and Dali in Yunnan, via Muse-Ruili at the Chinese border. Beijing has already invested 1.9 billion dollars in the Dali-Ruili section in Yunnan.

Map 2.4
High-speed Rail Corridors of the GMS and their Connections with China and India

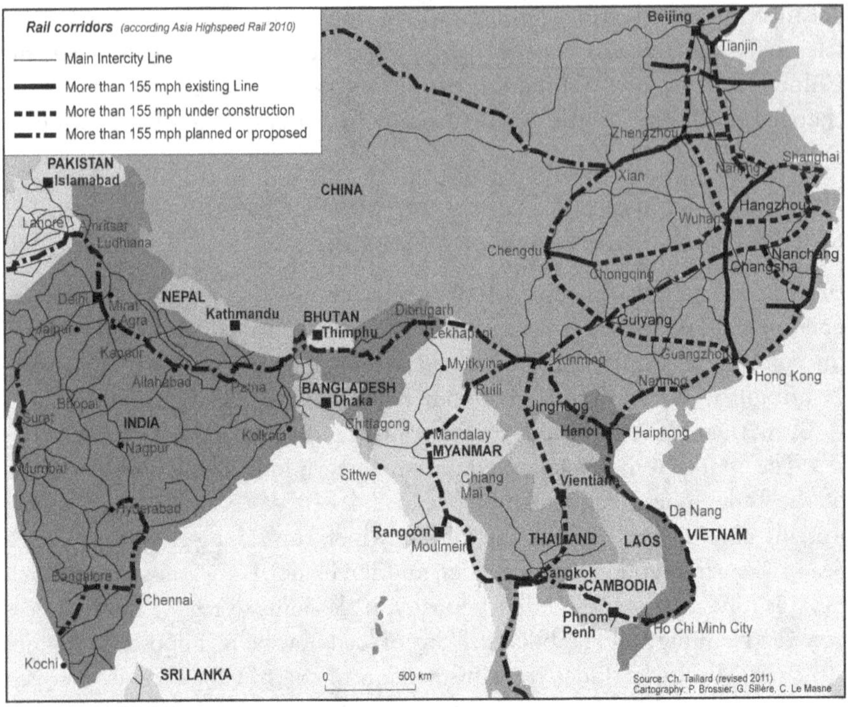

Source: Ch. Taillard, P. Brossier, G. Sillère, C. Lemasne, revised (2011).

The commitment of China to financing the Dali-Lashio railway does not guarantee the modernization of all the tracks in the western axis. Indeed, this new railway anticipates a second geopolitical stake beyond the Greater Mekong Subregion. For China, this is the first stage of an important trans-Asian project which will in the future link Yunnan and Assam in the context of a future Sino-Indian partnership.

A Central High-speed Railway to Strengthen Chinese Leadership in the Greater Mekong Region

Breaking with the first approach that emphasized the connection of existing rail lines in the eastern and western parts of the peninsula, China has committed itself to constructing a network of high-speed lines, which makes the most of national technology inspired by the Japanese Shinkansen and the Siemens and Alsthom TGV. In the past five years, 7,500 kilometres of railway have been built. The network should reach 45,000 kilometres in 2015 and 85,000 kilometres in 2020 (see Map 2.4). This network begins to project towards neighbouring countries. At the end of 2010, agreements were signed with Lao PDR and Thailand to build a high-speed line by 2015, whose Laotian section in itself is estimated at over 7 billion dollars.[3] This cost is commensurate with the scale of Chinese plans to invest 450 billion dollars in its network of high-speed lines by that date. The new line linking Kunming to Bangkok, constituting a section of the ASEAN-China railway, will be opened to passenger traffic and also to cargo freight. It will allow China to promote the free trade agreement signed in 2002 with ASEAN that has gradually been implemented since January 2010.

The route chosen for this first high-speed line is surprising because it takes the opposite view of the projects previously studied, that favoured, as we have seen, the western and eastern coastlines of the Peninsula. Indeed, it considers a route in the centre of the peninsula, but not the expected one. The topographical constraints and the economic vitality of the provinces crossed should have led to focus on a route starting from Jing Hong and running along the eastern branch of the North-South road corridor in Laos, and thence to Chiang Rai in the north of Thailand. Instead, the chosen option takes a more rugged route that runs via Muong Xai, Luang Prabang, Vangvieng and Vientiane, following the first section of the Central North-South Corridor in Lao PDR (Route 13) and subsequently joining the north-east corridor between Vientiane and Bangkok, via Udon Thani, Khon Kaen and Nakhon Ratchasima.

In doing so, China avoids the richest provinces of its two main competitors in the Indochinese peninsula, by leaving aside the meridian axis of Thailand, and the coastal axis of Vietnam that connects the Red River and the Mekong deltas. On the contrary, it reinforces the development potential of the poorest provinces of both Lao PDR and Thailand by promoting the sections of two road corridors of the second decade currently under construction. Therefore, this choice appears more geostrategic than economic. It is sufficient proof to measure the importance of crossing infrastructure on the Laotian section between Boten on the Chinese border and Vientiane: 190 km of tunnels and 90 km of bridges are planned out of 421 km of road!

Difficulties Encountered by the Projected Chinese Kunming-Bangkok Railway via Vientiane

China has encountered many difficulties in promoting its projected high-speed railway linking it to ASEAN countries, which was supposed to open in 2015 with the completion and implementation of the AFTA. These difficulties are the result of unfavourable internal Chinese circumstances and the reticence of its partners Lao PDR and Thailand.

The Kunming-Bangkok railway project via Vientiane was defined in 2010, at a time when Chinese investment dynamics in high-speed railways was undergoing a short-lived moratorium. This moratorium was decided in 2011 following the dismissal of the minister for railway concessions after the enormous debts of his ministry were revealed (317 billion dollars with an annual interest of 2.4 billion on the debt) and the derailment of a high-speed train in Wenzhou responsible for about forty deaths. To these internal difficulties were added the withdrawal of the Chinese company that was the partner in the joint venture on account of the project's lack of profitability. This company was supposed to provide 70 per cent of the investments, with the remaining 30 per cent the responsibility of the Laotian government which provided the land and took charge of the compensation of displaced persons.

The first draft of this project failed because of the conditions imposed by China which envisaged such an unequal partnership with Lao PDR and Thailand that it was perceived as a type of hegemony. From the financial as well as the industrial point of view, the turnkey project designed and constructed by the Chinese associated the other two countries as co-financers but not as technical and industrial partners. Lao PDR saw this project as an opportunity for training the skilled workforce that it lacked whereas Thailand wanted its own firms to take part in both planning and construction. With regard to land, the Chinese demanded a 200-metre concession on either side of the railway line, a condition refused by both the Laotian and Thai governments. This condition guaranteed the profitability of the investment for the Chinese firm until commercial railway traffic took over!

As a result of these problems in China, Peking transferred the Kunming-Vientiane railway project from the ministry of transport to the ministry of trade. It withdrew from the funding of the project, proposing only a sovereign loan granted by the State-owned Import-Export Bank, on the condition that the line was built by a Chinese firm with Chinese materials. The new proposed draft aims to make the Laotian government bear the entire responsibility for the project's risks, while limiting the financial risks for the Chinese partner. The loan is guaranteed, in addition to the future income from the railway and the assets linked to it, by Laotian royalties derived from joint Lao-Chinese mining ventures (exploiting gold, copper, potassium and in future bauxite and iron). This new draft is more like a "compensated exchange", a practice generalized by the Chinese with countries to the south to ensure supplies of raw materials for its industries.

According to Keith Barney, an Australian academic specializing in the development of natural resources, the seven billion-dollar loan proposed with a ten-year grace period and a two per cent compound interest rate over thirty years would have necessitated the payment of 3.3 billion dollars in interest. Principal and interest together would thus represent the most of Lao PDR's GDP. It is understandable that, in these conditions, the IMF and the ADB drew the attention of the Laotian government to the risks of such an investment, which would burden with debt most of its mining resources for thirty-eight years until the debt was being repaid, even if the income that would be derived from taxes on Chinese railway traffic in Laos is not known.[4] Apparently, the Laotian government has resumed negotiations with China, since the loan agreement was not signed, as announced, at the ASEM conference held in Vientiane in November 2012.

The Lao PDR, a country that currently counts less than 4 km of track (located at the mouth of the Friendship Bridge downstream from Vientiane), is now in a difficult position. It has resumed the project at its own expense, while decreasing the commercial speed of trains in this very mountainous country from 200 to 160 km/h for passengers and 120 km/h for goods. By doing this, it hopes to retain the advance achieved by this high-speed Kunming-Bangkok line crossing its territory, rather than the alternatives studied by neighbouring countries, so that it is on the future main axis of continental exchanges between China and ASEAN. To this end it has also undertaken a feasibility study for a road of international dimensions following the railway line in order to bring the section of the North-South corridor crossing northern Laos up to the standard of the other corridors of the Greater Mekong Subregion. Lao PDR is not however in entire control of all the conditions necessary for its construction since its profitability depends on the building of the Thai section, 615 km long, which will follow the route of the Nongkai-Bangkok railway and is estimated at nearly five billion dollars, and its subsequent extension to Malaysia to open up these two countries to Chinese exports.

Thailand is proceeding differently, giving preference to national imperatives for the construction of its high-speed rail network. Priority is given to the building of sections starting from Bangkok on lines towards Phitsanulok to the north, Hua Hin to the south, Eastern Seaboard ports to the east, and in last place the line from Nakhom Ratchasima to Lao PDR, which will only be extended in a second phase. Vietnam, another competitor for high-speed rail services, is taking the same pragmatic approach, and initially, according to a project studied by the JCA with the probable aid of loans from the Japanese government, intends to build the first sections (Hanoi-Vinh and Ho Chi Minh City-Nha Trang) of the line linking the two metropolises. This plan also sets to one side the missing link between Vinh and Nha Trang in central Vietnam and the connection with China and Thailand via Cambodia until a future date.

This analysis reveals that, at the beginning of the third decade, we are a long way from the approach stipulated by the ADB for the road corridors of the Greater Mekong Subregion: agreement to the project prior to the financial set-up, uniting all the countries crossed, partial funding by the regional bank and co-financing by the dominant partners of missing sections in the less developed countries that they cross, especially Lao PDR and Cambodia. Unless the ADB resumes its key role, the funding of interconnections between national networks may not take place for some time. This resumption should enable a renewal of the regional dynamics and cooperation that characterized the Greater Mekong Subregion during the first two decades. The commitment of the ADB, both as facilitator and financer, appears necessary for the third decade to become that of transnational rail corridors, since their inclusion in the trans-Asian railway plan defined by the ESCAP has not so far enabled the dynamics of regional cooperation initiated by the Greater Mekong Subregion to be maintained.

5. THE NEW ARCHITECTURE FOR THE GREATER MEKONG SUBREGION

A New Regional Integration within a Network

The Greater Mekong Subregion, despite the mountainous shape of remote boundary confines between China and the peninsular countries, is no longer an isolated region today. This vast region was never isolated, even during the pre-colonial era (Ch. Taillard 1989 and Map 2.2). Located on the hydrographic network fanning out from the Himalayan Mountains, Yunnan takes up a linkage position for long-distance trade, connecting the ancient southern Silk Road between China and India with the network of caravan routes which reached ports in the great deltas of the peninsula. The permanence of the caravan networks was underestimated for a long time by researchers who always highlighted the importance of the maritime trans-Asian routes. However, high value goods were traded on these caravan routes, with a simultaneous diffusion of civilizations which, when assimilated with native civilizations, served to forge the cultures of this “Indo-China”, enjoying relative autonomy with respect to these two continental states.

Since the end of the Cold War, which paved the way for the Greater Mekong Subregion, these networks have been reconstituted in a new context of transnational integration within East Asia which is part of regionalization dynamics related to the process of globalization. The first difference rests in linkage today: various corridors no longer form networks that juxtapose; they are integrated into a singular reticular organization (see Map 2.2). They all originate however from the periphery, regardless of whether this involves port or hinterland cities. The second difference arises from the position of these metropolises, which, previously isolated from one another, are now found on the main axis of integration.

The number of economic corridors commanded by the metropolises provides a measure of each one's structuring role. Kunming and Bangkok detach themselves; the meridian North-South corridor that links them constitutes the anchor for regional architecture. Hanoi and Nanning will shortly form with Bangkok a second main meridian axis. The new high-speed line which also occupies a central position in the Indochina Peninsula will enhance the emergence of a central meridian zone arranged in two beams, each corresponding to a trans-Asian itinerary connecting Singapore and Bangkok, on one side to central China and the north through Kunming and Chongqing, and on the other to southeast China through Nanning and Guangzhou (see Map 2.4). The southern metropolises, in the west (Rangoon with Moulmein) and in the east (Ho Chi Minh City with a relay to Phnom Penh) are well away from the central meridian zone.

The principal hubs of the regional reticular arrangement are found at the main intersection of the corridors based on three situations. They can first be found at diverging points on the two central bundles, not far from their command centres; Jing Hong and Chiang Rai south of Kunming on the North-South corridor, Ayutthaya and Nakhon Ratchasima north of Bangkok on the North-South and the Northeastern corridors. They also indicate the main junctions between the meridian and transversal highways: Phitsanulok and Khon Kaen on the North-South, Northeastern and East-West corridors, Mandalay and Haiphong on the western and eastern corridors. Lastly, Phnom Penh in Cambodia and Vientiane in Lao PDR will soon detach themselves from the secondary hubs to join this group as new corridors develop.

With the linkage of meridian and transversal corridors, the peninsula will operate once more as an integrated network, in which transnational axis structure a polycentric shaped arrangement spreading out from the coastal or continental periphery, where the command centres are situated. We point out on the one hand, the persistent character of the meridian structure forced by the division of the hydrographic basins and by the east and west coastal faces of the peninsula, and on the other by the intermediary position occupied by the Greater Mekong Subregion between China, India and insular Southeast Asia. This meridian compartmentalization and this in-between situation at the Angle of Asia, to borrow an expression by Elisée Reclus rendered famous by Paul Muss, forbids the development of polarized models.

The Lao PDR at the Crossroads of the Greater Mekong Subregion

Twenty years after the launch of the Greater Mekong Subregion, the vision of a return to a networked peninsula is becoming a reality, placing Lao PDR at the crossroads of the corridors that are increasingly structuring the regional linkage (see Map 2.2). Instead of being the battlefield of regional rivalry, Lao PDR is becoming the point of convergence of donors for funding the middle link in the new corridors. Unlike the buffer state separating potential enemies it was during the Cold War (Taillard 1989), Lao PDR is now in a position to take advantage of the flows that transit across its territory, by becoming a logistics actor on these corridors and an economic operator in the special economic zones set up at border crossings. Another decade will probably be needed for Lao PDR to benefit fully from its intermediate position in regional linkages.

However, it needs a clear-sighted territorial development policy that will combine national integration, structured around the meridian axis of the Mekong and Highway 13, with development of peninsular corridors of the Greater Mekong Subregion in order to receive a fair share of the revenues generated by the linkages between the region's economies. These two aims, apparently contradictory for a state with a meridian structure, will nevertheless determine Laos' integration into the new peninsular reality (Taillard 2005). Patient pursuit of simultaneous national and regional integration is therefore necessary.

To achieve these goals, the choice of a regional division that includes a medium-term perspective determines the effectiveness of spatial planning to be introduced in the coming years. From this point of view, the solution adopted in the master plan of the northern region, arranged with Chinese assistance, is not considered satisfactory (Bounthavy Sisouphanthong in this book). Dividing Lao PDR into two regions — the north, which includes all provinces between Vientiane province and northern borders with China and Vietnam, implying a south that would encompass all provinces between the Borikhamxay and the Cambodian border — appears too reminiscent of the regional division between Upper and Lower Laos from the colonial period. Today, in the context of the Greater Mekong Subregion, such a division appears to reflect very well a Chinese vision, and appears completely outdated for at least two reasons.

First, the need to link regional integration (by transverse corridors) and national integration (by the meridian axis of Route 13 which is the Laotian section of the Central North-South corridor) argues for a division into four regions (cf. Sisouphanthong and Taillard 2000). Each is organized around a structural corridor: the Laotian branch of the North-South corridor in the north, the North-East corridor uniting Xiang Khoang and Xamneua provinces in the North-East, the East-West corridor for the region controlled by Savannakhet, and the new East-West corridor for the one controlled by Pakse in the South. Each of the four regions will have transnational corridors that link Lao PDR with two of its neighbours: China and Thailand or Vietnam and Thailand for the northern and northeastern regions, and Vietnam and Thailand for the two other regions further south.

The second reason is that all countries in the region are redefining their urban metropolitan region by expanding their scope of planning to neighbouring provinces. Therefore the Vientiane Capital (the former Vientiane Prefecture) should be included in an urban capital region designed on the scale of national construction. Indeed, Vientiane province, which borders the Vientiane Capital in the north, controls the upstream territories of the Mekong valley, much of the Vientiane plain and the northern section of the national road 13 leading to Luang Phrabang and China. Borikhamxay province, which borders the east side of the Vientiane Capital, controls access to Paksane, the Laotian section of the North-East corridor leading towards the plateau of Xiang Khoang, Xamneua and Hanoi. It also controls access to the downstream area of the Mekong valley by the southern section of Route 13 that leads to Savannakhet, Pakse and Cambodia. It is at this scale that we therefore need to consider the major facilities that will allow Vientiane to count among the other metropolis of the Greater Mekong Subregion.

6. THE GREATER MEKONG SUBREGION: A CHALLENGE FOR THE PENINSULAR AND ASIAN POWERS

The Greater Mekong Subregion corridor linkage has whetted the appetite of major peninsular (see Map 2.2) and Asian powers (see Maps 2.3 and 2.4) to assume control or leadership of these transnational integration dynamics acting within the East Asia regionalization process.

Rivalry between Thailand and Vietnam

The first sign of competition between the peninsular powers, Thailand and Vietnam, is manifested by the blooming of peninsular organizations of varying geometries to strengthen or establish preferred relationships. Thus Prime Minster Taksin Sinawatra, wishing to assuage Thailand's neighbours resentful of the country's central position and economic domination (as attested by anti-Thai riots in 2003), launched the Irrawady-Chao Praya-Mekong Economic Strategy (ACMES) with an annual investment of 9 to 10 billion baht controlled by the Export-Import Bank (Exim) of Thailand. This bank finances, alone or with other regional donors (China or the ADB), roads, bridges and border business zones in Myanmar, Lao PDR and Cambodia. Thailand is trying to extend the model of the first Sino-Thai cross-border zone to the outlets of the other corridors on its territory. It is taking this opportunity to consolidate its economic dominant position over its close neighbours, enabling deployment by Thai companies and Thai foreign investment, by relocating their low labour-cost industries. The ACMES organization, piloted by Thailand, briefly excluded and then included Vietnam.

Hanoi, reviving its regional ambition and ready to compete with Thailand, suggested to its Indochinese partners, with the support of Japanese cooperation, the creation of a Cambodia-Laos-Vietnam Development Triangle. Initially a limited area, this triangle was extended to all provinces in the southeastern plateau of the peninsula. Vietnam later chose a series of projects, mainly hydropower resources and transportation, for which sources of financing were approached, for, unlike China and Thailand, Vietnam had no financing capacity. The financing of the section between Attopeu and the Ho Chi Minh highway, the new East-West corridor traversing the plateaus, has provided it with access to the Xe Kong basin and mainly to the Xe Kaman 3 dam (250 MW), on one of its tributaries. This dam costing 273 million dollars, completed in 2010, has been financed for the first time in Lao PDR by the Vietnamese firm Song Da Construction which will hold a twenty-five-year concession. It is the biggest Vietnamese investment overseas. The bilateral Lao-Vietnamese agreement signed in 2003, for the supply of 1,500 MW from 2010, requires the construction of the Xe Kaman 1 dam (322 MW and 441 millions dollars of Vietnamese investment, due for completion in 2014). This agreement will loosen a thirty-year Thai monopoly over the purchase of Laotian electricity and Lao PDR will attempt to negotiate better power prices for future dams with its top customer.

Rivalry between the two countries is gaining in momentum. By expanding its logistical and economic infrastructure, and taking advantage of its central location, Thailand intends to sustain profitable exchange with Vietnam and Myanmar, two former rivals, now left behind. But Vietnam, having embarked on economic reforms since 1986 boosted by strong growth, could well catch up with its main rival by taking advantage of the problems Thailand has been facing due to its political crisis during the past five years.

Competition between Yunnan and Guangxi

The Greater Mekong Subregion has not only become the field that displays the rivalry between Thailand and Vietnam, the two peninsular powers, it is also indicative of the competition that has emerged between the two Chinese provinces, Yunnan and Guangxi. Both provinces have become the intermediaries of Beijing's strategy towards its southern neighbours. The distribution of tasks between them by the central government, mentioned earlier, appears less balanced than it seems at first: namely, making Yunnan the Chinese hub controlling access to the Greater Mekong Subregion, and Guangxi the hub of relations with ASEAN countries (Colin's contribution).

Indeed, the decision taken in late 2004 to include the North-East corridor Bangkok-Vientiane-Hanoi-Nanning (which then extends to Guangdong and Hong Kong) in the regional network will soon offer Guangxi a second axis by which to structure relations between China and the Greater Mekong Suregion, adding to the first, the North-South corridor Bangkok-Kunming which benefits Yunnan. Moreover, the relative proximity of Nanning and Guangzhou-Hong Kong may allow Guangxi province to attract investors from those two cities of southern China more easily than does Kunming. Therefore, these metropolises might play an important role in the future of the Greater Mekong Subregion.

This may also explain how Kunming, in addition to its role as the Chinese link to the Greater Mekong Subregion, seeks to develop a regional function on a larger scale towards South Asia, as does Nanning towards ASEAN. China has already rebuilt its 600 km expressway between Kunming and Ruili and is financing another 500 km of the Stilwell Road in Myanmar to Ledo in Assam via Muse, Bhamo and Myitkyina. The building of the railway between Dali in Yunnan and Lashio in Myanmar can also be seen as a means of constructing (via the Lashio, Mandalay and Myitkyina section of the Burmese network) the missing link to connect the latter city with the Indian railway network at Likhapani in Assam (see Map 2.3). For Yunnan as well as Guangxi, the Greater Mekong Subegion may represent a first step toward a larger-scale deployment of China's economic ambitions in South Asia and Southeast Asia.

Competition between Chinese Hegemony over the Meridian Corridors and Japan's Dominance over Transversal Corridors

While China has made Yunnan and Guangxi platforms to develop exchanges with ASEAN, the Greater Mekong Subregion has also become a test of confidence of the new Chinese strategy activated by the ASEAN-China free trade agreement signed in 2002. With its proposal of multilateral negotiation instead of strictly traditionally bilateral relations imposed on its neighbours according to the adage "divide and rule", China has sent a strong message to its neighbours. Its recent decisions to freeze the second and third phases of dredging operations in the upper course of the Mekong to ease mercantile traffic has been well-received by its partners. Hence the Greater Mekong Subregion has become a vital mutual interest for both sides.

Japan signed an in-principle agreement with the regional organization in 2002. In 2008, Japan signed the ASEAN-Japan Comprehensive Economic Partnership planned with the first six members of the association for 2012 and with the last four members for 2017 (Cambodia, Lao PDR, Myanmar and Vietnam). It is also the top foreign donor in the Greater Mekong Subregion. Leaving China an open field on the meridian North-South corridor, it had very early supported the Cambodia-Laos-Vietnam Development Triangle. It has invested mainly on transversal corridors, the East-West and Southern ones. It could also become an important financing partner for the Vientiane-Xamneua section of the new Northeastern corridor. But now Japan will have to contend with China which has reserved 20 million dollars through the ADB to develop the three countries of this triangle. For the moment there appears to be a geographic division of tasks between China and Japan which will impact their own regional strategies as they compete for leadership in the Greater Mekong Subregion and also in East Asia.

In conclusion, it is clear that the Greater Mekong Subregion holds major strategic implications for the three leading Asian as well as other main peninsular powers. In this transnational regional integration, the various corridors, cross-border free development zones and twin border cities are ideal sites for observation, both on the peninsular and Asian scales. Located at converging points of state players, central and territorial administrations, national as well as local businesses, international donors and diverse transnational collaborations, they make relevant subjects for study to unravel the complex dynamics operating in the Greater Mekong Subregion. They also allow us to track the emergence of new centralities in the peripheral regions or in intermediate positions within their respective national constructions, and thus offer new opportunities to cities and provinces hitherto marginalized.

The way in which they manage to take advantage of these new opportunities will show the sustainability of the momentum induced by ADB-led economic corridor network strategy in the Greater Mekong Subregion. I hope that this book, with its comparative approach to continental corridors of the Greater Mekong Subregion and maritime corridors in Malacca Straits, can provide some answers to these important questions for the future of the transnational integration process in Southeast Asia.

Notes

1 This introductory text is an enlarged and updated version of my contribution "Corridor Linkage in the Greater Mekong Subregion: Newly Implication for Peninsular and Regional Power", in *New Dynamics between China and Japan in Asia*, edited by Guy Faure (Singapore: World Scientific, 2010), pp. 187–231.

2 This route has a variable standard. It has two lanes and two lateral strips for emergency (equivalent of a third lane). It has been increased to three lanes without lateral strips on the steepest slopes and reduced to two lanes only (without lateral strips) in some sections of narrow valleys.

3 Due to the collision between two Chinese high-speed trains in Wenzhou on 23 July 2011, the Chinese programme has been frozen until the causes of the accident can be identified. Therefore, it will fall behind schedule. The construction site of the Kunming-Vientiane-Bangkok railway has also been delayed.

4 As observed by Calving Zhang in the *Global Times* of 9 January 2013, in an article entitled "China unfairly targeted over railway", in response to an article that appeared in the *New York Times* on 2 January 2013, entitled "China builds railroad and Laos bears the cost".

References

Asian Development Bank. *Economic Cooperation in the Greater Mekong Subregion: Toward Implementation*. Manila: ADB, 1994.

———. *Economic Cooperation in the Greater Mekong Subregion: An Overview*. Manila: ADB, 1999.

Chia Peng Theng. "Sino-Thai Trade on the Mekong River: Leading Growth of the Economic Quadrangle, in Center for European Studies". *Regional Economic Cooperation: EU and GMS Development Strategy.* Regional conference, Chulalongkorn University, Chulalongkorn Printing House, Bangkok, 2004.

Evans, Grant, C. Hutton and K.E. Kuah, eds. *Where China Meets South-East Asia: Social and Cultural Change in the Border Regions*. Bangkok/Singapore: White Lotus and Institute of Southeast Asian Studies, 2000.

He Sheng Da. "Friendly and Wealthy Neighbours, Stable Nation: Yunnan Participation in the GMS". In *The Mekong Arranged and Rearranged*, edited by M.S.I. Diokno and Nguyen Van Chinh. Chiang Mai: Mekong Press, 2006.

Ishida Masami. *Toward a Single Market of the GMS: The Potential of the Three Economic Corridors.* Japan: Institute of Developing Economies, Japan External Trade Organization, 2007.

Kaosa-ard, Mingsarn and John Dore. *Social Challenges for the Mekong Region.* Chiang Mai: Social Research Institute, Chiang Mai University, 2003.

Kermel-Torrès, Doryane, ed. *Atlas of Thailand.* Chiang Mai: Silkworm, 2004.

Kunihiko Shinoda. "The Effect of Regional Trade and FDI Projects/Policies on Economic and Industrial Development: Its Achievment and Issues". In *Regional Economic Cooperation: EU and GMS Development Strategy*. Regional Conference 2004. Bangkok: Center for European Studies, Chulalongkorn University, Chulalongkorn Printing House, 2004.

Kwan Chi Hung and Osamu Yasuda. "The Indochina Economic Zone". In *Asia's Borderless Economy: The Emergence of Sub-regional Zones*, edited by E. Chen and C.H. Kwan. St Leonards: Allen & Unwin, 1997.

Pholsena, Vatthana and Ruth Banomyong. *Laos: From Buffer-state to Crossroads?* Chiang Mai: Mekong Press, 2006.

Sisouphanthong, Bounthavy and Christian Taillard. *Atlas of Laos: Spatial Structures of the Economic and Social Development of the Lao People's Democratic Republic.* Chiang Mai: Silkworm-NIAS, 2000.

Taillard, Christian. *Le Laos, stratégies d'un Etat Tampon* [Laos, Strategies of a Buffer-state]. Paris: GIP Reclus-La Documentation Française, 1989.

———. *Le Laos à la croisée des corridors de la région du Grand Mékong* [The Lao PDR at the Corridors' Crossroad of the Greater Mekong Subregion]. In Le Laos, doux et amer, 25 ans de pratiques d'une ONG [The Laos sweet and sour, 25 years of an NGO expérience], edited by D. Gentil et Ph. Boumard. Paris: Comite de Coopération avec les Laos-Karthala, 2005.

———. "Un exemple réussi de régionalisation transnationale en Asie orientale: les corridors de la Région du Grand Mékong" [A Transnational Integration Success: The Continental Grid of Economic Corridors in the Greater Mekong Subregion]. *L'Espace Géographique*, 2009/1 (2009): 1–16.

———. "Corridor Linkages in The Greater Mekong Subregion: New Implication for Peninsular and Regional Powers". In *New Dynamics between China and Japan in Asia: How to Build the Future from the Past?*, edited by G. Faure. Singapore: World Scientific, 2010.

Taillard, Christian and Doryane Kermel-Torres. "La Région du Grand Mékong, un espace transnational péninsulaire en Asie du Sud-est continentale" [The Greater Mekong Subregion: A Transnational Spatial Integration in Continental South East Asia]. In *Intégrations régionales en Asie orientale* [Regional Integration in Oriental Asia], edited by Ch. Taillard. Paris: Les Indes Savantes, 2004.

Vu Tu Lap and Christian Taillard. *Atlas du Viêt-nam* [Atlas of Viet Nam]. Paris: Dynamiques du territoire, Reclus-La Documentation Française, 1993 (in French, English and Vietnamese).

Zhao Guangzhi. "A Model of Decentralized Development: Border Trade and Economic Development in Yunnan". *Issues and Studies* (1996): 85–108.

3

MARITIME CORRIDORS, PORT SYSTEM AND SPATIAL ORGANIZATION IN THE MALACCA STRAITS

Nathalie Fau

The term "straits" designates a narrow maritime passage between two land masses that links two seas. Like an isthmus or a canal, it represents not only a strategic axis but also a structural axis in global organization. This unique and complex zone of land-to-sea interface has become a major factor in the twentieth century, as coastal nations have exercised their political will by enlarging their sphere farther and farther out to sea and by nationalizing the seaways. Although the Conference of Montego Bay in 1982 put an end to decades of tensions and even conflicts by establishing rules to regulate this appropriation and to delimit the maritime borders, international straits remain insufficiently defined geographical areas. Often considered solely as a narrowing of the maritime region in which navigating conditions frequently become difficult, straits are rarely considered as separate territories structured by longitudinal as well as latitudinal flow (Picouet and Renard 2002).

The originality of the Straits of Malacca is precisely the fact that it is both a zone of major exchange and transit in international commerce, in which the nations situated along its coasts have always been deeply integrated, and a region in itself, shaped, despite the borders that separate the opposing coasts, by a tight web of commercial and cultural relations between the two coasts (Fau 2003*a*, p. 34). A multi-level approach is necessary in order to understand the functioning of these straits, which are given structure both by the latitudinal flow, for the most part transoceanic, and by the transversal flow linking the two coasts.

This multiple reality of the Straits of Malacca will be studied in this chapter under the angle of an analysis of the maritime shipping networks and of the different levels of organization in the port systems. In order to understand the articulation of the different maritime flows and their impact on the spatial organization of the coasts, the straits are considered successively as a corridor of intense global maritime circulation, as a natural border between Malaysia and Indonesia, and lastly as an inland sea in which the local shipping flow has been given a new dynamic by the opening of the borders.

1. THE STRAITS AS A GATEWAY TO THE OCEANS: BETWEEN COOPERATION AND COMPETITION

Straits are corridors of intense maritime traffic. Since they are unavoidable passageways, they concentrate ever-increasing flows and constitute sensitive links in the maritime pathways. They are truly “gateways to the ocean” in the well-chosen expression of Jean-Claude Lasserre (2003), and this is reinforced by the increasingly seafaring nature of the global economy, the explosion in the trade of mid and high value-added manufactured goods, and the resulting boom in the transport of shipping containers. The Straits of Malacca are a major axis in worldwide maritime traffic and a vital artery of intraregional commerce. They are located at the juncture of the transoceanic merchandise shipping lines between the Indian Ocean and the Pacific, intra-Asian lines and East-West circumterrestrial maritime routes (see Map 3.1).

The volume of shipping traffic through the Straits is determined by trends in international seaborne trade. As it is a sea lane of communication (SLOC), the Straits also holds tremendous strategic value not only to the littoral states but also to the international community. East Asian economies such as China, Japan and Korea depend on the passage provided by the Straits to carry huge volumes of their imports, especially energy commodities from the Middle East. The Straits facilitates the transportation of an estimated one fourth of global seaborne trade and one third of its crude oil. In testimony to the commanding role it plays as a key global trade sea lane, the Straits recorded around 100,000 movements of vessels of all types in 2011 (Marine Department Malaysia, 2012).

Map 3.1
The Straits of Malacca: A Major Sea Line of Communication

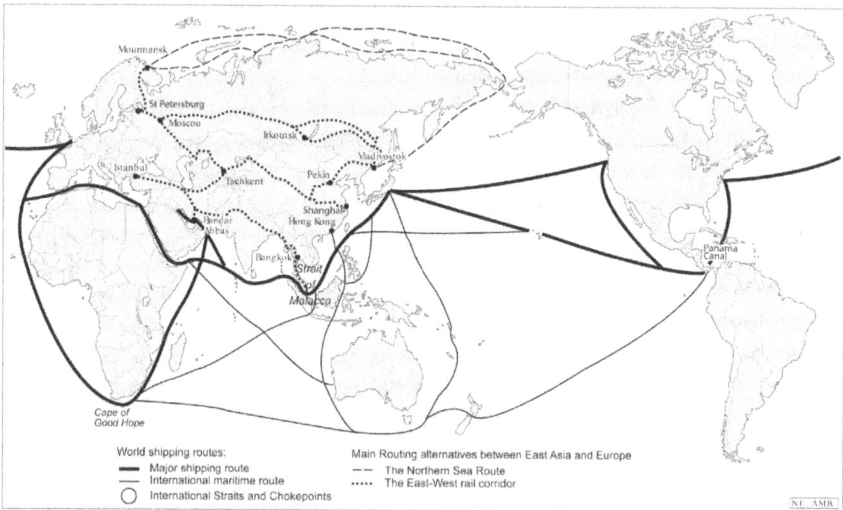

Source: Fau Nathalie, Ph.D. thesis, 2003*a*.

Necessary Cooperation: Guaranteeing Navigation Safety and Security

Such trends in the volume of traffic have an impact on the safety of navigation in a restricted area like the Straits of Malacca. The straits represent not only a stake for the bordering nations but also for the nations that use it, some of them quite distant, and the shipping companies. A closure of the Straits for whatever reason, even temporarily, would result in grave consequences for global trade and its economy, as an estimated one-third of world trade and half of its oil shipments are transported through the Straits. According to the Montego Bay Convention (UNCLOS) of 1982, the bordering nations are required to adopt the regime of "transit passage" and to respect the rights of passage of the ships (Bateman 2007). This text also specifies their obligation concerning the maintenance and financing of navigation infrastructures in the straits. Regardless of rivalry and political and economic disputes (Mak 2006), the coastal nations are under an obligation to set up a transnational management system for this international zone in order to guarantee the safety of maritime traffic.

In the Straits of Malacca, the border nations (Singapore, Malaysia and Indonesia) declared themselves responsible for the management and maritime safety of the straits in a joint press release on 16 November 1971 (Leifert 1978, p. 76). This trilateral cooperation follows two main lines. The first concerns navigation safety. In February 1977, Malaysia, Singapore and Indonesia signed an agreement establishing a traffic separation scheme for the Straits, including the establishment of sea lanes and navigational and safety aids in two particularly risky areas for navigation. This system has become progressively more sophisticated with the use of new information technologies. Since 1998, the straits of Singapore and Malacca have been equipped with a mandatory ship reporting system for ships in transit (see Map 3.2). Its goal is to strengthen navigational safety, to facilitate the passage of ships through dangerous zones and to protect the maritime environment (Fau 2003*a*, pp. 67–573; Abdullah 2008).

Map 3.2
Safety to Shipping in the Straits of Malacca

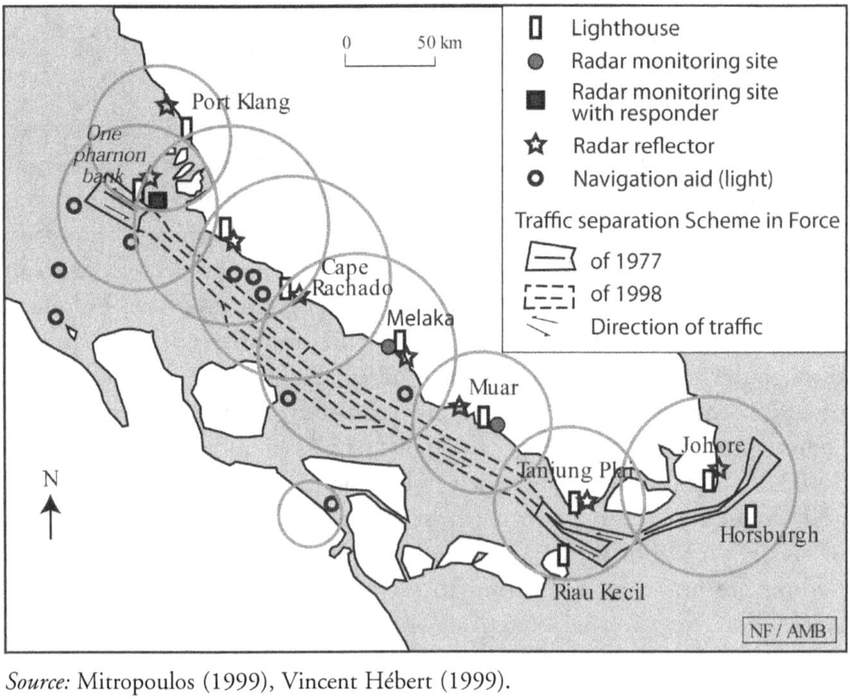

Source: Mitropoulos (1999), Vincent Hébert (1999).

The second goal aims at fighting piracy, which has been widespread in the straits since a fresh outbreak in the beginning of the 1990s. In 1992 the three border nations signed a series of bilateral agreements aiming at coordinating surveillance missions, and in 1993 they created, in cooperation with the International Federation of Ship Masters, a Regional Piracy Centre (Frécon 2003). Furthermore, the September 11 attacks have spurred the maritime sector to reassess the Straits' vulnerability to potential acts of terrorism, hostility and sabotage and so have required the littoral States to sign several multilateral agreements. On 20 July 2004, the navies of Indonesia, Malaysia and Singapore signed an initiative MALSINDO (Malaysia-Singapore-Indonesia) to combat piracy in the form of coordinated patrols. Equally important is the joint surveillance by the littoral states with the programme called "Eyes in the Sky" (EIS); commencing in September 2005, it features a combined maritime air patrol by the armed forces and maritime enforcement agencies of the littoral states and invited international participating nations (Bateman 2006). The launching of the EIS gives a clear signal to the international community that the littoral States are serious in ensuring safety and security in the Straits.[1]

Beside the three littoral states, other countries are involved in the security of the Straits of Malacca. Over the last thirty years, Japan, through the Nippon Foundation and the Japanese Shipowners' Association, has provided assistance by conducting hydrographic surveys, providing buoy tenders and installing navigational aids, and has contributed more than US$150 million to enhance safety in the Straits (Khalid 2007; Leong 1997; Tsuneskawa 2010). Moreover, at the IMO meeting in Singapore in September 2007, the three littoral states with the support of the international maritime community, such as user states, ship owners and shipping industries, launched the concept of Cooperative Mechanism. The Mechanism is a new framework conceptualized by the IMO council to provide a platform for the littoral states to engage in regular dialogue with other states and users for the maintenance of safety and security in the Malacca Straits. It will enable also the stakeholders to undertake joint projects and make voluntary monetary contributions towards maintenance efforts. For example, China has replaced navigation aids damaged by the 2004 tsunami and the United States has conducted a survey of incidents involving hazardous and noxious substances in the Straits (Khalid 2009; Mak 2006).

Although the financing of these costly measures is borne by the three border nations, they do not have an equal share in the dividends. For many years the benefits of international navigation benefited mainly Singapore, while profits remained marginal for the port economies of

Malaysia and Indonesia. At each of the international conferences on the Straits, the Indonesian and Malaysian delegates repeatedly declared their opposition to this injustice, which imposes on the border nations alone equal shares of the financial burden of the construction and maintenance of shipping facilities.[2] Faced with the obligation imposed by article 43 of the Montego Bay Convention, the governments of Indonesia, and especially Malaysia, set up new port management strategies fifteen years ago in order to attract a greater share of the international shipping in transit in the Straits.

Competition between Port Facilities

In the Straits of Malacca, worldwide maritime container route, which are by nature essentially transoceanic, are dominated by the port of Singapore. This port, spread over several sites, is a global hub of maritime traffic; it is linked with more than 600 ports spread over 123 countries. In 2011, its four container terminals handled 29.37 million TEU (Twenty Feet Equivalent Unit), which represents a fifth of worldwide container traffic and almost half of that of Southeast Asia.[3] Because of the confines of the Singapore market, the majority of this traffic consists of trans-shipping. This port platform is located at the centre of a complex bundle of regional maritime interconnections that provide it with a vast international hinterland. It is a well-known fact that within the "hub and spokes" container port system in the Southeast Asian region, there is a well-established order of container shipping traffic and container flow. Being a mega-hub port and the world's largest trans-shipment port, Singapore Port acts as the hub port in this scheme, while regional ports like Port Klang, Laem Chabang (Thailand), Manila Port (Philippines), Muara (Brunei) and Tanjung Priok (Indonesia) act as "feeder ports" along the "spokes" that connect to the "hub". Singapore Port is also the preferred destination of many main line operators, which are always looking to minimize their cost by calling at as few ports as possible, due to the availability of comprehensive one-stop shipping-related services at the port, which include bunkering, ship repair, legal support, financing, crewing and many others. Singapore prides itself on being an International Maritime Centre, a concept propagated by the city state to enhance its attraction to shipping lines as a global maritime hub (Fau and Khalid 2013). The other ports in the Straits that represent mere subsidiary branches of this network are accordingly restricted to a relatively secondary role (see Map 3.3) (Chia et al. 2003).

Map 3.3
Port System in the Straits of Malacca

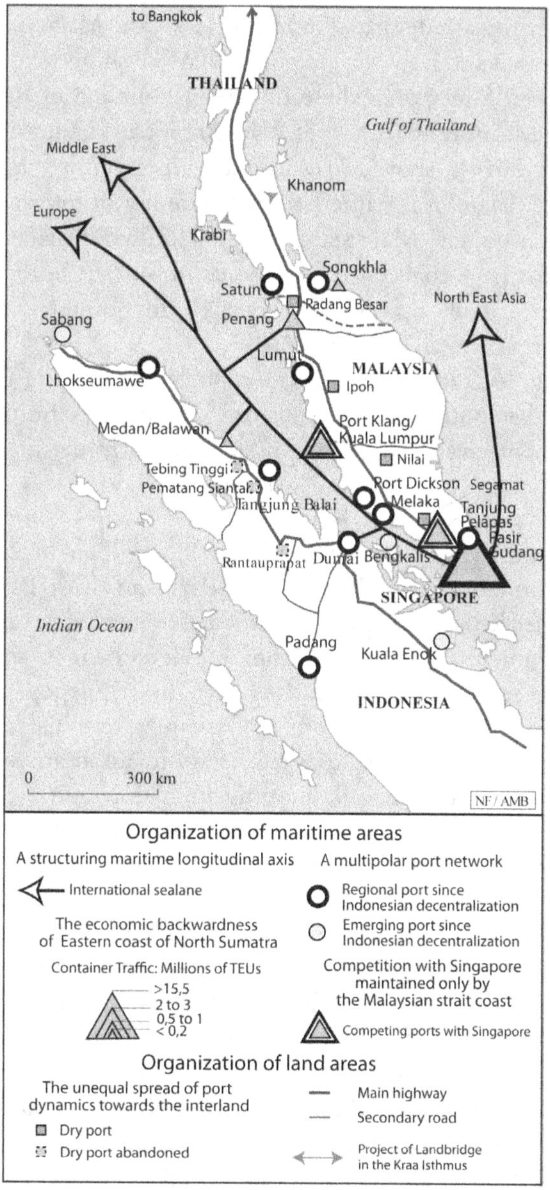

Source: Fau (2003*a*).

For the government of Malaysia and more recently for Indonesia, this situation was politically unacceptable. In response, they have undertaken massive projects to modernize and expand their port facilities to win a part of Singapore's maritime trade (Fau 2003*b*). In 1995, the Malaysian government introduced measures actively encouraging Malaysian shippers to use Malaysian ports rather than Singapore. These include a doubling of duties on laden trucks leaving Malaysia and a levy of RM100 on trucks entering the country (Hanizah et al. 2010). Above all, Malaysia set a target over a decade ago to attract a greater share of the international shipping in transit in the Straits. During the past decade, Malaysian ports, especially Port Klang and Port of Tanjung Pelepas have challenged Singapore's hub status in the region. They have to certain extents managed to wrestle trans-shipment volumes from Singapore Port, as shown in Map 3.3.

Port Klang, located 40 km from its capital, and Tanjung Pelepas (PTP), implanted at the southernmost tip of the Malaysian peninsula, have since established themselves as new regional port platforms in the Straits of Malacca.

Container traffic in Port Klang increased from 1.1 million TEU in 1996 to 3.2 million in 2000 and reached 9.6 million in 2011. Its three container terminals (North Port, South Port and West Port) have been entirely modernized and progressively privately operated since 1986. North Port invested more than RM500 million on redeveloping its existing wharves at Container Terminal 1, upgrading two multipurpose wharves and upgrading container yard operations at Container Terminal 2 and 3 (Jamal S. 2010). The quality of its facilities has begun to attract investments on the part of the major international shipping lines: for example, in 2000 the Hutchison International Terminal group, a subsidiary of the Hong Kong-based Hutchison Whampoa Group, which handles 10 per cent of worldwide container traffic, bought 30 per cent of the capital of the operator of the Westport terminal (Fau 2003*a*). The rapid growth of Port Klang can be explained by its capacity to attract an ever-increasing share of trans-shipment: 37 per cent of its traffic in 1999, more than 59.5 per cent in 2009. In order to achieve this growth, the main Malaysian container shipping operator, Malaysia International Shipping Corp. (MISC) entered into one of the broadest worldwide alliances between ship operators, the "Grand Alliance" comprising Hapag-Lloyd (Germany), Nippon Yusen Kaisha (Japan), Orient Overseas Container Lines (Hong Kong) and P&O-Nedlloyd (Great Britain). It deliberately chose not to join the "New World" alliance, which includes the main shipping operator of Singapore, the Neptune Orient Lines (NOL) along with Hyundai (South Korea) and Mitsui OSK (Japan).

In the south of the Straits of Malacca, the port of Tanjung Pelepas, built in close proximity to the second bridge linking the State of Johor to Singapore, was planned exclusively as a future direct rival of the port of Singapore. Although the official opening of the port was just nine years ago, in 2000, its growth has been spectacular: 418,000 TEU in 2000, 2.6 million in 2002 and 7.5 million in 2011! The port of Tanjung Pelepas is particularly competitive: it is capable of serving as the Port of Singapore Authority (PSA) for the latest-model container ships at less than 50 per cent of costs elsewhere, and because of its land reserves it can offer preferential wharves to the biggest ship-owners. The quality of its services has allowed it to attract international shipping operators, to the detriment of Singapore: in 2000, the Danish ship-owner Maersk Sealand, the world's number-one container transporter, transferred from Singapore to PTP the entirety of its shipping, and it was followed in 2002 by the Taiwanese ship-owner Evergreen Marine and again in 2004 by the Japanese companies Japan's K Line and Mitsui OSK Lines (MOL). In June 2008, MISC moved their hub from Singapore to PTP and introduced new services between PTP and Australia, New Zealand and the Philippines (Hanizah et al. 2010, p. 54). The phenomenal growth of this port could also be attributed to the greater number of services for imports and exports made available by shipping lines making new calls at PTP. Since its transfer, Maersk Sealand also furnishes 30 per cent of the capital of PTP. In just a decade of operation, PTP has emerged as the 17th largest container terminal in the world (*Containerisation International*, 2011) and the busiest in Malaysia, in terms of container throughput handled. An estimated 95 per cent of its throughput comes from trans-shipment containers, although its gateway throughput has been on the increase.

The Awakening of Indonesia's Ports?

In contrast to the well-developed port facilities on the Malaysian and Singaporean sides of the Straits of Malacca, Indonesia's port system is highly inefficient and in urgent need of upgrading. However, in the last few years, the Indonesian government has announced plans to build two new container transhipment terminals in Northern and Eastern Sumatra along the Straits of Malacca (Fau and Khalid 2013).

The first project is the Port of Sabang in Aceh Province on the island of Pulau Weh at the northern entrance to the Straits. Sabang is strategically placed, it is like a gate to the Straits from the Indian Ocean and the Andaman Sea, and it is a protected natural deepwater port. Located on the western tip of Indonesia, Sabang was one of busiest ports in the Dutch colonial era but its importance declined after World War II due to the destruction of the harbour facilities (Fau 2003*a*). Sabang also went through a golden age between 1970 and 1985 when the Indonesian government established it as a free port and free trade zone. The government, however, revoked this status in 1985. In 2000, after the government's decision to revive Sabang's status as a free trade zone and free port, the local Government of Banda Aceh Province, represented by the Sabang Development Authority (SADA), proposed to develop Sabang as a container trans-shipment hub for tankers and cargo ships unable to pass the Malacca Straits. In order to achieve this goal, SADA has signed a MOU with the Dublin Port Company Limited (DPCL), the world's twelfth largest port operator. Under this agreement, DPCL would handle all the port's operations and improve facilities, including the container yard, storage facilities, warehouses and cargo terminals. According to SADA, the port will start operating in 2014.

The second project is Tanjung Sauh, a deepwater port near Batam and Bintan and directly faces the Port of Singapore and Tanjung Pelepas Port in Malaysia. The port will have 2 km of pier and an estimated trans-shipment capacity of 4 million TEU. It aims to be a major transhipment point for Indonesia and a direct competitor of Singapore and Port of Tanjung Pelepas. Its construction will begin in 2013 and finish by 2016. This project is included in a nationwide freight transport programme introduced by the Indonesian government in June 2012 to improve the movement of goods in Indonesian waters and to reduce the cost of maritime logistics. The programme, named *Pendulum Nusantara* (Pendulum of the Archipelago), aims to create a single sea corridor connecting the eastern and western regions of Indonesia. It involves the ports of Belawan, Batam, Tanjung Sauh, Tanjung Priok, Tanjung Perak, Makassar and Sorong which will serve as transit points for smaller feeder ports all across the archipelago. They will also provide more frequent services for main line operators through the main seaports of Indonesia. With this project, Indonesia will no longer have to rely on Singapore Port to facilitate its export and import activities.

The Reaction of Port of Singapore Authority: Foregoing Its Regional Dependence

Faced with this increase in competition, PSA has multiplied its branches throughout the world (Fau 2010). In addition to its capital, PSA offers its expertise and logistics capabilities to more than 28 ports located in

16 countries. Its European, American and Asian terminals (other than Singapore), are capable of handling 76 million TEU per year; thanks to its 66 kilometres of wharves. Since 2006, PSA has handled more containers in its foreign terminals than in those of Singapore; the total numbers of containers handled in Singapore and elsewhere are, respectively, 16.8 million TEU compared to 7.7 in 2002, 20.6 compared to 12.5 in 2004, 24 versus 27.3 in 2006 and, finally, 27 versus 32 million TEU in 2007 (PSA Annual Report 2008). Its principal zones of investment are without doubt Europe and China. In Europe, its terminals are dispersed between Belgium (Antwerp, Zeebrugge), the Netherlands (Rotterdam), Italy (Genoa and Venice), Portugal (Sines) and Great Britain (Great Yarmouth). In China, PSA is involved in the construction of container terminals in Dalian, in Fuzhou Qingzhou and in Fuzhou International, in Guangzhou, in Tianjin, and it even invested in 17 of the 24 container wharves in the port of Hong Kong. Its most recent investment zones are Central and South America (Argentina and Panama), India (4 ports), Pakistan and Turkey. PSA has thereby strengthened its role as an Asian port platform by exerting greater influence over the direction of the flow of trans-shipment, while protecting itself from competition on the part of ports in the Straits and from the ups and downs of the economic conjuncture in East Asia. The port of Singapore has thus surmounted its regional dependence.

The competition among ports along the Straits to attract shipping lines and cargo is rising and this makes it impossible to develop complementarities between ports. This can result in non-optimal allocation of resources and even redundancies in services. At a time when ships are getting bigger and their owners are looking at minimizing the number of port calls, they will only consider berthing at ports which have the features, facilities and cargos to handle and accommodate their big ships. In the race to attract these ships by building bigger ports, which not all the littoral states can afford, there could be an absence of efforts to ensure that the ports in the Straits' region complement each other in terms of their facilities and services.

2. THE STRAITS AS A NATURAL BORDER: AN ASYMMETRICAL PORT SYSTEM

There are many straits that serve as borders, for they represent *a priori* a "natural" limit in a way that is similar to rivers or mountain chains. Since the Treaty of London in 1824, the Straits of Malacca have served as a linear border separating peninsular Malaysia from the Indonesian archipelago. Superimposed on this colonial division of the Malaysian region into two areas of political influence — the islands situated to the south of Singapore were under the dominion of the Dutch while the regions to the north were the reserve of the British — has been a gradually increasing demographic and economic gradient between the two coastlines. The Straits of Malacca link a city state that is perfectly integrated in the world economy (Singapore), the most economically developed coast of Peninsular Malaysia (the west coast) and an island at the extremity of the Indonesian archipelago, located at the periphery of its national territory (Sumatra) (Fau 2004, p. 296). However, a comparison of the port systems of the west coast of Malaysia and the east coast of Sumatra reflects not so much their economic disparities as their divergent maritime strategies: the Malaysian and Indonesian governments each see the role of the Straits with regard to their own interior land management.

The Port Systems: A Reflection of National Territorial Construction

In Indonesia, each successive government since its independence has striven to integrate the whole of its territory, first politically and then economically, into the framework of a centralized system centered on Jakarta and the island of Java. During the period of the New Order (1965–98), the government of Jakarta, fearing the rise of centrifugal forces and secessionist movements, adopted a policy of the closure of its national borders. The Straits of Malacca, although they are centrally located with respect to international navigation, are nevertheless on the periphery of Indonesian territory. As a result, immediately after its independence, the government of Jakarta adopted a law that centred trade once again around the Java Sea: as recently as 2007, out of the 5,085 million TEU handled in the ports of Indonesia, 2,658 million were handled in the terminals of Jakarta and 4,156 million in those of Java, as opposed to only 0.320 million in Belawan (North Sumatra) and 0.82 million in Palembang (South Sumatra) (Patunru et al. 2007). Its strategy also consisted in deviating a portion of the traffic in Malacca towards the straits located entirely in Indonesian territorial waters (particularly those of Lombok and the Sunda) and in restricting access of foreign ships to Indonesian ports: the maritime plan of 1982 authorized four departure points, Belawan (Medan), Tanjung Priok (Jakarta), Tanjung Perak (Surabaya) and Ujung Padang (South Sulawesi), through which all Indonesian transoceanic shipping was required to transit (Fau 2003*b*).

This Indonesian centralism was materialized spatially on the east coast of Sumatra by a concentration of port shipping in Belawan, the port of

Medan. In the nineteenth century, however, during the formation of the *Oostkustgebied*, the eastern belt of large plantations, the Dutch used the many ports located at the mouths of the rivers, which traded freely with the opposite shore of the straits. Shipping to and from Sumatra, however, has been reoriented towards Batavia, the former name of Jakarta, since the territorial unification of the Dutch Indies. Belawan played a role as collector for Tanjung Priok and gradually developed direct relations with the international circuits (Fau 2003*a*, pp. 580–90). Furthermore, at the end of the 1970s, in a context of administrative centralization, the government built the Trans-Sumatra Highway, a road axis 2,500 km in length that links the northern city of Banda Aceh to Bakauheni, the port serving the ferries that link Sumatra to Java (see Muriel Charras in this volume). This roadway not only anchors Sumatra to Jakarta, but also contributes to the concentration of international maritime shipping exclusively in the port of Belawan: the small coastal ports stretching from the north to the south have declined in inverse proportion to the progression of this north-south highway. Although Indonesia lays claim to its status as an archipelago nation, priority has always been given to road facilities.

On the west coast of Malaysia, however, Malaysian federalism and the importance of its international maritime commerce have contributed to the preservation of its multipolar port system. In keeping with the policy and the spatial perception of the former sultanates of Malaysia, the Malaysian federal government has given priority to the sea: 90 per cent of Malaysia's international commerce transits through its ports. The policy of the federal government has always had as a goal to create a multipolar port network (Fau 2003*b*). In 1970, Malaysia disposed of only two well-equipped ports, those of Penang, situated on the island of Georgetown, and Port Klang, the maritime access port of the federal capital, Kuala Lumpur. Starting with the second five-year plan, the national plan for ports (NPP) declared as its goal not only the renovation and development of its existing facilities, but also the construction of new ports that would support and stimulate regional growth. Johor became operational in 1977, Port Dickson in 1982 and Tanjung Pelepas, to the south of Johor, in 2000.

In addition, since the beginning of the 1990s the minor ports of Peninsular Malaysia have also benefited from improvement programmes aimed at stimulating the growth of these major ports. According to the Marine Department Malaysia, there are a total of 60 minor ports in Peninsular. The opening of ports on the east coast, however, began much later and has been less systematic: the port of Kuantan, the most important on the coast, dates only as far back as 1984, following the discovery of petroleum in the region. Minor ports in Malaysia play an important role in enabling producers from the hinterlands of the country to feeder their goods through these ports to the nation's main seaports (Khalid et al. 2011). Most of these ports handle conventional cargo, with the exception of a few that are established for specific types of commodity: for example Port Dickson in the State of Negeri Sembilan, Sugai Udang terminal in Melaka and Kemaman Port in Terengganu handle oil and petroleum-related cargos. Several minor ports are growing fast and have very good features and facilities which enable them to cater to large vessels and handle bigger volumes of cargo from year to year by handling international import and export trade. Lumut Port, which is a private port project initiated by the State Government of Perak, can handle Panamax-class vessels and serve trade internationally with Southeast Asia, Bangladesh, India, Sri Lanka, Pakistan, Middle East and also European countries (ibid.).

Difficulties with Public Financing in Indonesia, Privatization of the Ports in Malaysia

The lack of dynamic growth in Belawan must be considered in light of the lack of performance throughout the entire Indonesian port system. The principal and most competitive port in Indonesia, the Jakarta International Container Terminal, is ranked in all port classifications as the least competitive and the most expensive port in all of Southeast Asia (Carana 2004). This situation can largely be explained by a chronic lack of financing for the renovation and improvement of its port facilities. Although the private sector now plays a preponderant role in the construction, development and management of container terminals, investing 33 billion dollars in the ports of developing countries, of which 44 per cent are located in East Asia, Indonesia has remained on the fringes of this tendency. Law No. 21/1992, which granted four state-owned companies (Pelindo I, II, III and IV) a monopoly in managing the twenty-five strategic ports of Indonesia, rendered impossible the participation of private operators. Now these companies, which cumulate the functions of port regulator, operator and even warehouser, are no longer capable of assuming alone all of these prerogatives (Ray 2009). The new maritime law of 2008 should put an end to this monopoly within the port sector by separating the functions of regulator from those of operator and by authorizing the participation of the private sector (Howard 2008). However, its implementation is associated with the new 2009 maritime plan which calls for a reduction in the number of ports authorized to maintain direct international links: while Belawan would remain open, the flourishing ports of Dumai and Bandah Aceh would be closed.

As early as 1983, the Federal Government of Malaysia was already extensively considering privatization as an option to increase port efficiency in the country with the port authorities undertaking the role of trade facilitator, supervisor and regulator of the operations (Hanizah et al. 2010; Khalid 2007*b*). However, it in no way implies the complete withdrawal of the government, but rather a progressive redistribution of tasks between the public and private sectors, which in the port sector has resulted in the increased specialization of the wharves. For example, the privatization of Port Klang, which was managed until 1986 exclusively by the public Port Klang Authority (PKA), took place in stages: first the privatization of the container terminals of North Port in 1986, then the rest of North Port in 1992 and finally West Port in 1994; only South Port, which is specialized in coastal navigation and is closed to foreign ships, is still managed by the PKA. As for the construction of the new port of Tanjung Pelapas (PTP), it was planned and supported by the federal government and the government of Johor, but was entirely financed, managed and regulated by the private company Pelabuhan Tanjung Pelepas. In addition, the ship operator Maersk disposes of its own terminals and organizes its own handling and logistics operations.

The Unequal Impact of Ports on Local Economic Dynamics

Despite the quality of its location and its transport infrastructures, Belawan remains only a transit port. The development of container shipping has merely served to improve the pre-existing land-to-sea interface; contrary to many other Asian ports, it has not allowed Belawan to develop an export industry for manufactured goods. Its container shipping remains dominated by the export of raw materials (rubber, coffee, tea and palm oil) (Fau 2003*a*, p. 612). The surplus costs engendered by its geographical location at the periphery of Indonesia have not encouraged foreign companies to invest there. In the northern region of Sumatra, this surplus cost is not due to a relative lack of facilities, for Belawan is the third port of Indonesia, nor to the weakness of the interior market or the availability of manpower, for Medan is the fourth largest city in the country, but rather to the absence of decision-making power on the part of the local government. The outside services on which the companies must rely to guarantee their operation are all located in Jakarta.

On the peninsula, on the contrary, the construction or the improvement of port facilities is systematically accompanied by the development of industrial export zones or duty-free zones. Johor Port was the first Malaysian Port to have a free zone. Furthermore, each port is specialized in one type of industry: Port Klang in the manufacturing industries, PTP in packaging and Kuantan in the oleo-chemical industries. The port is the pivotal element of a subcontracting economy working for the multinational and global markets (see maps in Fau 2010). The construction of a new port complex in Penang in 1974 on the site of Butterworth was accordingly accompanied by the opening of two industrial duty-free zones (Prai and Prai Wharf). The quality of its facilities has made it a genuine logistics hub for Penang State, but also for the entire western part of Peninsular Malaysia. Likewise, from the moment of its conception, the new port of Tanjung Pelepas was linked with the contiguous construction of a commercial and industrial duty-free zone where goods that are directly re-exported to foreign countries are de-taxed. The creation of this port corresponds to the political will to stimulate growth at the southern end of the peninsula. Far from being an isolated project, PTP is integrated in the regional development programme called Iskandar. Furthermore, a minor port like Lumut is located near the Lumut Port Industrial Park (LPIP), an industrial estate for investors involved in activities such as manufacturing, boat building, bio-diesel and petroleum products (Khalid et al. 2011).

The Spread of Port Dynamics towards the Interior

Belawan has failed in its function as a catalyst for development (Fau 2003*b*). The rate of penetration of container transport in its traffic, which grew from 13 per cent in 1985 to 21 per cent in 1986, favoured the outfitting of a new container terminal, distinct from the other port facilities. In 1985 the Asian Bank and the governments of the Netherlands and West Germany financed the construction of the Gabion terminal, which handled 321,664 TEU in 2007 compared to 207,808 in 2000. With the exception of this facility, all of the other facilities intended to strengthen the links between the port and the hinterland have failed. The import service for Full Container Loads (FCL) which provides door-to-door service without offloading was set up in 1990, but the demanding administrative procedures, due to the centralized Indonesian government, have hindered its spread. The attempt to develop dry ports disposing of warehouses under customs control has also failed: the dry port of Tebing Tinggi was abandoned only four years after it opened in 1990 (see Map 3.3); the dilapidated state of its railroad facilities rendered the cost of transportation prohibitive in comparison to trucking.

In Peninsular Malaysia, on the contrary, since the middle of the 1960s the road and railway facilities have no longer been used only to carry export goods from the plantations towards the ports; they also favor diffusion and growth towards the country's interior. The priority of the Malaysian government, since the fifth five-year plan, has been the improvement of productivity in the port sector by means of the development of intermodal transport (Khalid 2004), the improvement of links between each port and its surrounding sphere of influence and the multiplication of dry ports (see Map 3.3), such as Ipoh, Padang Besar, Nilai or Segamat, built after the beginning of the 1990s, which play a role as secondary continental poles, collecting and redistributing shipments to and from the major ports (cf. Nazeri Khalid and Abdul Rahim et al. contributions in this volume).

Port Klang thus benefits from the quality of the Malaysian intermodal network. It is linked to the maritime ports and to the dry ports from Sungay Way to Petaling Jaya, from Prai to Penang and from Ipoh Cargo Terminal to Ipoh by ten railroad shuttles per day. In addition, the collaboration between the Malaysian and Thai railroad companies led to the creation in June 1999 of a railway bridge for containers between Bang Sue (Bangkok) in Thailand and Port Klang in Malaysia, resulting in a savings in transit time and cost evaluated by the firm Nedllyod's at 10 per cent of the total transport cost. This rail land bridge linking Port Klang to Bangkok is intended as the first stage of a rail link extending into southern China (Kunming).

Tanjung Pelepas also disposes of an excellent integration with the surrounding land facilities. The Malaysian government has invested 120 million dollars in the construction of 31.5 km of railroad linking the port of PTP to Kempas, the junction station on the main line along the west coast of the peninsula. In 2003, the private company Pengangkutan Timur Pantai Landbridge opened a landbridge between PTP and Hat Yai in the south of Thailand, the short-term objective being to extend it to Bangkok. Similarly, Malaysian Railways (KTM) plans to build a new rail landbridge on the east coast of the peninsula, linking Tanjung Pelepas to Bangkok, in order to relieve traffic congestion on the west coast.

3. THE STRAITS AS AN INLAND SEA: A POLYCENTRIC SYSTEM CRISS-CROSSED BY MULTIDIRECTIONAL EXCHANGES

Considering the straits as an inland sea allows us to go beyond the uniquely land-based territorial logic which places the coastline at the periphery of the nation, and which attributes to it the function of a boundary. Before we consider the straits, and more generally the maritime zone, as a point of rupture, they should be considered first as a link. In the Malay worldview, the sea has never been seen as an obstacle. This conception of the straits requires us to reconsider the notion of proximity and to go beyond a longitudinal sectioning of the straits, separating them into distinct bands, the hinterland, the coastline, the sea, and once the border has been crossed, once again the sea, the coastline and the hinterland; on the contrary, it considers the straits as composed of transversal zones playing complementary spatial roles and favouring the bringing together of the two sides. The Straits constitute a transversal passage point which can serve either to link with, or on the contrary to isolate, the territory located on the other side: this doorway between two land masses can indeed be maintained open, be closed or even be opened selectively according to the effectiveness of the border as a boundary.

During the Period of Closed Borders: Structuring by the Illicit Networks

Transnational shipping does not only exist when and where the political powers favour it; it also develops in the framework of a strategy of circumvention of the nations and their borders. Relegated to an informal, hidden and illegal sphere, this shipping possesses its own hubs, its own logic and traces a geographic map distinct from that of the official circuits. On the east coast of Sumatra, the principal hubs of illegal shipping are the port city of Tanjung Balai and the Riau islands (Batam, Bintan and Karinum) (Fau 2003*a*, chap. 6).

Tanjung Balai is a zone in an economic slump, but is leaving its position of economic marginalization on the sidelines, by establishing itself as a major pole in the contraband traffic. The focalization of international shipping on the port of Medan and the absence of a link between the city and the Trans-Sumatra Highway have transformed this export port, which was in competition with Belawan during the colonial period, into a small regional port. Tanjung Balai now serves only the small, isolated cities on the east coast of Sumatra and the Riau islands, and its international role is limited to supplying a link to Port Klang for the shipment of fresh fish and fruits. But behind its facade of economic decline is an active city that makes its living from contraband, and is ruled by local potentates who control not only its economic activities but also its society in general. Contraband shipping uses the same circuits and the same fruit and vegetable boats as the circuits used for illegal migration.

The second zone of contraband activity, the islands of the Riau archipelago, owes this specialization to its turbulent history. Joined to the east coast of Sumatra in a single province since 1958, this island zone is oriented historically, culturally and economically towards the west coast of Peninsular Malaysia and towards Singapore. The participants in the contraband activities are by no means adventurers operating at night in order to escape notice. They are businessmen engaged in, to borrow the expression used by the local population, "legal contraband", and they benefit from the support (in exchange for remuneration) of customs officials and the army. The Riau islands are specialized in three major types of contraband products: wood, petroleum and powdered pewter. For the smugglers of Riau these activities are not illegal, but are simply the pursuit of their role as intermediaries, the reflection of commercial continuity in a setting of political change.

These two "gray zones" function in tandem and cover each other's activities: the smugglers of Tanjung Balai indicate the Riau islands as their final destination and vice-versa. This system takes advantage of a legislative weakness of which the customs officials complain regularly. These two zones share the shipment of contraband. Tanjung Balai is specialized in contraband oriented towards Malaysia and in particular towards Port Klang, whereas Bintan and Batam are integrated in far more widespread illicit networks that have developed on the scale of East Asia and are based in the city state of Singapore. Singapore serves all the more easily as a "trade turntable" for contraband traffic as the shipments escaping the Customs taxation of Indonesia are not labelled illicit in the duty-free city state (Hastings 2010). According to this liberal logic, it is the banning of trade of certain goods itself that has created the contraband. However, in order not to put the Indonesian government in a difficult position, the government of Singapore does not publish statistics concerning its commerce with Indonesia, contenting itself with quoting Indonesian sources in all its official publications. This therefore makes it impossible to quantify this illicit trade.

In Malaysia, most traffickers uses the ungazetted entry points or "*pelabuhan tikus*" (rat ports) to enter the country. Smugglers from Java enter via Johor and from Sumatra via Penang and Negeri Sembilan and those from Thailand and Myanmar enter via the northern part of the Straits (Permal 2009).

During the Period of Open Borders: New Forms of Economic Cooperation

The north-south configuration of the Straits, which are 1,000 kilometres in length, has led to a geographical organization comprised of two zones of cross-border cooperation, known since the beginning of the 1990s by the generic term "growth triangle" (Fau 2004). The southern triangle, SIJORI, is a zone of cross-border cooperation regrouping Singapore, the state of Johor in Malaysia and the islands of the Riau Archipelago in Indonesia. In 1996 the government of Indonesia proposed extending it to the provinces of Riau and West Sumatra, and in 1997 to all of southern Sumatra (Jambi, Bengkulu, South Sumatra and Kalimantan West). On the Malaysian side, the states of Melaka, Pahang and Negeri Sembilan have also been included since 1997.

When the northern triangle, or IMT-GT, was created in 1993, it regrouped four states of western Malaysia (Perlis, Perak, Kedah and Penang), southern Thailand and two provinces in northern Sumatra (Aceh and Sumatra North). It now includes all the provinces of the island of Sumatra, all the peninsular states except Johor and Pahang, and fourteen regions in southern Thailand, which represent close to 100 million inhabitants. These two zones function basically according to the same logic: they apply the principles of international labour division on a subregional scale. They take advantage of the geo-economic hiatus resulting from the national borders. Their mode of operation depends on the existence of economic gradients (the cost of labour, the level of industrialization and development of the service sector), demographics (the availability of labour) and politics (protectionism or free trade) between the countries bordering the Straits.

Barter Trade and Policies in Favour of Smaller Ports

The development of the "growth triangles" led the border nations to take into account the diversity of maritime shipping in the Straits rather than merely passing legislation on the security of international maritime shipping. The traffic separation scheme of 1 December 1998 is thus the first legal text to establish maritime zones reserved for coastal traffic and for transversal shipping. This scheme guarantees greater security for inshore navigation and for fishermen by separating international transit from inshore and transversal shipping. Between 2000 and 2009, 20 per cent of total Straits shipping comprised cross-straits traffic (Shahryari and Arshad 2011). There is considerable transversal shipping that links the two sides of the Malacca Straits: small capacity cargo ships, outrigger fishing boats, feeder container vessels, tugs and barges and even *sampans* (small local trading craft) making daily runs from Medan or Tanjung Balai to Penang and Port Klang, from Dumai to Melaka and to Port Klang, and from the islands of the Riau archipelago to Singapore. There is also extensive passenger ferry traffic across the Malacca Straits, especially in the Melaka-Dumai corridor (26 per cent of ferry traffic) and from Port Klang which comprises 61 per cent of total of ferry movements to various ports in Sumatra Island (see Map 3.4). In 2009, there are also 290 passenger ferry movements between Batu Pahat Port (Malaysia) and Selat Panjang Port in Indonesia and from Muar Port, there were more than 500 ferry movements to Bengkalis Port, 150 to Dumai Port and only two to Batam. Between 2000 and 2009, almost 39,000 ferry movements occurred between the two regions (ibid., p. 9).

Since the reopening of the borders in the Straits of Malacca, numerous ports have emerged and have multiplied the number of passageways across the Straits. Cross-Straits traffic facilitates trade, enhances regional economic integration and provides better access to global markets. The policies of the governments of Indonesia and Malaysia have contributed to this growth, in particular by favouring the development of small ports specialized in "barter trade" (commerce based on compensation or exchange). The goal is to allow the coastal populations, which are often isolated from the big cities of the interior, to supply themselves with basic goods and farming tools. This traffic is particularly well developed at the two extremities of the Straits, to the north between the province of Aceh and the states of Penang and Perak, and to the south between the province of Riau and the states of Melaka, Negeri Sembilan and Johor: the southern part of the Straits of Malacca accounts for the bulk of the cross traffic due to the existence of various islands and shallow water (see Map 3.4).

Based on statistics from the Maritime Departement Malaysia, there were 206,300 barter trade vessel movement between Peninsular Malaysia and Sumatra Island,[4] originating mainly from Port Klang, Melaka, Muar and Baru Pahat ports to jetties and barter trade ports in Sumatra (more specifically Dumai, Bengkalis, Tanjung Batu, Batam and Tanjung Balai) between 2000 and 2009 (Shahryari and Arshad 2011). The largest share of barter trade movement with Sumatra island was from Port Klang (51 per cent of total movements), followed by Batu Pahat (42 per cent) and Muar (see Map 3.4). Asia Niaga Barter Trade Jetty (Port Klang) handles barter trade from Sumatra and almost 100 per cent of barter trades in Port Klang are Indonesian flagged vessels. Goods traded by them are seafood, vegetables and rice in exchange for groceries such as Milo, milk and snacks (Khalid et al. 2011). The bulk of barter trade vessels are Indonesian-owned. Eleven ports open to "barter trade" in the province of Riau are linked to a dozen ports on the other side of the Straits (Helmi 2009): in each of these ports, the coastal populations are permitted to carry out commercial transactions valuing up to 600 ringgit (130 euros) per crossing, without customs duties or administrative

Map 3.4
Container Throughput at Major Ports in SOM, 2001–11

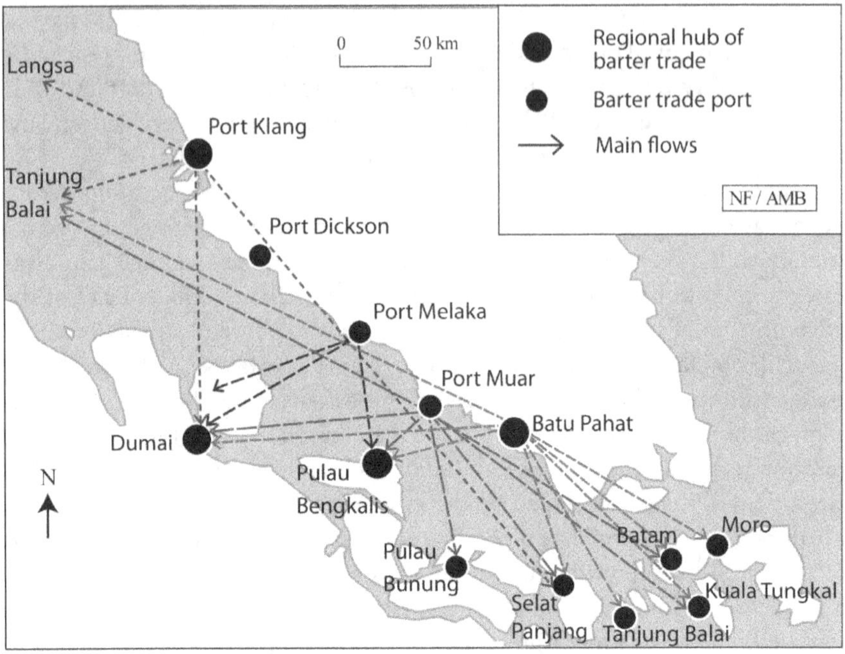

Source: Shahryari and Arshad (2011), Khalid et al. (2011), Helmi (2009).

authorization. The population of Riau is thereby able to trade fresh coconuts, copra, fresh fish and charcoal, as well as pineapples and bananas, in exchange for red and white onions, beans, cooking oil, herring and sardines, but also cans of beer and soda, snacks and baked goods. As a result of the economic crisis, activity in these small ports is expanding rapidly, to such an extent that the local authorities have asked their respective governments to raise the value of duty-free transactions to 6,000 ringgits (1,300 euros).

Despite flourishing, cross-Straits barter trade activities tend to be invisible to the authorities and are in danger of being crowded out, physically and financially. Physically, the importance of providing adequate facilities for the container and major bulk trades may result in a diminution of the space and facilities available for minor trade like barter trade. Financially, investment in facilities for barter trade is not attractive to major port operators, for example, as far as barter trade is concerned, there is a lack of investment in Port Klang's Southport, a terminal specializing in handling barter trade between

Indonesia and Malaysia. One other sensitive issue is the misperception of this trade. Barter trade between the two shores of the Straits is considered as an activity bordering on smuggling by some segments of the authorities, despite its economic advantages and political necessity. Given that it serves the needs of the traders and populations on both sides, there is a pressing need to improve facilities and to clarify the legislation governing barter trade to enable it to be carried out efficiently and to give it a more positive image (Fau and Khalid 2013).

CONCLUSION: GROWTH CORRIDOR AND THE MULTIPOLARITY IN THE MALACCA STRAITS

Port connectivity and economic corridors

This approach by "triangles" was enhanced at the beginning of the millennium by the definition of "growth corridors", a new development tool funded by the World Bank, not only in Asia (ADB 2007) but also in Latin America and in South Africa. Furthermore, in Southeast Asia, the introduction of the concept of "growth corridor" is currently associated with the improvement of "connectivity", a key concept in the policy debates on economic integration in the Association of Southeast Asian Nations (ASEAN), particularly since the adoption of the Master Plan on ASEAN Connectivity (MPAC) in October 2010. One of the objectives of the MPAC is to enhance regional connectivity by promoting subregional initiatives, which usually focus on less developed areas of the ASEAN region with less developed infrastructures (MPAC, p. 29). It is therefore of great importance that regional linkages among neighbouring countries in the region are strengthened, and also transport linkages in subregions within SEA, in order to facilitate trade, enhance regional economic integration and promote the development of the logistics sector to enable regional producers to have better access to the global market. These subregional initiatives are primarily supported by the Asian Development Bank (ADB) which promotes regional economic cooperation through the establishment of effective infrastructure linkages.

The Asian Development Bank (ADB) has included in its new development plan for IMT-GT for the period 2007–11 the creation of three transversal economic corridors, each one linking a pair or a trio of ports: Songkhla-Penang-Medan, Melaka-Dumai and Ranong-Phuket-Aceh (Banomyong 2006). The goal is to help twinned or sister ports to cooperate in the same way as twin cities on opposite sides of a land border. In order to encourage links between these major ports, priority has been given to the improvement of their infrastructures such as bridges and facilities to handle fast ferries and roll-on roll-off (RO-RO) ships. This can lead to mutually beneficial outcomes for the sister ports. For example, the establishment of a RO-RO ferry service between Penang and Belawan and between Dumai and Melaka would result in a decrease in offloading of passengers and an increase in freight traffic.

As with twinned cities, a balance of power is being established: Penang and in particular the North Butterworth Container Terminal port, the only one in the corridor that disposes of a terminal capable of receiving international containers, is taking its place as a new regional hub. The expansion of its hinterland towards the regions of Medan and Songkhla has resulted in the growth of container traffic: 950,000 TEU handled in 2007 in comparison to 720,000 in 2005, but also in the development of direct links with the ports of East Asia: 35 per cent of maritime shipping at present, whereas until 2000 almost all shipping transited through the ports of Kuala Lumpur or Singapore (Fau 2012). In addition, regarding Songkhla and Belawan, the quality of the facilities in Penang is an advantage that is helping the regions to surmount their peripheral status at the national level. The volume and value of their exports does not justify stopovers by the major ship operators. The proximity of Penang thus affords them a daily connection to worldwide commercial circuits without having to bear the cost. Being in a central position within the Medan-Penang-Songkhla corridor, Penang is in a good position to attract container traffic from North Sumatra and from the northern corridor region of Peninsular Malaysia and southern Thailand, and to become the economic centre of the entire corridor.

The twinning of Melaka and Dumai is not as well structured. Transport between the two ports is dominated by passenger transportation: the inhabitants of Riau regularly travel to the other side of the Straits for medical reasons, and exchanges are also favoured by cultural links. As for the last pair, its reality is non-existent for the moment, and its future remains unlikely insofar as Bandah Aceh has stronger relations with Penang than with Phuket.

The Multipolarity of Maritime Borders

This approach by port twinning and growth corridors nevertheless has the drawback of being the transposition into a maritime border of a spatial organization model specific to land borders. The fluid and relatively unconstrained nature of the maritime zone leads to a spatial organization in which "the exchanges are multidirectional, crossing the straits and seas in all directions and thereby creating a genuine network", in which "each link in the chain, each hub of the network, can benefit from several privileged links at a time with other hubs in all directions" (Charras and Franck 2000, p. 104). The twinned ports are not only connected with each other, they also multiply the exchanges across the Straits with ports located farther down in the port hierarchy, in a network that is more complex than a simple "hub and feeder" relationship.

The port of Penang is a good example of this complexity (Fau 2012): each wharf, specialized in one type of cargo, disposes of a different outer harbour within the Straits of Malacca. While the port's container terminal in Butterworth maintains an entirely "hub and feeder" relationship with the ports of Belawan and Songkhla, its wharf at Prai, with a large number of links in close proximity, is a pole for bulk trade; Prai is particularly well connected with the ports of Tanjung Balai and Belawan (North Sumatra) which supply Penang with fresh produce from the highlands of North Sumatra. It also has good linkages with the port of Kuala Langsa (Aceh) and Kuala Enok (Riau) since the flourishing of barter trade: agricultural products, raw materials from the forests and seafood from Sumatra are exchanged for machines and equipment from Malaysia. In 2008, the value of cargo handled through Prai Wharf involving barter trade was 104 million Ringgit and it is trying to expand its barter trade activities to other Asian countries (Shahryari and Ibrahim 2010). The fishermen of southern Thailand also participate in this barter trade, exchanging fresh fish for flour, sugar and powdered milk. As for Penang Port's wharf of Swettenham Pier (Georgetown), it specializes in handling luxury cruise liners. The wharf is now a mandatory stopover in the cruise circuit along the Straits of Malacca, along with the ports of Singapore, Langkawi and Melaka in Malaysia and Phuket and Krabi in Thailand. It is interesting to note that this circuit completely skips Sumatra owing to a lack of passenger terminal and cruise handling facilities there.

In the same way, most of the international freight in Belawan Port is transported via Singapore while the throughput handled at Port Klang only represents between 2 per cent to 5 per cent of the total volume of cargos transiting the Straits. There is only one weekly feeder service between Belawan and Penang, and Singapore Port plays an important role as a trans-shipment hub for North Sumatra owing to its facilities and connectivity.

Specificities of Maritime Borders

An analysis of port organization in the Straits of Malacca brings to light the differences and similarities between land and sea borders. The

Straits encompass a frontier delimiting state sovereignty: the proximity of an international seaway does not suffice as a catalyst for economic dynamism along the coasts that line it. It can perhaps be considered both as an opportunity to seize, the object of all sorts of rivalry, and as a crack in the national defense system; in the first case the coastlines take part in the flow of globalization; in the second they are geographical zones that are crossed.

The maritime border is, however, a very particular type of border; "whereas the dyad of a land border evolves on a two-dimensional plane, with a line and contacts on each side, crosswise, the maritime border functions as a triad, in three-dimensional space, since there is an additional lengthwise trajectory" (Pelletier 2004, p. 253). This third dimension plays a major role in the context of straits, since it participates directly in relations between the Border States that are forced to work together in managing this common zone. But above all, the fluidity of the maritime zone imposes not only the development of relations of proximity on each side of the frontier; the border is crossed in all directions and the links between two shores are neither limited nor constrained by the presence of mandatory points of passage. During open-border periods, the straits then become an axis that gives structure to a transnational zone comprising not only the coastlines of the two shores but also their hinterlands.

Notes

1. MALSINDO initiative was essentially a Malaysia-Indonesia public relations exercise to show to the world that the two littoral states were taking Straits security and were doing something about it so as to forestall possible foreign military intervention in the waterway (Mak 2006). See also Nik (2008) and Ho (2007).
2. For an analysis of the concept of burden sharing in the Straits of Malacca, see Hussin (2007). See also this observation by former Malaysian Prime Minister, Tun Dr Mahathir Mohamad. While foreign countries consider the Straits of Malacca as an international waterway, they are unwilling to bear the cost of keeping the waters clear of pollution, in particular oil, which affects the shores of our country. Tun Dr Mahathir Mohamad, Speech at IKMAL Honorary Fellowship Award, Kuala Lumpur, 20 March 2005.
3. Maritime and Port Authority of Singapore (MPA). <http://www.mpa>.
4. These do not include fishing vessels, illegal vessels, smuggling, human trafficking vessels etc.

References

Abdullah Abdul Aziz and Suppiah Rakish. "Safety of Navigation and Institutional Framework in the Straits of Malacca". In *Profile of the Straits of Malacca: Malaysia's Perspective*, edited by Ibrahim H.M. et Husin Hairil Anuar. Kuala Lumpur: Maritime Institute of Malaysia, 2008.

Amad Hamzad, ed. *The Straits of Malacca, International Co-operation in Trade, Funding and Navigational Safety*. Kuala Lumpur: Maritime Institute of Malaysia, Pelanduk Publication, 1997.

ASEAN. *Master Plan on ASEAN Connectivity*. Jakarta, ASEAN Secretariat, January 2011.

Asian Development Bank (ADB). *Indonesia-Malaysia-Thailand Growth Triangle Development Project, Overall Report*. Philippines, 1994.

———. *Indonesia-Malaysia-Thailand Growth Triangle, Building a Dynamic Future, A Road Map for Development 2007–2011*. Philippines, 2007.

Banomyong, Ruth. *Logistics Development Study of the Indonesia-Malaysia-Thailand Growth Triangle (IMT-GT)*. Bangkok: Thammasat University, 2006.

Bateman, Sam. "Regional Response to Enhance Maritime Security in East Asia". *Korean Journal of Defense Analysis* 18, no. 2 (Summer 2006): 25–53.

———. "UNCLOS and Its Limitations as the Foundation for a Regional Maritime Security Regime". *Korean Journal of Defense Analysis* 19, no. 3 (2007): 27–56.

Batman, Sam et al. "Shipping Patterns in the Malacca and Singapore Straits: An Assessment of the Risks to Different Types of Vessel". *Contemporary Southeast Asia* 29, no. 2 (2007): 309–32.

Carana Corporation. *Impact of Transport and Logistics on Indonesia's Trade Competitiveness*. Washington, 2004.

Chang, Young-Tae, Lee Sang-Yoong and Tongzon J.L Jose L. "Port Selection Factors by Shipping Lines: Different Perspectives between Trunk Liners and Feeder Service Providers". *Marine Policy* (2008): 877–85.

Charras, Muriel and Manuelle Franck. "Quarante ans d'introversion en Indonésie, l'éclipse de toute une région: l'Asie du Sud-Est insulaire dans les recompositions spatiales asiatiques" [Forty Years of Indonesian Introversion: The Fading of a Whole Region. Spatial Reconfigurations of Maritime Southeast Asia]. In *Indonésie contemporaine, Hommage à Jacques Leclerc* [Contemporary Indonesia, a tribute to Jacques Leclerc], edited by F. Cayrac-Blanchard, G. Defert and F. Durand. Paris, L'Harmattan, 2000.

Chia Lin Sien, Goh Mark and Jose Tangzon. *Southeast Asian Regional Port Development: A Comparative Analysis*. Singapore: Institute of Southeast Asian Studies, 2003.

Cleary, Mark and Goh Kim Chuan. *Environment and Development in the Straits of Malacca*. London: Routlege, 2000.

Economic Report 2008/2009. Kuala Lumpur: Ministry of Finance, 2009.

ERIA (Economic Institute for ASEAN and East Asia). *Geographical Simulation Analysis for Logistic Enhancement.* Research Project Report, no. 7–2 (March 2009). <http://www.eria.org/research/y2009-no7-2.html> (accessed July 2011).

Euan, Graham. *Japan's Sea Lane Security, 1940–2004: A Matter of Life and Death?* Oxon, UK: Routledge, 2006.

Fau, Nathalie. *Le nord de Sumatra: une périphérie indonésienne sur le détroit de Malacca; un espace partagé entre intégration nationale et recompositions transnationales* [Northern Sumatra: an Indonesian periphery on the Malacca straits. A shared space between regional integration and transnational reconstruction]. Ph.D. dissertation, Université de Paris X Nanterre, 2003*a*.

———. "Le rôle des infrastructures dans l'organisation spatiale du détroit de Malacca (Indonésie, Malaisie, Singapour)" [The role of infrastructure in spatial organization of the Straits of Malacca (Indonesia, Malaysia, Singapore]. *Cahier du GEMDEV* 29 (2003*b*): 163–84.

———. "Enjeux et organisations spatiales du détroit de Malacca" [Issues and Spatial Organization of the Straits of Malacca]. In *Intégrations régionales en Asie Orientales,* edited by Christian Taillard. NORAO vol. 2. Paris: Les Indes Savantes, 2004.

———. "Les zones franches en Asie du sud-est insulaire (Indonésie, Malaisie, Philippines et Singapour)" [Free Trade Zone in South-East Asia Maritime (Indonesia, Malaysia, Philippines and Singapore]. In *Atlas des zones franches dans le Monde* [World Atlas of Free Zones], edited by François Bost. Paris: CNRS-Libergéo, La Documentation française, 2010.

———. "Singapore's Strategy of Regionalization". In *Making the Global City: Singapore from Srivijaya to the Twenty First Century,* edited by J.L. Margolin, Karl Hack and Karine Delaye. Singapore: National University of Singapore Publishing, 2010.

———. "Penang: du global au local, une ville trans-détroit" [Penang: From Global to Local, a Cross-strait City]. In *Territoires de l'urbain en Asie du Sud-Est, métropolisations en mode mineur* [Urban territories of Southeast Asia and minor metropolisation], edited by M. Franck, Ch. Goldblum and Ch. Taillard. Paris, CNRS Alpha Editions, 2012.

Fau, Nathalie et Khalid Nazery. "Maritime Connectivity in the Straits of Malacca: Linking Regions, Connecting People". In *Port, Maritime and Hinterland Development in Southeast Asia,* edited by Muhammad Subhan, Sabariah Yaakub and Ahmad Bashawir Abdul Ghani. Malaysia: UUM Press, 2013.

Frécon, Eric. *Pavillon noir sur l'Asie du Sud-Est: Histoire d'une résurgence de la piraterie maritime* [Black flag on the Southeast Asia: history of a resurgence of maritime piracy]. Bangkok/Paris: "l'Harmattan", Institute of Research of Southeast Asia, 2003.

Hanizah, Idris et al. "Strengthening the Profile of Malaysian Ports in the National and Regional Port System". In *The Development of Maritime Sector in Malaysia,* edited by Hanizah Idris and Tan Wan Hin. Monograph Series 13. Kuala Lumpur: Institute of Ocean and Earth Sciences, University of Malaya, 2010.

Hastings, Justin V. *No Man's land: Globalization, Territory and Clandestine Groups in Southeast Asia*. New York: Cornell University Press, 2010.

Helmi, Achmad. "Peluang dan tantangan perdagangan Lintas Batas Indonesia-Malaysia" [Barter Trade opportunities and Challenges between Indonesia and Malaysia]. Paper presented in Kuala Lumpur workshop "Jaringan Strategik Malaysia-Indonesia dalam memperkarsakan Perdagangan, Perhitmatan dan Kepenggunaan", 18 March 2009.

Ho, Joshua. "The International Maritime Organization-Littoral State Meeting on Enhancing the Safety, Security and Environmental Protection of the Straits of Malacca and Singapore". *Maritimes Studies* (January–February 2007): 16–21.

Howard, Dick. "The 2008 Shipping Law: Deregulation or Re-regulation". *Bulletin of Indonesian Economic Studies* 44, no. 3 (2008): 333–406.

Hussin Abdul Rahim. "The Management of the Straits of Malacca: Burden Sharing as the Basis for Cooperation". In *Enhancing Security in the Straits of Malacca: Amalgamation of Solutions to Keep the Straits Open to All*, edited by Khalid Nazery. Kuala Lumpur: Maritime Institute of Malaysia, 2007.

Jamal, S. "Northport Ready to Unleash its Full Potential". *Bernama Online*, 30 August 2010.

Khalid, Nazery. "Whit a Little Help from My Friends: Maritime Capacity-building Measures in the Straits of Malacca". *Contemporary Southeast Asia: A Journal of International and Strategic Affairs* 33, no. 3 (December 2009): 424–46.

———. "Maritime Trade and Development". In *Profile of the Straits of Malacca: Malaysia's Perspective*, edited by Ibrahim, H.M. and Husin, H.A. Kuala Lumpur: MIMA, 2008.

———. "Yen and the Art of Malacca Straits Maintenance". *Maritime Affairs*, vol. 3, no. 2 (winter 2007*a*): 50–62.

———. *Global Trend Affecting Port Development and Their Implications for Malaysian Ports*. Kuala Lumpur: Maritime Institute of Malaysia, 2007*b*. <www.mima.gov.my/mima/htmls/papers2007>.

———. "ASEAN Initiatives on Trade and Transport: Implications for Malaysia and Its Maritime Sector". MIMA Research Report (unpublished), 2006.

———. *The Emergence of Multimodalism in the Straits of Malacca Region*. Kuala Lumpur: Maritime Institute of Malaysia, 2004. <www.mima.gov.my/mima/htmls/papers/2004>.

Khalid, Nazery and Ang Abu Hasan. "Promoting Minor Ports to Complement Major Ports in Malaysia". *MIMA Sea Views*, 2 June 2011.

Lasserre, Frédéric et Claude Comtois. "Les détroits maritimes des enjeux stratégiques majeurs" [Maritime Straits: Major Strategic Issues]. *Etudes Internationales* 34, no. 2 (2003).

Lasserre, Jean-Claude. "Le Pas-de-Calais et le détroit de Gibraltar. Quels enjeux géopolitiques? [Pas-de-Calais and the Straits of Gibraltar: What Geopolitical Issues?]. *Etudes Internationales* 34, no. 2 (2003): 195–213.

Leifert, Michael. *International Straits of the World: Malacca, Singapore and Indonesia*. Alphen aan den Rijn, the Netherlands: Sijthoff and Noordhohh, 1978.

Leong The Kong. "The Revolving Fund: A Unique Facility". In *The Straits of Malacca: International Co-operation in Trade, Funding and Navigational Safety*, edited by Hamzah Ahmad. Kuala Lumpur: Maritime Institute of Malaysia, 1997.

Lubis Sjafruddin and Isnaeni Dharmowijoyo. "Multimodal Transport in Indonesia: Recent Profile and Strategy Development". *Proceedings of the Eastern Asia Society for Transportation Studies* 5 (2005): 46–64.

Mak, J.N. "Unilateralism and Regionalism: Working Together and Alone in the Malacca Straits". In *Piracy, Maritime Terrorism and Securing the Malacca Straits*, edited by Graham Gerard Ong-Webb. Singapore: Institute of Southeast Asian Studies, 2006.

Ministry of Transport (Indonesia). *Perencanaan teknis physical distribusi angkutan barang strategis di pulau Sumatra* [Physical planning of the distribution of strategic good transports in Sumatra], 2008.

Nik Ramli Hj and Permal Sumathy. "Security Threats in the Straits of Malacca". In *Profile of the Straits of Malacca: Malaysia's Perspective*, edited by Ibrahim H.M. and Husin Hairil Anuar. Kuala Lumpur: Maritime Institute of Malaysia, 2008.

Patunru Aianto, Nurridzki Nanda and Rivayani. *Port Competitiveness: A Case Study of Indonesia*. Philippines: LPEM et ADB Institute, 2007. <http://www.adbi.org/files/2007.06.26.indonesia.sea.ports.competitiveness.pdf>.

Pelletier, Philippe. "L'Extrême Océan, enjeux territoriaux dans les mers entourant le Japon" [Ocean Far, territorial issues in the seas surrounding Japan]. In *Intégrations régionales en Asie Orientale* [New Regional Organisations in Oriental Asia: Regional Integration in Oriental Asia], edited by Christian Taillard. Paris: Les Indes Savantes, 2004.

Permal Sumathy. "Addressing the Issue of Trafficking of Humans through Malaysia's Sea". MIMA Bulletin 16, no. 3 (2009): 7–13.

Picouet, Patrick and Jean-Pierre Renard. "Les détroits: de nouveaux territoires, l'exemple du Pas-de-Calais" [Straits: new territories, the example of Pas-de-Calais]. *Mappemonde* 65, no. 1 (2002): 7–11.

Ray, David. "Indonesian Port Sector Reform and the 2008 Shipping Law". In *Indonesia beyond the Water's Edge: Managing an Archipelagic State*, edited by Robert Cribb and Michele Ford. Singapore: Institute of Southeast Asian Studies, 2009.

Republic of Indonesia, Coordinating Ministry for Economic Affairs. *Master Plan for Acceleration and Expansion of Indonesia Economic Development 2011–2025*, 2011.

Shahryari Mansoureh and Arshad Mohd. "Crossing the Straits: Risks and Responses". *MIMA Bulletin*, vol. 18, no. 3 (2011): 5–10.

———. *The Ship Carrying Capacity in the Straits of Malacca*. Kuala Lumpur: Maritime Institute of Malaysia (MIMA), 2010.

Taillard, Christian, ed. *Intégrations régionales en Asie orientale* [New Regional Organisations in Oriental Asia: Regional Integration in Oriental Asia]. NORAO vol. 2. Paris: Les Indes Savantes, 2004.

Trace, Keith, Barend Frielink and Denis Hew. *Maritime Connectivity in Archipelagic Southeast Asia: An Overview*. Southeast Asia Working Paper Series, no. 1. Asian Development Bank, August 2009.

Tsuneskawa Jun. "Japan Perspectives on the Straits of Malacca". In *Proceedings of the 6th MIMA International Conference on the Straits of Malacca: Charting the Future, Kuala Lumpur, 23–24 June 2009*, edited by Khalid Nazerih. Kuala Lumpur: Martime Institute of Malaysia (MIMA), 2010.

4

COMPARING CORRIDOR DEVELOPMENT IN THE GREATER MEKONG SUBREGION AND THE INDONESIA–MALAYSIA–THAILAND GROWTH TRIANGLE

Ruth Banomyong

INTRODUCTION

The development of logistics services and communication technologies has revolutionized production and distribution processes, and has created the "global" market. It is within this competitive environment that shippers and consignees require efficient logistics services that can move their goods in the right place, at the right time, in the right condition, and at the right price (Banomyong and Faust 2010).

According to Fujimura (2008), it is therefore of great importance that regional linkages among neighbouring countries in Southeast Asia are strengthened in order to facilitate trade, enhance regional economic integration and develop logistics for better access to the "global" market. It is acknowledged that having an efficient logistics system in Southeast Asia can help not only support regional economic integration but also improve connectivity with the rest of the world (ERIA 2010).

In Southeast Asia there exist a number of subregional initiatives that focus on economic corridor development to help accelerate economic integration such as the Greater Mekong Subregion (GMS) or the Indonesia–Malaysia–Thailand Growth Triangle (IMT-GT). The purpose of this chapter is to present a comparative analysis of corridor development level between these two subregional initiatives.

The chapter begins with a general background description of the GMS and the IMT-GT in order to provide a better understanding of the context involved. The following section introduces the conceptual framework and explains the various types of economic corridor development. Section 3 describes the results of the comparative assessment while Section 4 is the summary.

1. BACKGROUND OF SOUTHEAST ASIAN SUBREGIONAL INITIATIVES

The Greater Mekong Subregion

The Greater Mekong Subregion (GMS) is composed of Cambodia, Lao People's Democratic Republic, Myanmar, Thailand, and Vietnam, as well as Yunnan Province and Guangxi Zhuang Autonomous Region of the People's Republic of China (PRC). In 1992, with the Asian Development Bank's (ADB) assistance, the six countries entered into a programme of subregional economic cooperation, designed to enhance economic relations among the countries.[1] The GMS countries are illustrated in Map 2.2.

This figure provides a graphical illustration of the key economic corridors under development in the GMS. There are currently 3 main corridors in the GMS:

1. The Southern Economic Corridor (SEC) which links Thailand to Southern Vietnam via Cambodia.
2. The East-West Economic Corridor (EWEC) that links Myanmar to Vietnam via central Thailand and Lao PDR.
3. The North-South Economic Corridor (NSEC) which links Yunnan province in China with Bangkok, the capital of Thailand, via Lao PDR and Myanmar, as well as sub-veins that link Yunnan to Haiphong in Vietnam, and Nanning in China to Hanoi, the capital of Vietnam.

According to trade statistics compiled by the ADB, it is interesting to note that there has been a constant increase in trade value between GMS countries but this value is still considered marginal for certain GMS countries which mostly trade with North America, Northeast Asia or Europe. This is particularly true in the case of Vietnam, Thailand and China as, on average, intra-GMS trade represents less than 20 per cent of total trade value. However, for Yunnan and Guangxi province, as China's GMS members, their main trading partners are in fact ASEAN countries such as Vietnam, Myanmar, and Thailand, with an average of around 40 per cent of provincial trade value. This complexity with regards to China can create confusion as China is only a GMS member through its two provinces, but many of the decisions still have to be made at the central government level as not all authority related to GMS cooperation has been delegated to Yunnan and Guangxi (i.e. Colin in this book).

Cross-border trade, in the GMS, is generally undertaken in two forms: formal and informal. It plays an important part in trade between GMS countries. Formal border trade refers to trade transactions conducted through appropriate customs procedures at the border in accordance with rules, regulations and agreements of the Governments involved. Where applicable, customs tariffs are collected. Informal cross-border trade involves transactions that bypass or evade appropriate customs procedures. This type of informal border trade is relatively significant in this subregion as it is much greater than formal border trade. Customs and Commerce agencies in the GMS have guess-estimated informal border trade to be as much as at least two to three times the amount of formal border trade. Nonetheless, these agencies are hopeful that with the establishment of the ASEAN single market and the full implementation of the ASEAN-China Free Trade Agreement, border traders will not need to evade customs procedures as tariffs would have been eliminated.

The Indonesia-Malaysia-Thailand Growth Triangle

The Indonesia-Malaysia-Thailand Growth Triangle (IMT-GT) is a classic growth triangle drawing upon strong economic complementarities, geographical proximity, and close historical, cultural and linguistic ties. Subregional cooperation under IMT-GT is an initiative by the governments of the IMT countries to accelerate economic development in less developed provinces. It began in 1993, and has since grown to include 32 provinces and states, with population exceeding 70 million.

ADB has been a Development Partner of IMT-GT since 2007.[2] IMT-GT is distinct from the other regional cooperation initiatives supported by ADB in important ways: it is a grouping of sub-national entities; the cooperation is private-sector led; and, only one of the three national borders is land-based, the others are nautical boundaries.

The Indonesia-Malaysia-Thailand Growth Triangle (IMT-GT) consists of:

1. Fourteen provinces in southern Thailand: Krabi, Nakhon Si Thammarat, Narathiwat, Pattani, Phattalung, Satun, Songkhla, Trang, Yala, Chumphon, Ranong, Surat Thani, Phang Nga, Phuket;
2. Eight northern states of Peninsular Malaysia: Kedah, Kelantan, Melaka, Negeri Sembilan, Penang, Perak, Perlis and Selangor;
3. Ten provinces of Sumatra, Indonesia: Aceh, Bangka-Belitung, Bengkulu, Jambi, Lampung, North Sumatra, Riau, Riau Islands, South Sumatra, and West Sumatra.

The IMT-GT Road Map identified in 2008 five economic corridors as key to strengthening regional infrastructure to support increased intra- and extra-regional trade, investment and tourism. These are (i) the Extended Songkhla-Penang-Medan Economic Corridor; (ii) the Straits of Malacca Economic Corridor; (iii) the Banda Aceh-Medan-Pekanbaru-Palembang Economic Corridor; (iv) the Melaka-Dumai Economic Corridor; and, (v) the Ranong-Phuket-Bandah Aceh Economic Corridor.

The economic corridor approach has been identified as a core strategy for accelerated development of the IMT-GT, and development of these economic corridors was declared a Flagship Project of IMT-GT at the 13th Ministerial Meeting of IMT-GT in Selangor in September 2006.[3] The GMS programme started in 1992 and the IMT-GT started in 1993 but more progress has been made in the GMS than in the IMT-GT in terms of cooperation and subregional integration through the use of economic corridors as a driver for enhanced connectivity, development and integration. Map 1.1 of this book describes the IMT-GT corridors.

In the IMT-GT countries, Malaysia has substantial trade with both Thailand and Sumatra. There is quite an important flow of goods between Thailand and Malaysia. This trade is bilateral, cross-border and transit as well with a number of goods from southern Thailand utilizing Penang Port as port of loading to other markets. There is also a strong level of trade between Malaysia and Sumatra with Port Klang playing an important role. However, trade between Thailand and Sumatra is not very developed as Thailand only plays a marginal role in Sumatra's export and import trade. Trade linkage between southern Thailand and Sumatra must be further encouraged.

Malaysia plays a central role for trade flows in the IMT-GT. Thai exporters from the south of the country use Malaysia as a transit platform while Sumatra also sees Malaysia as a very important trading partner.

However, Singapore is shown to compete up to a certain extent with Malaysia with regard to transit goods.

2. DEFINING ECONOMIC CORRIDOR

Spatial accumulation tends to occur in productive places, when there is adequate land supply and accessibility, in this context corridors are an obvious choice of regional structure (Llanos 1997). The concept of economic corridor recognizes that development entails infrastructure integrated with other economic opportunities (Brooks and Hummels 2009). These include trade and investment, and efforts to address social and other impacts from enhanced connectivity. However, it is important to better understand corridor characteristics before allowing all corridors can be understood as "economic".

Transport Corridors

The purpose of a transport corridor is to physically link areas that were not previously connected within a country or a region. Transport corridors are characterized by having infrastructure linking nodes within the corridor (Beresford 1999). The most common infrastructure is a road that connects from origin to destination. The construction of bridges and tunnels are considered part of the necessary infrastructure for the establishment of a transport corridor (Beresford and Dubey 1990; ESCAP 2005). People, vehicles and freight should actually be able to flow within the transport corridor when it is completed, as the necessary "hardware" has been put in place to support these movements.

Transport corridors are an expression of spatial structures (Hoyle and Knowles 1992). They have a shape, offer distribution, and are subject to regulations. Wherever economic activities are distributed in space, transport corridor transactions are supported by distribution systems (Black 1993). This also means that whenever the economy is developing, the transactional demand will grow. Transport is a key factor that reinforces spatial inequality by linking *a priori* the most productive places. This is because when a set of economically concentrated areas interacts at the regional level, they reinforce the regional spatial inequality of accessibility by corridors of interactions (Rodrigue 1997). According to the World Bank (1994), transport corridors can provide two fundamental attributes for development: lower distribution costs and land supply for diverse activities.

Multimodal Transport Corridors

A multimodal transport corridor not only provides modal choice alternatives within the corridor but also supports the integration of different modes of transport under the multimodal transport approach. Multimodal transport is essentially an international through-transport combination with various modes of transport such as ship, rail, truck, aeroplane, etc., primarily through the use of containers. Containers will ensure the transport of unitized cargo from its origin to its final destination, with efficiency and least possible risk (UNCTAD 1994). In a multimodal transport corridor there is a need for intermodal transfer facilities to be established along the corridor to allow connectivity between the different existing modes of transport.

Multimodal transport is therefore a concept which places the responsibility for transport activities under one operator, who then manages and coordinates the total task from the shipper's door to the consignee's door, ensuring the continuous movement of the goods along the best route, by the most efficient and cost-effective modes and means, in order to meet delivery requirements.

A multimodal transport corridor is therefore a set of routes between hub centres where maritime, fluvial, land and air transportation systems converge (Fleming 1999). The structure of multimodal corridors lies within a set of interacting hub centres where modal interaction converges. Hub centres can be classified in four modal structures: maritime, fluvial, land and air corridors. Multimodal transport enables economies of scale within a transportation system where modes are used in the most productive manner. This is even more strongly reinforced by "just-in-time" production and "door-to-door" services that require low inventory level and movements between several points of origin and destination.

Logistics Corridors

A logistics corridor focuses not only on the physical connection but also on how the flow and storage of freight, people and vehicles is optimized in the corridor with the support of capable service providers and a facilitating institutional environment provided by relevant agencies.

The main stakeholders in a logistics corridor are shippers and consignees using the various routes along the corridors, service providers offering different types of logistics services, and government agencies involved in the infrastructure as well as the rules and regulations on movement and storage of freight along the logistics corridor.

The potential strength of logistics corridors lies primarily in the possibilities they offer in confronting the concerns and interests of all relevant stakeholders, public and private, who can focus on policies and initiatives to cater to specific routes and border crossings. They thus offer the possibility of tackling logistics issues in a holistic manner (institutional, administrative, and infrastructural), initiating and effecting changes that may otherwise be difficult to achieve at a wider national and/or regional level. The flows of people, vehicles and freight within a logistics corridor are therefore optimized.

Economic Corridors

In an economic corridor, economic development will not be solely concentrated in the large cities located along the corridor. Investment and economic development will need to reach smaller towns and rural areas along its route. This is the ideal situation in any economic corridor development. Incentives to attract private sector investments have to be reviewed and harmonized between different countries along the economic corridor to facilitate economic activities in less developed areas of the corridor. The success of an economic corridor will depend very much on the attraction of investment. Attraction of investment in turn relies on appropriate infrastructure and facilitation policies.

According to the Asian Development Bank,[4] economic corridors promote regional cooperation and have the following characteristics:

1. Cover smaller, defined geographic spaces, usually straddling a central transport artery such as road, rail, or canal;
2. Emphasize bilateral rather than multilateral initiatives, focusing on strategic nodes particularly border crossings between two countries;
3. Highlight physical planning of the corridor and its surrounding area, to concentrate on infrastructure development and achieve the most positive benefits.

Stages of Corridor Development

It is impossible to establish economic corridors at the outset. There is a gradual evolutionary phase that must be followed if their establish-

Table 4.1
Corridor Development Level

Stage	Corridor	Definition
Type 1	Transport Corridor	Corridor that physically links an area or region.
Type 2	Multimodal Transport Corridor	Corridor that physically links an area or region through the integration of various modes of transport.
Type 3	Logistics Corridor	Corridor that not only physically links an area or a region but also harmonizes the corridor institutional framework to facilitate the efficient movement and storage of freight, people, and related information.
Type 4	Economic Corridor	Corridor that is able to attract investment and generate economic activities along the less developed area or region. Physical linkages and logistics facilitation must first be in place.

Source: Adapted from Banomyong (2008).

ment is to be sustainable. The stages of development are shown in Table 4.1.

The assessment of corridor development level can be derived from the supply chain literature when the supply chain overall performance is based on its weakest link (Christopher 2004). The weakest link analogy refers to the fact that even though infrastructures have been developed to provide through transport along any given corridor, a missing link, either physical or regulatory, will hinder the overall capability of the corridor to connect from origin to destination. This would mean that the development level of corridors based on the typology presented in Table 4.1 would depend very much on the capability of each link and node within each targeted corridor.

3. COMPARING GMS AND IMT-GT CORRIDORS

In this section, a comparison of GMS and IMT-GT corridors will be made, based on the assessment framework explained in Section 2. In order to illustrate key differences, certain corridors will be selected to provide in-depth analysis of development level. Two corridors in the GMS will be compared with two IMT-GT corridors.

The GMS North South Economic Corridor (NSEC)

In the assessment of the NSEC, the infrastructure (hardware) along the corridor is still lacking but improving. Hopefully, the infrastructure will be in place and physically connected by 2015. Rules and regulations (software) such as the Cross-Border Transport Agreement are in place but have not been completely implemented. This may require a lot of effort in terms of implementation as some countries still cannot fulfil their contractual obligations. Signing or even ratifying an international agreement does not mean immediate implementation.

Border crossings are the weakest link in the NSEC corridors and special attention must be made to deal with required border issues. Time is of the essence as the infrastructure is moving faster than the institutional arrangements between countries. This may also be a reason why transit trade is currently minimal compared to border trade, though transit can become an important component of trade along the NSEC. Thus, border facilities must not be forgotten as these facilities support the expansion of border trade.

Level assessment conducted on the various branches of the NSEC, based on the proposed classification in Table 4.1, shows that there is currently no level 4, or economic corridors, in place yet along the NSEC. At present, therefore, the NSEC is still primarily a transport corridor.

Tables 4.2, 4.3, and 4.4 illustrate how the overall level assessment is determined by the weakest link in the corridor. Note though that logistics corridors do exist, but only within the boundary of a country, not at a regional level. However, if border crossings are improved with the completion of bridges and full implementation of AFTA framework agreement and the GMS Cross-Border Transport Agreement (CBTA) in 2015, then there is a strong possibility that transport corridors along the NSEC will evolve into logistics corridors.

The same conclusions could be applied to two other corridors: Haiphong-Kunming and Nanning-Hanoi.

Table 4.2
Level Assessment of Route No. 3, Bangkok-Kunming

From	To	Level
Bangkok (Thailand)	Chiangrai (Thailand)	3
Chiangrai (Thailand)	Mae Sai (Thailand) R3E	3
Chiangrai (Thailand)	Chiangsaen (Thailand)	3
Chiangrai (Thailand)	Chiangkhong (Thailand) R3W	3
R3E Mae Sai/Tachilek (Thailand/Myanmar)	R3E Mengla/Daluo (Myanmar/China)	1
R3E Tachilek	R3E Kunming (China)	3
Chiangsaen (Thailand): river	Jinghong (China): river	2
Jinghong (China)	Kunming (China)	3
R3W Chiangkhong/Huayxai (Thailand/Lao PDR)	R3W Boten/Mohan (Lao PDR/China)	1
R3W Mohan (China)	Kunming (China)	3
	Overall level	**1**

Source: Compiled by the author.

Table 4.3
Level Assessment of the Haiphong-Kunming Corridor

From	To	Level
Haiphong (Vietnam)	Hanoi (Vietnam)	3
Hanoi (Vietnam)	Lào Cai (Vietnam)	3
Lào Cai (Vietnam)	He Kou (China)	1
Hekou (China)	Kunming (China)	3
	Overall level	**1**

Source: Compiled by the author.

Table 4.4
Level Assessment of the Nanning-Hanoi Corridor

From	To	Level
Nanning (China)	Pingxian (China)	3
Pingxian (China)	Lang Son (Vietnam)	1
Lang Son (Vietnam)	Hanoi (Vietnam)	3
	Overall level	**1**

Source: Compiled by the author.

The GMS East West Economic Corridor (EWEC)

The overall assessment level of the EWEC is also based on the weakest link of the corridor (see Table 4.5). It must be noted that logistics corridors do exist, but only within the boundary of a country and not at the EWEC or cross-border level.

The current status of EWEC border crossings is still based solely on existing physical links as the institutional framework facilitating border crossing has not been totally implemented. There are two main veins that exist within the EWEC:

1. The route from Tak (Thailand) to Danang (Vietnam), which is part of the original EWEC route agreed upon, designated with support from ADB. Linkages with Myanmar cannot be considered due to international sanctions and embargos. The logistics assessment can therefore only be made between Thailand and Vietnam.
2. The EWEC routing developed by the private sector, which links Bangkok and its surrounding industrial estates in Thailand, with Hanoi and Ho Chi Minh City in Vietnam as its origin and destination points (see Figure 4.1).

Table 4.5
EWEC Corridor Assessment Level

From	To	Level
Mawlamyine (Myanmar)	Myawady (Myanmar)	1
Myawady (Myanmar)	Tak (Thailand)	1
Tak (Thailand)	Mukdahan (Thailand)	3
Mukdahan (Thailand)	Sawanakhet (Lao PDR)	1
Sawankhet (Lao PDR)	Dansavanh (Lao PDR)	3
Dansavanh (Lao PDR)	Lao Bao (Vietnam)	1
Lao Bao (Vietnam)	Danang (Vietnam)	3
	Overall level	**1**

Source: Banomyong, Sopadang and Ramingwong (2010).

Figure 4.1
Policy Driven EWEC vs. Actual EWEC

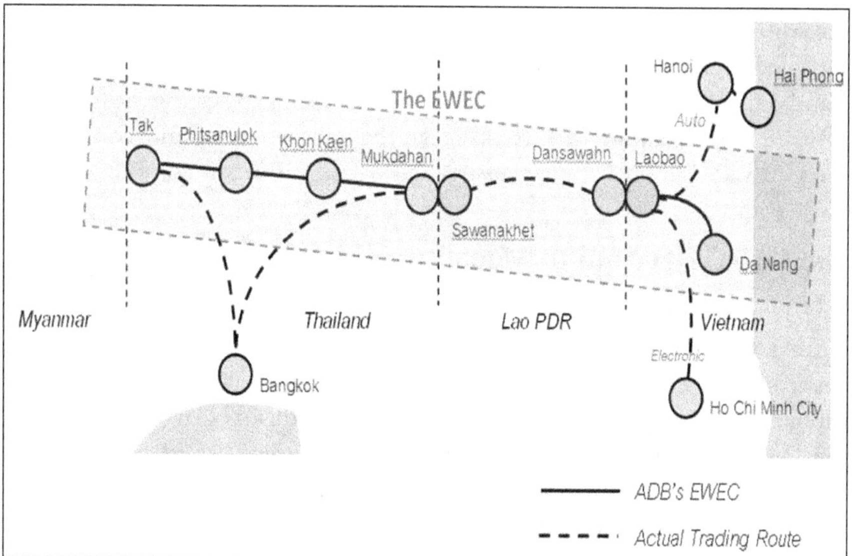

Source: Banomyong, Sopadang and Ramingwong (2010).

The issues in the EWEC are similar to those in the NSEC with border crossings being the weakest link in the system, thus affecting the overall connectivity within both corridors. The private sector also has a different take on the corridor with expanded routing that includes origins and destinations not included in the EWEC agreed on by member countries.

Currently, commodity flows on ADB's EWEC are almost non-existent (Than 2005). This is not surprising as there are no real economic or industrialized hubs along this corridor and trade flows between cities in the EWEC are scarce. On the Thai section of the EWEC, the connection between Tak, Phitsanulok, Khon Kaen and Mukdahan is mostly with Bangkok and its surrounding area. Danang in Vietnam is not sufficiently industrialized to attract higher levels of cargo flows. The main existing product flows within EWEC are mostly from/to Bangkok/Laem Chabang port in Thailand and from/to Hanoi/Haiphong port (automotive products) and Ho Chi Minh City (electronics products) in Vietnam. Current existing flows are illustrated by the "doted" line in Figure 4.1.

However, when appraising freight flows on ADB's version of the EWEC, there is in reality no flow of goods between Danang and Tak province at the border with Myanmar. Myalamine in Myanmar is not even taken into consideration due to the difficult transit situation in the country.

The Southern Economic Corridor (SEC) has not yet been assessed according to the described typology but, based on the existing issues for border crossing between Thailand and Cambodia as well as between Cambodia and Vietnam, it would not be surprising to see this SEC receive a Type 1 classification. This would mean that all these economic corridors in the GMS are currently only transport corridors and nothing more.

The IMT-GT Hat Yai to Medan Corridor

The Hat Yai to Penang is a very important corridor for Thai exporters located in the South of the kingdom while Medan in Indonesia is not a destination. This is particularly true for Thai exporters who are exporting rubber-related products as well as those in agro-industry. Cargo flows within the IMT-GT region are mostly concentrated between southern Thailand and North Malaysia but are minimal with Sumatra. The estimate is that around 12 per cent of Penang cargo throughput originates from or is destined for the Penang area but Penang port is still congested and not well managed from the perspective of some users.

In Belawan port, near Medan, one of the biggest concerns is the access road to and from the port as the infrastructure is in a poor state and cannot cope with the number of trucks. The trucking cost between Belawan and Medan has been quoted at US$100 per TEU with an average transit time of one hour. The cost of actually transporting a container between Belawan and Medan is less than the actual cost of processing the entry of the imported goods. The most important non transport-related cost is the Customs license. If the goods need to be quarantined, then non transport-related cost can be as high as twice the transport cost. Speed money or "tips" do not seem so high compared to all the other related costs. Import cost is usually much higher than export cost, at an average of US$250 to 300.

The overall assessment in the Hat Yai-Penang-Medan corridor is that there are two distinct levels of logistics integration. The linkages between Hat Yai and Penang are quite well developed (i.e. Abdul Rahim and Amel Farhat in this book) while on the other hand the links between Penang and Medan are more on a traditional or "archaic" basis. This does not mean that there is no logistics linkage, it only means that the type of logistics services available is a reflection of the type of trade between Penang and Sumatra.

Table 4.6
Hat Yai to Medan Corridor Development Level

From	To	Level
Hat Yai (Thailand)	Sadao (Thailand)	3
Sadao (Thailand)	Bukit Hitam (Malaysia)	2
Bukit Hitam (Malaysia)	Penang (Malaysia)	3/4
Penang (Malaysia)	Belawan (Indonesia)	2
Belawan (Indonesia)	Medan (Indonesia)	3
	Overall level	**2**

Source: Compiled by the author.

The direct linkage between southern Thailand and Sumatra is almost non-existent or has to be made via Penang or Port Klang. Table 4.6 assesses the development level of the Hat Yai to Medan corridor and provides a description of the corridor level. This table is quite interesting as the overall assessment is only equivalent to a level 2 corridor which is a multimodal transport corridor. This corridor is assessed as a multimodal transport corridor because there is a need to combine road and sea transport in order to move the freight from origin to destination, from Hat Yai to Medan. This corridor is characterized with an origin on the Gulf of Siam side at Hat Yai and then crossing over to Penang in order to cross the Malacca straits.

The weakest point in this corridor is the weak physical linkage between Penang and Belawan port. This seriously hinders the integration of this IMT-GT corridor, even though within each respective IMT-GT country there are domestic corridors that are at a logistics and almost economic corridor development level such as within Malaysia (i.e. Nazeri Khalid in this book).

The IMT-GT Aceh to Palembang Corridor

The Bandah Aceh to Palembang corridor is the only designated domestic corridor in the IMT-GT. It runs from the North to the South of Sumatra Island. This corridor is considered important due to the fragmented nature of economic activities on the island. However, this does not mean that there is actually freight moving to and from Bandah Aceh and Palembang. Freight flows are mostly concentrated between Bandah Aceh and Medan. Medan is the main trading hub for the north Sumatra region. Most of the imported goods from foreign countries transit via the port of Belawan as ports in Aceh have limited capability for containerized freight.

This corridor is characterized by the poor condition of the roads in certain areas and is in need of rehabilitation and upgrading. Local forwarders acknowledged that cost and time may vary depending on the commodity carried as well as the volume. These service providers were also the first to explain that there is no traffic flow between Aceh and Palembang. They stated that traffic moved on shorter distances, Medan to Aceh, Dumai to Pekanbaru, which is the provincial capital of Riau province (i.e. Muriel Charras, Manuelle Franck in this book). In fact it is easier to access Port Klang in Malaysia or Singapore port via Medan and Dumai than actually moving freight from Medan to Dumai.

Table 4.7 is an analysis of the corridor development level between Bandah Aceh and Palembang. Currently there is just a road link within this corridor. There is no railway connection and the road infrastructure in certain sections is lacking. However, since this is a domestic corridor, there is no issue related to import and export regulations. The development level of the corridor is therefore higher than that of the other international corridors but this does not mean that this is the most advanced or developed corridor. It means that there is physical linkage within the corridor as well as a common institutional framework that regulates the movement of freight.

Table 4.7
Aceh to Palembang Corridor Development Level

From	To	Level
Banda Aceh	Medan	3
Medan	Dumai	3
Dumai	Jambi	3
Jambi	Palembang	3
	Overall level	3

Source: Compiled by the author.

It is important to note that the characteristics of the Aceh to Palembang corridor are different from those of the other corridors in the IMT-GT as it is only a domestic corridor. Even though it has achieved an initial assessment as a logistics corridor this does not mean that there is freight moving along the corridor from origin to destination. Comparison with other corridors does not make sense but is necessary to reflect the key difference between a domestic and an international corridor which is the common national institutional framework.

The performance of this corridor is weak when compared with other international corridors in the GMS or the IMT-GT region. The typology is helpful for classification purposes but further performance dimensions will need to be explored before being able to conclude which is the highest performing corridor.

CONCLUSION: INDICATIVE ECONOMIC EFFICIENCY OF GMS AND IMT-GT CORRIDORS

In order to put the comparison of corridors in perspective, the initial assessment results are compared. However, these comparisons are only indicative as each corridor is subject to different characteristics in terms of time, cost and distance.

From the benchmark tables given below, preliminary investigation indicates that actual transport cost ratios in the IMT-GT are somewhat higher than in the GMS corridors. Table 4.8 clearly shows that from a total cost perspective, transport costs in the IMT-GT corridor represent at least 57 per cent of the total cost from origin to destination. Corridors in the GMS are characterized by a higher ratio for non-direct transport activities with the minimum representing 44 per cent and the maximum representing 60 per cent of the total cost from origin to destination. The case of the Aceh-Palembang corridor is somewhat of an outlier as it is the only domestic corridor that is within the scope of the study. Comparatively speaking, this means that formalities and other fees are less than in the other non IMT-GT corridors.

In terms of direct transport time, the IMT-GT corridors do not seem to be as efficient as the others. A lot more time is taken during loading/unloading, administrative and customs formalities. It must not be forgotten that one of the main characteristics of the IMT-GT corridors is that short sea shipping is also involved and thus requires specialized equipment

Table 4.8
GMS and IMT-GT Cost Comparison

Route	Physical Transportation	Non Transport Activity
NSEC: R3W	42%	58%
NSEC: R3E	40%	60%
NSEC: Mekong River	Road 32% River 15%	53%
EWEC: Danang-Tak	56%	44%
IMT-GT: Hat Yai-Medan	67%	33%
IMT-GT: Trang-Melaka	74%	26%
IMT-GT: Melaka-Dumai	57%	43%
IMT-GT: Aceh-Palembang	100%	0%
IMT-GT: Ranong-Aceh	82%	18%

Source: Compiled by the author.

and facilities in order to provide door-to-door service which includes intermodal transfers at key nodes in the corridor.

Border crossing charges are high in the GMS corridors (Kieck 2010). These charges are composed of documents, processing, permit, license fees and the usual under the table payment to facilitate the border crossing of goods. Payment of under table is not mandatory but high levels of delays can be expected if such payment is not made as there will always be some documents or necessary permit missing. This is quite understandable as GMS border crossings suffer from a lack of information related to the necessary procedures related to the transit movement. Users are therefore inclined to pay to the required fees in order not to delay shipment at the border. Border crossing charges along the EWEC are less compared to the NSEC, but that is understandable as along the EWEC there is a pilot site for the Cross-Border Transport Agreement (CBTA). This pilot site is at the border between Lao PDR and Vietnam. The border posts are Densawanh (Lao PDR) and Lao Bao (Vietnam) where single window and single stop inspection are being implemented. This has resulted in higher levels of standardization and transparency during border crossing procedures.

Table 4.9
GMS and IMT-GT Time Comparison

Route	Physical Transportation	Non Transport Activity
NSEC: R3W	80%	20%
NSEC: R3E	85%	15%
NSEC: Mekong River	Road 32% River 54%	14%
EWEC: Danang-Tak	57%	43%
IMT-GT: Hat Yai-Medan	25%	75%
IMT-GT: Trang-Melaka	51%	49%
IMT-GT: Melaka-Dumai	8%	92%
IMT-GT: Aceh-Palembang	100%	0%
IMT-GT: Ranong-Aceh	35%	65%

Source: Compiled by the author.

The issue is somewhat different in the IMT-GT corridors as overall border crossing charges are less when compared to the actual transport cost. This does not mean that there is no under table for border crossing procedures but the border crossing system is already well established when compared to GMS border crossings. Procedures and protocols for border crossings are standardized and users are more familiar with the practices, thus reflecting the higher ratio of direct transport cost in the IMT-GT corridors' total cost.

In Table 4.9, the time dimension in terms of corridor performance is presented and it is very clear that in GMS corridors the majority of time taken is actually the transport from origin to destination. There is in fact an inverse relationship between the money paid at the border and the time taken to cross the said border. The more money is paid the less time it will take. In the IMT-GT corridors, delays occur at the main nodes due to port congestion issues. This is a reflection of the limited infrastructure capacity at key intermodal facilities and lack of frequent shipping services between the main IMT-GT ports.

It is therefore important for all subregional initiatives to have increased and enhanced cooperation. This is particularly true at border crossings where delays are the norm. Trade and transport facilitation can play a role in supporting economic growth within these subregions. However, it is important for all member countries to gain benefits from logistics integration and facilitation and not just a few selected members.

Physical infrastructure is not considered as a major issue, especially between Thailand and Malaysia, as there are already numerous projects in place. Infrastructure rehabilitation within Sumatra is still an important component of improved logistics in Sumatra. The linkages between Sumatra and the other IMT-GT areas also need to be accelerated in order to gain the benefits from such growth triangle. In the GMS infrastructure links are expected to be completed by 2015.

Penang and Port Klang in Malaysia can play a crucial role in becoming the logistics platform for the IMT-GT region with its strong connection with southern Thailand and Sumatra. The overall impression from empirical evidence is that there are two distinct level of logistics integration in the IMT-GT. The linkages between Southern Thailand and North Malaysia are quite well developed and the various corridors that link both countries can be considered as a logistics corridor, almost ready to be a full-fledged economic corridor with the various industrial development projects that are currently being put into place, such as the border dryport project in Bukit Hitam, Malaysia (i.e. Abdul Rahim and Amel Farhat in this book).

However there are still some important issues that need to be resolved, in particular those issues that are related to the current existing institutional framework. Transport rules and regulations should be relaxed in order to promote efficient cross-border logistics and transportation.

Activities within the GMS and the IMT-GT should foster closer public and private collaboration as well as advancing the role of provincial and state governments. Coordination and support among provincial and central agencies in the GMS and IMT-GT are critical as there are still "gaps" in the understanding and the implementation of related subregional agreements. Maybe it is also a question of capacity building from governmental officials' perspective. They need more support in order to understand better how to implement the various related facilitation agreements.

Even if all these supporting policies are implemented, the challenge remains on how to transform all of these corridors into fully fledged economic corridors that can attract investment and generate economic activities in remote areas such as border crossings and secondary nodes along the corridors located far from the subregional economic hubs.

Notes

1 <www.adb.org/GMS/> (accessed 15 January 2011).

2 <http://www.adb.org/IMT-GT/>.

3 For the IMT-GT, the corridor approach will (i) provide focus and serve as a catalyst to the development and growth of ancillary activities and areas within and around the economic corridors, (ii) generate investment opportunities in terms of commercial activities and infrastructure support services, (iii) achieve synergy through clustering of projects within and around the economic corridors (IMT-GT Roadmap of Development 2007)

4 <http://www.adb.org/GMS/Economic-Corridors/approach.asp> (accessed 3 September 2011).

References

ADB. *Connecting Greater Mekong Subregion Railways: A Strategic Framework.* Manila: Asian Development Bank, 2010.

Banomyong, Ruth. "Logistics Development in the Greater Mekong Subregion: A Study of the North-South Economic Corridor". *Journal of Greater Mekong Subregion Development Studies* 4 (2008): 43–58.

———. "Benchmarking Economic Corridors Logistics Performance: A GMS Border Crossing Observation". *World Customs Journal* 4, no. 1 (2010): 37–46.

Banomyong, Ruth and Peter Faust. "GMS Logistics Development Strategy". Asian Development Bank, RETA 6450 Enhancing Transport and Trade Facilitation. Unpublished Report. Manila, 2010.

Banomyong, Ruth, A. Sopadang and S. Ramingwong. "Logistics Benchmark Study of the East West Economic Corridor". *Business & Management Quarterly Review* 1, no. 2 (2010): 1–12.

Banomyong, Ruth, P. Cook and P. Kent. "Formulating Regional Logistics Development Policy: The Case of ASEAN". *International Journal of Logistics Research & Applications* 11, no. 5 (2008): 359–79.

Beresford, Anthony K.C. "Modelling Freight Transport Costs: A Case Study of the UK-Greece Corridors". *International Journal of Logistics: Research and Applications* 2, no. 3 (1999): 229–46.

Beresford, Anthony K.C. and R.C. Dubey. *Handbook on the Management and Operations of Dry Ports.* RDP/LDC/7. Geneva: United Nations Conference on Trade and Development, 1990.

Black, William R. "Transport Route Location: A Theoretical Framework". *Journal of Transport Geography* 1, no. 2 (1993): 87–94.

Brooks, Douglas and David Hummels, eds. *Infrastructure's Role in Lowering Asia's Trade Costs: Building for Trade.* Cheltenham: Asian Development Bank Institute and Edward Elgar Publishing, 2009.

Christopher, Martin. *Logistics and Supply Chain Management: Strategies for Reducing Cost and Improving Service*, 3rd ed. Harlow: Prentice Hall, 2004.

ERIA. *Comprehensive Asia Development Plan*. Jakarta: Economic Research Institute for ASEAN and East Asia, 2010.

ESCAP. *Free Trade Zone and Port Hinterland Development, Economic and Social Commission for Asia and the Pacific*. United Nations, New York, ST/ESCAP/2377, 2005.

Flemming, Douglas. "A Geographical Perspective of the Transhipment Function". Paper presented at International Association of Maritime Economics (IAME) Conference, Halifax, Canada, 14 September 1999.

Fujimura, Manabu. "Economic Integration in the Greater Mekong Subregion and Cross Border Transport Infrastructure". *Journal of Greater Mekong Subregion Development Studies* 4 (2008): 21–42.

Grant, D.B., D.M. Lambert, J.R. Stock and L.M. Ellram. *Fundamentals of Logistics Management*. European ed. McGraw Hill, 2006.

Hoyle, B.S. and R. Knowles, eds. *Modern Transport Geography*. London: Belhaven, 1992.

Kieck, Erich. "Coordinated Border Management: Unlocking Trade Opportunities Through One Stop Border Post". *World Customs Journal* 4, no. 1 (2010): 3–14.

Llanos, J. "The Linkages between Transportation Investments and the Regional Development: Flaws in the Macro-economic Approach". In *Proceedings of the 7th World Conference on Transport Research*, edited by Hensher D., J. King and Th. Oum. Oxford: Pergamon, 1997.

Rodrigue, J.P. "Transportation Corridors in Pacific-Asian Urban Regions". In *Proceedings of the 7th World Conference on Transport Research*, edited by Hensher D., J. King and Th. Oum. Oxford: Pergamon, 1997.

Than, Mya. "Myanmar's Cross-Border Economic Relation with the People's Republic of China and Thailand in the Greater Mekong Subregion". *Journal of GMS Development Studies* 2, no. 1 (2005): 37–54.

UNCTAD. *Multimodal Transport and Trading Opportunities*. Geneva: UNCTAD, UNCTAD/SDD/MT/5, 1994.

World Bank. *World Development Report 1994: Infrastructure for Development*. New York: Oxford University Press, 1994.

PART II

NATIONAL POLICIES RELATED TO REGIONAL INTEGRATION

5

THE PARTICIPATION OF YUNNAN PROVINCE IN THE GMS
Chinese Strategies and Impacts on Border Cities

Sébastien Colin

Since 2010, China has undoubtedly been considered by most observers as a leading actor and sponsor of the Greater Mekong Subregion (GMS) cooperation programme. A member of this programme since its inception in 1992, China has mostly increased its participation in the course of the 2000s, especially through greater involvement of the central authorities. The National Development and Reform Commission, the Ministry of Foreign Affairs and the Ministry of Finance have in this way co-authored in 2002, 2005 and 2008, three state reports on the participation of China in the GMS, in which they emphasized the importance for their country of cooperating with the States of the Indochinese peninsula in the development of transport infrastructure, the establishment of economic corridors, the facilitation of border crossing or the increasing of hydropower production.

However, the participation of China in the GMS is not limited to the central authorities. Yunnan Province and Guangxi Zhuang Autonomous Region, both bordering the Indochinese peninsula, are also two major Chinese actors. Although they did not have the power to negotiate directly and sign agreements with the central authorities of other Member

States of the GMS, they are nevertheless crucial to the implementation of projects and the establishment of some strategies, which they often try to defend at the central authorities' level. Crossing by the headwaters of the Mekong River (called *Lancangjiang* in Chinese territory), Yunnan Province was designated in 1992 by the Chinese central authorities for participation in the GMS. The integration of the Guangxi Autonomous Region is more recent and dates from 2005. It results from the desire of the Chinese central authorities to symbolize the more active involvement of China in the GMS. If this integration of Guangxi reflects a geographical expansion of regional cooperation, whose boundaries now largely exceed those of the Mekong River Basin, it raises the question of potential competition with the province of Yunnan. Timothy A. Summers (2008), for example, revealed that since the integration of Guangxi in the GMS, officials and academics from Yunnan are keen to present their province in their writings as "the main Chinese representative in the GMS" and as being "in the first line of cooperation".

The more active participation of China in the GMS required in advance and in parallel, in the course of the 1990s and 2000s, an improvement of political relations with the bordering countries, but also with ASEAN, which meanwhile was being geographically extended to the whole Indochinese peninsula. The Asian financial crisis of 1997–98 was also, in hindsight, a turning point because, far from scuttling the cooperation process, it raised awareness in China and other countries in the region of the need for developing mechanisms leading to cooperation and regional integration. Finally, the decision in 2002 by China and ASEAN to establish the China-ASEAN Free Trade Agreement (CAFTA), whose coverage will extend to the entire Indochinese peninsula in 2015, has also contributed heavily to boosting cooperation projects within the GMS.

The participation of Chinese actors in the GMS serves several national and international interests. It emphasizes first the strategic importance of the Indochinese peninsula as a land passageway between China and "the South Seas" (i.e., the South China Sea and the Indian Ocean) which could be a possible alternative to certain maritime routes. It is also motivated by the desire to increase Chinese influence in the Peninsula while reducing the role of Japan which has been in this area for several years through an active policy of development aid and its dominant position within the Asian Development Bank. In this sense, China's participation in the GMS is a kind of modern version of an older, long-term Chinese tropism toward Southeast Asia. Economically, greater integration between China and the countries of the GMS provides Chinese economic actors with a market exceeding 200 million people and an access to raw materials. More generally, for the Chinese central authorities, the GMS is a component of their "border opening policy" while at the scale of Yunnan and Guangxi it meets provincial strategy of socio-economic development and willingness to open up, especially through the establishment of North-South economic corridors. In summary, for Kunming and Nanning, as for Beijing, the main official objective is to generate economic development for raising "backward regions", mostly rural and populated by ethnic minorities, above the poverty line.

It is precisely this national policy of border opening as well as these provincial strategies, especially that of Yunnan, that I propose to review in the first two parts of this article because they are more likely to highlight the importance of the GMS to China. The choice of focusing my analysis on Yunnan province is explained by the length of its participation in the GMS, its geographical position bordering three States of the Indochinese peninsula and its greater dependence on the establishment of three main North-South road and rail corridors already existing or under construction (Kunming-Hanoi, Kunming-Bangkok and Kunming-Rangoon) unlike the autonomous region of Guangxi which has, for example, other alternative projects like the "Pan-Beibu (Tonkin) Gulf Economic Cooperation".

In the past few years these three main road and rail corridors have helped to transform the outlying areas that they cross, especially the border cities. In this context, it seemed important to me in the last part of this article to focus on these border towns. Indeed, the analysis of the participation of Yunnan Province in the GMS also requires taking into account the dynamics present across these cities which, because of their geographical position at the intersection of transportation corridor and border, are theoretically supposed to become important centres of activity within the GMS. So, what are the plans and ambitions of these cities? And what are the economic and spatial dynamics that occur there? To provide some answers to these questions, I have selected a series of border towns on the three north-south routes described above. They are Hekou, Ruili, Jinghong and the two border trade "ports" of Mohan and Daluo (see Map 5.1). More generally, this study of border cities enables the impact of Chinese border opening and GMS integration policies on the peripheral province of Yunnan to be analysed.

Map 5.1
Cities and Transport Network in Yunnan Province

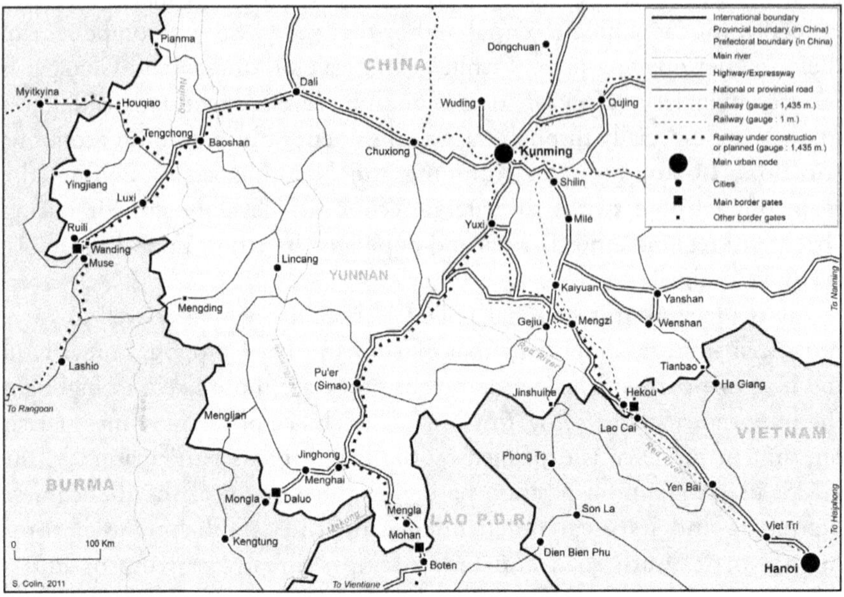

Source: Data compiled by S. Colin (2011).

1. GMS AS A COMPONENT OF CHINESE BORDER OPENING POLICY

For more than two decades, Yunnan border regions have been at the centre of several national and provincial development strategies whose objectives overlap and mostly correspond to those of the GMS. Among them, the most important is "border opening policy", one of whose main objectives is to promote cross-border economic cooperation. Started at the beginning of the 1980s with some provincial initiatives, it became really institutionalized by the Chinese central authorities in 1992, i.e. the same year as the launch of the GMS cooperation programme. The implementation of this policy and its development required the establishment of a favourable international context. Indeed, the improvement of bilateral relations between bordering countries played and still plays a determining role for both central and local authorities to ensure the success of different cross-border cooperation projects.

An Overall International Context Favourable to Border Opening and Cross-border Cooperation

The resumption of Sino-Soviet relations in the second half of the 1980s followed by the collapse of the USSR in 1991 represented together a turning point. They left the way open for China to resume its relations with Vietnam and Lao PDR. The redefinition of their common borderlines inherited from the end of the nineteenth century, which took place very quickly with Lao PDR (1991), and slightly more slowly with Vietnam (1999), is emblematic from this point of view.

Despite the resumption of trade links, relations between China and Vietnam wavered greatly during the 1990s. Hanoi constantly hesitated between permissiveness and protectionism, according to changes in its perceptions of security. However, although distrust and fear of Chinese domination have still not entirely disappeared in Vietnam, and although the question of disputed islands in the South China Sea remains a very sensitive subject, it must be admitted that bilateral relations, and especially economic interdependence between the two countries, have become more sustained since the beginning of the 2000s. Many meetings have taken place between the two Communist Parties to tackle common issues, such as the combination of Socialism and the market economy or the securing of power in periods of economic opening. From 2004 onwards, with the strengthening of the economic corridor strategy during the second decade of the GMS, at the time when Guangxi joined the transnational programme, discussions were held concerning the modernization and building of new transport infrastructures and the simplification of customs formalities. In May 2007, Hu Jintao and Nguyen Minh Triet reaffirmed at the highest level their desire to implement the bilateral "Two corridors, one circle" plan,[1] in view not only of speeding up the integration of their economies, especially in border regions, but also of meeting the requirements of the GMS and ASEAN (Do Tien Sam and Ha Thi Hong Van 2009). Efforts are currently being made to modernize trade on the Sino-Vietnamese border and encourage economic development thanks to the creation of cross-border economic cooperation zones (Cross-Border Economic Zone — CBEZ) at major border-crossing points such as Hekou-Lào Cai.

With Lao PDR, the strengthening of bilateral relations has considerably speeded up since Jiang Zemin's official visit in 2000. Chinese aid, via the buying up of debts and the construction of official buildings, is considerable. In the economic field, China increasingly competes with Vietnam and Thailand, Lao PDR's two traditional partners. There is an increasing

Chinese presence in the north of the country, which has contributed to making considerable changes in the local economy in the past few years. It can be seen in major investments in agriculture (rubber, sugar cane), trade, hydroelectricity, the mining sector and tourism (see Danielle Tan's paper in this volume).

Finally, in Myanmar's (formerly Burma), the Chinese Communist Party's progressive withdrawal of its support for the Burmese Communist Party during the 1980s initially led to a closer relationship with the military junta. Bilateral relations were then strengthened from the early 1990s onwards, when each of the two States in turn was subjected to international sanctions — Myanmar in 1988 and 1990 and China in 1989 — for crushing the democratic aspirations of part of their society. Since then, China has become Myanmar's chief arms supplier and one of its main economic partners. However, in spite of growing economic and logistic integration between the two countries, their bilateral relations are not without limits. In particular, Myanmar's continuing instability, both internally and along the border, continues to preoccupy the Chinese authorities and sometimes complicates border cooperation (see Ella Vignat's chapter). More recently, the resumption of a dialogue between Myanmar and the United States, coupled with some measures of openness and reform initiated by the Burmese regime, are new elements that could influence the Sino-Burmese cooperation process.

Opening Borders and Developing Outlying Regions

The border opening policy was launched at the national scale in 1992. It was indeed at this date that the State Council set up "National Level Economic Cooperation Border Zones" (*guojia bianjing jingji hezuo qu*), whose administration is directly dependent on the abovementioned Council, and promulgated the opening of "first category" (or national level) border trade "ports" (*yilei bianjing kou'an*).[2] The Chinese central authorities then opted for the decentralization of economic decision-making powers to the border provinces and to certain local authorities situated at sub-provincial administrative level. The latter are thus authorized to receive foreign investments, set up economic zones, manage border trade and sign contracts for border cooperation with neighbouring countries. Beijing also provides subsidies and reduced-interest loans, and also a certain number of tax privileges, to help them finance infrastructure projects, encourage trade and attract investments.

Since the beginning of the 2000s, this border opening policy has taken shape as a national programme (*Quanguo xingbian fumin xingdong guihua gangyao*), consisting of "Building prosperous border regions". Its objective is to distribute the benefits of economic growth beyond the coast towards outlying regions, to reduce social and regional inequalities, and, via the setting up of various partnerships with neighbouring countries, to work towards China's increasing integration into its regional environment. Its main themes include the development of transport infrastructures (roads, motorways, railways, airports) and telecommunications, and the construction of new energy installations such as hydroelectric power stations. At the same time, during the 2000s, the Chinese central authorities launched other national policies to reduce regional imbalances and social inequalities, whose spatial impacts in border areas such as Yunnan are considerable. They are the "Open up the West" policy,[3] of which the province of Yunnan is officially part, the "Building a New Socialist Countryside" programme,[4] made official in 2005, and an urban policy aiming to absorb the major part of the future rural exodus.[5] We can also add the November 2008 reflationary plan and the national rail development plan, whose objective is to extend the railway network to 120,000 km by 2020 and to 170,000 km by 2030, including in some border areas.

An Opening which Relies on Cities and the Development of Border Trade

In this framework of border opening policy, the provincial authorities quickly revealed themselves as important actors. With the help of provincial economic and academic actors, they have designed many cross-border cooperation projects with the local authorities of neighbouring countries, planning the building of transport infrastructures and trying to give prominence to the geographical, economic and cultural features of their provinces, in order to attract national and foreign investments. They also encourage specialized production with the aim of being complementary to bordering countries and dynamizing foreign trade.

The authorities rely mainly on the cities, especially provincial capitals, and above all, border cities. The first, where decisions are made, obviously play a primordial role in the opening policy, and are the most important from an economic point of view. They are relayed by border cities, especially those with economic zones and national level border trade "ports". These may be administrative centres of greater or lesser importance and, in demographic terms, often correspond to small or medium-sized towns. It should also be noted that the central and local authorities have also encouraged cross-border ethnic minorities to trade and renew transnational economic links with their counterparts on the other side of the border, so as to associate them with the development thus created.

The State Council also legislates on border trade, allowing customs duty and tax exemptions for small-scale trade between neighbouring populations. Thus, during the 1990s, alongside State and collective firms, residents began to take part progressively in both border trade, and barter trade. The year 1992 saw the halving of import tax and the end of export quotas for goods (with the exception of some "strategic" food and energy products: rice, soybeans, maize, petroleum products). During the same period, permission to sell imported products throughout the country also dynamized these trading activities.

The legislation recognizes four type of border trade. The first is small-scale trade, found in border markets (*bianjiang hushi dian*) and between residents living less than 20 km from the border. The second consists of trading activities, controlled by small firms in "open" border towns with a national or provincial level border trade "port". The third takes the form of occasional purchases made in neighbouring countries during tourist visits. The fourth, and last, is larger-scale trade carried out by companies based elsewhere in the province or the country, whose commercial activities may be linked to economic or technological cooperation, for example via joint ventures created with counterparts over the border.

However, these policies linked to border trade and its vitality vary greatly from one border to another according to the level of economic development of the border regions of neighbouring countries, the absorption capacity of local markets and, of course, the geopolitical situation. Overall, the structure of border trade is the same everywhere. Chinese exports are mainly agricultural and manufactured products (consumer and intermediate goods), whereas imports are mainly raw materials.

2. YUNNAN'S DEVELOPMENT STRATEGY: TURNING AN OUTLYING REGION INTO A CENTRAL AREA

In 2010, the province of Yunnan is still at the bottom of the national hierarchy from an economic point of view. Furthermore, on a provincial scale, the southern and eastern border regions, densely populated by ethnic minorities, are overall less urbanized and industrialized than the northeastern region of Yunnan, polarized by the provincial capital, Kunming. However, since the middle of the 1980s, the central and provincial authorities have implemented a joint strategy with the aim of transforming this outlying province into a new economic hub and reducing territorial economic disparities both within China and within Yunnan province itself. This strategy includes the redevelopment of old economic activities, such as mining and agriculture,[6] as well as the active exploitation of the province's high hydroelectric potential,[7] but also new initiatives directly concerned in GMS development (Gan and Chen 2004, p. 540; Wen Jie 1997): opening of borders to favour border trade and concrete implementation of the "bridgehead strategy" aiming to transform Yunnan into a province at the crossroads of trade flows between China, Southeast Asia and South Asia.

Opening and Revival of the Former "Southern Silk Road"

To contribute to the development of border trade with the Indochinese peninsula, the borders of Yunnan have been progressively opened. The starting point was in 1984 when the Chinese authorities allowed twenty-seven border districts in the province to establish trade relations with the neighbouring countries. This decision was followed the same year by the approval of a "Provisional ruling for the management of small-scale border trade", giving the local authorities full autonomy to carry out this type of trade (D'Hooghe 1994, p. 300). In 1991, the Yunnan authorities opened a series of border trade "ports" at provincial or second category level (*erlei bianjing kou'an*), mostly located along the Burmese border. Finally, between 1992 and 1993, the central authorities promoted six border areas in Yunnan to "First-category border trade ports" — Hekou, Wanding, Ruili, Mohan, Tianbao and Jinshuihe — while Xishuangbanna airport, located in Jinghong, and the river ports of Jinghong and Pu'er (Simao) were promoted in 1995 and 2001, respectively (see Table 5.1).

In the long term, as can be seen in numerous slogans,[8] the provincial authorities' strategy is above all to make Yunnan, still an outlying area within Chinese territory, into a bridge or crossroads between the three major markets of China, Southeast Asia and South Asia. This strategy consists of reviving the main branches of the former "Southern Silk Road" thus destroying several decades of enclosure, responsible, according to Kunming, for the province's backward economic development. During the last two centuries of the Qing dynasty (1644–1911), Yunnan province was crossed by important trade routes: some were located in the north and east of the province and connected to the rest of the Chinese empire via Sichuan and Guizhou, while others, going west and south, were cross-border routes linking the province to the Indochinese peninsula and beyond, towards South Asia.[9] This network was structured by five trading nodes controlled by the Chinese customs: Dali (Tali) and Kunming (Yunnan-fu), the two main

Table 5.1
Yunnan Trade "Ports"

Border Trade Ports	Type	Border Location	First Category (National Level) Date Opened	Second Category (Provincial Level) Date Opened
Kunming	Airport			
Tianbao	Road	Vietnam	1993/02	
Hekou	Railway and road	Vietnam	1993/05	
Jinshuihe	Road	Vietnam	1993/11	
Xishuangbanna	Airport		1995/12	
Jinghong	River	GMS countries	2001/06	
Simao (Pu'er)	River	GMS countries	2001/04	
Mohan	Road	Lao PDR	1993/12	
Wanding	Road	Myanmar	1992/06	
Ruili	Road	Myanmar	1992/06	
Houqiao	Road	Myanmar	2000/04	1991/08
Daluo	Road	Myanmar	2007/11	1991/08
Mengding	Road	Myanmar		1991/08
Pianma	Road	Myanmar		1991/08
Nansan	Road	Myanmar		1991/08
Menglian	Road	Myanmar		1991/08
Zhangfeng	Road	Myanmar		1991/08
Yingjiang	Road	Myanmar		1991/08
Lunyuang	Road	Myanmar		1996/09
Tianpeng	Road	Vietnam		1996/09

Source: Zhongguo kou'an nianjian 2001 [China's trading port yearbook], (Beijing: Zhongguo haiguan chubanshe, 2002), pp. 683–91; compiled by the author.

crossroads of the north and south roads, and three border posts, Tengchong, Simao and Mengzi, which were opened to western influence at the end of the nineteenth century.[10]

Bridgehead and Urban Development Strategies

Motivated by geopolitical and economic necessities, this "bridgehead" strategy was formulated in Kunming in the second half of the 1980s[11] before being politically and financially supported by the central authorities in the 1990s–2000s.[12] Its materialization necessitates the prior development of a real provincial network of modern transport infrastructures, well connected to the rest of China and neighbouring countries. Thus, after finishing the construction of the Kunming-Nanning[13] and Kunming-Dali railways during the 9th Five-Year Plan (1996–2000), in the 2000s the central authorities started to extend the motorway and rail networks toward border regions and the Indochinese peninsula. This extension was consistent with the plan prepared by several meetings of the GMS where Yunnan Province actively defended the setting up of "One region and three corridors" (Kunming-Mandalay-Rangoon, Kunming-Hekou-Hanoi-Haiphong and Kunming-Vientiane-Bangkok). In November 2004, during a conference of the United Nations Economic and Social Commission for Asia and the Pacific, held in Bangkok, the Chinese representative also signed a preliminary agreement on the construction of trans-Asian rail routes, including the connection of the Yunnan railway network to those (existing, under construction or planned) of Myanmar, Lao PDR and Vietnam (UNESCAP 2004).

The success of Yunnan's strategy can also be seen in the building of an urban network given hierarchy and structure by the main transnational transport corridors. The main effort was to transform Kunming, the head of the network, with a population of about 2.5 million, into a new regional centre for Southeast and South Asia. To do this, the provincial authorities launched an ambitious urban restructuring and development project for the 2008–20 period, similar to that carried out in the main cities along the seaboard such as Shanghai and Tianjin, or in Beijing.[14] Another recent example of this objective is the construction of a new international airport, the Kunming Changshui International Airport,[15] inaugurated in February 2011. It should be connected by a high-speed rail link to the centre of Kunming by 2015. It is supposed to become the fourth biggest air hub in the country, after Beijing, Shanghai and Guangzhou, before asserting itself as a regional hub for Southeast and South Asia.

At a lower level on the urban scale are a series of medium-sized or small cities, industrial and/or tourist centres, located along the main transport routes, with populations between 500,000 and one million inhabitants (Dali and Baoshan), or less than 500,000 (Kaiyuan, Gejiu, Mengzi, Hekou, Yuxi, Pu'er, Jinghong, Chuxiong, Luxi and Ruili) (see Map 5.1). Among these cities, all of which want to build up a partly internationalized economic network, Hekou, Jinghong and Ruili stand out because of their border position, now seen as a factor of internationalization and a "geographical advantage". They are supposed to attract capital, technologies and talents so as to become real economic growth centres, not only in Yunnan, but also at the level of mainland Southeast Asia (He and Sheng 2005, pp. 304–7). So what is the current status of their projects, ambitions and dynamics?

3. IMPACTS ON YUNNAN'S BORDER CITIES: EXAMPLES OF HEKOU, RUILI AND JINGHONG

On the scale of China's urban network, Hekou, Ruili and Jinghong are small cities governing a rural environment, mainly populated by ethnic minorities (see Table 5.2).

Hekou and Its Cross-border Cooperation with Lào Cai

Hekou is a small border town located on the Sino-Vietnamese border, and crossed by the Kunming-Hanoi corridor. It is the capital of the Yao autonomous district of Hekou, which depends, administratively, on the Hani and Yi autonomous prefecture of Honghe. Located at the confluence of the Red River, the main river route between Yunnan and Vietnam, and the Nanxi River, Hekou and its Vietnamese counterpart, Lào Cai, are "twin cities". The two urban areas are geographically contiguous, separated only by the two narrow, shallow rivers. This characteristic is illustrated locally by the expression "two countries, one city" (*liangge guojia, yige chengshi*), which is used, as we will see below, as the basis of an ambitious cross-border economic zone. The completion of the motorway between Kunming and Hekou at the end of 2008, the inauguration of a third bridge between the two sides of the Red River in 2009 and the building of the new Kunming-Hekou railway, due for completion in 2013, are recent vectors of urban change and give rise to high hopes for economic development.

The origin and development of Hekou date back to France's colonization of the Indochinese peninsula and its wish to increase its economic influence in Southern China. It was with the aim of opening up the Chinese market

Table 5.2
Demographic Characteristics in 2010 of the Main Prefectures and Municipalities Responsible for the Cities Studied

	Surface Area (in km^2)	Density (inhab./km^2)	% of Rural Pop.	% of Urban Pop. (Cities and Small Towns)	% Han	% Ethnic Minorities
Dehong Autonomous Prefecture	11,526	105.1	65.8	34.2	51.9	48.1
Municipality of Ruili	1,020	177	48.2	51.8	57.6	42.4
Xishuangbanna Autonomous Prefecture	19,700	57.5	65	35	30	70
Municipality of Jinghong	7,133	72.9	55	45	38.6	61.3
Honghe Autonomous Prefecture	32,930	136.7	64.2	35.8	42.8	57.2
Hekou District	1,313	79.6	55	45	41.4	58.6

Sources: — *2010nian Dehong zhou diliuci quanguo renkou pucha zhuyao shuju gongbao* [Bulletin of main data of the sixth national census of the population of Dehong Prefecture], *Dehong zhou tongjiju*, 16 May 2011; *2010nian Honghe zhou diliuci quanguo renkou pucha zhuyao shuju gongbao* [Bulletin of the main data of the sixth national census of the population of Honghe Prefecture], *Honghe zhou tongjiju*, 17 May 2011; *2010nian Xishuangbanna zhou diliuci quanguo renkou pucha zhuyao shuju gongbao* [Bulletin of the main data of the sixth national census of the population of Xishuangbanna Prefecture], *Xishuangbanna zhou tongjiju*, 19 May 2011, available on <http://www.luqyu.cn/zhuanti.asp?zid=469> (accessed 22 September 2011).
— *2010nian Ruili shi diliuci quanguo renkou pucha zhuyao shuju gongbao* [Bulletin of the main data of the sixth national census of the population of Ruili municipality], *Ruili shi tongjiju*, 10 June 2011, available on <http://www.rlr.gov.cn/Zhengwu/tongzhi/2011-6/22/RuiLi@1162215234171.html> (accessed 22 September 2011).
— *2010nian Jinghong shi diliuci quanguo renkou pucha zhuyao shuju gongbao* [Bulletin of the main data of the sixth national census of the population of Jinghong municipality], *Jinghong shi tongjiju*, 25 May 2011, available on <http://kmwh.xxgk.yn.gov.cn/canton_model57/newsview.aspx?id=1731723> (accessed 22 September 2011).

via a southern front that French colonial forces occupied Lào Cai in 1886, then finalized the definition of a border in two treaties signed with the Qing dynasty in 1887 and 1895. The second treaty allowed Hekou to be opened up to international trade, followed two years later by diplomatic and border missions. Next, the building of the Yunnan railway by the French between 1903 and 1910 quickly made the Hekou-Lào Cai pair of border cities the main trade route between Yunnan and French Indochina. During the Vietnam War, it became an important supply route for North-Vietnamese troops. The deterioration of relations between China and Socialist Vietnam during the second half of the 1970s led to a strong decrease in trade flows, before armed conflict between the two countries in 1979 interrupted them altogether, leading to the closure of the Hekou-Lào Cai border. The resumption of relations with Vietnam in 1991, and the redefinition of the border in 1999 provided a new start. Official relations between Hekou and Lào Cai and the transport of goods by railway were resumed in May 1993 and February 1996. In 2000, the two cities proclaimed themselves "Friendship Cities", and this proclamation was symbolized by the inauguration the same year of a new road bridge over the Nanxi River, parallel to the old railway bridge[16] (see Map 5.2). A year later, in December 2001, the first border trade fair (*bianjing jingji maoyi jiaoyihui*) was organized in Lào Cai. It has been held alternately in each of the two cities ever since. One of the last was held in Hekou from 25–30 November 2010, and attracted about 200 Chinese, Vietnamese but also Thai firms.

Cross-border cooperation between Hekou and Lào Cai has been given new impetus since the construction of the Kunming-Hanoi transport corridor. It is the cornerstone of the brand-new Beishan (Hekou)-Kim Than (Lào Cai) cross-border economic zone (CBEZ)[17] (see Map 5.2). This project, initiated in 2004, is aimed at facilitating and promoting trade and investment, not only between the two cities, but also between the provinces of Yunnan and Lào Cai. Its ambition is to turn them into a node on the north-south economic corridor and also a centre for the ongoing economic integration between China and ASEAN, especially in view of the broadening of the CAFTA.

The first project agreement was signed in 2005 between the authorities of Honghe prefecture and those of Lào Cai province. A common vision, with a map and a few temporary administrative rules, was proposed by both parties. The projected zone would be 10 km^2 (5 km^2 on each side of the new motorway bridge). Working groups have been set up on the Chinese side. They consist of members of the local and provincial authorities but also of

Map 5.2
Hekou: CBEZ and Urban Development

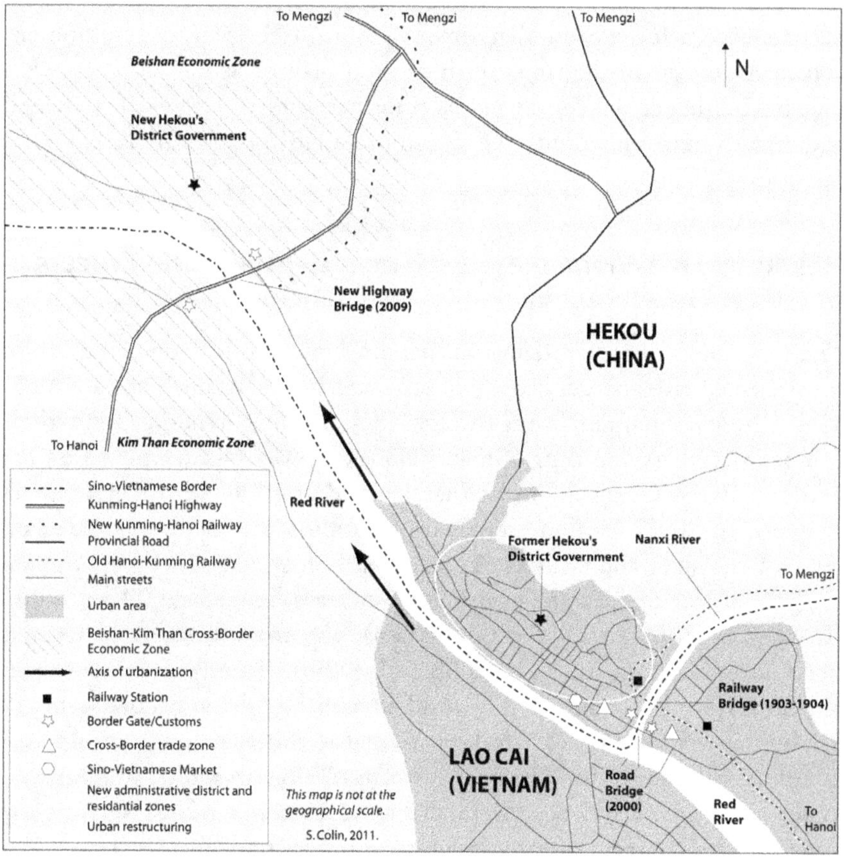

Source: Data compiled by S. Collin (2011).

the Ministry of Trade, the State Council and the National Development and Reform Commission.[18]

This project, which is supported financially by the Asian Development Bank and the UNDP, is now in the hands of the central authorities of both countries. A common launch meeting was held in Nanning (Guangxi) in November 2010, followed by three preliminary studies performed in November 2010 and March and July 2011, whose main aim was to set up a logbook. A final meeting was organized on 15 and 16 July 2011 in Lang Son, where representatives of both governments discussed the conclusions and recommendations of the studies, before agreeing to undertake bilateral negotiations, which should result in the signing of a MOU concerning the development and management of CBEZs. The main tasks will then be to create a joint Sino-Vietnamese Committee for the development and management of the zone, to draw up a plan of action and locate key investments in sectors considered as priorities, and to create a better legal and fiscal environment for the development of cross-border economic activities.

Hekou: A Sino-Vietnamese Exchange Centre in the Process of Urban Development

The twin cities of Hekou-Lào Cai, pending the possible concretization of the projected cross-border economic zone, are an important centre for Sino-Vietnamese exchanges, even though they are less dynamic than their competitors Pingxiang-Lang Son and Dongxing-Mong Cai, situated on the border between Guangxi and Vietnam (see Map 2.2 in Christian Taillard's chapter). In 2008, with an estimated 1.5 million tonnes of goods, Hekou was, for the sixth consecutive year, the first border trade "port" in Yunnan province. After a rapid and continuous increase between 2000 and 2007 (from 770 million to 8.6 billion Yuan), the total amount of foreign trade nevertheless dwindled in 2008 and 2009 (5.1 and 4.7 billion Yuan, respectively) on account of the world financial crisis, and now seems to be on the rise again (3.15 billion Yuan for the first seven months of 2010). At the end of the 11th five-year plan (2006–10) it totalled 30 billion Yuan, more than 2.6 times the total for the 10th plan (2001–05). At the same time, the number of firms involved in border trade increased from 123 in 2005 to 193 in 2010. At this date, Hekou's foreign trade represented 80 per cent of that of Honghe prefecture and 10 per cent of that of Yunnan province.[19]

The actors involved in this foreign trade — nearly all of it border trade — are varied. There are many Han from other regions of Yunnan or other Chinese provinces, especially Sichuan, who arrived in the 1990s, either alone, in partnership, or with their families, and set up firms to sell a wide range of Chinese manufactured goods to Vietnamese traders. Overall, these firms are modest in size and Chinese in origin: no foreign firms were registered in Hekou at the end of 2008. From the economic point of view, Hekou district is, on the scale of Yunnan province, an important rubber and tropical fruit production centre. It is also highly dependent on tertiary activities (56.4 per cent of total local GDP in 2010), headed by border trade, but also tourism. In 2009, 1.3 million tourists visited the border town. Finally, Hekou is also a centre for the informal economy. Apart from a great deal of smuggling, the city is known for its brothels, especially its "Vietnamese market" where local products and specialities can be seen side by side with young prostitutes from Vietnam (He Zhixiong 2003). Some of the landlords of these establishments are mixed Sino-Vietnamese couples (see Caroline Grillot's chapter).

The opening of the Sino-Vietnamese border to bilateral and transnational cooperation, as well as the country's town planning policy, has created a process of urbanization that may be geographically confined, but is in no way negligible.[20] During the last few years, the city has extended along the Red River towards the new bridge, which is now a point of anchorage around which the new district of Beishan is developing, consisting, apart from the projected cross-border economic zone, of housing, public buildings (a hospital, schools, etc.) and an administrative centre, including the new head office of the local authorities (see Map 5.2). Hekou, a market for Sino-Vietnamese exchanges, seems to have turned into a small town. The commercial district, where exchanges between Chinese and Vietnamese traders take place, near the two bridges spanning the Nanxi River, has been undergoing redevelopment since 2010–11, to leave room for residential areas, among others, as if small-scale border trade had become of secondary importance next to the ambitious cross-border trade zone project.

Ruili: A Sino-Burmese Trading Place

The city of Ruili is one of the four administrative subdivisions of the Dehong Dai and Jingpo autonomous prefecture. It is across the border from the Burmese border town of Muse, and crossed by the Kunming-Rangoon road. It should be linked by rail to Kunming by 2016 then connected to the Burmese rail network via an extension as far as Lashio (see Map 5.1). It also shares a national airport with the city of Luxi (Mangshi), about 60 km away. Starting in 2013, the arrival of the Sino-Burmese oil and gas pipeline will make it a strategic location.

As a border city, the history of Ruili only really starts in the 1950s. For a long time, the region's main border city was Tengchong, where the Chinese set up a customs post in 1790, until the pressure of the British colonial forces, which completely subdued Burma in 1886, forced the

Imperial authorities to move the customs post several dozen kilometres further south to Yingjiang in 1885. The region around Ruili, Wanding, Muse and Namkhan, already inhabited by a large Han mining community, was bitterly disputed in the Sino-British border negotiations from 1894 to 1897. It was once more a subject of the treaty signed between the Republic of China and the British in 1941. In fact, unlike that of Hekou, the interruption of the building of the British-built Rangoon-Kunming railway at Lashio, in northern Burma, did not really allow the emergence of a town on the border at this time. On the contrary, the construction of the famous "Burma Road" from Lashio to Kunming in 1937–38 by the British forces and the Kuomintang, at that time resisting the Japanese invasion, provided what is still today the main route between the two countries. At the beginning of the 1950s, the Chinese authorities set up two foreign trade agencies in Wanding and Ruili, which at the same time became two official border posts. Nevertheless, because of volatile Sino-Burmese relations in the 1960s and 1970s, cross-border trade between Ruili and Muse did not really develop until the middle of the 1980s. In 1990, the city was promoted "Open Tourist Zone" by the authorities. Two years later, it became a first-category trade "port", like its neighbour, Wanding, until the latter's subsequent administrative incorporation into the municipality of Ruili in 1999.

Currently, the city of Ruili is a centre for trade and tourism within a widespread Sino-Burmese cross-border area and is also a relay station between Kunming and Rangoon. Its development is mainly due to the precious stone trade with Myanmar, which took off in the second half of the 1990s. The self-proclaimed title of "Jewel City", as well as serving to attract investors and tourists, is a good illustration of this phenomenon. In total, this trade probably employs more than 35,000 people, or more than 40 per cent of the city's working population! In 2000, the State Council approved the creation of a free zone of 2.4 km^2 in Jiegao, a small border village between Ruili and Muse (see Map 5.3). This is at once a trade, tourist and industrial area. There are processing industries for raw materials imported from Myanmar (precious stones, ore, agricultural products, wood, etc.) as well as factories for the assembly of manufactured goods for the Burmese market (textiles, foodstuffs, household appliances, machines, etc.). The tourist attractions mainly consist of shops selling precious stones and jewellery,

Map 5.3
The Twin Cities of Ruili-Muse

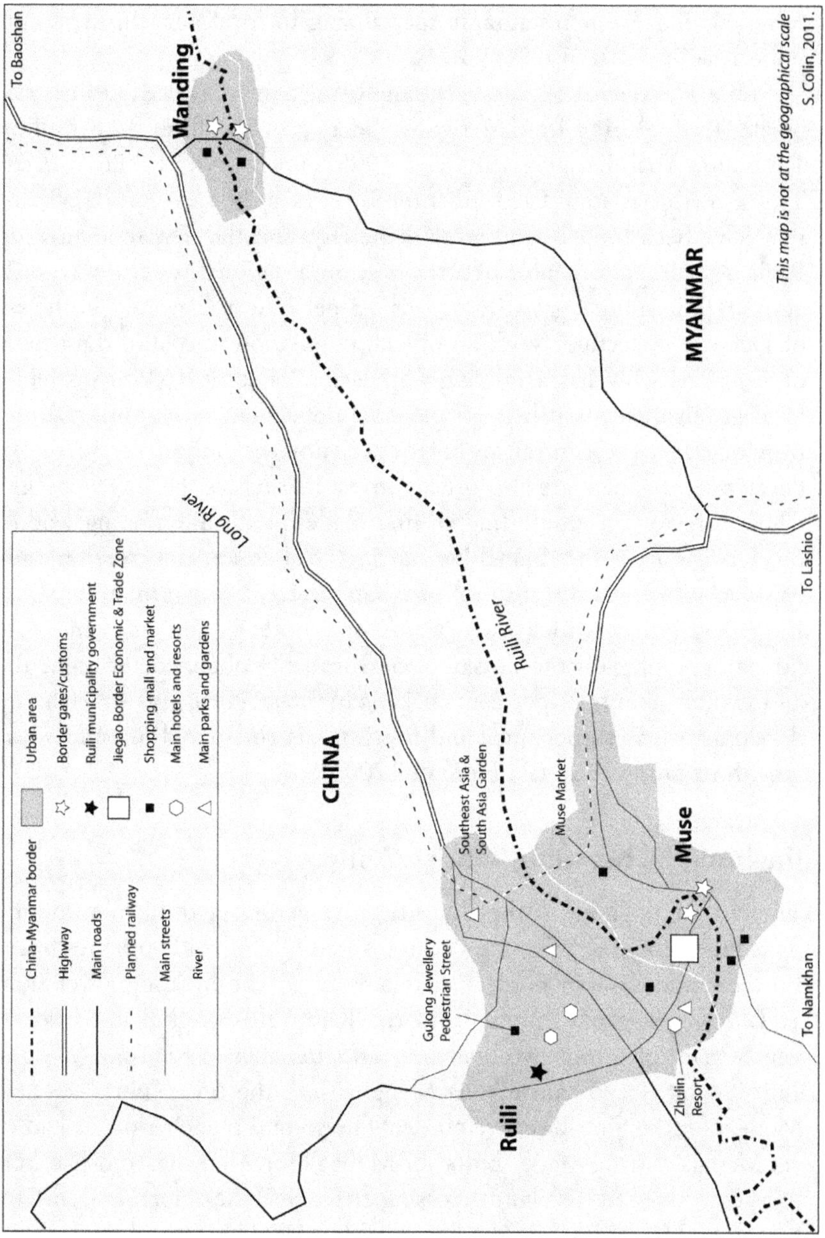

Source: Date compiled by S. Collin (2011).

"ethnic" villages, or restaurants specializing in local and Burmese cuisine (Guo Xiaolin 2010, pp. 91–93).

Like Hekou, there has been, in Ruili, a very noticeable increase in border trade during the last two decades, growing from 44.8 million to 1.5 billion Yuan between 1985 and 1995, and reaching 6 billion in 2008. This growth has stimulated urbanization, as can be seen by the new roads that have been built in and around the city, and the new residential areas. Ruili is an important pole of attraction on a regional scale. Between 2000 and 2010, its annual growth rate was 2.3 per cent, the greatest in the whole of Dehong prefecture. With an official population of 3,000, the new zone of Jiegao actually has a floating population of some 16,000 people. The local government, which is 60 per cent dependent on subsidies from the province and the central authorities, is not the only actor in the city's development. There are also private investors such as the Jingcheng group, whose director, an ethnic Jingpo, first made his fortune trading wood and precious stones before branching out into real estate, construction and the banking sector, via the creation of a bank specializing in micro-credit. The Jingcheng group, which owns a four-star hotel in the city centre and a holiday complex on the outskirts, has probably obtained the right to use and exploit about 10 per cent of the city area. Finally, as in Hekou, the development of border trade and tourism has gone hand in hand with the growth of prostitution (Pu Hongyan 2006).

Jinghong: A Booming Tourist Centre

The city of Jinghong is the administrative centre of the Dai autonomous prefecture of Xishuangbanna. Located about a hundred kilometres from the Burmese and Laotian borders, it is crossed by the Mekong River (known as *Lancangjiang* in Chinese) and the Kunming-Bangkok road corridor, which then splits into two branches. The first goes to Kentung (Myanmar) and Chiang Rai (Thailand) via Menghai and the trade "ports" of Daluo-Mongla on the Sino-Burmese border. The second branches off to Lao PDR via Mengla and the trade "ports" of Mohan-Boten on the Sino-Lao border, then continues to Thailand, crossing the twin cities of Huay Xai-Chiang Khong on the Mekong at the Lao-Thai border. This second section should be duplicated by the Kunming-Jinghong-Vientiane-Bangkok railway (see Map 2.4 in Christian Taillard's chapter). Once it is operational, this rail-road corridor will make Jinghong, which already has an international airport and river port, an important multimodal logistic centre.

The city of Jinghong mainly owes its existence to the creation of the Xishuangbanna autonomous prefecture by the Chinese authorities in 1953. Before that, the site was occupied by a series of villages depending on the Thai principality of Sipsong Panna, which the Qing authorities progressively, and not without difficulty, absorbed into Chinese territory (Giersch 2000; Atwill 2002). The trade route that crossed this region was controlled by the Chinese customs post at Simao, located to the north. From the 1950s onwards, the Communist authorities made Xishuangbanna a centre for rubber production,[21] before "ethnic" and "natural" tourism became identified as a major regional development factor in the 1980s.[22] Coupled with this new Chinese state tourist policy, the building of Xishuangbanna airport in 1990, then its proclamation as an international airport (the second largest in Yunnan after Kunming) in 1995 has completely changed the city's appearance. Directly linked to the big Chinese metropolises and to the Thai cities of Bangkok and Chiang Mai, while its road and rail connections were out of date or non-existent, Jinghong, a small conurbation of *Dai* villages along the Mekong, quickly became a major tourist centre on a provincial and national scale. Classified as a "famous site at provincial level" in 1982, it was given the status of municipality in December 1993 when the first signs or urbanization began to appear, then was promoted to "Tourist city of excellence at national level" by the central authorities in 1999.[23] For the Han Chinese middle classes, it has become a well-known and much admired tourist destination.

In Jinghong, and in Xishuangbanna in general, tourism is the trademark of a municipality whose projects and urban development have been directed since 2008 by a master plan designed by the Institute for research into urban planning projects of the Shanghai Tongji University (*Shanghai tongji chengshi guihua sheji yanjiuyuan*). Over a twelve-year period, this scheme especially plans to redevelop the former *Dai* village of Menghan, located in the southern part of the municipality, into a "small tourist town" (*Jinghong shi Manghan lüyou xiaozhen zongti guihua*, 2008–20) (see Map 5.4). In the coming years, the city of Jinghong should also acquire new structural installations: a new bridge over the Mekong, shopping centres, new luxury hotels and resorts as well as a conference centre, supposed to host, among other events, meetings or conferences of the GMS, or even ASEAN

Map 5.4
North-South Corridor and Tourism Development in Xishuangbanna Prefecture

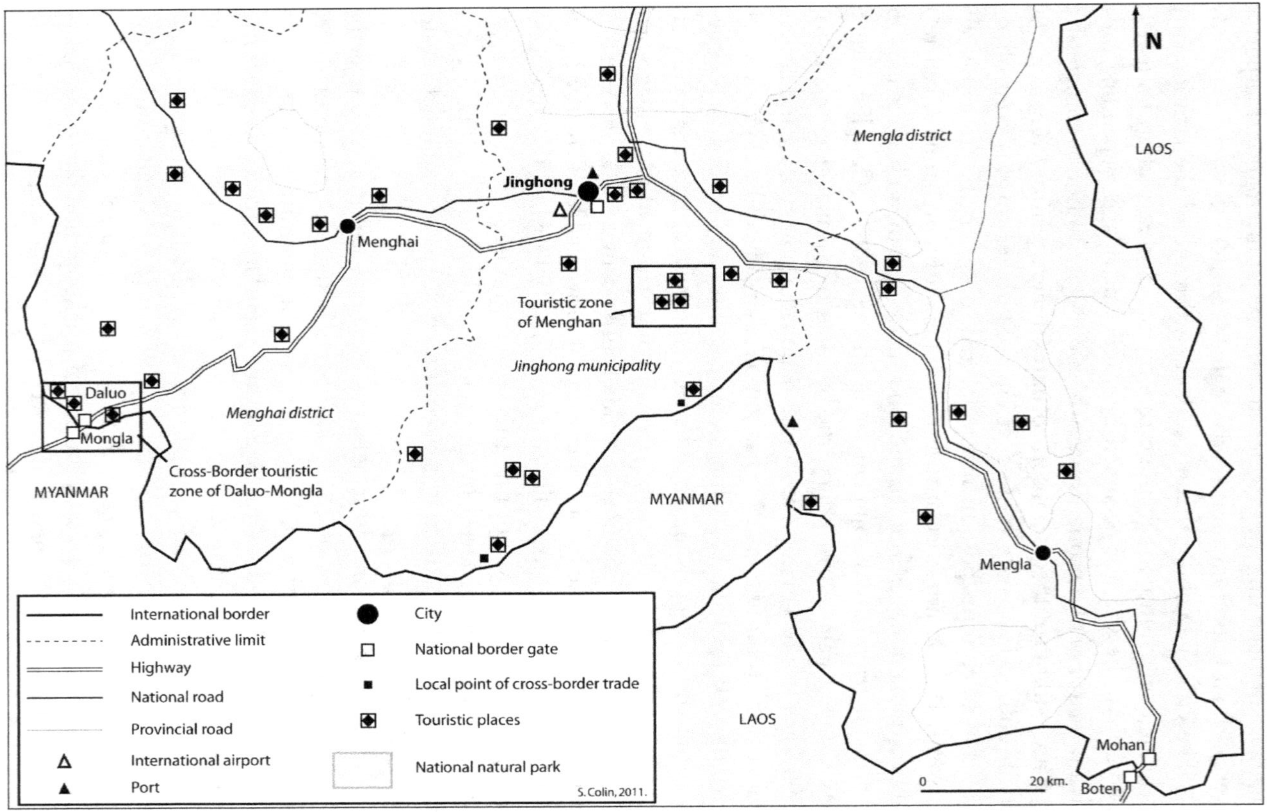

Source: Data compiled by S. Collin (2011).

Map 5.5
Jinghong: Tourism as a Driving Force behind Urban Development

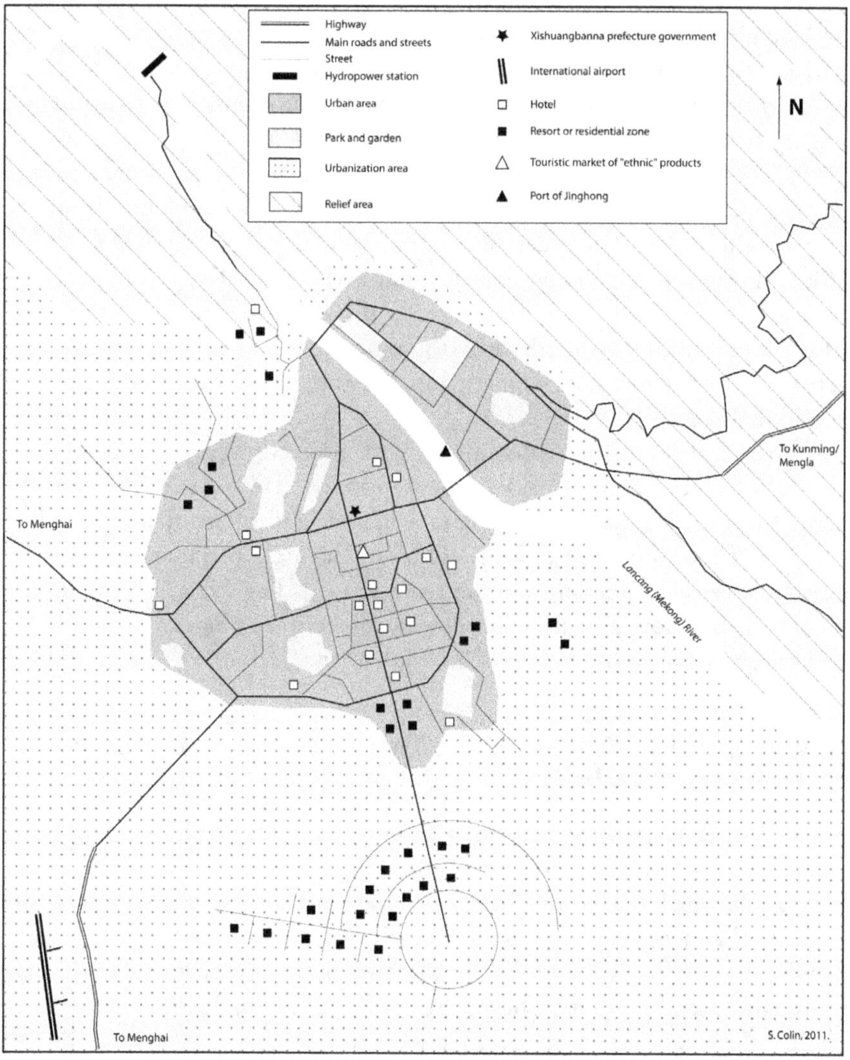

Source: Data compiled by S. Collin (2011). This map is not at the geographical scale.

(see Map 5.5). The plan also recommends the rehabilitation of the Mekong, with the development of the waterfront and the creation of an eco-tourism centre in the surrounding countryside.

As in Hekou and Ruili, profound changes in urban structure took place in Jinghong in the 1990s and 2000s, with the appearance of numerous hotels and residential areas built with Sino-Thai, Taiwanese, Hong Kong or Shanghainese capital (see Map 5.5). Apart from these there are theme parks, traditional *Dai* dance shows, restaurants, karaoke bars, night clubs, gambling dens, massage parlours and a few brothels. Prostitution has greatly developed, making Jinghong a veritable "*piaocheng*" (prostitution city). The city attracts numerous tourists, nearly all men drawn by the erotic reputation of *Dai* women (who are mostly Han migrants from the two neighbouring provinces of Sichuan and Guizhou, or from further afield, dressed in traditional *Dai* costumes) (Hyde 2007). During the last ten years, there has been a simultaneous process of destruction/promotion of the local culture, symbolized by the modern apartment blocks that have replaced the former *Dai* villages, but are topped with reproductions of traditional roofs.

Apart from tourism, which, directly or indirectly, makes up nearly half the local GDP, Jinghong is also an important centre for trade with the GMS countries. There is an extensive market and many stalls, all selling products from Southeast Asia, most often Burmese precious stones. This border trade does not only use the river, towards Lao PDR, Myanmar and Thailand, but now also uses the new roads between the city and the two first-category border trade "ports" of Mohan and Daluo.

Jinghong: A Pair City with the Two Trading and Tourist Ports of Mohan and Daluo

Starting in Jinghong, the new road links to Myanmar and Lao PDR serve many tourist sites scattered around Menghai, the capital of Pu'er tea cultivation, which has recently become a local tourist attraction, Mengla, where residential and holiday areas with evocative names such as "New Vientiane City" are growing up, and finally the two border trade ports of Mohan and Daluo (see Map 5.4). These new roads are leading to the formation of "pair cities" connecting a command centre, Jinghong, with two border ports located on each branch of the North-South corridor (see Manuelle Franck's chapter and Christian Taillard's conclusion in this volume).

Officially promoted in March 1992, the trading "port" of Mohan (about 50,000 inhabitants) located close to the Laotian border, has acquired border trade and tourist zones and an international logistics centre (*guoji wuliu zhongxin*). The completion of the motorway between Kunming and Mohan at the end of 2008 and the future arrival of the Kunming-

Vientiane-Bangkok railway, which the city is preparing for via the construction of a rail logistics centre (*tielu wuliu zhongxin*), give the local authorities grounds for believing that Mohan could become an important centre for cross-border exchanges with Lao PDR and Thailand. For the time being, we can just observe that cross-border flows of people (traders and tourists) and goods have strongly increased since 2003 and that infrastructures are progressively being built (see Table 5.3). The little country town has also begun to change, with many stalls and a few housing estates on either side of the road to Lao PDR. Mohan also counts on the development of tourism, which was at one time associated with the special economic zone of Boten a few kilometres south, in Laotian territory, but this has recently been closed (see below). However, this closure makes the current prospects of development still uncertain.

Daluo, which was promoted to provincial level trade "port" in August 1991 and to first-category port in November 2007, is still a small country town (less than 20,000 inhabitants, 94 per cent of them were farmers in 2007). Its promotion to trade "port" on the Burmese branch of the North-South corridor has brought it to the fore. Like Mohan, it is at the centre of various provincial development plans.[24] One of the main projects consisted of creating a cross-border tourist area with the Burmese town of Mongla.[25]

Table 5.3
International Traffic on Mohan-Boten Border between Yunnan and the Lao PDR, 2003–07

	Passenger Volume (Number of Persons)	Cargo Volume (10,000 Tons)	Number of Vehicles from Yunnan to Lao PDR	Number of Vehicles from Lao PDR to Yunnan
2003	8,961	3.56	2,114	6,679
2004	27,841	3.89	2,408	6,088
2005	39,254	4.75	4,152	6,153
2006	58,361	7.34	6,279	12,138
2007	95,182	41.03	15,356	29,306

Source: Transport and Trade Facilitation in the Greater Mekong Subregion: Time to Shift Gears, Evaluation Study of the Asian Development Bank, Reference Number: SAP: REG 2008–86, December 2008, p. 180.

Created after widespread opium poppy eradication campaigns in the 1990s, this area included about twenty sites in 2009, mostly in Burmese territory around Mongla, and accessible without a visa to tourists accompanied by a guide (see Map 5.4). Unlike Mohan, the town of Daluo, which was connected later than Mohan by the transport corridor, had not substantially changed in 2009. Despite the presence of a few groups of Chinese or Thai tourists, trade and human flows across the border are still very slight. The total value of border trade was only 551 million Yuan in 2007.[26]

CONCLUSION: YUNNAN'S BORDER CITIES: NEW NODES ON TRANSNATIONAL CORRIDORS OR CITIES MENACED BY MARGINALIZATION?

Undeniable, but Still Insufficient and Incomplete Economic Development

Benefiting from the various national policies and Yunnan development strategies, in a context of regional cooperation, the cities of Hekou, Jinghong and Ruili have clearly become more central. The changes taking place in each of these three cities have numerous common features. Border trade has increased everywhere. Although it mainly consists of small businesses, larger-scale trade is developing and is supposed to grow in the future with the development of infrastructures and the simplifying of border checks. It has led to the development of numerous activities in each area: markets, small businesses, wholesalers, transport and import-export services, hotels and restaurants. The three localities also depend on tourism, to a great extent for Jinghong and Xishuangbanna, which adapts its urban development policy accordingly. More ambitious projects, parallel to the setting up of transport corridors, such as cross-border zones, seem to show a desire to transcend a model that has enabled temporary survival, but is insufficient to create genuine industrialization and more sustained socio-economic development.

From an economic point of view, the GDP and GDP per capita of these border cities are still well below those of Kunming, the primary sector is over-represented, and, on the contrary, the secondary sector is generally weak (see Table 5.4). The important share of the tertiary sector, however, displays a strong specialization in trade and tourism. For the moment, the development of Yunnan border cities is not accompanied by an important industrialization process. Nearly all the manufactured goods exported to Vietnam, Lao PDR, Thailand and Myanmar via these cities are made outside

Table 5.4
Economic Characteristics of Kunming and Border Regions Studied in 2009

	Kunming Municipality	Ruili Municipality	Hekou District	Xishuangbanna Prefecture
Population (in millions)	6.28	0.2	0.1	1.13
GIP (in billion Yuan)	212.04	2.5	1.5	16.09
Primary sector (in % GIP)	5.7	20.7	19.9	27.6
Secondary sector (in % GIP)	45.3	22.3	23.7	29.6
Tertiary sector (in % GIP)	49	57	56.4	42.7
GIP per inhab. (in Yuan)	33,550	12,500	14,284	14,905

Source: Hong Kong Trade Development Council, 2011 <http://www.hktdc.com>.

Yunnan, mainly in Guangdong and Fujian provinces. However, in this context of projects managed and funded initially by public and international donors, the role of industrial companies also deserves to be further analysed, even if they are currently often presented as less enthusiastic and unimportant, because things may be changing rapidly.

Three Cities Whose Economic Opportunities Differ According to the Political Development of Neighbouring Countries and Their Place in Yunnan Urban Hierarchy

The development of border cities is also dependent on the economic vitality and the political situation in neighbouring countries. In this respect, Ruili and Daluo are faced with certain difficulties on account of Myanmar's weak economic development, and the return of considerable political and military instability to Shan State since 2008–09, both of which tend to paralyse cross-border cooperation. Relations between Ruili and Muse, where the border trade fair has already been cancelled several times, now seem strained, to the extent that the main centres of exchange with China are now further south, in Lashio and Mandalay. Similarly, the border between Mongla and Daluo was closed several times in 2009 and 2010.

The further development of these three urban centres will also depend on the state of their relations with Kunming and the medium-sized cities in the interior, often with a higher administrative status and/or with industrial activities (see Map 5.1). Thus, Hekou's future depends as much on Mengzi, the administrative centre of Honghe prefecture, to which it belongs, as on Kunming. In the same way, Ruili is directly concerned by the ongoing changes in Luxi or in Baoshan, a city with a much larger population which could become a junction when the Kunming-Baoshan railway is extended to Myitkyina, in Myanmar, or even further, to India. The danger threatening Hekou and Ruili with the development of new transport infrastructures and the easing of border controls may well be the loss of that which has ensured their success up to now — small-scale border trade — without any guarantee of developing alternative activities. After all, it is perhaps cities like Baoshan and Mengzi that will benefit the most from the new transport corridors and GMS cooperation. In this context, Jinghong appears to have a real advantage. At the head of the urban hierarchy of Xishuangbanna, structured by two branches of the corridor, the city is now a well-known national and regional tourist centre. Its position as a logistic junction and its status of main town in the prefecture make Jinghong likely to become an important centre on the Kunming-Bangkok route, to the detriment of Mengla, Menghai, and especially Mohan and Daluo, which will have trouble emerging close to a centre of this kind, especially since their relations with Boten and Mongla are still limited.

Cities Faced with Persistent Non-traditional Security Issues

Another of the main limits, and one that has been the source of failures and difficulties in the last few years, is the continuing existence, on both sides of Yunnan's border, of non-traditional security issues (illegal immigration, drug trafficking, gambling, Aids, etc.). Most of the heroin and synthetic drugs sold in China cross, for example, the borders with Myanmar and Lao PDR. Prostitution is widespread in the cities of Hekou, Jinghong and Ruili, where, despite increasingly severe preventive measures, it is still responsible for spreading Aids (Hyde 2000). Finally, the trafficking of women and children and also kidnappings linked to gambling debts have increased in recent years, especially in Xishuangbanna. Consequently, Yunnan has the highest addiction and HIV-positive rates in the country, and, with Fujian, is the province most severely affected by the scourge of people trafficking (Chu 2011).

These numerous non-traditional security problems have, of course, been taken into account for several years by Beijing and Kunming. Whether in the context of the GMS, ASEAN or ASEAN+3, or bilaterally with Vietnam, Lao PDR and Myanmar, China is actively involved in international cooperation against drug trafficking and other non-traditional security issues (Sheng 2006). On a nationwide scale, in April 2005 Beijing ordered the launch of a "People's war against drugs and Aids" in order to build a "Peaceful and orderly Yunnan". This "war" is backed up by new measures against poverty in the context of the "Building a New Socialist Countryside" policy.

Beijing and Kunming's desire for stability has recently been responsible for the successive closure of the casinos in Mongla, in Myanmar, and the casino-hotel complex in the Boten economic zone in Lao PDR, even though they were at the centre of tourist projects in association with the Chinese border areas of Daluo and Mohan, respectively. These projects, initially developed to offer some economic alternatives in the anti-drug campaigns, and mainly funded by the same Hong Kong company, quickly ran into difficulties. The increase in the crime rate, especially the kidnapping of Chinese nationals who had got into debt, finally exasperated the Chinese central authorities, and in the context of the "Peaceful and orderly Yunnan" campaign, they began discouraging tourists from going there before putting direct pressure on the company and the Burmese and Laotian authorities to terminate these two projects.

Since 2005, China has also reacted to complaints by the Burmese military junta about the "illegal exploitation" of natural resources, directly contracted by Chinese firms from the armed ethnic groups of Shan State and Kachin. The Chinese border police have increased border checks in recent years in an attempt to prevent these "illegal" mining and forestry activities. This measure has prompted the return to Yunnan of 20,000 Chinese workers and led to a decline in illegal trade, although it has failed to eradicate smuggling, which is still rife.

Yunnan Threatened by Guangxi's Inclusion in the GMS?

Finally, on another geographical scale, we must return to the integration of Guangxi into the GMS. With two entry points into the Indochinese peninsula, China has now strengthened its position of hegemony within the GMS and its influence on CLMV countries, but is not this choice by the authorities contrary to the interests of Yunnan, which would now face more competition at national level in its participation in the GMS? Even if Guangxi seems more interested in the value of its opening to the South China Sea and the activation of connections with the countries of maritime Southeast Asia, as evidenced by its active participation in the *Pan-Beibu* project, it also heavily involved in the realization of its portion of the Nanning-Bangkok corridor and in the establishment of CBEZ at Pingxiang-Lang Son. Geographically, Guangxi is also crossed by the Hong Kong-Guangzhou-Singapore axis, of which it could take advantage economically. It is very important to take this situation into account because Guangdong Province would also be interested in participating in the GMS, which could further marginalize the position of Yunnan. Accordingly, on a different geographical scale, we can highlight the importance to Yunnan Province to develop links with its hinterland, represented by the Chengdu-Chongqing urban area, as well as its just developing connections with India.

Notes

1 The two corridors correspond to the Kunming-Hanoi and Nanning-Hanoi routes, whereas the circle corresponds to the *Pan-Beibu* (Gulf of Tonkin) economic development zone, created in 2006 on a local level and officialized by the Chinese authorities in 2008.

2 A border trade port known as "first category" or "national rank" theoretically allows the passage of nationals from a third country, whereas a "second category" or "provincial rank" port is only open to nationals of both countries involved.

3 The aim of this "Open up the West" policy (*Xibu da kaifa*) is above all to disenclose the widespread western regions and integrate them with the rest of Chinese territory via the development of transport infrastructures and a call for investment from public and private firms based in coastal regions. In the long term, the authorities express the official wish that development will improve the living standards of local populations, mainly consisting of ethnic minorities, some of which, such as the Uigurs, and especially the Tibetans, contest more or less actively China's annexation. For an analysis of this policy, see David S.G. Goodman (2004); Barry J. Naughton (2004).

4 The "Building a New Socialist Countryside" policy (*jianshe xin shehui nongcun*) aims to solve the socio-economic problems of rural areas on a national scale. It takes a special form in outlying regions with the building of "New Socialist villages" along new transport infrastructures, in view of assembling local populations often widely spread out in mountainous enclaves.

5 The urbanization rate of the country, which was 46.6 per cent (i.e., 620 million city-dwellers) at the end of 2010, should reach 52 per cent in 2015 and 65 per cent in 2030. During the 12th five-year plan (2011–15), this rate should increase by 0.8 to 1 per cent each year, representing the arrival of a rural population of more than ten million in the cities.

6 Yunnan's main agricultural products include rubber, Pu'er tea, sugar cane and bananas.

7 The exploitation of Yunnan's hydroelectric potential via the building of numerous dams on the Salween (*Nujiang* in China) and Mekong (*Lancangjiang*) speeded up in the 1990s before becoming one of the province's priorities in the "Open up the West" policy. Many studies in the past few years have analysed the mechanisms of Yunnan's hydroelectric policy and the political, social and environmental consequences of the construction of multiple dam building. See, among others: Philip H. Brown and Xu Kevin (2010); Chen Lihui (2008); John Dore et al. (2007); Darrin Magee (2006); Yang Rongxin (2006).

8 See, for example: "*Zhongnan bandao* (Indochina Peninsula) economic circle", "New Southern Silk Road", "Cooperative Bridge between China, Southeast Asia and South Asia", "Connecting to Coastal Provinces in the East, and Southeast Asia in the West", or the most recent, "Bridgehead strategy", displayed in 2008 in the context of a new provincial development plan (Timothy A. Summers 2010; Margaret Swain 2002).

9 Although the presence of long trade routes is attested in the region from the Han dynasty onwards, it is only under the Ming, and especially the Qing, that trade volume increased and the regional network started to develop. The old Yunnan trade routes have been the subject of several studies in recent years: Prasertkul Chiranan (1989); Harald Bockman (1991); Andrew D.W. Forbes (2000); C. Patterson Giersch (2011); Nanny Kim (2011).

10 Two French consuls were appointed in Kunming and Mengzi, while a British consul was appointed in Tengchong. The objective of the colonial powers was to develop trade and modern transport infrastructures, as can be seen in the building of two rival railways from Hanoi and Rangoon. Although the colonial administration of French Indochina achieved its end, by building the Hanoi-Kunming line between 1903 and 1910, this was not the case for the British administration in Burma, which put an end to its plan after reaching Lashio. On this point, see, among others: J. Chandran (1971).

11 The idea of connecting Yunnan to the Burmese ports via Lashio and Bhamo, for example, emerged in the middle of the 1980s (Pan Qi 1985).

12 During a visit to Yunnan in July 2009, Hu Jintao gave a reminder of Yunnan's importance as a bridgehead in China's relations and cooperation with Southeast and South Asia. From the financial point of view, State investments in fixed capital destined for Yunnan have increased greatly since 1990, from 398 million dollars in 1995 to 7 billion in 2005 and 18 billion in 2007 (Freemen and Thompson 2011, p. 67).

13 The Kunming-Nanning railway enables Yunnan, beyond Nanning, to be connected to the ports of Beihai and Fangchenggang, situated on the Gulf of Tonkin, and also to Guangzhou and the Pearl River delta.

14 For more details on the urban development programme, see *Kunming chengshi guiha ju* (Kunming urban planning office). Available at <http://ghj.km.gov.cn>.

15 When the building of the airport began in 2009, there was some question of naming it Zheng He, after the famous Ming dynasty navigator, originally from Kunming. The expansion of the Ming navy under his command in the South China Sea and beyond the Malacca Straits in the Indian Ocean was an excellent illustration of the contemporary provincial policy of assertion in Southeast and South Asia, but perhaps with an overly hegemonic connotation, which is probably why the idea was abandoned.

16 This is a narrow bridge directly connecting the two city centres. It is mainly used by pedestrians, non-motorized vehicles (bicycles, carts) and low-tonnage trucks. Chinese and Vietnamese customs are situated on either side of the bridge, and, in 2009, still carried out separate checks.

17 A second Sino-Vietnamese cross-border economic zone is also projected between Pingxiang (Guangxi, China) and Lang Son (Vietnam).

18 Representatives from the Land and Resources, Finance and Foreign Affairs Ministries; and from the Central Development Bank and the Confederations of Public and Private Enterprises have also sometimes been invited to these working groups.

19 Hekou xian shangwuju (Business Office of Hekou County), "Shiyiwu hekou kou'an duiwai maoyi chengxiao xianzhu" (Result of the foreign trade of the border port of Hekou during the Eleventh Five Year Plan), 29 March 2011, available at <http://www.ynf.gov.cn/canton_model1/newsview.aspx?id=165393> (accessed 22 September 2010).

20 The Vietnamese city of Lào Cai is also undergoing a similar process although at first sight it appears less intense. Since 2003, it has been at the centre of a new urban scheme aiming to quadruple the surface area of the city. On this point, see Marie Mellac's chapter in this volume.

21 Exploited only by State farms in the Maoist era, rubber has also been grown since the decollectivization of the early 1980s by independent farmers, mainly *Dai,* who soon internationalized their production to Lao PDR thanks to the reactivation of ethnic links made possible by the opening of the border. Also, in the last few years, agricultural production in Xishuangbanna has become more varied, with rice growing in the lowlands decreasing in favour of more profitable crops such as sugar cane, and especially bananas. For detailed analyses of rural changes in Xishuangbanna, specifically in rubber production, see Janet C. Sturgeon (2011); Yin Shaoting (2009); Zhang Deshang et al. (2009).

22 Since the 1980s, Xishuangbanna has been one of Yunnan's main tourist centres, with Kunming, Dali and Lijiang. Apart from the presence of ethnic

minorities, the basic element of "ethnic" tourism, the setting up of State nature reserves is also supposed to promote the development of eco-tourism (Wen Jie 1997).

23 "Jinghong shi" (Jinghong Municipality), *Xishuangbanna daizu zizhizhou zhengwuwang* (Xishuangbanna Dai Autonomous Prefecture Governmental Affairs Information Network), 7 March 2006, available at <http://www.xsbn.gov.cn/situation/ShowArticle.asp?ArticleID=2605> (accessed 1 December 2010).

24 This is the "Provincial long-term development Plan (2000–2020)" of the town of Daluo and the "General plan of the economic development zone of Daluo trade port" (2009–2021). "Daluo kou'an" (Daluo Trading port), *Menghai xian zhengwu xinxiwang* (Menghai County Governmental Affairs Information Network), 30 April 2010, available at <http://www.ynmh.gov.cn/channels/351.html> (accessed 1 December 2010).

25 "Xishuangbanna jingji fazhan zhuangkuang" (Situation of Xishuangbanna's Economic Development), *Xishuangbanna daizu zizhizhou zhengwuwang* (Xishuangbanna Dai Autonomous Prefecture Governmental Affairs Information Network), 7 March 2006, available at <http://www.xsbn.gov.cn/situation/ShowArticle.asp?ID=2603> (accessed 1 December 2010).

26 "Menghai xian Daluo zhen" (Daluo Township of Menghai County), *Menghai xian zhengwu xinxiwang* (Menghai County Governmental Affairs Information Network), 16 July 2009, available at <http://www.ynmh.gov.cn/contents/5/1267.html> (accessed 1 December 2010).

References

Atwill, David G. "Trading Places: Resistance Ethnicity, and Governance in Nineteenth-century Yunnan". In *Dragons, Tigers and Dogs: Qing Crisis Management and the Boundaries of State Power in Late Imperial China*, edited by Robert J. Antony and Jane Kate Leonard. Ithaca, N.Y.: East Asia Program, Cornell University, 2002.

Bockman, Harald. "Yunnan Trade in Han Times: Transit, Tribute and Trivia". In *Asia Trade Routes: Continental and Maritime*, edited by Karl Reinhold Haellquist. London: Curzon, 1991.

Brown Philip H. and Xu Kevin. "Hydropower Development and Resettlement Policy on China's Nu River". *Journal of Contemporary China*, issue 19, no. 66 (September 2010): 777–97.

Chen Lihui. "Contradictions in Dam Building in Yunnan, China: Cultural Impacts versus Economic Growth". *China Report*, issue 44, no. 2 (April–June 2008): 97–110.

Chiranan Prasertkul. *Yunnan Trade in the Nineteenth Century: Southwest China's Cross-Boundaries Functionnal System*. Bangkok: Institute of Asian Studies, Chulagongkorn University, 1989.

Chu Cindy Yik-yi. "Human Trafficking and Smuggling in China". *Journal of Contemporary China*, issue 20, no. 68 (January 2011): 39–52.

Colin, Sébastien. *La Chine et ses frontières*. Paris: Armand Colin, 2011.

Dehong Daizu Jingpozu zizhizhou gaikuang [General situation of the Dehong Dai and Jingpo Autonomous Prefecture]. Beijing: Minzu chubanshe, 2007.

D'Hooghe, Ingrid. "Regional Economic Integration in Yunnan". In *China Deconstructs: Politics, Trade, and Regionalism*, edited by David S.G. Goodman and Gerald Segal. London; New York: Routledge, 1994.

Do Tien Sam and Ha Thi Hong Van. "Vietnam-China Relations and Building the 'Two Corridors, One Economic Belt'". In *Towards Pax Sinica? China's Rise and Tranformation: Impacts and Implications*, edited by Yeoh Emile Kok-Kheng. Kuala Lumpur: Institute of China Studies, University of Malaya, 2009.

Dore, John, Yu Xiaogang and Li Kevin Yuk-shing. "China's Energy Reforms and Hydropower Expansion in Yunnan". In *Democratizing Water Governance in the Mekong Region*, edited by Louis Label and et al. Bangkok: USER; Chiang Mai: Mekong Press, 2007.

Gan Chunkai and Chen Zhilong. "Yunnan". In *Developing China's West: A Critical Path to Balanced National Development*, edited by Yeung Y. M. and Shen Jianfa. Hong Kong: Chinese University Press, 2004.

Giersch, C. Patterson."The Sipsong Panna Tai and the Limits of Qing Conquest in Yunnan".*Chinese Historians* 10, no. 17 (Fall 2000): 71–92.

———. "Cotton, Copper, and Caravans: Trade and the Transformation of Southwest China". In *Chinese Circulations: Capital, Commodities, and Networks in Southeast Asia*, edited by Eric Tagliacozzo and Chang Wen-Chin. Durham: Duke University Press, 2011.

Goodman, David S.G. "The Campaign to 'Open Up the West': National, Provincial-level and Local Perspectives". *China Quarterly*, no. 178 (June 2004): 317–34.

Guo Xiaolin. "Boom on the Way from Ruili to Mandalay". In *Burma: Inside Challenges, Outside Interests*, edited by Lex Rieffel. Washington D.C.: Konrad Adenauer Foundation, Brookings Institution Press, 2010.

He Shengda. "Friendly and Wealthy Neighbors, Stable Nation: Yunnan's Participation in the GMS". In *The Mekong Arranged and Rearranged*, edited by Maria Serena I. Diokno and Nguyen Van Chinh. Chiang Mai, Thailand: Mekong Press, 2006.

He Shengda and Sheng Lijun. "Yunnan's Greater Mekong Sub-Region Strategy". In *ASEAN-China Relations: Realities and Prospects*, edited by Saw Swee-Hock, Sheng Lijun and Chin Kin Wah. Singapore: Institute of Southeast Asian Studies, 2005.

He Zhixiong. "Migration and the Sex Industry in the Hekou-Lao Cai Border Region between Yunnan and Vietnam". In *Living on the Edges: Cross-Border Mobility and Sexual Exploitation in the Greater Southeast Asia Sub-Region*, edited by Muhadjir Darwin and Anna Marie Wattie. Yogyakarta: Gadjah Mada University, 2003.

Hekou Yaozu zizhixian gaikuang [General Situation of the Hekou Yao Autonomous District]. Beijing: Minzu chubanshe, 2007.

Honghe Hanizu Yizu zizhizhou gaikuang [General Situation of the Honghe Hani and Yi Autonomous Prefecture]. Beijing: Minzu chubanshe, 2007.

Hyde, Sandra Teresa. "Selling Sex and Sidestepping the State: Prostitutes, Condoms, and HIV/AIDS Prevention in Southwest China". *East Asia: An International Quarterly*, issue 18, no. 4 (Winter 2000): 108–36.

———. "Sex Tourism and Lure of the Ethnic Erotic in Southwest China". In *China's Transformations: The Stories Beyond The Headlines*, edited by Lionel M. Jensen and Timothy B. Weston. Lanham: Rowman & Littlefield Publishers, 2007.

Li Chenyang and He Shengda. "China's Participation in the GMS Cooperation: Progress and Challenges". In *China-ASEAN Sub-Regional Cooperation: Progress, Problems and Prospect*, edited by Li Mingjiang and Kwa Chong Guan. Singapore; Hackensack, N.J.: World Scientific, 2011.

Magee, Darrin. "Powershed Politics: Yunnan Hydropower under Great Western Development". *China Quarterly*, no. 185 (March 2006): 23–41.

Naughton, Barry J. "The Western Development Program". In *Holding China Together: Diversity and National Integration in the Post-Deng Era*, edited by Barry J. Naughton and Yang Dali L. Cambridge, England; New York: Cambridge University Press, 2004.

Pan Qi. "Opening to the Southwest: An Expert Opinion". *Beijing Review*, 2 September 1985, pp. 22–23.

Pu Hongyan. "Burmese Migrant Sex Workers in Ruili, Yunnan". In *Cross-Border Sexuality in the Greater Mekong Sub-Region: Executive Summary*. Nakhon Pathom, Thailand: Southeast Asian Consortium on Gender, Sexuality and Health, 2006.

Sheng Lijun. "China-ASEAN Cooperation against Illicit Drugs from the Golden Triangle". *Asian Perspective*, issue 30, no. 2 (2006): 97–126.

Sturgeon, Janet C. "Rubber Transformation: Post-Socialist Livelihoods and Identities for Akha and Tai Lue Farmers in Xishuangbanna, China". In *Moving Mountains: Ethnicity and Livelihoods in Highland China, Vietnam, and Laos*, edited by Jean Michaud and Tim Forsyth. Vancouver: UBC Press, 2011.

Summers, Timothy A. "China and the Mekong Region". *China Perspectives*, no. 3 (2008): 68–77.

———. "Narrativizing Yunnan's International Borders". Paper presented at the Conference on China's Policies on Its Borderlands and Their International Implications. University of Macau, 11–12 March 2010.

Swain, Margaret. "Looking South: Local Identities and Transnational Linkages in Yunnan". In *Rethinking China's Provinces*, edited by John Fitzgerald. London; New York: Routledge, 2002.

UNDP. "Enhancing China-ASEAN Economic Integration: Cross-Border Economic Cooperation Zones at the China-Vietnam Border". Beijing: UNDP China, 2007.

UNESCAP. "Report of the Regional Meeting for Drafting the Intergovernmental Agreement on the Trans-Asian Railway". Bangkok, 22–23 November 2004.

Wen Jie. "Tourism and the Regional Development of China: Its Role and Experience of Yunnan". In *China's Economic Growth and Transition: Macroeconomic, Environmental and Social/Regional Dimensions*, edited by Clement A. Tisdell and Joseph C.H. Chai. Commack, N.Y.: Nova Science Publishers, 1997.

Xishuangbanna fazhan luntan wenji (neibu) [Proceedings of the Forum on the Development of Xishuangbanna (Internal document)], edited by Zhonggong Xishuangbanna zhouwei xuanchuanbu [Propaganda Office of the CPC of the Xishuangbanna Autonomous Prefecture] and Xishuangbanna Daizu zizhizhou jingji xuehui [Economic Association of the Xishuangbanna Dai Autonomous Prefecture]. Jinghong, 2008.

Yang Rongxin. "Yunnan's Water Ressource Development: Problems and Prospects". *China Report*, issue 42, no. 1 (January–March 2006): 25–39.

Yin Shaoting. "Rubber Planting and Eco-Environmental/Socio-Cultural Transition in Xishuangbanna". In *Good Earths: Regional and Historical Insights into China's Environment*, edited by Abe Kenichi and James E. Nickum. Kyoto: Kyoto University Press; Melbourne: Trans Pacific Press, 2009.

Zhang Deshang, Fu Guohua and Liu Zhiyong. "Hainan, Yunnan and Guangdong State Farms' Rubber 'Going Global' Strategy and Cooperation with ASEAN Countries". In *China-ASEAN Relations: Economic Engagement and Policy Reform*, edited by Yeoh Emile Kok-Kheng et al. Kuala Lumpur: University of Malaya, Institute of China Studies, 2009.

6

VIETNAM, AN OPENING UNDER CONTROL, LÀO CAI ON THE KUNMING-HAIPHONG ECONOMIC CORRIDOR

Marie Mellac

Created in 1992 with the assistance of the Asian Development Bank by the five mainland Southeast Asian countries and China, the Greater Mekong Subregion (GMS) is officially presented as "a programme of subregional economic cooperation, designed to enhance economic relations among the countries" (ADB 2009). The will to concentrate on economic matters is not surprising in the regional context of economic renovation and opening of the early 1990s. It is also to be understood as a pragmatic approach, taking into account previous integration experiments (the International Mekong Committee and ASEAN for instance) which all show the necessity of putting aside political constraints to emphasize concrete realizations. The GMS mainly concentrates on trade and operates following annual sector-based financing plans supported by varying members and donors. During the first decade (1992–2002), the GMS gave priority to transport infrastructures and developed a "corridor by corridor" approach based on five economic corridors which retraced the former caravan networks within the peninsula (Taillard GMS, this volume). Since 2002, the GMS has developed a more integrated approach to corridors which became development corridors, by emphasizing the linking of infrastructure improvement to the signing of specific free trade and free circulation agreements. It has also reinforced its territorial approach by creating three new corridors and enlarged its fields to the creation of industrial and commercial zones, interconnectivity improvement, human resources building and environmental conservation.

As suggested in its name, this new "Three Cs" strategy — connectivity, competitiveness and community (ADB 2009), links economic dynamics to the development of a greater sense of community. However this link is not evident — is a sense of community a precondition or a result of economic dynamics? — and ASEAN experiments tend to show that it is highly difficult to go beyond decades of isolation and disputes. This communication emphasizes the role of territorial representations resulting from such contentious situations. Such representations are considered as subjective but long-lasting factors which vary according to the level (or scale) and the actors considered. My hypothesis is that these representations impact on regional construction and transnational territorial construction in various, diffuse but significant ways. In order to track down their impact, this communication explores different actors perspectives at different scales. It focuses on the only institutional (political and administrative) actors who are considered as key figures of the regional construction regulatory and material framework conception and implementation.

The country studied is Vietnam, with a starting focus on Lào Cai city and province[1] on its northern border. At this level, the major developments observed during the last decade show an apparent discrepancy between the evolution of the territory as planned by local civil agents and the actual economical and social dynamics. This situation is then questioned by replacing it within the Vietnamese internal territorial dynamics and by confronting the various representations of China as expressed by civil servants at different levels.

1. LÀO CAI, A LOCALITY UNDER CONSTRUCTION

Lào Cai city, like most Vietnamese cities, has not escaped dramatic construction and huge urban extensions. Such metamorphoses are linked to Vietnam's recent rapid economic growth and to an optimistic belief in the future. In Lào Cai city, it must also be understood as the result of its special location on the Chinese border.

A well-situated border province along the Kunming — Haiphong corridor[2]

Facing China and located at the confluence of Nam Ti and Red River (see Map 6.2), Lào Cai city is halfway between Kunming (Yunnan county town) and Hanoi, on the Kunming shortest route to the sea (see Map 6.1). As for Lào Cai province, it shares 183 km of border (48 km of which is land border and 135 km river and stream border) with China (Nguyen Hong Thao

Map 6.1
Lào Cai, along the Kunming–Haiphong Corridor

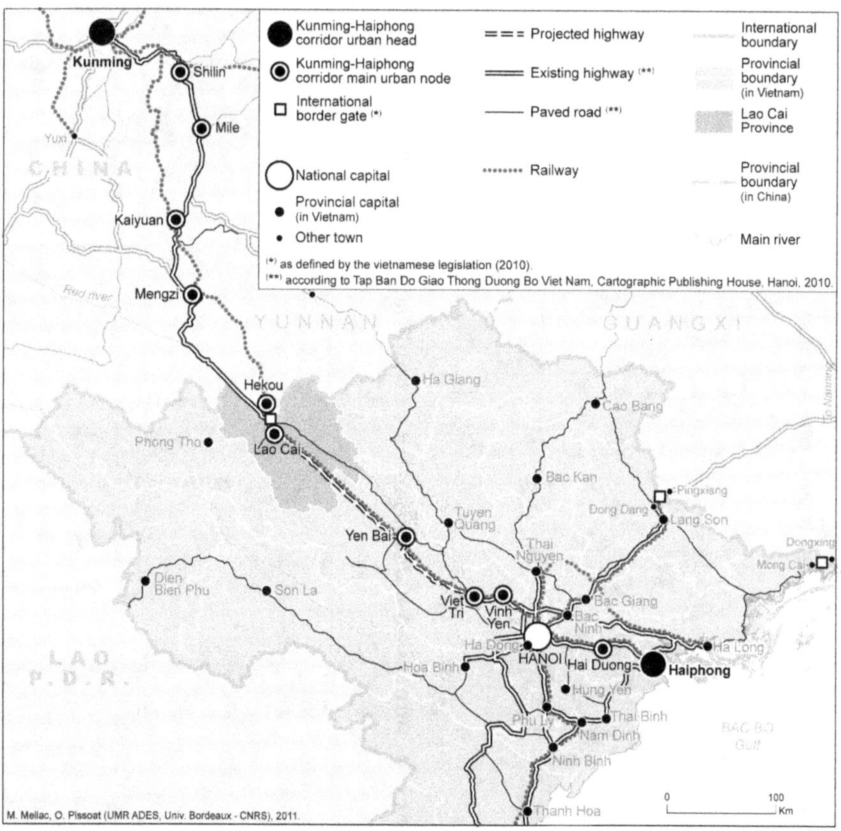

Source: Asian Development Bank, *Greater Mekong Subregion: Kunming-Haiphong Transport Corridor — Noi Bai-Lào Cai Highway*, 2005; *Vietnam Road Atlas* [Tap Ban Do Giao Thong Duong Bo Viet Nam], Cartographic Publishing House [Nha Xuat Ban Ban Do], 2010.

Map 6.2
Lào Cai–Cam Duong Urban Area Development Plan 2003–13

LAO CAI and CAM DUONG URBANIZED AREA (2003)

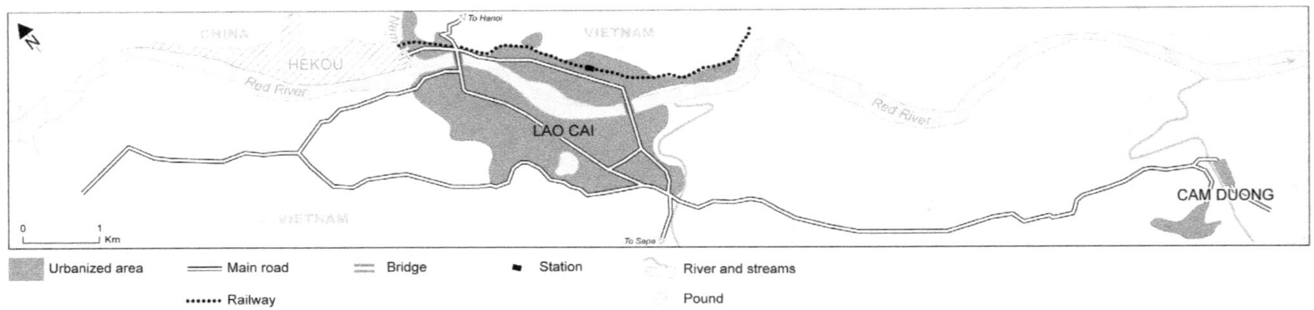

LAO CAI CAM DUONG NEW URBAN AREA DEVELOPMENT PLAN (2003-2010)

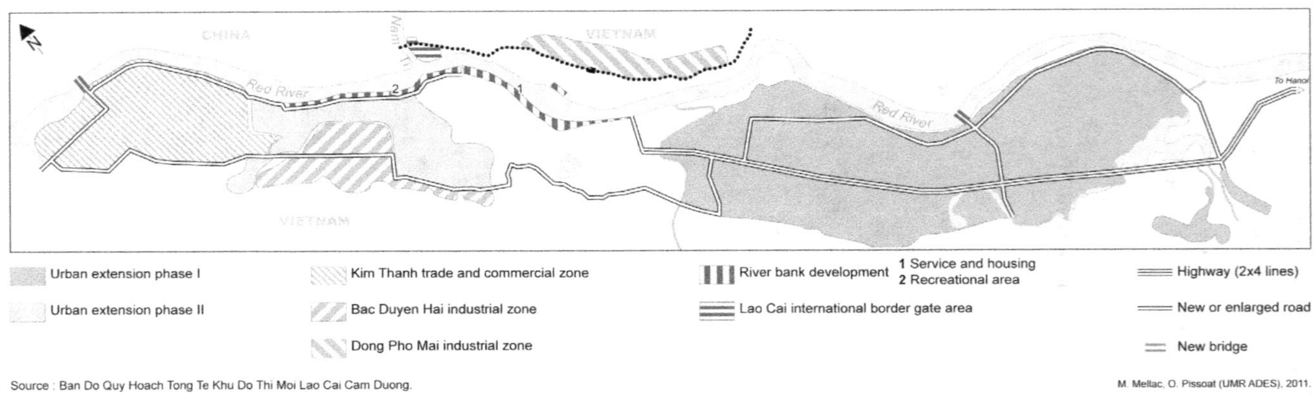

Source: Lào Cai Cam Duong New Urban Area Development Plan (Ban Do Quy Hoach Tong Te Khu Do Thi Moi Lào Cai Cam Duong).

2009). It is crossed from north to south by the Red River course which is an important penetration axis into China. Both Lào Cai city and province undoubtedly have a good location. Such advantages depend however on relations between Vietnam and China and their long history of dispute and separation has long placed Lào Cai city in a "*cul-de-sac*" position on a national scale. For a long time, Lào Cai was the last *Kinh* (or Viet) position along the Red River and was no more than a fortress that was disputed by both local and national powers.

At the time of French penetration in the mid-nineteenth century, the region was scarcely controlled by the Hue court and was agitated by the migration of ethnic minorities coming from China (Roy 2001). French people, who settled in the Red River Delta and wanted to control as much territory as possible, entered the city quite early (in 1886) and set up a military post facing the Chinese troops in Hekou.[3] Subsequently, the French spent a great deal of energy and money to maintain their position and benefit from this access to China. They invested in roads and bridges, and partly transformed the Red River waterway in order to sail further every year with bigger ships than the traditional junks. They mainly concentrated on railway development with the principal idea of connecting Yunnan-sen (Kunming) directly to the Gulf of Tonkin in Haiphong. They finished their project quite quickly. Lào Cai city was connected to Haiphong in 1906 and to Yunnan-sen in 1910. Thanks to these infrastructures, Lào Cai competed with Yen Bai as the delta's northern supply centre and attracted both activities and populations, including French people. The city then became a modern colonial centre which was an important circulation node for both goods and people. This situation also benefited the province territory. During the 1920s, tourism developed in Sapa, a colonial hill station located 30 km from Lào Cai city. The French also modernized the exploitation of mineral resources, for which they "imported" many coolies from the delta. Benefits, however, were concentrated on few people and places, and the province remained mainly an agricultural one, peopled by ethnic minorities (except in cities and mines) living in quite poor conditions and in relative isolation.[4]

In Vietnam, this isolation was reinforced during the following fifty years by the successive wars and economic choices associated with the socialist experiment. Concerning China, years of tensions ended in an armed conflict (1978) followed by ten years of isolation. All these factors partly explain why Lào Cai demographic profile remains typical of mountainous province: low population (593,600 inhabitants), low density (95 hab/sqm), low urban population proportion (20 per cent) and a multiethnic mosaic (33 per cent Kinh, 26 per cent Tai-Nung, 21 per cent Hmong). With an active population mainly employed in the primary sector (78 per cent in 2004) and an economy still dependent on the state sector (38 per cent of total income; 75 per cent of industrial income in 2007),[5] Lào Cai province is also typical of rural Vietnam and does not significantly differ from the neighbouring provinces.

However, *Doi Moi* (1987) and bilateral normalization of relations with China (1990), gave Lào Cai the opportunity to benefit once again from its good location! With Lang Son and Quang Ninh at the East, Lào Cai city is one of the first nine International Border Gates Economic Zones (BGEZ) of Vietnam[6] whose creation was authorized by the State in 1998 (see Map 6.1). Nowadays, Lào Cai city still remains the only international border gate with Yunnan and the only important one on the western side of upper North Vietnam. Lào Cai city is also a strategic place on a regional scale. It is located on one of the GMS first-generation economic corridors (1992–2002). As it was in the French period, this corridor is planned to link Kunming to Haiphong through Hanoi by both highway and railway. Well supported by China on its Chinese side, this corridor is largely funded by the Asian Development Bank (ADB) on its Vietnamese side. In 2008, for instance, ADB agreed to provide US$1.1 billion to build a 244-kilometre highway linking Hanoi to Lào Cai city which made this project the most expensive one to be funded by the ADB at the time (Thayer 2008). Vietnam, for its part, invests in the renovation of old transportation and border infrastructures such as bridges, roads, rails and border gates, and recognized this corridor as its national highest priority for infrastructure investments in 2009.

Unlike other GMS countries, Vietnam does not allow Chinese loans to finance its infrastructure. This creates a dissymmetry between the two national corridor sections (Zhu 2010). In the Chinese part of the corridor, the last section of the highway (the southern one) was opened in February 2008 and is now fully operational (ibid.). In the Vietnamese part, a discontinuous highway connects Haiphong to Yen Bai through Hanoi (see Map 6.1)[7] and the old colonial road which was hopefully renovated in 2009–10 is the only way to reach Lào Cai city from Yen Bai to date. Works for the 244 km expressway agreed by ADB in 2008 (cf. supra), had only just begun in 2010 and are planned to be finished in 2012. The highway is invested in by a Vietnamese company from bond issuance and a loan from the ADB. It is designed by a Japanese company and its different sections are contracted to be built by South Korean, Chinese and Vietnamese companies. In its current design (phase 1), the highway will have four lanes in the Noi Bai-Yen Bai section (123 km) and two lanes in the Yen Bai-Lào Cai section (121 km). When finished (phase 2), the highway will allow vehicles coming from Haiphong to avoid crossing Hanoi. It will run from Noi Bai (Hanoi airport) to Yen Bai and will double the current highway in a northern trajectory (see Map 6.1). On its northern section, between Yen Bai and Lào Cai, the highway will run on the right bank of the Red River, unlike the present road which goes along the left bank.

The railway improvement project operation started in 2005 but very little has been done to improve the quality of the rails to date. Regularly announced, the improvement project is still at the stage of technical choice discussions (new high-speed line, enlargement or simple renovation). As a consequence, most of the passenger traffic is now concentrated on the road which is faster and financially competitive for individuals and small trade activities.

Since 2004, the Kunming-Haiphong economic corridor is also part of the "two corridors and one circle policy" under ASEAN+3 which made Lào Cai theoretically included as such in the pan-Beibu Bay (or Bac Bo gulf) economic area. As far as I know, this new cooperation programme between China and Vietnam has still not had any impact on the North-South corridor, when some decisions were already taken in Quang Ninh province and Haiphong city.[8] But it reinforces its strategic weight and gives it the opportunity to remain connected to the development of the whole Bac Bo gulf area.

City and Province under Ambitious Functional Projects

As mentioned previously, large infrastructure investments are also locally visible. Since 2003, a new urban development plan (see Map 6.2) fully reorganized and resized the Lào Cai urban area. According to this plan which runs until 2010, the city is projected to accommodate 200,000 people (Nguyen Duc Muoi 2003), twice the present population,[9] in an area four times larger than in 2003, running continuously for 15 kilometers (9.32 miles) and merging with Cam Duong urban area. In the future, the Lào Cai Master Plan (Decision No46/2008/QD-TTg) envisages that Lào Cai municipality will develop in order to become a grade-II urban centre, in other words, a city of more than 300,000 inhabitants with an agricultural labour force accounting for less than 20 per cent.[10] The new 2003 city plan organized the urban area into several functional zones which are located following the new role given to the border by the local authorities. In 2009, buildings and infrastructures were far to be finished but the main "architecture" was already visible.

Heading north, a new road bridge (inaugurated in September 2009) links Vietnam with China. This bridge is the first one to be constructed on the river border since the colonial period. On both sides of the bridge, both Hekou town and Lào Cai city defined new trade zones (Kim Thanh commercial zone on the Vietnamese side and Bac Son in China). The Vietnamese zone (250 hectares) will offer "the most favourable environment for trade promotion and import-export activities between Vietnam and China" (ibid., p 224) and will comprise border infrastructures, a tax free International Trade Centre, a conference hall (where the annual bilateral trade exhibitions will be held), hotels, entertainments (like casinos), insurance and banking services. This area will progressively take over some of the commercial and border activities of the historical "Lào Cai International Border Gate Area".

This historic area never appears on planning documents and is thus not represented on the map in figure 2. Small (fifty hectares), it mainly comprises the "border-gate management and control" building and the Biti's building[11] which houses a supermarket, hotel and office accommodation. This area was the only "international border gate area" until 2009, thanks to the two colonial rail and road bridges, Ho Kieu Bridges 1 and 2, respectively.[12] It was and still is a very lively neighbourhood because of permanent movements of trucks and people during day time. But the old colonial quarter will be, according to the plan, dedicated to tourism and business services.

In very close proximity to the two border gate areas, industrial activities will be concentrated on two specific zones. The smaller one, Dong Pho Moi (146 ha planned) is located on the left bank of the Red River in the close vicinity of the current railway station and the old bridges and gate. It is mainly dedicated to storage activities for the railways. Bac Duyen Hai Industrial Area (300 ha planned) is progressively emerging south of the new Kim Thanh commercial zone and will be linked to China by the new highway and road bridge. This area is dedicated to what is called "green industries" i.e., light ones. These two areas already house industrial activities and/or storage buildings, and they are already serviced with roads, electricity, water supply and drainage.

Three "residential and service" zones are planned in the city. The central one corresponds roughly to the original urban area (as for 2003).[13] This area does not constitute a real "zone" (planned area) since it is never mentioned in planning documents or outlined on maps. Because of its past, this area is not specialized and comprises most of the current residential (small) buildings, administrative buildings from different periods, traditional markets, urban parks, hotels and all buildings typical of old city centres. According to different local informants, this area will be progressively emptied of most of its present residents and activities which will be redirected to the new urban area. Thanks to its location and architecture, the area will be then dedicated to high value activities such as company headquarters, recreation and tourism services, in relation with the border gate economic development. This new orientation (a "window on China") will be materialized by impressive improvement works already in progress along the Red River bank. In the northern part of the city, the river bank is turned into a "promenade" (which was partly built in 2010) and will offer a pleasant view of Lào Cai city from China. In the centre of the new urban area, the bank will be used to set up an alley of 6–7 storey buildings which will be — when finished — the highest in this part of the city. Building height is not yet restricted in this area.

Local administrations and new residential areas are planned to be concentrated into the Lào Cai — Cam Duong urban zone (urban extension phase I) in the southern part of the city. This new area is the most impressive one: it is planned to occupy 2,350 ha located along a fifteen kilometre-long four-lane highway whose dimensions, already visible in 2010, were also quite impressive (sixty metres wide)! This zone is designed to concentrate the entire administration in nine brand-new buildings (8–9 storeys) placed on artificial hills and theoretically[14] surrounded by gardens and recreational pounds. The area will also contain a 500-bed hospital, fifty-six hectares of lakes, cultural and sports complexes, a technical and industrial college and several residential high-standard quarters (ibid.) which had already started to emerge in 2010. It will also comprise a large set of collective buildings for middle-class households. In this zone, as in the Kim Thanh trade area, most of the planned buildings will be owned or financed by provincial public funds. Phase II of urban extension, located along the northern part of the Red river bank, is still not described on planning documents. The only map available indicates a large set of buildings looking like hotels and/or high standard residences.

Taken as a whole, the new structure of Lào Cai city is surprising. The new trade area is located far from the main urban extension and is described by its promoters as being both an isolated trade zone and a buffer zone between China and Vietnam (ibid., p. 223). On the contrary, the main urban extension is located away from the border. Even though they are described as "green", the two industrial areas are also, astonishingly, located in the direct vicinity of residential areas and/or the historic centre of the city which is planned as an area attracting tourists and businessmen. It seems then that everything is done, in a manner which can hardly be considered to be unconscious, to dedicate the border to economic activities and isolate it from the residents and local administrations. In other words, Lào Cai city plans to benefit from the border but not to be used or managed by it!

The situation is quite similar on the provincial scale. Lào Cai province does not tend to develop the border area, including the Muong Khuong national border gate (located in the eastern part of the province) which is not significantly different to the other rural localities. Considered as a whole, the province is mainly dedicated to traditional agriculture and activities and is not undergoing rapid change, with the exception of two localities. One is Tang Loong industrial complex (officially created in 2001, thirty kilometres south of Lào Cai city) which is dedicated to extraction (apatite) and heavy industries (apatite ore processing, chemical industries) in an area planned to reach 2,000 hectares in 2025, 650 of which will be for industries. The area, which was already exploited during the French period (and probably before), is planned to create jobs for 4,000 workers and be the home of 50,000 people in 2025 (ibid.). The second one is Sapa hill station which is well known for its French architecture and its ethnic minorities (Hmong and Yao people). The city is the "focal point" of tourist sector development which was itself considered as the key economic sector in the province for 2005–10 (ibid.). With an increasing number of foreign tourists visiting the area, the city is home to huge investments in both luxurious hotels and small guest houses. Nowadays, the city also attracts an increasing number of national tourists[15] whose specific demands help to progressively transform the city.

A Decentralized Institutional Context Favourable to the Development of Lào Cai

All these changes are made possible by a large set of superstructures and institutions (administration and legal system) which were created or transformed during the last twenty years. Many changes were made, for instance, to permanently enlarge and reinforce Lào Cai city. In 1992, one year after being created from Hoang Lien Son province, Lào Cai province assembled eight districts (rural units) and two municipalities (urban units), an old one, Lào Cai and a brand new created one, Cam Duong. In 2002, the two municipalities were merged by the government into a single one,

Lào Cai municipality, which was enlarged again two years later when it was upgraded to the administrative rank of provincial city. Since then, Lào Cai municipality has comprised five (rural) communes and twelve (urban) wards. Such a process (enlargement and upgrading) is typical of Vietnamese cities since Doi Moi. It is yet not specific to Lào Cai city, but these changes come together with those concerning border areas and give Lào Cai city a specific configuration.

Lào Cai province border management agenda is coherent with that at national level. The provincial border management started with the opening of a single gate limited to locals in 1990. Since then, the number of gates increased and the categories of people allowed to use them were broadened.[16] In 2001, three years after the national authorization (cf. supra), the province officially created the Lào Cai Border-Gate Economic Zone (BGEZ) and its management board which assembled all the services related to border management (customs, border guards, quarantine, etc.) with the aim of improving (and facilitating) border control and managing all related investments and commercial activities.[17]

Latter, in 2008, the province issued an important decision in order to regulate operation within the zone[18] and define its exact boundaries. The zone consists of two areas located around the two provincial international border gates (Lào Cai city and Muong Khuong)[19] and is divided into several functional zones partially coherent with the ones of the 2003 new urban area city plan. The key element of this new configuration is the "trade-industry zone" which lies adjacent to the border[20] in Lào Cai city and which constitutes a kind of territorial enclave within Vietnamese territory, with a counterpart in Chinese territory on the other side of the bridge. This zone is designed to facilitate investments[21] and local movements between the two countries. It is fully separated from the other functional zones by solid fences and is equipped by two control stations: one located at the border (the head of the bridge) aiming at controlling temporary passes and cards, and another one located in its outer limits playing the role of international border gate.

Out of the BGEZ, many institutions were created and much money spent to attract national and foreign investments.[22] In 2007, the province created the Centre for Promotion of Trade, Investment and Tourism in Lào Cai Province whose main tasks consist of organizing the biannual trade fair with China in Lào Cai city and representing Lào Cai in all exhibitions in Vietnam and China. The centre also publishes information books in three languages (Vietnamese, Chinese and English), it organizes business tours and provides information for all visitors.[23] Investors can also ask for assistance at the "Investment and Trade Facilitation Centre" which is in charge of all administrative and financial procedures. This centre is assisted in each functional zone by specific management boards run by local officers under the authority of the province and/or the municipality. These boards aim at facilitating investments and, in particular, are in charge of all tasks concerning land (clearance, renting, provision of services, etc.). As such, they take the place of the "Land Funds Development Organization" in the zone.[24] In this field, as in the field of development in general, Lào Cai provincial and municipal authorities leave little room for private and individual initiatives.

All these realizations would be ineffective without specific agreements with the Chinese side in order to speed up the border economy. Many bilateral agreements were issued on different scales (city/district, province, and prefecture) and on different subjects (education, health, investments, etc.). The most important ones are still quite recent. They involved Lào Cai province and Hong Ha prefecture in large programmes. In 2005, for instance, Lào Cai and Hong Ha agreed on the development of a "Common Economic Cooperation Zone" (Kim Thanh — Bac Son border trade special zone) which included, on the Vietnamese side, the Kim Thanh commercial zone which was designed in the Lào Cai city 2003 urban master plan. The two also agreed to build a new common border gate and a new share-funded bridge. On the Vietnamese side, the bridge was financed by national funds, like most projects concerning the border.[25] The two also agreed to organize alternate trade fairs and in 2007, they decided, in agreement with national decisions, to facilitate border crossing and trade by local permit recognition, 'one stop' procedures, and an extension of the permitted circulation perimeter for vehicles.[26]

For everybody passing through Lào Cai city nowadays, the efforts made by the authorities to develop both the city and the border area are immediately visible. In coherence with all measures taken to attract investments and boost border activities, all the officers I talked to gave an enthusiastic account of the realizations and the potential of Lào Cai. They all mentioned the strong involvement of the province and justified the dramatic investments made by the province, which will anticipate the Chinese "Go west" policy or "great reclamation of the West" (Nguyen Duc Muoi 2003).

2. THE OTHER SIDE OF THE SCREEN

Visible official enthusiasm is therefore consistent with the scale of the investments made. But officers — and the documents they provide to visitors — insist much more on projections than on concrete realities. In other words, they avoid mentioning the "other side of the screen" which is that the present situation is not as encouraging as the projected future, and that their plans are partly discordant with the ones made at a higher level.

An Impressive Economic Growth Tempered by Its High Spatial Concentration along the Corridor

Concerning the economic situation, one must first confess that the province met with undeniable economic success during previous years. Yet most of the growth indicators are impressive[27] and are higher than the national ones. It was the case, for instance, of the GDP average growth rate which was 11 per cent per year between 2000 and 2007 (Lào Cai Statistical Office 2008) when the national rate never exceeded 8.5 per cent during the same period.[28] Such data reveal dramatic changes in the province. But they are also due to a very low initial development level.

Indeed, the province is still one of the poorest provinces in the country, in both absolute and relative (to the population) value. In 2010, for instance, GDP per capita in Lào Cai was only US$365 when it was already US$1,156 at national level.[29] Lào Cai, whose population belongs to a wide range of ethnic groups, is also in third place (just after Dien Bien and Lai Chau) of provinces having the highest rate of poor households in the country (Ministry of Labour 2011). The province can still benefit from all the poverty task force programmes and poverty eradication policies. But it gives it an image of backwardness that is hard to pass over. Despite the economy is still dominated by the state sector and by primary resources production (cf. supra). There is also a majority of small firms (30 per cent have less than 10 employees)[30] and foreign direct investments, even though they are more developed than in the other northern border provinces, are slight in the industrial sector.[31]

To better understand how economic success can be linked to in Lào Cai's location at the border point of the corridor, it is also useful to compare the province to the neighbouring ones in different situations. The Table 6.1 provides some economic indicators for five northern provinces located on the Chinese border. In this panel, Lang Son is the only other one located on a GMS corridor.

According to these statistics, Lào Cai's economy is comparable to that of Lang Son with an industrial output value higher than that of agriculture and a substantial registered capital of FDI. But the province is also characterized by a higher level of poverty and social inequalities than in Lang Son (Minot et al. 2003), probably because of the share of industrial activities (mainly mineral extraction) in the composition of the GDP, the high level of isolation due to elevation, and a poor road network in most of the rural

Table 6.1
Economic Indicators in Five Northern Mountainous Provinces of Vietnam, 1998–2010

	Agricultural Output Value		Industrial Output Value			Foreign Direct Investments		Poor HH	GDP
	current prices	growth	current prices	growth	part of FDI	capital[(a)]	projects[(b)]	in total HH	per capita
	2008	1998–2008	2008	1998–2008	2008	2009	1998–2008	2010	2004–05
	Billion Dongs	%	Billion Dongs	%	%	US$ Million	nber	%	US$
Northern midlands and mountain areas	**16,531**	**46.3**	**54,394**	**5.38**	**5.5**	**158.9**	**262**	**28.4**	–
Cao Bang	691	15.3	1,498	–15.75	1.1	2.4	12	38.06	228
Ha Giang	908	52.0	766	7.40		3.9	6	41.80	154
Lai Chau	362	–13.4	231	8.96	0.8	–	–	46.78	212
Lang Son	1,145	33.9	1,491	3.77	1.3	25.3	29	28.34	237
Lào Cai	**732**	**47.3**	**4,203**	**8.16**	**0.5**	**18.7**	**36**	**43.00**	**321**

Notes: (a) registered capital; (b) licensed projects; HH = Households. The Vietnam Dong exchange rate for 31/12/2008 was 17,494.3 for one US$ (<http://www.exchange-rates.org/>).
Sources: General Statistical Office, 2011; Ministry of Labour, 2011.

communes. Lào Cai's prosperity, when considered in its spatial distribution (ibid.) is concentrated on the main infrastructure axis along the Red River course. There is a higher rate of poverty along the border, with the exception of Lào Cai municipality. Lào Cai's economic development therefore seems to be largely affected by its position along a corridor. The province, as for Lang Son, is in a relatively better economic situation than comparable provinces not located on corridors. But a very narrow portion of its territory along the Red River course and at the head of the corridor (Lào Cai city) benefits from this situation. It is typical of a tunnel effect characteristic of the corridors.

Imbalance of Commercial Relations with China

Another reason to question the dimension given to the infrastructures set up in Lào Cai is the uneven and imbalanced commercial relations with China. This situation is not specific to Lào Cai. Vietnam, whose leading trade partner is China, has a very negative commercial balance that increases every year and reached US$9.146 million in 2007 (Ha et al. 2009). Vietnam, in particular, was only the second partner of Yunnan within GMS in term of exports in 2002. It contributes to 9.4 per cent of Yunnan exported value, far behind Myanmar (20.4 per cent) (Poncet 2006). Lào Cai province itself has a huge trade gap with China which has rapidly increased since 2000[32] due to a low level of exports and a high and uneven level of imports (because of electricity). In 2007, according to the provincial Department of Trade and Industry,[33] China absorbed 81 per cent of Lào Cai's exports and provided close to 100 per cent of its imports.

Ha et al.'s (2009) paper tempers this picture a little. It shows that Lào Cai province was the main destination for Chinese FDI in Vietnam and represented close to 20 per cent of the total Chinese FDI capital in 2007. It is followed by Quang Ninh province and the cities of Ho Chi Minh, Hanoi and Haiphong respectively which means that the main cities on the Kunming-Haiphong corridor are of great interest to China.

The nature of imported and exported goods is yet another problem for Lào Cai. Except for (plastic) shoes which are the local industrial flagship (with the well-known Bitis' firm), Lào Cai mainly exports primary resources at a low price (iron, copper and rice). Such resources, in the case of minerals, are mostly extracted, traded and transported by the Chinese, reducing the income of the Vietnamese side. The Chinese, more generally, invest in sectors of importance to their own market (light and heavy industries, infrastructures, services, etc.) and have a lower impact on the Lào Cai economy than the amount of investments would let us hope. In the tourism sector for instance, the Chinese invest in infrastructures designed for Chinese tourists which means that Chinese tourist expenses return to China. For its part, Lào Cai mainly imports agricultural products like fruit, vegetables and tobacco leaves which compete with local products due to a lower price level in relation to a much more intensive level of production. China-Vietnam trade relations are also unbalanced in the domain of traders' businesses. Vietnamese businesses are only involved in retail trade on a local scale. Approximately four hundred traders gather in a market — called the Vietnamese Market — and hundreds of others run small shops and hostels in Hekou (Grillot, in this volume). The Chinese, on the contrary, specialize in wholesale trade in all sectors, and thereby take out the main profits of the high level trade relations. They do not take part in retail trade and let Vietnamese small traders sell Chinese products in Lào Cai's shops and markets. It partly explains why Chinese citizens are absent from Lào Cai city,[34] an unusual situation for a border city.

To complete the picture, one must mention that the efforts made to facilitate trade and investments are neither complete nor immediately "logical". Considering infrastructures, for instance, Lào Cai city transformations are much rapid than the upgrading of the infrastructure giving access to the city and the province. While the Chinese have already finished the highway linking Kuming to Hekou, the one planned on the Vietnamese side is in its first stage and Lào Cai city is still 9–10 hours drive from Hanoi. The situation is worse concerning rail and even air transport construction, which is endlessly under discussion.

Another question concerns border crossing. It is still difficult and is mainly practised by pedestrians. In 2009, the "one stop" procedure was still not in use and the overall procedures were, even if improved, still time-consuming. Locals are allowed to cross the border on a daily basis, with a simple "pass".[35] But the amount of goods they can trade is small[36] and most of them cross the border on foot to avoid the procedures linked to the use of a vehicle. As mentioned earlier, vehicles cannot leave the province (or Hong Ha prefecture) and their use is not of great interest except to reach Sapa station or Tang Loong industrial complex. In addition, the railways have different gauges in the two countries. Finally, most of the goods cross the border in arm wagons and most of the people are on foot. The border bridge still acts as a barrier except for very local movements.

This does not stop border activities. According to the Lào Cai Border-Gate Economic Zone Management Board (BQLKKTCKTPLC 2006), around 30,000 vehicles (22,000 were Chinese), 1,200 trains (operating the junction) and one million people crossed the Lào Cai border in 2005, and the value of goods exchanged was close to US$430 million the same year. In 2008, Chinese sources were much more optimistic and announced that the number of passengers who passed through Hekou trading port exceeded 3 millions people when the total volume of import and export trade exceeded US$1 billion per year during the last three years (2007–09) (Zhu 2010). But these statistics related to present trade can hardly explain or justify the very ambitious local development plans. Trade with China remains deeply unequal and is only of slight benefit to the local economy. The huge investments made are concentrated in a border city, but still not properly connected to Hanoi.

Lào Cai is therefore in an ambiguous situation. On the one hand, local plans are not justified by the present economic situation. They appear to be the result of a political will. On the other, the institutional framework is still incomplete and there is a surprising gap between local investments made to attract Chinese investments (and the dimension given to the new urban area) and the effective "physical" closure of the border. This ambiguity is mainly due to discordant visions of Lào Cai's future, in particular those of institutional actors between officers at provincial and central levels, who do not play at the same level of geographical scale.

3. VIETNAM'S INTERNAL TERRITORIAL DYNAMICS

Vietnam's territory, like that of most of the countries in the region, is historically constructed around lowland areas inhabited by rice growers organized into kingdoms or empires. In Vietnam, such lowlanders, *Kinh* people, were limited to the two deltas and to the tiny costal plains of the central region.

An Incomplete Decentralization

This historical construction was reinforced by the French colonial powers that settled in Hanoi and Saigon. It persists under the socialist regime in the context of a planned economy and a centralized power. As a result, people, activities, infrastructures and all types of powers are still highly concentrated in the lowlands, in particular in the two deltas, so that Vietnam is often compared to a "two-headed dragon".

The first important decentralization measures appeared in the 1990s after they became a key element of the Program of Administrative Reform (PAR) in 1995, reinforced in the 2001 programme. Local democracy progressively became an important political argument and today, nobody contests the reality of the process any more. But it is still limited and unfinished (Mellac 2009). The administrative and political hierarchy is confused and the legal system is weak, giving room to many different interpretations. Vietnam also has a strong and resistant socialist heritage of centralized planning. The Vietnamese government regularly (five and ten years) draws up economic plans which give the direction to follow for all regions and provinces. Land-use plans are designed following a top-down procedure where all plans must respect higher-level ones and can be reshaped by a higher authority (Mellac et al. 2010). Also, provincial master plans are modified and commented by the Prime Minister before being approved.

The provinces, however, are the only level to really benefit from decentralization thanks to the capacity they have to legislate and to their increasing resources. These resources, even if substantial, were not, for years, enough to support ambitious development plans. Since 2004, the provinces have been able to raise funds from land-funds. This is the case, for instance, in Lào Cai province which was strongly encouraged to do so in its ambitious 2008 Master plan (Decision No46/2008/QD-TTg). The province started even earlier to raise such funds and its huge investments coincided perfectly with the application of the last land law in 2004. Using land-funds is of great social and political significance on a local scale, and it activates a strong territorial competition at national level.

A Permanent Competition Between Administrative Units

For Gironde and Tréglodé (2011, p. 312) Vietnamese policy is made up of strained relationships between clans organized according to geographical origins, training networks and family. They feed competition which takes place at two levels: between provinces and between the provincial and central levels.

In spite of the recent financial autonomy given to the provinces by land-funds, their dependence on the State is still strong and is felt as such. Lào Cai officers often complain about the slowness of the central administration concerning both border infrastructure and institutions. If not openly criticized, the government's reticent attitude is sometimes opposed to Chinese willingness. As mentioned by an official of the Lào Cai Border-Gate Economic Zone Management Board,[37] "in general, Yunnan would like to accelerate the process of opening, but Vietnam (state) doesn't." Local officers described Lào Cai province as a very dynamic one, in advance of its time. But they also listed all the obstacles it encounters and they frequently mentioned the poor road between Yen Bai and Lào Cai city whereas Lang Son and Mong Cai are connected to Hanoi by highway. The situation is seen as unfair and is due to the inherited negative image of Lào Cai in the central administration (a backward and isolated province, inhabited by ethnic minorities) when Yunnan is "knocking at the door".

For their part, central level officers often adopt a superior stand *vis-à-vis* the provinces. They describe provincial officers as easily corrupted, weak and incompetent. This is the case in many domains, such as land (Mellac et al. 2010). Gironde and Tréglodé (2010) noted that provinces are used as scapegoats (or fuses) in corruption affairs and that local officers may be arrested in order to hush up scandals. Tensions between local and central levels are especially high where China is concerned, and are always presented as "questions of State".

> Border provinces are not in the position to understand what happens between the two countries. They look after their own interest and manage with the bordering units of the other side. Considering Lao Cai for instance, the province does not carry weight in front of Yunnan. That is why the government must keep the lead. It is true in all domains, but especially about border matters.[38]

A perfect example of competition in relation to the hyper-centralization of North Vietnam is the decision taken by the National Assembly to extend the Hanoi urban area by merging Ha Tay province with Hanoi province.[39] By this act, the central power tends to raise Hanoi to the rank of an attractive international metropolis (Musil and Labbé 2011). It also shows its capacity to distribute resources and thus that the PCV and the government still represent a political model for the future (Gironde and Tréglodé 2011). Also, the powers in Hanoi show their capacity to organize the Vietnamese territory and to do so around the capital.

Furthermore, Vietnamese development strategies are made of long-term projections intensifying competition. Vietnam has set a target that by 2020 (Decision No. 52/2008/QD-TTg) the total import-export turnover of goods and services through the borders with neighbouring countries will reach US$42–43 billion. The number of "border gate economic zones" (BGEZ) will rise to 30 and Vietnam will receive 7.8–8 million tourist arrivals through BGEZ. Provinces, like Lào Cai, are thus highly encouraged by the State to speed up their development and do to it quickly before their favourable position is weakened.

The Hanoi inflation strategy also affects many provinces and administrative capitals, and Vietnam is now covered by ambitious projects and plans. Local ruling elites, like the central ones, need to show their capacity to manage their territory and participate in national development. During interviews, they always insist on their will to surpass other border provinces and to be at the forefront of innovations and investments, given all local efforts in that direction and despite the handicap of poor transport infrastructures. Local officials then tend to reinforce and legitimize their position by projecting themselves as visionary leaders. It is thus possible to think that the dimension of Lào Cai city plans are linked to the leaders' need of legitimacy, in a context where economical choices (liberalization and privatization) challenge the PCV ideological foundations.

Land-funds and permanent competition may partly explain the dimension of Lào Cai's development plan. But this is not enough to understand why Hanoi has not sped up the process of border opening with China when it does not hesitate to do so with Laos and Cambodia. To fully understand the delicate situation of Lào Cai it is necessary to take into account the incessant strategic game between Vietnam and China.

4. VIETNAM, CHINA, AND THE BORDER

As everybody knows, Vietnam and China share a long common history, made up of competition and alliances, mutual enrichment, estrangement, open or underlying conflicts. This long, turbulent and still conflicting history, combined with the indisputable Chinese political and economical domination in the region, makes China a sounding question.

"An Uncommon Border"[40]

The year 2009, with the signature of the "Treaty on Vietnam-China Land Border Demarcation and Marker Planting" is a memorable date in Sino-Vietnamese relations. For Nguyen Hong Thao (2009) who wrote immediately after the treaty, it marked the end of a long period of tensions.

> [F]or the first time in history, the two countries have fixed a clear land borderline with a system of border markers, thereby laying a firm foundation for building a borderline of peace, friendship, cooperation and development between the two countries and ushering in a new era in the history of the relationship. (2009, p. 11)

But the honeymoon was very short. In 2010, the Chinese press revealed a Chinese Ministry of Defence attack plan against Vietnam and provoked renewed tensions (Gironde and Tréglodé 2010). Latter, a controversy about Chinese workers living in Vietnam occupied the front pages of the press and the government voluntarily let the press express virulent sentiments against Chinese attitude in the Paracels (Hoang Sa). Very recently (Spring 2011), a succession of disputes about the sovereignty of the two countries in the South China Sea caused anti-China demonstrations in big Vietnamese cities and an outburst of virulent patriotism in official media as well as in independent blogs and social networks.

How can we explain such violent reactions? Most of the reasons are to be found in the history of the two countries and in their territorial construction. First of all, one must remember that "Vietnam" was under the domination of China for the first ten centuries of our era. During this period, the limit of the *Kinh*[41] settlement was an administrative one. With Kinh autonomy, it then became a state border that specialists consider to remain globally stable until today. Stability, however, did not mean peace and quiet for *Kinh*. On the one hand, the border was characterized, as common in the Chinese vicinity, by the concept of a zone border, not by clearly demarcated borderline (Nguyen Hong Thao 2009). On the other, the Chinese refused to regard it as a border (but a limit) until the end of the nineteenth century. Vietnamese dynasties still had to fight permanently to retain the border's full present extent.

Later, the French found a very confused situation in the upper Tonkin area and signed a succession of treaties with China in order to stabilize their position and benefit from commercial advantages. In 1895, after long and perilous demarcation works engaged by both sides, all posts were planted and recognized by the two countries (Fourniau 1989). The situation remained stable overall until 1974, when the Socialist Republic of Vietnam and the People's Republic of China entered progressively into an open conflict. During this conflict, the boundaries were a crystallization point for tensions (Dauphin 1989) and it took a very long time, once peace was made, to solve the multiple tensions arising around them. The chronology of the resolution is very complex.[42] In an initial stage, the negotiations mainly concentrated on land demarcations. They ended with the 2009 treaty on Land Border Demarcation (cf. supra) which followed a long and controversial period of post planting. This treaty was negotiated until the last minutes. It is thus not consensual and gives rise to great suspicions about the way the border line was negotiated.

Suspicions are permanently reactivated by the remaining disputes on sea boundaries, in particular the matter of sovereignty over the Spratley (Truong Sa) and Paracel islands. These islands have been occupied by China since 1974 and are both a strategic (on account of their location) and an economic stake (oil). In 2008, two years after the first joint patrol in the Gulf (April 2006), Thayers reported that "in contrast to the Greater Mekong Subregion, Vietnam and China have been able to cooperate successfully in the Gulf of Tonkin" (2008, p. 17). According to him, demarcation and circulation problems in the Gulf were then solved. But the disputes over the Spratleys and Paracels did not end and provoked new tensions in the Gulf. Border problems between China and Vietnam are still active and this two thousand-year-old issue is still of great importance in the way the two countries deal with each other.

China, Not an Ordinary Neighbour

Land and sea demarcations are not only a territorial (spatial and economic) stake. Boundaries have, as Dauphin (1989) remarked concerning land boundaries, a very symbolic dimension. In the case of Vietnam, the symbolic force of the issue is proportional to the multiple roles given to China in internal affairs. China has long been a political model for the Vietnamese rulers even if they do not recognize it (Dosch et al. 2008). It is the case, for instance, of the current democratization process experienced in Vietnam (Gironde and Tréglodé 2011). China is also used by the Hanoi administration to demonstrate its capacity to act as a strong state by guaranteeing national territorial integrity and security. About the recent tensions in the South China Sea, Timberlake (2011)[43] estimated that "the difficulty will be for the government not to provoke too much Beijing without looking too weak to its people". Relationships with China are also manipulated to solve internal disputes. In 2010, the Vinashin, Vedan and Jet Star affairs[44] allowed the Prime Minister's opponents to reproach him for being weak (and probably corrupt) in his dealings with China. As stated by Dosch and Vuving (2008), the complexity of Sino-Vietnamese interactions in the post-Cold War era is closely linked to the internal division of the Vietnamese ruling elite in the period (conservatives versus reformers). For a long time, it was a political tradition in Vietnam to oppose China without showing it. But this tradition was recently superseded when China was officially described as "inconvenient" in the Third Vietnamese Defence White Book (Gironde and Tréglodé 2011).

It would be wrong to say that China only has a negative image in Hanoi. The Chinese "sister country" also means prosperity, Asian success and power. But the feeling that China is pushing at the door is very strong and was expressed in many interviews. Hanoi officers may used images like "numerous ants" and frequently used terms like "rush", "dependence", "pressures" to describe Chinese attitude. The country is also described as a colonial power already expanding in Laos and Cambodia. When speaking of the different highways linking Hanoi to China, some officers expressed the idea that these will open the delta to Chinese pressure and will allow the Chinese to siphon the country's heart.

This ambivalent feeling partly explains the dispositions taken to keep China at a distance. Since the 1999 Treaty on Land Borders, Vietnam and China have signed 54 state-level agreements and 59 bilateral cooperation documents and they have exchanged more than 100 delegations at all levels (Thayers 2008). China, moreover, benefited from a different treatment to Laos and Cambodia concerning border management. If considering International Border Gates, there are only four with China (for a 1,281 km long border) (see Map 6.1) whereas there are seven with Cambodia (1,228 km) and seven also with Laos (2,130 km). In the 2008 "Master Plan on Vietnam's Border-gate Economic Zone Development up to 2020" (Decision No. 52/2008/QD-TTg) China is also the only country proposed to "associate economic development with protection of national sovereignty and border security on the basis of resolving social problems and improving people's material and cultural life". Whereas the question of security at the Chinese border is mentioned several times in the plan, it is only mentioned once for Laos and Cambodia, which are only offered "solutions on consolidation of defence and security in BGEZ".

In another important decision (Decision No. 1151/2007/QD-TTg) approving the up-to-2020 planning on building of the Vietnam-China border region, the Vietnamese government plans to displace households living in the vicinity of the border (around 1,000 in Lào Cai) to create a buffer zone and for the population distribution to "form a chain system in support of economic development and security as well as defence maintenance" (article 1). Such dispositions are also to be funded in local decisions such as the 2008 "Master plan on socio-economic development of Lào Cai province until 2020" (Decision No. 46/2008/QD-TTg). In this decision, two long paragraphs mention security and defence. It is planned, for instance, "[to] build permanent border stations and posts; to build border ring roads and border patrol paths, consolidate and develop comprehensively the economic-defence zone in Bat Xat district; to firmly build provincial and district defence areas in the province... To launch mass movements to involve the people in protecting national security...."

But such territorial decisions are much more a concern in Hanoi than in Lào Cai. Local officers consider security as an important and serious matter and mentioned problems keeping the border under control. But they are very reluctant to speak about illegal traffic and border crossings and they never stress the point of security during interviews. About the future highway coming from Hanoi, they insist on the opportunity it represents for Lào Cai's development and they surprisingly do not worry that the city may be crossed by trade and migration flows rather than being a stop. They spend time in China and do not consider it as a distant body. They learn the Chinese language and use the Chinese health system, they buy plenty of materials and products in China, and their children study in China. All of them frequently cross the border; they have all been to Kunming at least once and some to other cities in the south or even further away in China. China is part of both their private life and their function, both a daily concern and a political one. Having nothing to prove but plenty to gain both collectively and personally, they insist much more on proximity with China than on distance. That is probably why they concentrate most of their efforts on developing the city and the province to adapt to Chinese presence.

Multilateralism as a Solution

Hanoi is not completely reluctant to opening, and develops strategies to mitigate its tensions with Beijing. One of these strategies consists of focusing and giving much media coverage to non-problematic questions. This was the case of the 2006 memorandum of understanding on the 'two corridors, one circle' cooperation. The two corridors consist of developing transport routes in the corridors (Nanning-Lang Son-Hanoi and Kunming-Lào Cai-Haiphong). The circle (or belt) refers to the joint development of the Gulf of Tonkin (including joint patrols). This memorandum is a significant advance in terms of symbolic and practical cooperation (funding procedures for instance). But it does not contain any decision solving the problematic questions as shown by the rapid reappearance of tensions in the Gulf (cf. supra).

Another strategy consists in managing bilateral relations in a multilateral context. Both countries participate in several multilateral organizations including some on a regional basis, for example, ASEAN+3 and GMS. In so doing, they committed themselves to modifying the nature of their relations. According to Thayer (2008), the GMS provides a unique cooperation framework between Vietnam and China to develop their respective border regions. Since the early 1990s, both China and Vietnam have included the development of border regions in their national development plans. And GMS led to substantial development of energy, communications and transport infrastructures. On the contrary, all discussions on Mekong waters have been rendered fruitless by the absence of China in the Mekong Committee. Vietnam does not react to China's rise in the region by pursuing an external balance of power strategy. Its strategy consists in using all multilateral organizations in order to manage and codify its relations with China through understanding reached at a higher level (ibid.). This process is facilitated by the Chinese will to foster the separation of economic, strategic and environmental issues in different organizations. Its effectiveness is weakened by the absence of a global solution — and the possibility that specific tensions may reappear. But the Vietnamese choice not to come into head-on conflict with China is a pragmatic positioning. Vietnamese strategy consists also in having a paternalistic and cooperative attitude *vis-à-vis* Laos and Cambodia in order not to appear isolated or give a weak internal and external representation of itself. This is made possible by classic bilateral cooperation but also by the multidimensional nature of cooperation within the GMS. The GMS is therefore an effective organization to make Vietnam develop as a strong state faced with a much stronger one.

CONCLUSION: AN OPENING UNDER CONTROL

This chapter has addressed the question of how to explain the paradoxical local situation of Lào Cai. Lào Cai can be compared to an impressive trade and business system (under construction) waiting for the border to open and the road corridor to be finished. This local situation is the result of paradoxical (Vietnamese) representations of China, varying according to the focus and the level of actors. Vietnam manages its delicate position as China's neighbour and former colony by trying to keep it at a distance without entering into conflict. Its position is favoured by multilateral organizations such as the GMS but also necessitates strong territorial control. Vietnam reacts to China's power by pursuing "a strategy of internal balancing by a modest effort to modernize its military forces and by maintaining national unity (ibid., p. 25). By doing so, Hanoi sent contradictory information to Lào Cai and led to local territorial dynamics becoming temporarily disconnected from current reality. This is an particular form of integration caused by a weak link at the national level.

Factors of scale, as shown in this communication, are not only a question of geographical distance and dimension of the phenomena. They are also crucial to the way actors play and interact in relation to the way they apprehend their complex context. Representations do not form an independent and ideal sphere either at each level or in the material context. They are dependent upon each other and are accompanied (they produce and are produced) by acts such as legal decisions and funding choices which impact on infrastructures and on flows of material and population, at all levels.

For how long will Lào Cai's development be checked by Hanoi's fears? And for how long will the Chinese complain about the poor activity on its Yunnanese border?[45] Maybe not for long! Lào Cai was, according to "The Vietnam Provincial Competitiveness Index 2010", the second most competitive province of Vietnam after Da Nang. "Despite many other difficulties confronting the province, positive firm perception of its economic governance has clearly signified a critical success of Lào Cai for the past years" (ibid., p. 12). Lào Cai's strong will and GMS choices (combined with strong economic changes all over Vietnam) may counterbalance Hanoi's weakness in the near future. I would not conclude this communication without remembering the social cost of both Hanoi's fears and Lào Cai's huge plans. For the present, Lào Cai's development rests on local land funds resulting from expropriations and produces a very standardized urban (and tourist) model which provides no room for the weak local population, especially minorities. The future of these populations in the context of the ADB corridor grid strategy in general and of the Kunming-Lào Cai-Haiphong corridor in particular, is a matter of concern which has never been questioned, either by central officers, or by local ones.

Notes

1 Lào Cai is both the name of a province and of its city-capital. When not specified, "Lào Cai" will refer to the province.

2 Apart from published data, this communication is based on interviews conducted in Lào Cai and Hanoi in 2009 and early 2010 with public officers.

3 Hekou (or Ha Khau) is the Chinese town facing Lào Cai on the other bank of the two rivers. Hekou depends on Honghe (or Hong Ha) autonomous prefecture within Yunnan province.

4 Such isolation, as emphasized by Michaud and Forsyth (2011), is relative to the national territory only (and to its economic and political centre located in Hanoi) as long as local people maintained — until today — cultural and trading links across borders.

5 All data in the paragraph are for 2007 (Lào Cai Statistical Office 2008), except the ethnic composition which is for 2009 (General Statistical Office 2011).

6 The first nine were created in 1998 and were located in Lào Cai city (Lào Cai province), Dong Dang Lang Son (Lang Son province), Mong Cai (Quang Ninh province), Cau treo (Ha Tinh province), Lao Bao (Quang Tri province), An Giang city (An Giang province), Moc Bai (Tay Ninh province), Bo Y (Kon Tum province), Dong Thap (Dong Thap province). In 2008, the government decided to increase their number to 30 by 2020 (Decision No. 52/2008/QD-TTg).

7 In 2010, this highway was mainly a four-lane one except in short sections located in rural areas. These sections have also been recently renovated.

8 If the Haiphong-Kunming corridor is part of the "two corridors and one circle policy", the "Tonkin Gulf coastal economic belt section in the Vietnamese territory covers Quang Ninh province and Haiphong city" (Decision No. 34/2009/QD-TT, article 1).

9 Lào Cai city's average population was 94,192 people in 2007 (Lào Cai Statistical Office 2008).

10 According to Decree No. 42/2009/ND-CP on the Grading of Urban Centres, administrative units are classified into six grades of "urban centers" (from Special Grade to Grade V) which are defined both by socio-economic functions, population scale and density, infrastructure, architectural and landscape development situations.

11 The building was fully funded by the Binh Tien Consumer Goods production Company (Biti's) (Interview, Lào Cai Province Centre for Promotion of Trade, Investment and Tourism, 22 April 2009).

12 The road bridge was fully reconstructed during the 1990s.

13 It corresponds approximately to the whole area already built in 2003 and comprises, on the east bank, part of the Pho Moi quarter and, in the other bank, the French quarter around the cathedral and the socialist period extensions to the south (around the Nhac Son park).

14 The Lào Cai Promotional Book (Nguyen Duc Muoi 2003) praises the "green" quality of the plan and indicates for instance that "for trees, there has been no cutting down and moving" (p. 213). However, a rapid observation demonstrates that all "natural" elements including trees and land forms were completely eradicated.

15 Tourists spending at least one night in Lào Cai province were reported to be around 100,000 in 2007 (Lào Cai Statistical Office 2008). They come from all over the world (80 nationalities) but local and Chinese tourists increase every year.

16 Today, the province border infrastructure is composed of: four "Local Border Gates" where only locals from the adjacent district can cross the border with a "pass" (they are allowed to stay one day in the close vicinity of the border);

one "Local International Border Gate" (Muong Khuong) where all (but only) Chinese and Vietnamese can cross the border with a visa; two "International Border Gates" located in Lào Cai city (the "historic" one on the Nam Ti and the new one opened close to the Kim Thanh bridge in 2010) open to all foreigners possessing passports and visas.

17 Interview, Lào Cai Province Industrial Zones Management Board, 22 April 2009.

18 It did so by Decision No. 44/2008/QD-TTg of March 2008 promulgating the regulation on operation of Lào Cai Border-Gate Economic Zone.

19 The part located in Lào Cai city is called the "Lào Cai International Border-gate Area". It covers six wards of Lào Cai municipality and two communes of Bao Thang district. The "Muong Khuong Border Gate Area" comprises the only Muong Khuong commune.

20 I did not have a chance to see any map of this zone. It seems to correspond to the "Kim Thanh Trade Area" which is now the only area surrounded by fences.

21 According to Decision No. 44/2008/QD-TTg, all organizations and individuals who invest in the BGEZ are allowed to mortgage their land use right to all credit organizations operating in Vietnam. All investment projects are granted with best incentives applied for localities with special socio-economic disadvantages (Corporation Income Tax — CIT — rate of 10 per cent for fifteen years from their date of operation; CIT exemption for four years since earning taxable income; tax reduction of 50 per cent for the nine following years). Chinese and Vietnamese citizens residing in the districts adjoining the border are permitted to enter and stay in the BGEZ for fifteen days if they have a pass issued by their respective authorized agencies. Foreigners and Vietnamese holders of foreign passports and their family members will be granted long-term visa to enter the EZ.

22 Interview, Lào Cai Province Department of Planning and Investment, 21 April 2009.

23 Interview, Lào Cai Border-Gate Economic Zone Management Board, 22 April 2009.

24 Land funds (or Land Fund Development Organizations) were created by the government in 2004 (by an application decree of the 2003 land law) in order to organize investors' access to the land. They are in charge of land clearance and to all procedures needed to invest. They are also supposed to manage the funds they put together from their services and rent, and to compensate the previous land users. These organizations whose status is not clear (half public, half private), allow the State (and the provinces) to keep the sensitive question of land at a distance (Mellac et al. 2010).

25 The bridge funding is not very clear. It differs from one informant to another. But the Lào Cai promotion book (Nguyen Duc Muoi 2003) mentioned that "the work of Kim Thanh bridge having a political significance related to the national border is invested by the central budget capital" (p. 224).

26 On the Vietnamese side, the perimeter of permitted circulation was enlarged to the whole of Lào Cai province. Previously it was 30 km and only enabled vehicles to reach Sapa and Tang Loong industrial complex. Inside the perimeter, the circulation is restricted to vehicles having special permits. Outside the perimeter, the circulation of foreign vehicles is forbidden. On the Chinese side, the perimeter corresponds to Honghe prefecture (Interview, Lào Cai Department of Communication and Transports, 23 April 2009).

27 Between 2000 and 2007, all total growth indicators were impressive. The total growth of tourist numbers was 514 per cent, the total growth of export and import values (in current US$) were respectively 368 and 3,164 per cent and the total growth of the GDP at constant price was 232 per cent (Lào Cai Statistical Office 2008).

28 Indexmundi, 2010, <http://www.indexmundi.com/vietnam/gdp_real_growth_rate.html> (accessed April 2011).

29 To compare with US$2,800 in Ho Chi Minh city or US$1,950 in Hanoi in 2010. All data from the Committee for Ethnic Minorities <http://cema.gov.vn/> (accessed March 2011).

30 All data from LCSO (2008).

31 As a result of these weaknesses, Lào Cai is the Vietnamese province that, every year, has the lowest industrial gross output of foreign invested sector per year in Vietnam since 2000.

32 The balance was positive in 2000, it was US$7 million in 2005, US$20 million in 2006, US$91 million in 2007 (Lào Cai Statistical Office 2008) and US$410 million in 2008 (General Statistical Office 2011). In the official statistics, all trade registered until 2007 was carried out by local and national state sectors.

33 Interview, Department of Trade and Industry, 23 April 2009.

34 One officer of the Department of Industry and Trade (interview, 23 April 2009) noted that this situation is the complete opposite of that of Mong Cai in Quang Ninh province where Chinese traders are numerous on the Vietnamese side and Vietnamese traders are few in China.

35 Chinese citizens residing in districts bordering Lào Cai province may enter "Lào Cai Border-Gate Economic Zone" with border passes and are allowed to reside temporarily in Lào Cai (Article 8, Decision No. 44/2008/QD-TTg). The reverse exists on the Chinese side.

36 It was 2 million Vietnam Dong per day and per adult in 2009 (Interview, Lào Cai Province Department of Industry and Trade, 23 April 2009).

37 Interview, Lào Cai Border-Gate Economic Zone Management Board, 22 April 2009).

38 Interview, Ministry of Planning and Investment, 4 May 2009.

39 The new Hanoi urban area comprises the previous Hanoi municipality, Ha Tay province, the Me Linh district (Vinh Phuc province) and four communes (Dong Xuan, Tien Xuan, Yen Trung et Yen Binh) of Luong Son district

(Ho Binh province). The area of the capital was tripled and Hanoi then became the largest capital in Southeast Asia, ahead of its southern rival, Ho Chi Minh city.

40 In reference to the cover title of the *Vietnam Law and Legal Forum*, vol. 15, no. 176 (April 2009).

41 During this period, *Kinh* settlement (Giao Chi) was limited to the Red River delta.

42 The main episodes are: 1990 — normalization of bilateral relations; 1992 — Vietnam and China began their boundary negotiations; 1993 — agreement on the basic principles for settlement of territorial and border issues. The following negotiations were conducted over a six-year period and ended in 1999 with the "Treaty on the land border" (but not on the maritime ones); 2000 — Agreement on the Demarcation of Waters, Exclusive Economic Zones and Continental Shelves in the Gulf of Tonkin; 2005 — Agreement on the conduct of joint naval patrols in the Gulf of Tonkin; 2006 — negotiations on demarcation of maritime waters outside the Tonkin Gulf.

43 Ian Timberlake, "'Vietnam: l'effet Facebook' enflamme les tensions avec la Chine", *Agence France Presse*, 10 June 2011, <http://blog.vietnam-aujourdhui.info> (accessed June 2011).

44 Vinashin (Vietnam Shipbuilding Industry Group) was a successful public enterprise which was generously funded by the government until major financial difficulties were revealed to the press in 2009. Vedan Vietnam is a Taiwanese chemical firm which was convicted of seriously polluting the Thi Vai river. Jet Star Pacific is a Vietnamese low-cost airline company (partly funded by Qantas since 2008) whose directors were convicted of failing to respect air safety procedures. Both cases were widely covered by the press and were an opportunity to discuss the responsibilities of various high officials (some under arrest), including Prime Minister Nguyen Tan Dung. As such they are supposed to be used to settle political disputes.

45 "We are sad to see that even after 12 years development since GMS was set up in 1992, cross-border trade and tourism in Hekou and Lào Cai, Guliu port group, are still in a low level. Hekou has river, railway and highway transportation to connect with Vietnam. Its facilities are rare in all ports in Yunnan Province" (Zhang Bing and Deng Gang 2004, p. 247).

References

Asian Development Bank. *Greater Mekong Subregion: Kunming-Haiphong Transport Corridor — Noi Bai-Lào Cai Highway*. Technical Assistance Project number: 33307, Report and Recommendation of the President to the Board of Directors. Sri Lanka: ADB, November 2005.

———. *Greater Mekong Subregion*. Updated 27 February 2009. Available at <http://www.adb.org/gms/> (accessed October 2009).

BQLKKTCKTPLC (Ban Quan Ly Khu Kinh Te Cua Khau Thanh Pho Lào Cai). *Report on the Construction and the Results of the Lào Cai Border Gate Economic Zone* [Bao cao tinh hinh xay dung va ket qua hoat dong Khu Kinh Te Cua Khau Lào Cai]. Lào Cai: UBND Tinh Lào Cai, 2006.

Dauphin, André. "La frontière sino-vietnamienne de 1895-1896 à nos jours" [The Sino-Vietnamese Border from 1895–1896 until present]. In *Les frontières du Vietnam, Histoire des frontières de la Péninsule Indochinoise* [Vietnamese Borders, An History of Borders in the Indochinese Peninsula], directed by P.B. Lafont. Paris: L'harmattan, coll. Recherches Asiatiques, 1989.

Dosch Jörn and Alexander L. Vuving. *The impact of China on governance structures in Vietnam*. Discussion paper 14. Bonn: Deutsches Institut für Entwicklungspolitik, 2008.

Fourniau, Charles. "La frontière sino-vietnamienne et le face à face franco-chinois à l'époque de la conquête du Tonkin" [The Sino-Vietnamese Border and the Sino-French Face to Face at the Time of the Tonkin Conquest]. In *Les frontières du Vietnam, Histoire des frontières de la Péninsule Indochinoise*, directed by P.B. Lafont. Paris: L'harmattan, coll. Recherches Asiatiques, 1989.

Gironde, Christophe and Benoît de Tréglodé. "Vietnam, Une année à l'ombre du Congrès" [Vietnam, a Year under the Congress' Shadow]. In *L'Asie du Sud-Est 2011*, directed by Leveau, Arnaud and Benoît de Tréglodé. Paris: Les Indes Savantes, 2011.

———. "Vietnam". In *L'Asie du Sud-Est 2010*, directed by Leveau, Arnaud and Benoît de Tréglodé. Paris: Les Indes Savantes, 2010.

Ha Thi Hong Van and Do Tien Sam. *Vietnam-China Trade, FDI and ODA Relations*. Institute of Developing Economies, Japan External Trade Organization, 2009. Available at <http://www.ide.go.jp/English/Publish/Download/Brc/.../01_vietnamandchina.pdf> (accessed March 2011).

Mellac, Marie. *La gestion déléguée des terres forestières au Viêt Nam: jeux d'acteurs et d'illusions* [The Delegated Management of Forest Land in Vietnam: Actor Games and Illusions], 2009. Available at <http://halshs.archives-ouvertes.fr/halshs-00378969/fr>.

Mellac, Marie, Frédéric Fortunel and Tran Dac Dan. "Land Reform in Vietnam: The Analysis of the Roles Played by Different Actors and Changes Within Central and Provincial Institutions". Paris: AFD/MAE, 2010. Available at <http://www.foncier-developpement.fr/travaux-recherche/>.

Michaud, Jean and Tim Forsyth. "Rethinking the Relationship between Livehoods and Ethnicity in Highland China, Vietnam, and Laos". In *Moving Mountains, Ethnicity and Livehoods in Highland China, Vietnam and Laos*, directed by Jean Michaud and Tim Forsyth. Vancouver: UBC Press, 2011.

Ministry of Labour, Invalids and Social Affairs. *Results of the Survey on Poor and Nearly Poor Households, Year 2010* [Ket qua tong dieu tra ho ngheo, ho can ngheo nam 2010], 2011. Available at <http://www.molisa.gov.vn/> (accessed June 2011).

Minot Nicholas, Bob Baulch and Michael Epprecht. *Poverty and Inequality in Vietnam: Spatial Patterns and Geographic Determinants*. International Food Policy Research Institute/Institute of Development Studies, 2003. Available at <http://www.isgmard.org.vn/.../Poverty%20Mapping%20Final%20Report-e.pdf> (accessed February 2011).

Musil, Clément and Danielle Labbé. "The Extension of Hanoi's Administrative Limits: Territorial Restructuring in Tension". *Cybergeo: European Journal of Geography*, 2011.

Nguyen Duc Muoi, ed. *Lào cai van hôi moi* [Lào Cai new opportunities]. Cong Ty Co Phan Hop Tac Truyen Thong Viet Nam, Ha Noi: Nha Xuat Ban Chinh Tri Quoc Gia, 2003.

Nguyen Hong Thao. "A line runs through it, Vietnam and China complete boundary marking process". *Vietnam Law and Legal Forum* 15, no. 76 (April 2009): 7–13.

Poncet, Sandra. "Economic Integration of Yunnan with the Greater Mekong Subregion". *Asian Economic Journal* 20, no. 3 (2006): 303–17.

Roy, Alexandre. *La pénétration des technologies occidentales dans la province de lào Cai (Vietnam nord) pendant la colonisation française, de 1886 à 1945* [The Progression of Western Technologies in Lào Cai Province (North Vietnam) during the French Period, 1886–1945]. Master Degree Dissertation, Department of Geography, University Bordeaux III, 2001.

Thayer, Carlyle A. "The Structure of Vietnam-China Relations, 1991–2008". Paper for the 3rd International Conference on Vietnamese Studies, Hanoi, 4–7 December 2008. Available at <http://www.viet-studies.info/kinhte/Thayer_Sino_Viet_1991_2008.pdf> (accessed 2010).

Timberlake, Ian. "'Vietnam: l'effet Facebook' enflamme les tensions avec la Chine", *Agence France Presse*, 10 June 2011, <http://blog.vietnam-aujourdhui.info> (accessed June 2011).

Zhang Bing and Gang Deng. "GMS Development as Viewed from the Cross-border Tourism and Trade in Hekou Town, Yunnan". Regional Economic Cooperation: EU and GMS Development Strategies, Regional Conference 2004. Bangkok: Centre for European Studies, Chulalongkorn University, 2004.

Zhu Zhenming. "Construction and Development of Kunming-Hanoi-Haiphong Economic Corridor". *The Ritsumeikan Economic Review* 58, nos. 5–6 (2010): 731–41. Availabe at <http://www.ritsumeikeizai.koj.jp/koj_pdfs/58504.pd> (accessed May 2011).

7

INTEGRATION OF GREATER MEKONG SUBREGION CORRIDORS WITHIN LAO PLANNING, ON NATIONAL AND REGIONAL SCALES
A New Challenge

Bounthavy Sisouphanthong

Since the establishment of Lao People's Democratic Republic (Lao PDR) in 1975, the Lao economy has been through a series of transformations. With the introduction of the New Economic Mechanism in 1986, Lao PDR has gradually increased its integration with the subregional, regional and global economy. Some of the milestones include membership of the Greater Mekong Subregion cooperation (GMS) in 1992, the Association of Southeast Asian Nations (ASEAN) in 1997 and now WTO in 2013. The country has also been an active member of many regional and subregional cooperation initiatives, creating the opportunity for both investment and foreign trade. The economy has grown at a remarkable rate of 6.7 per cent a year over the past two decades. Despite the global financial crisis, the Lao economy managed to grow by 7.6 per cent in 2009 and 7.8 per cent in 2010. The GDP per capita rose from US$329 in 2001 to US$1,069 in 2010. In the social field, progress has also been observed. The literacy rate for both males and females, the primary completion rate as well as the total enrolment rate show a gradual improvement in education. Health indicators, such as infant mortality rate and life expectancy, show some improvements. There has been a significant progress in poverty eradication. The poverty headcount ratio at national poverty line fell from 46 per cent in 1992/93 to 27.6 per cent in 2007/08 and is estimated at 20.4 per cent for 2010.

This chapter aims to review the integration of Lao PDR into the regional and subregional economy. It will first summarize the evolution of Lao economic integration. The chapter will then outline how and why Laos needs to integrate with the region. In the last part, it will discuss the need for a better integration at different levels, including a discussion of the opportunities and challenges arising from the integration. In the conclusion the concept of connectivity is introduced.

EVOLUTION OF LAO PDR'S INTEGRATION IN THE REGION

The introduction of a comprehensive economic reform, entitled the New Economic Mechanism (NEM or *Jin Ta Na Karn Mai* in Lao) in 1986, marks the beginning of the transformation from a planned economy to a more market-oriented economy. The government begins with the guarantee of free participation by private enterprises in the production and the distribution process. Some major reforms as part of the NEM include price liberalization reform to allow prices to move in accordance with the forces of supply and demand, agricultural production reform to modernize production and improve productivity, reforms of state-owned enterprises to improve management and introduce some privatization, as well as other changes in the legal and institutional framework. Production, consumption and investment decisions could be made in accordance with the market signal and private incentives.

Regional Integration Initiatives

The series of economic reforms mentioned above has led to increased confidence of foreign investors and trading partners, resulting in increased foreign trade and investment. This gives Lao PDR the opportunity to gradually increase economic integration with the region. The first major integration took place in 1992 when Lao PDR became a member of Greater Mekong Subregion cooperation (GMS). The Greater Mekong Subregion is a subregional cooperation programme supported by the Asian Development

Bank (ADB), aiming to enhance economic relations, building on shared histories and cultures and covering nine priority sectors (ADB 2011). Some outstanding ADB-GMS projects include the upgrade of East-West and North-South economic corridors, contributing to the transformation of Lao PDR from land-locked to land-linked status.

The membership of the Association of Southeast Asian Nations (ASEAN) in 1997 reconfirms Lao PDR's determination to deepen integration with the region. ASEAN is a comprehensive regional bloc working towards the establishment of an ASEAN Community by 2015. The main aims and purposes of ASEAN are acceleration of economic growth, social progress and cultural development; promotion of peace and stability; promotion of active collaboration and mutual assistance; provision of assistance in the form of training and research; collaboration for better utilization of resources and capacity in priority sectors; promotion of Southeast Asian studies; and promotion of close and beneficial cooperation with other existing international and regional organizations. The commitment to establish the ASEAN Community by 2015 is reaffirmed in the Cebu Declaration of 2007 during the 12th ASEAN Summit. The ASEAN Community consists of the ASEAN political-security community, the ASEAN Economic Community (AEC) and the ASEAN Social-Cultural Community. The ASEAN Free Trade Area (AFTA) is among the most observable outcomes of ASEAN cooperation, gradually removing trade barriers between member countries.

Other regional integration initiatives include the Cambodia-Laos-Vietnam (CLV) Development Triangle and the Irrawaddy-Chao Phraya-Mekong Economic Cooperation Strategy (ACMECS). These subregional cooperation initiatives answer the need for better and more effective regional integration especially for Lao PDR. They will contribute to the reduction of the development gap between old and new ASEAN members, and to expedite development in the peripheral areas. The CLV aims to accelerate economic growth, eradicate hunger, reduce poverty and advance cultural and social progress in the Development Triangle Area by exploiting the potential and advantages of each nation (Vietnam Embassy 2011). Improving infrastructure such as transport, electricity and information communication technology (ICT) is considered as the priority area for regional development, with the support of the Japanese government in the framework of Japan's initiative for Mekong Region Development (MOFA Japan 2011). The ACMECS is a cooperation framework between Cambodia, Lao PDR, Myanmar, Thailand and Vietnam to utilize member countries' diverse strengths and to promote balanced development in the subregion. The main objectives of the ACMECS are to increase competitiveness and generate greater growth along the borders; facilitate relocation of agricultural and manufacturing industries to areas with comparative advantages; to create employment opportunities and reduce income disparity among the four countries; and enhance peace, stability and shared prosperity for all in a sustainable manner.

Significant Progress in Terms of Trade Openness

The continuing integration of the Lao economy has led to expansions in cross-border trade and foreign direct investment. Both exports and imports have increased gradually. The growth of exports increased from an average of 7.4 per cent per year during 2001/02 to 2005/06 to 36.9 per cent during 2006/07 to 2010/11. The growth of imports also increased from 5.2 per cent to 23.6 per cent during the same period. A more rapid growth in exports leads to a shift from trade deficit to trade surplus. For instance, the trade balance progressed from an average of –10 per cent of GDP during 2001/02 to 2005/06 to a trade surplus of 1.74 per cent of GDP in 2010/11. Recent external shocks have shown the Lao trade sector is sensitive to external vulnerability. The value of both imports and exports decreased during the Asian financial crisis and the recent global economic and financial crisis.

The integration of the Lao economy with the region is confirmed by the significant role of the region in foreign trade. Thailand represents Laos' most significant trade partner with trade volume accounting for more than 40 per cent on average of total trade. Vietnam and China became the second and third largest Lao trade partners respectively. Their bilateral trade volume with Laos covered on average respectively 11 per cent and 8 per cent of the total (see Table 7.1 and Figures 7.1 and 7.2). However, among the three trade partners, Laos-China bilateral trade achieved the highest growth rate at 34 per cent yearly while the growth rate for Thailand and Vietnam was 26 per cent and 1 per cent respectively. Thailand and Vietnam were important not only as bilateral trade partners with Laos, but also in the AFTA scheme. Similarly, France, Germany and the United Kingdom are Laos' top three trade partners in the European market (with a total share of 8 per cent of total Lao trade on average). It is also interesting to note that the market for Lao exports is rather concentrated. Trade with Thailand, Vietnam, China and France accounts for more than 77 per cent of total trade volume on average. On the other hand, the import share from Thailand and China alone accounts for more than 87 per cent of total imports.

Table 7.1
Laos PDRs' Major Trade Partners
(Unit: % of Total Trade Volume)

	2000/01	2005/06	2010/11
Thailand	26.85	25.21	42.60
Australia	0.03	6.39	11.40
China	5.78	5.24	7.37
Vietnam	6.24	10.05	6.96
Japan	0.45	0.50	2.01
United Kingdom	1.16	3.49	1.34
France	2.67	1.54	1.31
Korea (South)	0.52	0.32	1.14
Germany	1.44	1.89	1.00

Sources: Ministry of Industry and Commerce.

Figure 7.1
The Value of Export and Imports During 2000–10 Period
(Unit: US$ million)

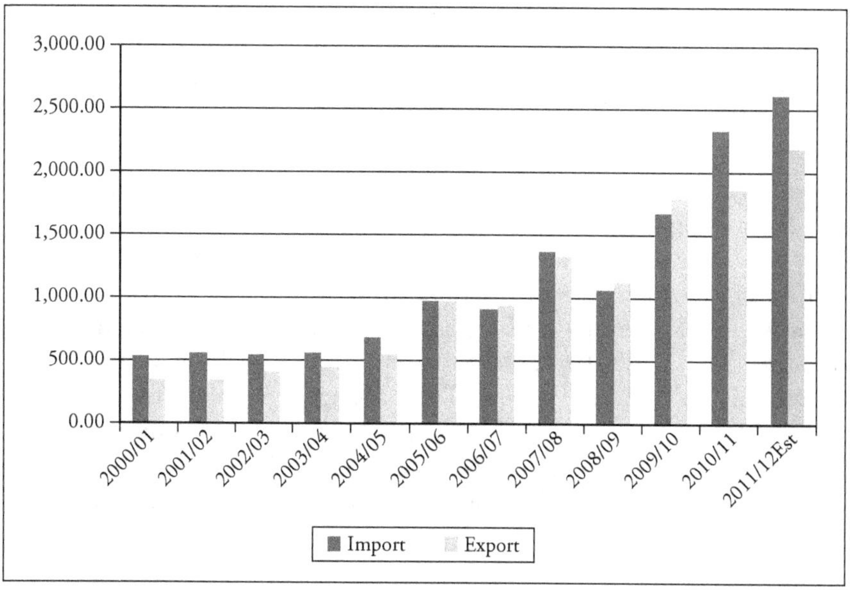

Source: Ministry of Commerce and Industry, Lao PDR.

Figure 7.2
Lao's Trade Balance During 2000–10 Period
(Unit: US$ million)

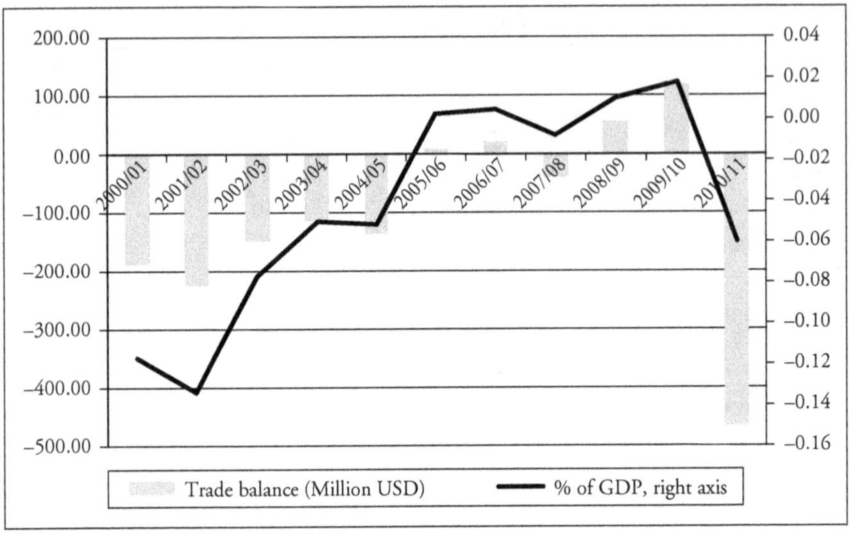

Source: Ministry of Commerce and Industry, Lao PDR.

Expansion in FDI shows another dimension of Lao PDR's integration with the regional economy (see Figure 7.3). The value of FDI surged from a marginal level of US$4 million in 1989 to over US$4 billion in 2009. The value of FDI inflow already grew very strongly before the Asian financial crisis in 1997. After slowing down during the crisis, FDI resumed a rapid growth again in early 2000 fuelled by the boom in electricity generation and the mining sector (see Figure 7.3). On average, the FDI inflow into Laos grew 156 per cent a year during 2000 to 2011. The fastest-growing sectors are electricity generation, mining and, in more recent years, services. The FDI inflows into Laos are largely from the neighbouring countries. During 2001 to 20011 the top investors in Laos were Vietnam, China and Thailand (see Figure 7.4). On average, FDI inflow from Vietnam covers 32 per cent of total FDI. Some 24 per cent is from China and 21 per cent is from Thailand. Apart from the neighbouring countries, some other important investors include France, Japan, South Korea, India and Australia.

Figure 7.3
The Value of FDI Approvals in Laos
(In US$ million)

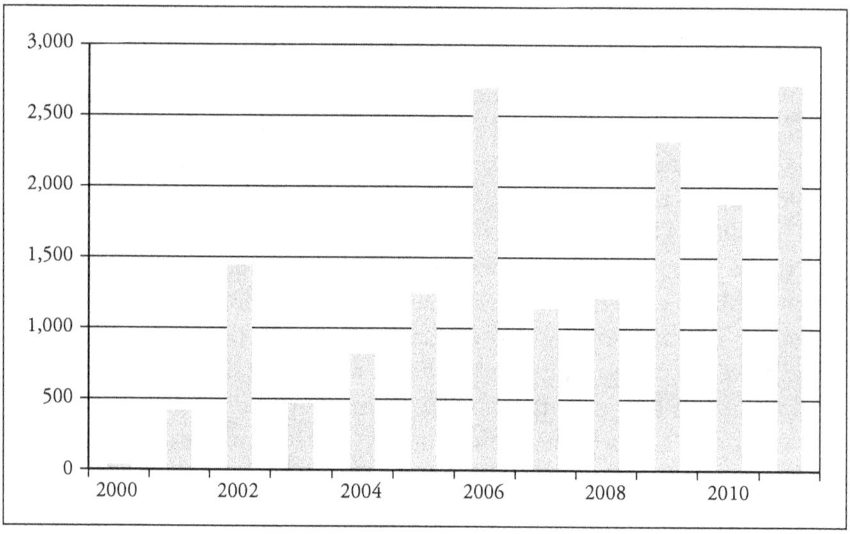

Source: Ministry of Commerce and Industry, Lao PDR.

Figure 7.4
Top Foreign Direct Investors in Laos

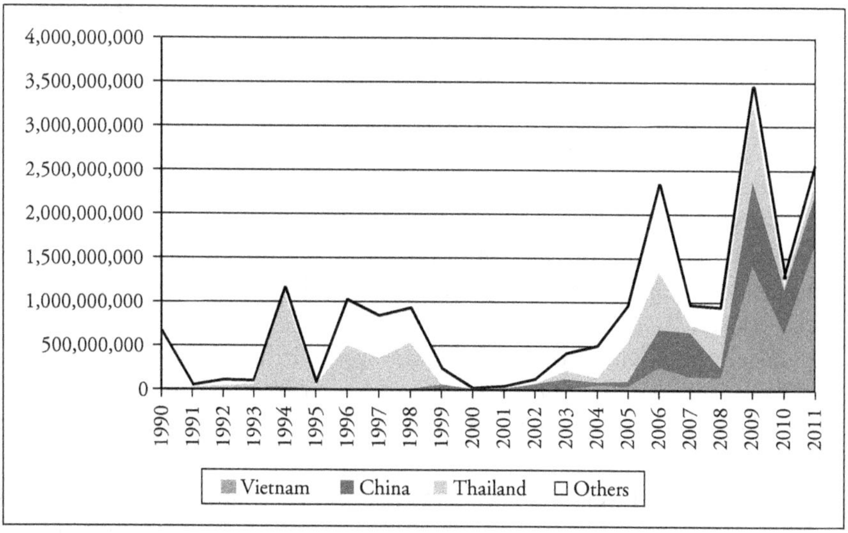

Source: Department of Investment Promotion, Ministry of Planning and Investment.

2. POLICY ENVIRONMENT, OPPORTUNITIES AND CHALLENGES FOR LAO PDR'S ECONOMIC INTEGRATION

Policy Environment for Lao PDR's Integration

The integration of the Lao economy with the region is well acknowledged in the framework of Lao PDR's socio-economic development planning. The Seventh Five-Year Socio-Economic Development Plan (NSEDP-7) sets out policy directions for the development of Lao PDR over the period of 2011 to 2015. The NSEDP-7 is considered to be the most research-driven plan that comprehensively reviews both external and internal environments, presenting both opportunities and challenges for development over the next five years. The aims of the Lao government (GOL) in the next five years are to move the country towards a modern and industrialized society; enjoy rapid economic growth with stability; visibly improve the living standards of the people; achieve the Millennium Development Goals (MDG) by 2015 and significantly contribute to the achievement of the 2020 overall goal to graduate from Least Developed Country (LDC) status.

Specific directions concerning the regional and global integration of Lao PDR are clearly discussed in the 7th plan within various major frameworks. In the ASEAN framework, achieving the objective of building the ASEAN Economic Community (AEC) by 2015 is the priority. Eleven priority sectors in this plan are expected to be integrated further, such as: agro-based products, air travel, automotives, E-ASEAN, electronics, fisheries, healthcare, rubber-based products, textiles, tourism and wood-based products. Achieving the establishment of the AEC has important implications for the integration of Lao economy into the region. Opportunities include the inflow of goods, technology and capital to fuel higher economic growth, increased access to goods and services, and the development of trade, logistic and tourism hubs. However, the AEC could also have several challenges for Laos such as increased competition for local businesses, in particular those in the import-competing sectors, fierce competition in the labour market and labour migration (NERI 2011).

Under the GMS cooperation framework, integration will focus on 'GMS Development' within the Vientiane Action Plan for 2009–12 to foster economic growth, alleviate poverty, promote social development and protect the environment. Priority projects in transportation, energy, agriculture, investment, trade facilitation, human resource development, environment and tourism will be promoted. Some major infrastructure projects under the GMS framework include the 4th Mekong Bridge in Bokeo Province; upgrading the Luang Prabang-Tha Ho (Vinh province of Vietnam) road; improving the corridor between Vientiane and Xamneua via Paksane and Phonesavan; and constructing the second East-West linkage route in Khammouane Province (Thakhek to Ha Tinh). The GMS framework will play an important role in improving the connectivity of Laos with the region. However, an important task was to promote the local absorption capacity so that the country could gain from this improvement in transport infrastructures.

Membership of the WTO in early 2013 marks the new landmark of Lao PDR's integration with the global economy. The membership package was approved by WTO's General Council on 26 October 2012. The formal ratification by the National Assembly was completed on 2 February 2013. WTO accession is explicitly promoted in the 7th Plan. To prepare for accession to the WTO, the 7th plan calls for more coordination between ministries and sectors in working on all required measures and reforms. There should be more discussions with the relevant sectors to make progress in drawing up legal acts, and provide answers (and also make policies) in response to the WTO's questions. The regulatory framework should be revised in accordance with the regulations of the WTO. Markets for export goods should be further developed and backward and forward linkage industries should be strengthened. The 'rules of origin of the WTO' should be strictly observed. Finally, awareness of integration into the world economy and accession to the WTO should be further raised among public servants, scholars, business people and the general public, so they may be prepared through training schemes, workshops and various forums.

Like other economic integrations, membership of the WTO could have both benefits and costs to the economy. The challenges include the capacity of Lao exports to meet the quality requirements of the WTO. The quantitative assessment of WTO accession, using the computer general equilibrium model, concludes that the Lao economy will gain only marginally from being a member of the WTO (particularly from tariff reduction) while trade deficit will increase and output in some vulnerable sectors will decline (Kyophilavong et al. 2010). The reason for such low economic return from tariff reductions as a result of accession to the WTO could include the fact that the Lao economy is relatively small and the tariffs applied by Laos and its trading partners are already low. Laos will still be able to gain from other indirect benefits of this multilateral trade agreement. Potential benefits could include the opening of Lao exports to wider markets and increasing the confidence of the international community, including foreign investors, in the Lao economy. In addition, WTO accession creates an opportunity for Laos to reform its trade policy in order to increase competitiveness of local industries (Lord 2010). This will have longer term benefits to the Lao economy.

Opportunities and Challenges of Increasing Integration of the Lao Economy

The integration of Lao PDR into the regional economy comes with both opportunities and challenges. The country's integration is blessed with favourable domestic environments such as political stability; the strategic location of Lao PDR at the heart of the GMS subregion; abundant natural resources and preferential treatments from many trading partners in relation to the LDC status. These provide much potential for increased integration for the Lao economy. Recent integration of Lao PDR into the GMS, ASEAN and other global frameworks has led to increases in the inflow of capital, technology and know-how. Our neighbouring countries, especially Vietnam and China, are growing rapidly, together with other ASEAN members. This leads to higher demand for goods and services from Laos. Also, with Lao PDR's fulfillment of AFTA and other AFTA+ measures, our exports will be able to expand to other countries and increase the product lines. At the same time, this will make investing in Laos more attractive to foreign investors.

However, some characteristics of the Lao economy remain obstacles that limit the gains from this improved integration. The transformation from subsistence-based agriculture to industrial production creates a challenge, especially for labour force adjustment. Most domestic private sectors are of micro, small and medium size with low levels of competitiveness, coupled with a low level of technological intensity. These make it difficult for the domestic private sector to absorb the benefits of improvement in connectivity in our economy. The expansion of the private sector and exports still relies on imported technology and inputs, and the natural resource sector. In addition, despite improvements, a lot more needs to be done to improve infrastructure, especially transport linkage within the country, and transport costs remain high. Finally, the progress of soft connectivity such as the implementation of the Cross-Border Trade Agreement is relatively slow, limiting some of the benefits from hard connectivity such as transport infrastructure development.

3. A NEED FOR A BETTER INTEGRATION OF THE LAO ECONOMY AT DIFFERENT LEVELS

For a more effective integration of the Lao economy, the chapter proposes an analysis of the ASEAN contribution to the integration process at regional and global level, and a focus on the advancement of recent cooperation initiatives including the Northern region development strategy, road economic corridors and the Kunming-Singapore high-speed railway via Laos.

ASEAN Facilitation of Lao PDR's Integration at Regional and Global Level

In recent years, there has been a large increase in the number of Preferential Trade Agreements (PTAs) and the formation of regional trading blocs. At the same time, the WTO has found it very difficult to achieve agreement during the recent trade rounds. Such developments raise concern whether the expansion of PTAs supports or hinders the progress of the WTO, especially for removal of global trade barriers. The question of how PTAs affect the progress of the WTO, which is often referred to as multilateralism, has always been an important and inconclusive issue for trade and political economy theorists. Viner (1950), a pioneer of the static analysis of PTAs, argues that PTAs could have negative effects on both the member countries and world welfare. Viner explains that the effects of PTAs depend on whether they lead to trade diversion or trade creation. Moreover, Krugman (1989) develops a model to the Regional Trade Agreement (RTA) at the time when there was a concern that rapid increase of PTAs, mainly RTAs, could hinder multilateral trade liberalization and reduce global welfare. Krugman concludes that RTA divides the world into many small trading blocs and therefore reduce welfare even though each bloc aims to maximize the welfare of members. These concepts have become important tools for many subsequent studies on the costs and benefits of PTAs.

The integration of Lao PDR with ASEAN facilitates the efforts of Lao PDR to further liberalize trade on a global scale. Lao PDR's integration into ASEAN creates a platform for the country to further liberalize its trade sector by removing both tariff and non-tariff barriers to trade with ASEAN and also, gradually, with non-ASEAN members. Lao trade with ASEAN and non-ASEAN members has increased during the past decade, concluding that the trade diversion argument is not observed in Lao PDR's economic integration. From the recent gravity model by the IMF (2010), it is found that ASEAN integration promotes intra-ASEAN trade and trade with the rest of the world. In addition, one of the old arguments against the formation of the regional trade agreement is that RTA is established so that insufficient multilateral trade liberalization appears not to be applicable to ASEAN. The new wave of regionalism including ASEAN is designed to facilitate entry into the multilateral trade liberalization system. For instance, many members of ASEAN become members of the WTO after membership of ASEAN, including Cambodia, Vietnam and recently Laos.

Northern Region Development Master Plan

The Northern Region development strategy was developed to further develop the regional economy. The Northern provinces development cooperation plan aims at promoting socio-economic development, trade and investment and poverty reduction for the Northern provinces of Laos. This master plan comprehensively examines the potential, challenges, development models and options, shaping the key development direction for the Northern region until 2020. The master plan covers nine provinces: Phongsaly, Luang Namtha, Bokeo, Oudomxay, Huaphanh, Xiang Khoang, Luangprabang, Xayaboury and Vientiane Province. The Northern Master Plan proposes the "1334" development concept. Luang Prabang will be developed as the economic center of the region, in recognition of its strategic location and its status of economic powerhouse. It is proposed to develop three economic corridors in this region. The first economic corridor runs from the Chinese border to Nongkhai in Thailand via Oudomsay, Luang Prabang and Vientiane province. It is part of the central North-South GMS corridor which links Kunming to Phnom-Penh. Two other horizontal economic corridors link Thailand to Vietnam via the Northern provinces of Laos. The second economic corridor runs from the Thai border facing Huayxai to the Vietnamese border, via Luang Namtha and Muong Xai. Finally, the third proposed economic corridor links Huaixay to Tha Ho at the Vinh province border of Vietnam, via Hongsa, Luang Prabang, Phonesavan and Nonghaed.

In addition, the master plan proposes the development of major road infrastructures including 3 longitudinal roads, 5 transverse roads and 8 connecting roads. One of the longitudinal roads runs from the Vietnamese border of Huaphanh province to Paksane in Borikhamsay province, via Xamneua and Phonsavan. This road is the main Laotian section of the Xamneua-Vientiane route which is part of the North-East GMS corridor linking Nanning to Bangkok via Vientiane. These road infrastructures will substantially improve the connectivity of the Northern region with the rest of the country as well as with neighbouring countries. The challenges in implementing this Northern master plan include resource mobilization and sequencing the development of many priority projects. The master plan suggests, for example, the development of three industrial focal areas in Vientiane Province, Oudomxay and Xiang Khoang, and four border trade areas in Luang Namtha (Lao-Chinese), Bokeo (Lao-Thai-Myanmar), Xiang Khoang (Lao-Vietnam) and Xayabury (Lao-Thai).

Corridor Development

Following the government's policy of transforming Lao PDR from land-locked to land-linked status, there have been substantial investments in road infrastructures. Such initiative is driven by development partners, mainly the Asian Development Bank (ADB), through the Greater Mekong Subregion (GMS) framework and bilateral cooperation between Laos and its neighbours. This gives birth to the economic corridor development concept, focusing on developing transport infrastructure to connect Lao PDR with her giant neighbours. The major corridor developments include the East-West Economic Corridor (EWEC) and the North-South Economic Corridor (NSEC). The developments are part of the ADB GMS programme whose objectives are promoting the development of GMS members by accelerating regional development through linking two rapidly growing economies via Lao PDR in order to facilitate trade and investment in the region, and helping to reduce poverty in Lao PDR through enhancing market access, providing services, improving income and employment opportunities and enhancing overall development potentials.

Despite the massive investment in hardware, together with gradual reforms in trade and investment regulations, the expansion of trade and investment in the subregion, and in Lao PDR in particular, remains below expectations and the transport cost has not reflected this development. The development of the Special Economic Zones has also been delayed. Some of the reasons leading to this observation include the lack of know-how, slow development of necessary infrastructure, the clarity of related regulations and low local absorption capacity.

Vientiane-Kunming High-speed Railway

The construction of the Mohan/Boten-Vientiane railway is the latest mega infrastructure development project pursued by the Lao government, aligned with the "land-linked" policy. The project was recently approved by the National Assembly. The Mohan/Boten-Vientiane section is part of the Singapore-Kunming Rail Link (SKRL) project and part of a Trans-Asian Railway in the Southeast Asia initiative, which is a broader railway development cooperation driven by the UNESCAP. The Kunming-Singapore Railway would connect Southwest China and Southeast Asia by a railway running from Kunming to Singapore via Laos, Thailand and Malaysia.

The Lao section has a total length of 420 km with an estimated construction cost of US$7 billion. The project will be entirely financed by the Lao government with a loan from one of China's commercial banks. Thirty-one stations are planned to be opened, including seven major stations in Vientiane Capital, Vientiane Province, Luang Prabang, Oudomxay and Luang Namtha. The completion of SKRL will reduce the travel time between Singapore and Kunming to about ten hours by train, contributing to increasing cross-border activities in the region. This railway project will also substantially improve the connectivity of Laos and make Vientiane Capital one of the important subregional cities and hubs on this axis (Kunming, Khon Kaen, Bangkok) as well as other main hubs on other corridors of the GMS grid, like Hanoi, Ho Chi Minh and Rangoon.

CONCLUSION

Lao PDR has increased integration with both the regional and the global economy. Integration has led to increases in both trade and foreign direction investment, fuelling the economic progress of Lao PDR. Increased integration with both the regional and global economy is explicitly promoted in the Seventh-Five Year Socio-economic plan. The key targets in integration include further development of the linkage infrastructure in the GMS region and the formation of the ASEAN Community by 2015. Despite the ongoing debates over the cost and benefits of forming regional trade agreements, Lao PDR's integration with ASEAN appears to support further trade liberalization in Laos and further integration with the rest of the world.

Integration of the Lao economy has created both opportunities and challenges. In order to optimize the benefits of integration and minimize the losses, the integration of the Lao economy should be systematically promoted. The development of Special Economic Zones and cross-border cooperation areas should be priorities. Development of the Savan-Seno Special Economic Zone which takes advantage of the East-West Economic Corridor should be the priority and could be a model for SEZ development in other areas. At the subnational level, the experiences of the Northern Region

Development Master Plan could be an example for regional development models in other parts of the country. At the subregional level, Laos should continue participating actively in the GMS framework to further utilize the economic corridors already developed. Local absorption capacity should be further improved so that the country will gain more from this improvement in connectivity. At the regional level, Laos will deepen its integration with ASEAN in line with the regional plan, such as the formation of the ASEAN Economic Community by 2015. Finally, at the global level, local businesses should be supported so that they gain more from accession to the WTO.

The ultimate goal of increasing economic integration will be to improve the connectivity of Laos with the region and the world. Unless necessary measures are taken, Laos will not gain much from increased economic integration and improvement in connectivity. These measures will include active and effective investment in the transport infrastructure such as regional roads and railways. To materialize all the expected benefits from the hardware, there is also a need to improve the soft measures of connectivity, especially reducing the cost and time of cross-border transactions. As a whole, these improvements in connectivity will allow Laos to realize the government vision of changing the country's status from "land-locked" to "land-linked".

References

ADB. "Greater Mekong Sub-region (GMS)", 2011. Available at <http://www.adb.org/gms/>.

ITC. *Trade Map-Trades Statistics for International Business Development.* International Trade Center, Geneva, 2010. Available at <http://www.trademap.org>.

Khonthapane, S. "The Integration of Lao Economy: Opportunities and Challenges". A Presentation at the Seminar on Planning and Economic Monitoring for Champasack Province Government Officials, July 2010.

Kyophilavong, P., S. Takamatsu, and J. Ko. "The Impact of Laos's Accession to the Word Trade Organization". Paper presented at the 13th Annual Conference on Global Economic Analysis, 15 April 2010.

Krugman, P. *Is bilateralism Bad?* National Bureau of Economic Research Working Paper 2972. Cambridge, MA, 1989.

Lord, M.J. "Implications of WTO Accession for Selected Domestic Industries of Laos". MPRA Paper No. 41151, 2010. Available at <http://mpra.ub.uni-muenchen.de/41151/>.

MFA Thailand. Available at <www.mfa.go.th/internet/emerald/ACTIONPLAN.doc>.

Ministry of Planning and Investment. *The Seventh Socio Economic Development Plan of Lao PDR 2011–2015*. 3rd Draft. MPI, Vientiane, 17 May 2010.

Ministry of Industry and Commerce. *Official Trade Statistics.* Ministry of Industry and Commerce, Vientiane, 2010.

MOFA Japan. 2011. Available at <http://www.mofa.go.jp/region/asia-paci/clv/joint0411.html>.

NERI. "Estimating the Impacts of AEC on Lao PDR". Research to Support the Implementation of the 7th Five Year Plan 2011–15. Draft Report for NERI Technical Seminar on November 2011. National Economic Research Institute, Vientiane, 2011.

Nolintha, V. "Preferential Trade Agreements as Stepping Stones towards Global Free Trade". Research Essay. Master of International and Development Economics, Australian National University, Canberra, 2008.

———. "Economic Sub-corridors and Potentials for Regional Development in Laos". In *Emerging Economic Corridors in Mekong Region*, edited by Masami Ishida. *BRC Research Report*, no. 8. Bangkok: IDE-JETRO, 2012.

Ornelas, E. "Trade Creating Free Trade Areas and the Undermining of Multilateralism". *European Economic Review* 49 (2005): 1717–35.

Tumbarello, P. "Are Regional Trade Agreements in Asia Stumbling or Building Blocks? — Implications for the Mekong-3 Countries". IMF Working Paper, No. WP/07/53. Washington D.C., 2007.

Vietnam Embassy. 2011. Available at <http://www.vietnamembassy-brunei.org/vnemb.vn/tin_hddn/ns061206095000>.

Viner, J. *The Customs Union Issue, Carnegie Endowment for Peace*. New York, 1950.

8

SHAN STATE IN MYANMAR'S PROBLEMATIC NATION-BUILDING AND REGIONAL INTEGRATION
Conflict and Development

Ella Vignat

Myanmar's (formerly Burma[1]) strategic geographic location at the tri-junction of South Asia, Southeast Asia and China has been attracting foreign countries as it is a dynamic crossroads linking Southeast Asia, Western China and the Indian subcontinent and serves as an economic gateway to a colossal potential market of more than two billion consumers as well as a primary gateway to energy supplies and trans-shipment. In the context of energy security, Myanmar's emergence as a significant regional exporter of natural gas including its immense hydropower potential has become a pivotal stake to energy-hungry superpowers such as India and particularly the economic powerhouse, China.

Indeed, Myanmar's geostrategic location is of critical significance for China as it is a "land bridge" for the economic development of its south-western provinces, a strategic corridor allowing China to secure access to the Indian Ocean, trans-shipment of oil and source of natural gas. Currently, China is building oil and gas pipelines linking Kunming, the capital of Yunnan Province, to the deep-water port of Kyaukphyu on the Bengal coastline to tap Myanmar's huge gas reserves and to provide an alternative route for its crude oil imports from the Middle East and Africa, reducing shipping time and its dependence on traffic through the Straits of Malacca.

However, Beijing's influence in Myanmar has been challenged since the November 2010 national elections in Myanmar. Following the elections, the junta, the State Peace and Development Council[2] (SPDC) was dissolved by the parliament and a new civil government — mainly composed of newly retired generals and military officers — was formed. The elections aimed to institutionalize the junta's political role and gain domestic and international legitimacy, and respectability with its new civil government. The end of the military rule was perceived as a move to attract foreign investment. At first, the handover of power by the junta was greeted with scepticism, since then, the international community has been seeking engagement and adopting a more flexible approach with the new government after the April 2012 by-elections that marked the return of the main opposition National League for Democracy (NLD) party into the Myanmar political system. The reasons for this reversal are, first, the long-awaited elections have taken place, second, Myanmar's emergence as a vital regional exporter of natural gas and hydroelectric power, third, Myanmar's pivotal position in the development of the Greater Mekong Subregion, and last, but not least, countervailing China's influence over the region.

While Myanmar's emergence as a significant regional energy exporter and a potential crucial transport hub in Asia has revived its role in the region and strengthened its regional integration, Myanmar has been constantly challenged, since independence, in the process of its nation-building by ongoing ethnic tensions on its peripheries, in particular in its eastern region, Shan State,[3] where clashes[4] have continued to occur despite ceasefire agreements concluded by the new government with most[5] ethnic armed groups operating in Shan State.

The Shan plateau with its 160,000 square kilometres occupies one fifth of the total surface of the Republic of the Union of Myanmar,[6] and it is home to a complex mix of ethnic groups — including the Lisu, Danu, Intha, Palaung, Pao, Lahu, Akha, Kachin, Wa, Kokang Chinese — dominated by the Shan majority, the second largest ethnic group of the country, accounting for 7 per cent of the total population of Myanmar according to the 1931 census,[7] the last proper census taken in Myanmar. Due to its geostrategic location bordering China to the north, Lao PDR to the east and Thailand to the south, Shan State is the recipient of huge investments including China's multi-billion transnational oil and gas pipeline project. Furthermore, it is the point of entry of the North-West corridor (Kunming-Mandalay-Rangoon) and the North-South corridor (Kunming-Kengtung-Chiang Rai-Bangkok) of the Greater Mekong Subregion Economic Cooperation Program (GMS ECP) through its border towns:

Muse in northeastern Myanmar which has been acting as China's gateway to central Myanmar, Southeast Asia and the Indian Ocean, and Tachilek in southern Shan State at the Thailand-Myanmar border (i.e., Map 2.2 in this book).

This chapter attempts to identify the major development projects implemented in Shan State and their impacts on local, national and transnational scales in a context of ethnic issues intertwined with economic stakes associated with foreign investments in the border areas. Thus, the understanding of the origins of Myanmar's longstanding ethnic conflict and its complexity is necessary. This chapter is based on the available sources produced by scholars and researchers working on Myanmar, and the collection of observations and interviews conducted in Myanmar from 2001 to 2011.

The first part of the research will present the development strategies carried out by the Myanmar government in favour of the borderlands and their impact on Shan State regional and local stakeholders. The second part will examine Myanmar's role in regional integration with a particular focus on Shan State's development, and it will raise questions concerning conflicting issues between regional and national integration and development. Finally, the conclusion will expose the human impacts of the development projects in the Shan State.

1. IMPACT OF NATIONAL DEVELOPMENT STRATEGIES ON SHAN STATE REGIONAL AND LOCAL STAKEHOLDERS

Shan State has played a prominent role throughout Myanmar's history, particularly its post-colonial history, in its nation-building and regional integration strategies. This feature is more true than ever now that Shan State represents a major stake for Myanmar national and regional integration. Furthermore, recent resumption[8] of hostilities has brought back to light the intractable deep-rooted ethnic issues in Shan State where a myriad of state and non-state actors including local, national and regional stakeholders have influenced the course of events to protect their vested interests.

Shan State Paradox: From the Frontier Areas to Panglong, from the First Attempt at Nation-building to Ethnic Conflict

The Shan (*Tai* in Shan) belong to the same ethno-linguistic family as their Thai and Laotian neighbours with whom they share a similar language and culture. During the pre-colonial period, they were organized under fiefdoms ruled by hereditary chiefs referred to as *saophas* or *sawbwas* (in Burmese). These princes were vassals of their strongest neighbours such as the Burmese, Thai and Chinese sovereigns to whom they paid tribute. That traditional power structure did not favour political unity among the kingdoms.

Nevertheless, at the end of the Second World War, after British power was restored, together with Chin and Kachin leaders, the Shan were one of the first peoples of the Frontier Areas to join Burma represented by the Burmese nationalist leader, General Aung San — after reaching a consensus at the Panglong conference in February 1947 — and ask for independence. The Panglong Agreement is the basis of post-war relations between the central Burmese authorities and the frontier peoples. The Panglong Agreement guaranteed the Frontier Areas "full autonomy in internal administration" (Article 5) and the use of democratic "rights and privileges" (Article 7). Furthermore, a democratic constitution was compiled, that current pro-democracy activists would agree to use as a basis for Myanmar's new constitution despite its imperfections. Crucially, this Constitution conceded to the Shan ethnic group the right to secede after a ten-year term and allow their traditional rulers to keep their feudal authority to a large extent. After Burma's independence in January 1948, a Shan, Sao Shwe Thaike became the first president of the Union of Burma. However, two major events and their impacts ended the Shan experiment within the Union of Burma.

In 1949, U.S.-backed Kuomintang (KMT) remnants invaded Shan State from southern Yunnan in the wake of Mao Zedong's communist victory in China. This foreign invasion was fraught with consequences for Shan State's socio-economic and political landscape. Ever since, Chinese networks have been crucial to the borderlands economy and linked to the illicit opium trade now operating on a global scale. Opium and insurgency in Shan State have always been two sides of the same coin as ethnic conflict and the narcotics trade are intertwined.

These events in the Frontier Areas alarmed the central government as the Shan states had the constitutional right to secede from the Union in 1958. In response to this fear, in 1962, the commander-in-chief of the Burmese Army, General Ne Win, staged a coup that ended the Federal Movement and the era of Burma's parliamentary democracy. In 1988, while the international community focused on democratic issues launched by the National League of Democracy (NLD) in Rangoon, in 1989 another major event announced the beginning of a new era for the periphery and the centre. Following the collapse of the Communist Party of Burma (CPB), Burma's largest insurgent force, developing conflicts in Shan State became more complex with the increase in the number of stakeholders.

The Ceasefire Agreements: A New "Institution"?

From the implosion of the CPB, four new armed groups emerged, including the present powerful United Wa State Army (UWSA). The central government in Rangoon exploited the mutiny and reached ceasefire agreements with the new forces. The breakaway groups were promised they could control their territory, get involved in any economic activities, have development assistance from the government, and retain their weaponry if they did not help the other insurgent groups fleeing Rangoon in the aftermath of the August-September 1988 urban uprising. Thus, between 1989 and 1994, most ethnic rebel groups made similar deals with Rangoon. These forces included the Shan State Army (SSA) and Kachin, Palaung, Pao members of the National Democratic Front (NDF), the Kayan New Land Party, and Shan State Nationalities Liberation Organization. By 1994, the entire Shan State, except part of southern Shan State, was nominally under central government control. In 1996, after Khun Sa's Mong Tai Army (MTA) surrendered to the Burmese government, the Shan United Revolutionary Army (SURA), a faction of MTA, under the command of Colonel Yawd Serk, refused to accept the MTA's unconditional surrender, and it has been carrying on fighting[9] against the Tatmadaw (Burma Army) in southern Shan State under the name of Shan State Army-South (SSA-S) ever since.

Foreign involvement in Shan State territory — such as KMT invasion and counter-insurgency interventions from China (People's Liberation Army), influx of Red Guards into Shan State to support the CPB, and Thailand's "buffer-state" policy toward the pro-federal ethnic armies on Shan State southern border — sustained the conflict.

The ceasefire agreements announced a new era in Shan State developments. In many parts of the Shan territory, truces had put an end to decades of civil war, and offered state and non-state actors opportunities to reconstruct and develop former war-torn zones. Moreover, 1988 was also a turning point in Burma's development with the repression of the uprising of the democratic movement in urban central and lower Burma, and the open-door policy adopted by the military regime to develop bilateral and multilateral relations with its neighbouring countries. The fallout from those events has had serious consequences on the Shan State's national and regional integration ever since.

Since Myanmar's independence, the actors of the Shan arena — including indigenous villagers, hereditary leaders, state agencies, ceasefire and active armed opposition forces, religious leaders, traders, business people, loggers, Non-Government Organization (NGO) personnel, spies, drug dealers, smugglers, human traffickers — have been living in a conflict-affected environment where insurgency has become "a way of life" that regulates their survival. Decades of warfare in the Shan State have created war economies depending on external resources and connections out of reach of the central authority's control. The ceasefire agreements initiated in 1989 by the junta have not overwhelmed these networks. On the contrary, the multifaceted networks of actors that have emerged, have an impact on the reconstruction, development and national integration of the war-torn Shan State.

Government Implementation Strategies for Border Areas Development

In the context of "neither war nor peace, but a kind of post-civil-war, not quite peacetime environment",[10] have emerged complex and multilayered networks on the local, national and transnational scale dominated by a myriad of state and non-state stakeholders. Governance in Shan State is characterized by the "mosaics of power". Their predominance fluctuates across time and space. Yet, political power in the Shan State belongs to those who have military power because military capacity is the main asset for local autonomy (political and economic) and control of territory and natural resources. Ethnic groups with powerful armies enjoy greater autonomy and have better leverage in negotiations with the central power.

In the ceasefire agreements, the Myanmar military regime, the SPDC, has promised development for armed conflict-affected areas. Following the truces, the SPDC's strategy focused on the Border Areas Development Program (BADP) by sending out its regional commanders and bureaucrats from various ministries such as the Ministry for the Development of the Border Areas and the National Races, to build roads, bridges, dam infrastructures, Myanmar language schools and health centres. Nevertheless, health facilities lacked proper equipment and staff, and the quality of education was poor because of the appointment of low-skilled teachers. Sixty-five per cent of the budget of the BADP projects went to road infrastructure with a dual-purpose target, it would link outlying areas to the national economy but it also ensured rapid expansion of the junta troops into ceasefire areas.

Furthermore, what was not acknowledged was that most of this road infrastructure was built by local labourers who were required to fulfil the *corvée* and "donate" days' labour; lands on which the projects were launched were confiscated from farmers who were displaced and forcibly relocated. Moreover, these projects were financed by taxes levied on local citizens and deals from foreign companies from China and Thailand looking for natural resources.

The implementation of the BADP projects by state agencies varies geographically over time and space as the relations between the government and the ceasefire groups are not homogeneous. For example, in the Wa region in northeastern Shan State where the authority of central power is limited, government troops have the authorization to enter Wa territory only after prior notification. At the other extreme, in ongoing conflict-affected patches in southern Shan State, the central authority is an occupation power. However, in Pao territory in southern Shan State and Kachin controlled area in northern Shan State, government agencies, ceasefire groups, traditional and religious leaders, traders, and NGO personnel used[11] to coexist.

Relay of Non-State Actors: Ceasefire Armed Groups and FDI

As the central government did not provide sufficient and sustainable development projects for the former conflict-affected areas, the ceasefire groups have to find alternative sources of income to finance the reconstruction and the development of their territory. Given that access to legal trade and business is in the hands of the central power, ceasefire armed groups have to rely on the shadow economy and informal economic activities such as taxation of black market commodities, extraction of natural resources in logging and mining, gambling, drug and human trafficking. Since the fighting has stopped, political and economic functions of the border areas have changed, and the outcome is the growth of trans-border trade with the neighbouring countries. The majority of the ceasefire groups as well as the government projects have prioritized the development of infrastructures such as the construction of roads and power stations.

Other projects have been conducted by foreign companies (mostly Chinese) in exchange for logging concessions. The United Wa State Party (UWSP) and the National Democratic Alliance Army (NDAA), better known as the Mongla group based in Eastern Shan State, have invested in various businesses including a brewery, a paper mill, lighter and cigarette factories, tea plantations, mines, hotels, karaoke bars, casinos[12] and airlines (Yangon Airways[13]). Many of these undertakings were not successful as they were barred from entering Chinese and Myanmar markets by high Chinese import taxes and central government's trade restrictions (registration procedure and export regulations). Monoculture plantation projects for sugar cane, tea and rubber trees financed by Chinese capital have been set up in Kokang and Wa regions to replace opium cultivation. They depend on markets in China as access to markets in central Burma is restricted.

As the ceasefire groups face huge difficulties to finance themselves, they have taken to relying on the sale of their natural resources, mainly hardwoods and mining. Large-scale logging and mining concessions have been given mainly to Chinese and Thai companies whose governments carried out bans on logging in the late 1980s, and these have put the environment of the Shan State at risk during the last decade. Furthermore, the central state's stranglehold on natural resources has pushed many ceasefire groups to intensify the drug trade. Armed conflict and exploitation of natural resources have been intertwined in Myanmar for decades since the pre-colonial period, battles were fought for control over the territory's natural resources.

Booming Cross-border Trade

Since 1988, after the junta seized power in the aftermath of the repression of the August-September nationwide uprising for democracy, its military regime, the State Law and Order Restoration Council (SLORC), later replaced by the SPDC, decided to engage towards a "market-oriented" democracy with its "open-door" policies by formalizing border trade with its neighbouring countries, namely China, Thailand and India, hitherto an illegal activity in Myanmar. Developing bilateral and multilateral relations with its neighbours was the main focus of Myanmar's regional integration strategy. On 6 August 1988, China was the first nation that signed an agreement with Myanmar initiating official trade across the common border. On 17 March 1996, Thailand signed a border trade agreement between the two countries through the border trade posts of Mae Sai, Mae Sot, and Ranong on the Thai side, and Tachilek, Myawaddy, Kawthaung, and Myeik (Mergui) on the Myanmar side. Likewise, on 21 January, 28 May 1994 and 6 December 2000, Myanmar signed a border trade agreement with India, Bangladesh and Lao PDR, respectively. Furthermore, in 1992 Myanmar joined the Great Mekong Subregion (GMS) Economic Cooperation, and with the support of China and Thailand, Myanmar became a member of the Association of Southeast Asian Nations (ASEAN) in 1997. Since then, Myanmar's cross-border trade has significantly grown and flourished as an outcome of the ceasefire agreements initiated in 1989.

Because of the economic sanctions imposed by the West in 1997, Myanmar has relied more on its neighbouring countries, and its trade relations with China, Thailand and India have changed and been reinforced. Thereby, trade through cross-border points has become more dynamic. It is estimated that the value of border trade (including legal and undocumented trade) accounted for 25 per cent of Myanmar's total value of trade in 2006. Myanmar's trade with China and Thailand accounts for 47 and 51 per cent of its total border trade, in terms of legal border trade value. In 2008, the value of trade with China, Thailand and India accounted for 70 per cent of Myanmar's total trade with the world. Thus, these three countries are Myanmar's main cross-border trade partners.

Cross-border trade with China and Thailand is carried out mostly by land through border checkpoints. The value of trade between Myawaddy border town in Myanmar's Kayin State facing Mae Sot in Thailand, accounts for 24 per cent, 19 per cent of cross-border exports and 31 per cent of cross-border imports, while Muse border gate in northern Shan State, facing China's Ruili border town, is the recipient of the most valuable border trade, accounting for 39 per cent of Myanmar's total documented cross-border trade with China's Yunnan province, and its share in documented cross-border exports represented 41 per cent and 38 per cent in cross-border imports in 2008. These two border gates are Myanmar's most important border-crossing points in terms of value. Four out of eleven border gates are located in Shan State including three on the Chinese border and one on the Thai border, Muse, Chin Shwe Haw, Mongla and Tachilek respectively. Muse ranks first whereas Tachilek ranks fifth in terms of cross-border trade value.

Muse, Myanmar's Showcase: A Model of Success

The border regions which flank areas of the world's fastest economic growth have undergone drastic changes in the nature of their functions that have evolved from outlying regions to dynamic centrifuge nodes. At the beginning of the 1980s, relations between Rangoon and Beijing improved as the latter diminished its support for the CPB. Chinese leaders expressed their intention of finding an outlet for trade from China through Myanmar to the

Indian Ocean. The situation changed in 1987, and Chinese goods started to penetrate the Myanmar market. The border trade grew faster after 1988 following the ceasefire agreements. The first border trade checkpoints with the Chinese border were officially set up in Shan State in Muse, Namkham, and Kyukok in October 1988.

Since the beginning of the 1990s, cross-border trade has mushroomed. Trade between Yunnan and Myanmar accounts for half of total bilateral trade between the two countries. According to Chinese official figures, Myanmar-China bilateral trade hit US$4.44 billion in 2010, a 53.2 per cent increase over the previous year, with China standing as Myanmar's second largest trading partner after Thailand. As part of plans to further promote trade between the two countries, special economic or free-trade zones are being set up. The border crossing between Muse and Ruili is the biggest trade route at present. The Yunnan provincial authorities have lately proposed to set up a Ruili cross-border economic cooperation zone, whereas the Dehong County proposes to create a Ruili-Muse free-trade area. It is anticipated that in future development, transit trade to third countries would be launched through the two countries' border points to effectively implement the normal trade. Construction of a new railroad to link Myanmar's border town of Muse with China's Jiegao, Yunnan province, to facilitate commodity flow in bilateral border trade has started. Once the railroad is completed, transportation should become rapid and help to boost bilateral border trade. Commercial banks from Myanmar and their counterparts from Ruili and Jiegao will establish a direct banking system to facilitate bilateral border trade payments, which were settled in cash during the past decade. Since 2001, a Myanmar-China border trade fair has been held annually, alternately in Muse and Ruili. In 2012, the border trade fair was held in Myanmar's Muse 105th Mile Border Trade Zone, the first and largest border trade zone in Myanmar, upgraded in 1996 for trans-shipment with full import-export equipment.

Muse's strategic function keeps growing. Muse, the gateway between China's Yunnan province and Myanmar, is on the northern economic corridor on the former "Burma Road" that runs south from Kunming, capital of Yunnan Province (Kunming-Baoshan-Mangshi-Ruili-Muse-Lashio-Mandalay) to Mandalay, capital of central Myanmar. Whereas a second road, the western branch of the North-South economic corridor links Yunnan with Thailand through northeastern Myanmar (Kunming-Simao-Jinghong-Daluo-Mongla-Kengtung-Tachilek-Chiang Rai). Both corridors enter Myanmar through Shan State border towns, Muse and

Tachilek, respectively. Muse border town is located in northern Shan State, 178 km from Lashio, 458 km from Mandalay, and 1,449 km from Rangoon; though it is home to a population of approximately 200,000 predominantly ethnic Shan. The cross-border boom attracts a large number of ethnic *Bamar* and Chinese migrants. The past two decades have seen life in the borderlands transformed and enjoying a higher standard of living than in Myanmar's major cities, and local economy in the border areas experienced an economic boom. Similarly, boosted by the economic growth of the border towns, business in cities like Lashio and Mandalay is flourishing.

After 1989, the ceasefire brought relative peace to the borderlands and some degree of economic prosperity to the local communities. Due to western economic sanctions, the formal and informal cross-border trade has tremendously increased on the available market where a large number of various stakeholders have revitalized and structured economic activities in Myanmar's rugged borderlands. The regions to have experienced a cross-border economic boom are Myanmar's eastern borderlands along the Thai border, and particularly the Chinese border. Business are run by a array of state and non-state actors on national, local and regional scales — such as Myanmar government and its civil and military representatives, ceasefire and non-ceasefire groups, local militia, high-profile and local businessmen, traders, hereditary leaders, indigenous villagers, loggers, smugglers, drug dealers and human traffickers — including foreign stakeholders, mainly from neighbouring countries, China, Thailand and India respectively, but also from other Asian nations such as Hong Kong, Singapore, South Korea and Malaysia.

During the first decade, the government and most ceasefire groups prioritized the development of infrastructures such as the construction of roads, bridges and power stations. In the 1990s, the logging trade, banned in the neighbouring countries, financed by Chinese and Thai networks, caused widespread deforestation in the eastern borderlands. All actors were involved in this business. Since 1989, strategic, money-spinning natural resources, especially oil, gas and jade, have been taken under the central power's control, while other sectors such as construction, gold and gemstone mining have been left to non-state patronage.

Nonetheless, the recent growth of large-scale development projects in hydropower and oil and gas trans-shipment infrastructure conducted by Chinese state-owned companies backed by the Myanmar government has fuelled ethnic tensions[14] in the eastern borderlands.

2. REGIONAL INTEGRATION, NATIONAL INTEGRATION AND SHAN STATE'S DEVELOPMENT: CONFLICTING ISSUES?

Due to its strategic geographic location at the tri-junction in Asia, Myanmar sees itself more and more as a regional transport hub connecting Southeast Asia, Western China through Yunnan, and the Indian Subcontinent with a huge potential market of over two billion consumers. Given Shan State's strategic position in Myanmar, at the crossroads of two main regional stakeholder states — China and Thailand — some of the development projects directly concern Shan State, such as the road network, hydropower dams and mines, and some only cross the Shan space, such as the energy transportation corridor, the China-Myanmar pipeline. The development projects implemented by the central power have fuelled tensions and conflict. Resentment among local residents toward the Myanmar government over its regional development strategy is significant as the projects have generated negative social, economic and environmental impacts on the local community.

Regional Integration Precedence over National Integration and Local Development: Regional Rail Hub as an Alternative Route to the Straits of Malacca

In the past two decades, cross-border trade has been Myanmar's main driving force for its regional integration. Its role in the region has been revived in a changing geostrategic context. Beijing is the powerhouse of the massive infrastructure development across Asia, it will provide skills and funds for an integrated railway system that will connect all Greater Mekong Subregion (GMS) countries by 2020. Several routes linking Yunnan with ports in Southeast and South Asia, are to be built.

China will construct a rail link connecting Kunming with a new deep-sea port and special industrial economic zone under construction at Kyaukphyu on Myanmar's Arakan coast. On 27 April 2011, a Memorandum of Understanding on building the rail link between the Myanmar border town of Muse and Kyaukphyu was signed in Naypyitaw between China Railways Engineering Corporation and the Myanmar Union Ministry of Rail Transportation, and the route is expected to be finished by 2015. China will also be in charge of developing the port and the industrial zone which are part of its ambitious plan to develop a trade outlet for its landlocked southwestern regions, and an oil and gas trans-shipment point connected to oil and gas pipelines already under construction.

A second railroad will connect the 1,920 km between Kunming and Rangoon, would also link a railway connecting to the Tavoy (Dawei) project on the southern coast, including the construction of a new rail link between Tavoy and Bangkok. Furthermore, a potential route has been identified that would cross eastern Shan State connecting Kunming with Chiang Rai, and from there link into the Thai rail network. This link, together with a high-speed rail line in Lao PDR, planned to be built in 2015, will enable the shipment of goods by rail between China, Cambodia, Thailand and Singapore. Two other routes connecting the Chinese town of Dali in Yunnan with the Myanmar towns of Myitkyina and Lashio, are planned. Both Myanmar towns are large trading centres and railheads. China has also contributed to upgrading Myanmar's rail stock by donating engines from the rail transport department.

Myanmar plays an especially important role in Beijing's ambitious plans to construct a high-speed rail network linking China with Southeast Asia, South Asia, the Middle East and Europe. Much of this network, especially through Myanmar, corresponds to the 14,000 km Trans-Asian Railway Project, a United Nations plan to create an integrated freight railway network across Europe and Asia. If completed as planned, it will represent the largest infrastructure project in history. The exact routes of the lines are still undefined, but they will follow three general directions. As far as Southeast Asia is concerned, a western route will link Yunnan to India through Myanmar. A middle route will go through Lao PDR and Thailand, and an eastern route will extend through Vietnam and Cambodia to Thailand (i.e., Map 2.4 in this book). Currently, China's only rail connection to Southeast Asia runs to Vietnam. One major obstacle for the project is Beijing's insistence on the use of the same gauge of tracks as its domestic high-speed network. The rail systems in Myanmar and other Southeast Asian nations use different gauges, meaning it will be necessary to replace current tracks or lay new ones parallel to the existing ones, to make the connection. Due to the high cost of the project, observers are sceptical that the network could be completed over the next ten or twenty years. Up to now the only portions of China's planned network actually under construction are, firstly, a rail link between Yunnan province and northern Myanmar, and secondly, a high-speed route to Bangkok through northern Lao PDR and northeast Thailand.

Besides supplying export routes for Chinese goods from its land-locked southwestern region, the railways will also secure the transportation of energy resources imported from the Middle East and Africa. The new ports and rail links will allow alternative routes for China's crude imports reducing shipping time and its dependence on traffic through the Straits of Malacca, where currently as much as 80 per cent of China's imported energy travels. In Myanmar, new deep-sea ports at Kyaukphyu and Tavoy with new rail links will almost halve the distance needed to transport oil from the Middle East and Africa by sea.

With Bangladesh, China has discussed establishing a rail link between Kunming and a new deep-sea port project under construction at Cox's Bazaar via Myanmar's rail network. The proposed 111 km route is planned to go through eastern Bangladesh to Gundum in Myanmar where it will either connect with Myanmar's existing rail network or with the new high-speed route. In July 2010, Dacca (Dhaka) announced plans to build a railway to the Myanmar border.

As for India, an extension of a broad track from Jiribam in southwestern Manipur state to Moreh on the border with Myanmar will connect the current railhead at Segyi in western Sagaing to the border gate of Tamu on the Myanmar-India border. A rail link between the two countries would enable more efficient shipment of goods between India and China. Even though they are competing for influence in Myanmar, both economies are interdependent, and China is now India's largest trading partner.

Due to its strategic geographic position, Myanmar has been acting as a regional transportation hub and playing a central role in the construction of the dual Asian transportation corridors. Myanmar will mainly provide a regional transit crossroads for energy and goods trans-shipment to neighbouring countries (see Map 8.1). These transportation corridor projects will deepen Myanmar's regional integration and revive its role in Asia. The main impact of Asian railway network projects on Myanmar will be the expansion of Tatmadaw's presence and its control of the periphery areas.

Map 8.1
Natural Resource Development Projects in Shan State

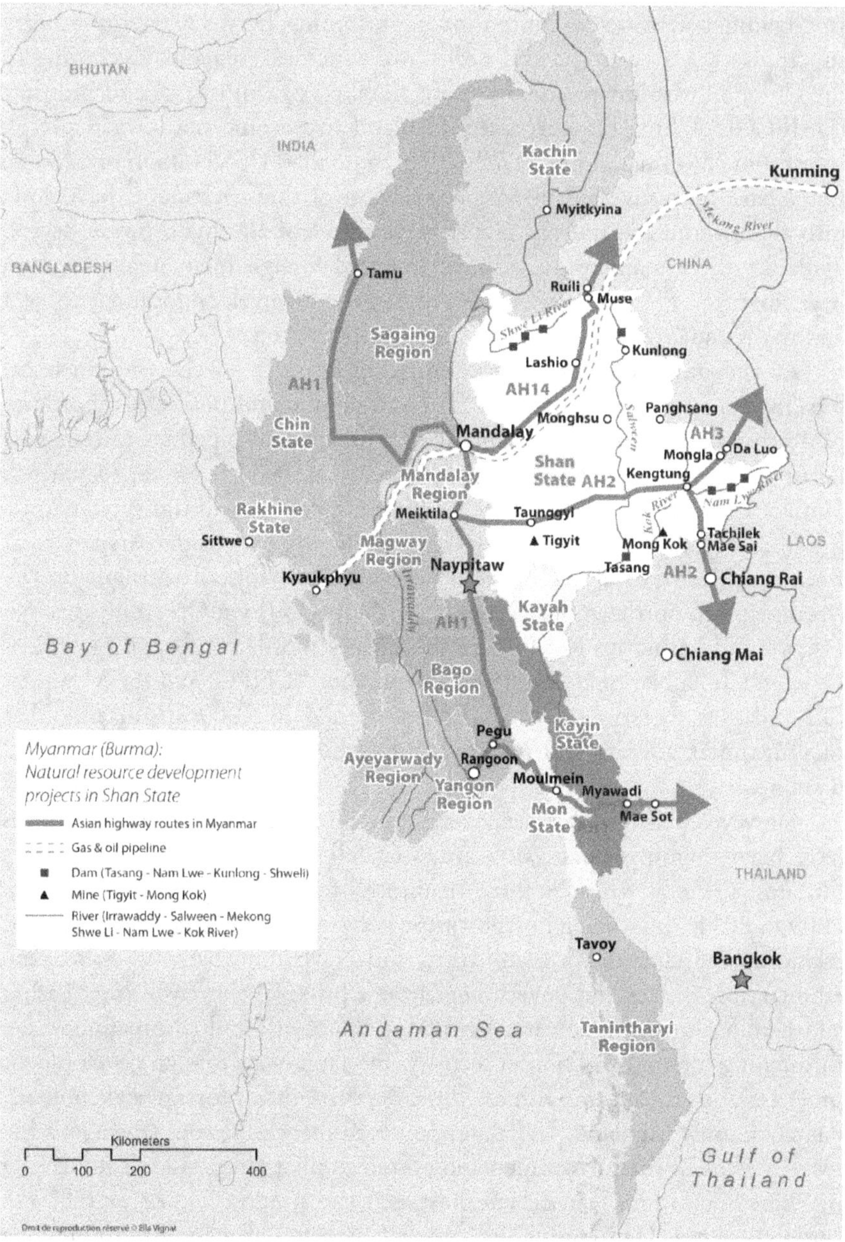

Source: AMEICC, Myanmar Information Management Unit, Salween Watch, Shwe Gas Movement, The Pa-Oh Youth Organization.

Regional Integration Projects Challenged by Instability in Shan State: Gas and Oil Pipeline Development

In Myanmar a strategic contention is unfolding between energy-hungry superpowers, China, India and new powers such as Thailand. According to the Union of Myanmar Federation of Chambers of Commerce and Industry (UMFCCI), China has surpassed Thailand to become the largest foreign investor in Myanmar. Since March 2010, more than US$8 billion of Chinese funds have been invested in Myanmar. Most of the investments have gone into the oil and natural gas sector, well ahead of the hydropower sector. Thus far, three major projects have attracted foreign main investments in Myanmar: the Shwe Gas pipeline Project, the Tavoy Development Project and the Kaladan Multi-modal Transit Transport Project.

At the moment, the most significant project — the Shwe gas-oil pipelines project — has stepped up the economic and political importance of northern Shan State. The construction of the dual pipelines to transport natural gas and oil from the Arakan coast across Myanmar to China's Yunnan Province has started. The 4,000-km pipelines will transport natural gas from Myanmar's offshore reserves and oil shipped to the Arakan coast from the Middle East and Africa. The pipelines will cut through central Myanmar and northern Shan State, from Maymyo (Pyin Oo Lwin) through Hsipaw and Lashio up to Muse on the Chinese border. A formal agreement between China National Petroleum Corporation (CNPC) and the Myanmar Oil and Gas Enterprise (MOGE) was signed in 2009 stipulating that the Myanmar government will guarantee the safety of the onshore dual pipeline.

However, tensions between Naypitaw[15] and the ethnic armed groups have been ongoing since 2005 after the fall of the ceasefire agreement's initiator, General Khin Nyunt, Myanmar's former prime minister. From 2009 to 2011[16] increasing hostility arose between most ethnic armed groups, including the United Wa State Army, and Naypyitaw. Skirmishes between ethnic armed forces and government troops broke out in Kyawkme, Lashio and Kutkhai townships in northern Shan State on the dual pipeline route, threatening internal and border security, and large-scale foreign development projects. With the resumption of ethnic tensions, major foreign stakeholders, namely China, Thailand, India and South Korea, raised concerns about the security issue of their investments in ethnic conflict affected areas, especially in Shan State. The last decade has seen the predominance of Chinese companies over Thai companies and the presence of western corporations in the development projects.

Regional Integration Projects Contested by Local Population: Hydropower Generation, Lignite and Coal Power Plants

China's involvement in Myanmar's hydropower sector is the most controversial issue. Following the recent event that took place in Kachin State, many blame the Chinese dam projects for triggering the strife. Among these numerous projects, the largest is the US$9 billion Tasang Dam on the Salween River.

Tasang Dam is the largest planned for the transnational Salween River, Southeast Asia's longest free-flowing river, and will submerge 870 square km in the heartland of Shan State. It is located in southern Shan State, 106 km from the Thai border. Tasang Dam is 228 meters high (the tallest in Southeast Asia), with 7,110 megawatts installed capacity and 35,446 GWh annual electricity production to be exported to Thailand. Initially, Thailand's MDX Group signed the deals with the government. Amidst rumours of dissatisfaction with MDX, the China Gezhouba Group Co. (CGGC), a large state-owned development company, won a contract for initial dam construction in early 2007. In November 2007, the Myanmar government gave CGGC a 51 per cent stake in the project, reducing the MDX share to 24 per cent. In mid-2008, Sinohydro, China Southern Power Grid Company, and China Three Gorges Project Corporation signed an agreement for the development of the Salween River Basin in Myanmar, mentioning the Tasang Dam in particular. The British firm Malcolm Dunstan Associates has also been involved in the project.

It is to be acknowledged that Tasang Dam is situated in the midst of a former conflict-affected area where the SSA-South fought the Tatmadaw for decades. On 9 May 2011, four Chinese engineers from the Chinese firm working on the project disappeared from the construction site. They were believed to have been abducted by an unknown militia. There are also several armed groups in the area, mostly backed by the government, and the expanded presence of the Tatmadaw is visible.

Chinese companies are planning to build the Kunlong Dam (or Upper Salween Dam); the dam site is situated on the mainstream Salween at Ton Gyap village in Kunlong Township, about 25 km from the border with China's Yunnan Province. On 24 February 2010, Myanmar's Ministry of Electric Power signed a Memorandum of Agreement with Hanergy Holding Group and Goldwater Resources Ltd. to develop the Kunlong Dam. The dam will have a total installed capacity of 2,400 MW. Feasibility studies have been conducted by a team of Chinese and Myanmar technicians.

The completion of the Shweli 1 Dam was in 2008 under Myanmar's first Built-Operate-Transfer (BOT) hydropower agreement with China. The 47 meter-high Shweli 1 Dam is located about 30 kilometres from the China-Myanmar border and is the first of a planned three-dam cascade that will generate 1,420 MW of electricity, most of which will either be exported to China or used for domestic military-owned mining operations. The main contract holder was the Yunnan Joint Power Development Company (YUPD) and the construction was carried out by Sinohydro Corporation. The installed capacity of the Shweli 1 Dam will be 600 MW. The actual power supply will be 174.8 MW, and the annual power output 4,033 GWh. The electricity will be transmitted to both Myanmar and China through 230 kV and 220 kV cables. The Shweli 2 and Shweli 3 hydropower projects are under construction near Mongmit Township. The Yunnan Machinery Equipment Import & Export Company Ltd. (YMEC) will be involved in these two projects. The Shweli 2 power plant will be located 20 km downstream from Shweli 1, and have an installed capacity of 460 MW. Shweli 3 will be 48 km downstream from Shweli 2 with an installed capacity of 360 MW. According to the BOT, the Chinese companies will get 80 per cent of the revenue generated by the project regardless of where the electricity is transmitted.

On 4 February 2010, the Yunnan Power Grid Corporation and the State Development and Investment Corporation (SDIC) Huajing Power Holdings signed a "formal stage agreement" to develop one dam on the Nam Kha, a tributary of the Salween River, and six dams on the Nam Lwe River, a tributary of the Mekong. Project pre-feasibility studies have already been conducted. The planned dam sites are former conflict-affected areas. In effect, the Nam Kha flows through the region under UWSA control, and the Nam Lwe runs through NDAA's territories. In 2011 tensions between the Tatmadaw and these two ethnic armed groups intensified and both sides reinforced their troops on their territory.

Saraburi Coal Co. Ltd., a subsidiary of Italian-Thai Power Company, has received concessions from Myanmar's authorities to extract 1.5 million tons a year for an initial ten years (renewable for five years at a time) at Mong Kok, located 40 km north of the Thai border in the mountains of eastern Shan State in a former conflict-affected area. The production of the open-pit coal mine is for export to Thailand. A lignite-fired power plant will be built by Italian-Thai Power Company to produce 405 MW. 369 MW will be sold to the Electricity Generating Authority of Thailand (EGAT) over a period of twenty-five years. Power is planned to come on line in 2016, the electricity will be distributed to Chiang Rai, in Thailand.

Local communities and Thai environmentalists have launched a campaign against the Thai lignite mine. They express concern about the disastrous environmental impacts of the lignite-fired plant on the Kok River which flows into Thailand through Chiang Rai Province and the Mae Ai district of Chiang Mai province before joining the Mekong River. The Tatmadaw had made a condition for Saraburi to build a new route across the border (despite the existence of a shorter 78 km route inside Myanmar's Tachilek to Thailand's Mae Sai), in exchange for the concession. Due to this proposed road plan, inhabitants from both sides are concerned that lignite shipment could destroy and affect local villages' life along the proposed route as well as endangering the environment and their safety. At a meeting held in Chiang Rai in May 2011, Saraburi Coal Company announced that it was considering handing over their products to China if people in Thailand went on opposing the project.

Tigyit is Myanmar's biggest open pit coal mine with a daily production of two thousand tons of coal. The Tigyit coal mine and coal-fired power plant are situated about 21 km from ASEAN heritage site, Inle Lake. The coal from the mine is conveyed to Myanmar's only operating coal-fired power plant in Tigyit. Most of the power from the plant is used at Pinpet Iron Factory operated by Russian and Italian companies. In 2002, the mine and the power plant were implemented by the China National Heavy Machinery Corporation (CHMC) and the Myanmar companies Eden Group and Shan Yoma Nagar. The plant produces 600 GWh of electricity annually, using 640,000 tons of coal per year from the Tigyit coal mine. Electricity from the Tigyit power plant mainly supplies power to companies that are further exploiting natural resources.

Myanmar's natural resources have been the major assets and the first factor for its regional integration before becoming vital to Asian road and rail transportation network schemes whose outcomes will only become visible in the medium term. As soon as the ceasefire agreements were signed in the peripheral regions, energy development projects were launched by the central power in the remote resource-rich borderlands without prior consultation with local communities. The local residents have strong resentment against these massive development projects as they do not obtain any share of the profits of this economic integration.

Strategic Mining Development Project: A Vector for National Strategies

Shan State is rich in gems and mineral resources, including iron, lead, silver, gold, copper, zinc, nickel, tungsten, wolfram, manganese and coal. The Namtu Bawdwin mines in northern Shan State were the world's largest source of lead, and one of the world's largest sources of silver before the Second World War. A major mining site, the Pinpet Projects, is the recipient of foreign investments (the Tigyit Project and the Mong Kok Coal Project mentioned earlier are excluded).

One of the most recent large mining projects in Shan State is at Mount Pinpet near Hopong in southwest Shan State, where Myanmar, Russian and Italian companies have begun mining iron. Located 11 km southeast of Taunggyi, the capital of Shan State, it has a long history of attempts to develop its mineral resources and contains the Myanmar second of the largest known iron ore deposit. The Pinpet deposit consists of reserves of hematite estimated at ten million tons with 56.4 per cent iron and reserves of limonite at seventy million tons with 42.6 per cent iron (the best grades of iron in the world contain 60 per cent iron). In addition, the mountain also has a deposit of an estimated thirty million tons of limestone, which can be used to make cement including copper, limestone, wolfram, and tin. The presence of uranium on the mountain has also long been suspected but not officially made public. Officially, the current plans are for iron ore extraction but there are suspicions among local villagers that uranium will also be mined at Mount Pinpet, in line with Myanmar's recently announced nuclear ambitions and partnership with Russia.

Since 2003, a third attempt to excavate minerals from Mount Pinpet has been moving forward. The iron ore will be extracted via open-pit mines. In 2006, a road was built on the upper west part of the mountain, to lead to two abandoned underground tunnels from exploration conducted in the 1960s. The area protected by the heavy presence of Tamadaw troops has been restricted and marked as a "military area". In 2004, the construction of Pinpet Iron Factory, located at the base of the mountain on its eastern side, began. Reports point out there will be various sources of power for the factory. In 2006, the Russian state-owned company Tyazhpromexport announced an agreement to supply US$150 million in equipment for the project. Another western company, Danieli & C. Spa is believed to be involved in the iron processing plan. Danieli has long been the only Italian company in Myanmar working on iron and steel. In 1979,

Danieli signed a turnkey contract with Mining Company No. 3 for an iron and steel factory in Maymyo (Pyin Oo Lwin). In 2007, the company confirmed in a press release that they operated in Myanmar in the steel-making industry but denied that they had any connection to the military industry or were supplying any products for armaments.

Tatmadaw's control over mining strategic resources is indeed a national strategy but by no means an integration strategy. On the contrary, that implies the use of coercion in dealing with the local population in the mining areas. While most development projects in Myanmar have involved Asian Transnational Corporations (TNCs), since western companies were prevented from investing in Myanmar by the sanctions, for its strategic mining development projects the junta has called in western expertise — namely of Russian and Italian firms — instead of its neighbouring partners.

3. HUMAN AND ENVIRONMENTAL IMPACTS OF THE DEVELOPMENT PROJECTS IN SHAN STATE

Impacts of Border Development Projects in Shan State

Related natural resource development projects do not significantly benefit the population. On the contrary, in the absence of environmental and social standards, these projects could have drastic impacts on the local communities and their environment. Common factors emerge from the natural resource development projects: heightened militarization around project areas, confiscation of farmlands, loss of livelihoods, forced relocation without adequate compensation leaving behind culture and ancestral homes, use of forced labour for the Tatmadaw and the development projects, forced conscription into an Army-associated militia as a part of the security apparatus in the area, and increased vulnerability of women due to an influx of male migrant workers.

Dams and mining projects radically alter the landscapes and ecosystems, causing deforestation, loss of animal habitat, pollution of water sources and intensive logging activities around development project areas. These human rights violations against the local communities are the common denominator of the development projects that take place in ethnic minority areas along Myanmar's borders with India, Bangladesh, China, and Thailand. In Myanmar's law, there is no public participation in decision-making, no environmental, social, or human rights impact assessment, and effectively no access to justice.

The Price of So-called Myanmar Competitiveness

The attraction of Myanmar for TNCs, in addition to its geographic location and natural resources, is largely linked to its lack of environmental labour and private property rights that allow this sort of environmental disregard. There is a line-up of Asian stakeholders — Chinese, Thai, Indian, Korean, Singaporean and recently Japanese TNCs, as well as many western TNCs — waiting to invest in Myanmar after most sanctions have been eased or lifted as a reward for the new government's political, social and economic reforms. Yet, since 24 December 2012, unprecedented air strikes launched by Myanmar armed forces against the Kachin rebels have raised doubts about the sincerity of President Thein Sein's commitment to peace and reconciliation. For decades ethnic minority people in the borderlands have put up with longstanding ethnic issues ignored by the international community.

Recently, resumption of armed conflict between the SSA, the KIO and the Tatmadaw has attracted international media coverage. Nevertheless, the outbreak of violent fights was hardly surprising for observers working on Myanmar issues. In point of fact, heralds of violence were already there. Since February 2011, a series of bombings has hit the country, altogether fourteen bombs exploded in different cities of Myanmar. The bombings targeted government buildings, military and police check-points, public areas such as train stations and markets in central Myanmar and in the ethnic region. In 2010 seven bomb blasts occurred, including one casualty at the Myitsone hydropower in Myitkyina in Kachin State, one blast in Muse, the China-Myanmar trade gate, several victims in Laukkaing in Kokang region, three bombs exploding at the Thaukyaykhat hydropower dam project in Taungoo Township, and the deadly water festival bomb blast in Rangoon. Bombing regularly hits Myanmar cities but it has escalated since the late 1990s. Whenever there is an incident such as a bomb explosion in the country, the authorities invariably accuse ethnic armed groups through their mouthpiece, the state-owned media. Unless someone claims responsibility, nobody will find out who is behind the attacks.

"A War of Business that Transcends Ethnicity"

Nevertheless, the ceasefire agreements, first initiated in Shan State in 1989, are only military arrangements with no political resolution. Since the economic and geostrategic realities have been changing, the status quo cannot fulfil its role anymore. Many believe that the resumption of armed conflict is caused by economic interest, by "greed"-based motivations. In effect, the roots of Myanmar ethnic conflict are intertwined with genuine political grievances and economic motivations. Insurgent ethnic armed groups have genuine grievances that have not been addressed by the central power since the ceasefire agreements. They have been asking for equal rights, and they are deeply unhappy about their status in the new Constitution. In the development projects issue, the ethnic communities have been affected by the authorities' lack of transparency. Hence, they have strong feelings against the large-scale development projects because they do not benefit from the financial fallout.

Since the end of 2012, armed conflict between the Tatmadaw and the KIA has escalated after Myanmar forces launched airstrikes on rebel headquarters in Kachin State. Religious, civil society groups have called for an immediate ceasefire and peace talks between the antagonists. They also launched public appeals to the international community, including Myanmar's "new ally", the U.S. government, and opposition leader Aung San Suu Kyi to intervene and pressure the Myanmar government to stop its assault. UN Secretary-General Ban Ki-moon and the U.S. government expressed "concern" about the conflict in Kachin State but did not condemn the assault. Many Kachins are distressed and disappointed by Daw Aung San Suu Kyi's silence and her refusal to help to stop the bloodshed. While she condemned the violent crackdown on mine protesters in Monywa, her silence over Rohingya violence in Rakhine State and attacks in Kachin State has cost her support in ethnic areas. Furthermore, tensions in the northern region have spilled over into China, Myanmar's and the insurgents' ally, with an influx of refugees flooding into its territory. China has deployed troops to the border between Yunnan Province and Kachin State and has complained to the Myanmar government about several bombs landing on its territory from Myanmar air strikes on KIA. As the main stakeholder with vested interests in Myanmar, China has asked both sides to end the fighting and find a peaceful solution. Unless a peaceful resolution is reached, widespread civil war will threaten the country's internal and border security. It will also put at risk what has been gained in the last two decades and will endanger the stability of neighbouring countries with the influx of refugees. Last but not least, there will be an increase in drugs and smuggling as armed forces need to finance their armies.

CONCLUSION

Myanmar and the Difficulty of Reconciling National and Regional Integration Processes

On 30 March 2011, in his inaugural speech, the president, ex-General Thein Sein, addressed the issue of national unity focusing on the ethnic nationalities, and declared the government's top priority was to build national unity in order to build "a peaceful, modern and developed nation". He also said that the new government would engage in development in ethnic areas with concrete measures such as roads, railroads, bridges, health, education and socio-economic measures to improve the living conditions of the ethnic nationalities. This strong message was received positively by the ethnic communities. It has been twenty months since President Thein Sein assumed power, his envoys have met fifteen ethnic armed movements for peace talks and have concluded thirteen ceasefire agreements. Despite the positive outcome of the government's efforts in the peace process, ongoing clashes between the Myanmar armed forces and the Shan State armies — two of the peace agreement signatories — have repeatedly occurred in Shan State, meanwhile a civil war has flared in Kachin State. As a result, thousands of people have died and 100,000 people are internally displaced. This questions the credibility of the peace process launched by the government. There will be no development in the ethnic areas without peace. Peace must precede development and there will be no peace without rights.

In post-conflict environments, efforts in nation-building and economic reforms have implications on peace in the borderlands. National integration can be achieved only by convincing the ethnic peoples of the Frontier Areas to be part of the Republic of the Union of Myanmar. Economic development is complementary to the political and security factors of peace-building in the process of nation-building. In the context of transnational integration on the tri-regional scale, the integration of the periphery is the condition *sine qua non* of its viability.

China's Role and Influence over Myanmar and in the Region: Counterbalancing China's Expanding Influence?

To some extent, peace and political stability in Myanmar's border areas will be unlikely to be reached without China's involvement. Given that some of China's major investments are located in the border areas in Shan and

Kachin states sharing the 2,192 km common border with Yunnan province, stability in the borderlands is crucial to China's economic and strategic interests. Myanmar is critical to China, as are the borderlands to Yunnan province appointed as one of China's bridgeheads to mainland Southeast Asia by President Hu Jintao in July 2009. Furthermore, Yunnan is China's only gateway to India and the Indian Ocean via Myanmar.

The ethnic armed groups have requested China's help to be the "referee" and mediator in any potential peace negotiations between the protagonists. Though the ethnic nationalities appreciate China's role in advising the central government against military action, and to solve the differences through peaceful negotiations, they know through China's growing relationship with the junta that they are pawns in Beijing's Myanmar strategy, and China's support is provisional and urged by its interests. Furthermore, in present Myanmar ethnic tensions, Beijing and Kunming's divergent interests and approaches to border management and ethnic groups are put to the test.

While Beijing prioritizes the security and stability of the border regions, Yunnan authorities value border trade and close cross-border relationship with ethnic groups. Border trade between China's southwestern province and Myanmar represents more than three-quarters of Yunnan's total border trade. Formerly blessed by the central power, this special relationship has recently irritated China's leaders in the north. Beijing is concerned about the increase of informal and illegal cross-border trade — particularly in logging, mining, drugs, wildlife and human trafficking — that is taking place. Thus, local business activities profit from instability in the borderlands. Nevertheless, China's top priority is the security of its shared border because turmoil would jeopardize China's domestic stability and regional economic development. In order to consolidate its political and economic influence over Myanmar, it has stepped up high-level diplomatic visits, trade and investments.

To counterbalance China's preponderant influence over Myanmar and to diversify the sale of its energy resources, the junta has reinforced bilateral ties with India, the other regional superpower. India and Myanmar share a 1,640 km-long border, and Myanmar's location is pivotal to India's Look East Policy, energy security and counterbalancing China's influence in Southeast Asia. Though India has reshaped its policy towards its south-eastern neighbour to offset China's rising influence in the region and over Myanmar, it has lost the first rounds and suffered a setback in the energy diplomacy race with China — loss of its status as preferential buyer in the Shwe project — and with Thailand by losing out on the Tavoy mega-project. Yet, India is currently Myanmar's fourth-largest foreign investor and has sought to deepen its relationship with Naypitaw and to expand their economic and trade cooperation. Both countries have agreed to double bilateral trade from US$1.5 billion to US$3 billion by 2015. One major project emerging from this collaboration is the US$120 million Kaladan Multimodal Transit Transport Project — comprising a waterway and a roadway — that will create a trade corridor between Indian ports and Myanmar's western coastal Sittwe Port to transport goods from Northeast India to Southeast Asia.

However, China's growing influence with its large-scale investments in Myanmar has provoked resentment among the local population but also within the armed forces. The symbol of anti-Chinese sentiment is currently the controversial US$3.6 billion Myitsone Dam construction project. Myanmar's new regime has recently come under pressure over the potential environmental and social impact of the construction of the 6,000 mw Myitsone dam on a confluence of the Irrawaddy River in Kachin State. "Save the Irrawaddy" campaign and demonstrations have been organized inside and outside Myanmar by environmentalists and Myanmar opponents to protest against the project financed by the state-owned China Power Investment Corporation (CPIC). Heated argument about the controversial hydropower occurred in Naypyitaw as government ministers had different points of view on the matter. On 30 September 2011, a statement to Parliament on behalf of President Thein Sein announced the decision to halt the construction of the Myitsone dam because it was "contrary to the will of the people". This decision has surprised everyone at home and abroad, as CPIC President Lu Qizhou declared he was shocked as he heard of the suspension of the project through the media, and that it "will lead to a series of legal issues". Observers are cautious in analysing the implications of such a move from Myanmar's new regime. Furthermore, the government has recently found itself in a quandary facing a public outcry against a vast copper mining project run by the Myanmar military and a Chinese firm Wanbao Mining, a subsidiary of the North Industries Corporation (Norinco), a Chinese arms manufacturer, near Monywa, northwest of Mandalay in Upper Myanmar. Wanbao Mining is accused of confiscating 7,800 acres of land from 26 surrounding villages for the project, destroying cultivated fields and polluting water sources. The three-month protest was eventually broken up by riot police with 50 injured. This reflects growing resentment towards Chinese companies across the country, even within the military. Since late March 2011, President Thein Sein seems to have played an important role in a number of political, social and economic reforms that have been launched in the country. Nevertheless, questions have been raised whether President Thein Sein has control over the Myanmar armed forces, since his orders to stop fighting against ethnic minorities groups and the KIA were ignored twice by the Tatmadaw that seems to answer only to Commander-in-chief Gen Min Aung Hlaing. It means that ethnic concerns, which are a critical issue of national security, are still the responsibility of the Tatmadaw. In this context, in the coming months, particular attention will be paid to President Thein Sein's agenda for further economic reforms and national reconciliation. While Naypyitaw states its willingness to change to Myanmar citizens and the international community by undertaking a series of political, social and economic reforms that have little impact on people's daily lives, especially in the conflict-affected areas, intensive fighting has been going on in Shan and Kachin states.

Notes

1 In 1989, the junta regime changed the English name of the country, Burma, to "Myanmar, citizens were to be called "Myanmars". The name changes are not unanimously recognized by all countries. The junta adopted new English pronunciations, for cities ("Rangoon" became "Yangon", "Maymyo" became "Pyin Oo Lwin") and ethnic groups ("Karen" became "Kayin"). In this chapter, I use the term "Burma" when referring to a period prior to 1989. Likewise, from 1989, I use the term "Myanmar" in reference to the country and to the citizens of this country. I am not using "Myanmar" to make any kind of political statement.

2 Formerly named State Law and Order Restoration Council (SLORC), the SPDC was dissolved on 30 March 2011.

3 The term "state". Within Myanmar, the territorial and administrative units where many ethnic minority peoples live are called "states".

4 Since 2009 tensions have arisen between the authorities and the ethnic armed forces after the latter refused to join the Border Guard Force (BGF) plan implemented by the government based on the 2008 Constitution that requires "One Nation, One Army".

5 Since President Thein Sein assumed power in March 2011, the government envoys have met 15 ethnic armed movements and have concluded ceasefire agreements with 13, including major non-state armed groups present in Shan State, namely the United Wa State Army (UWSA), the National Democratic Alliance Army (NDAA), the PaO National Liberation Organization, the Shan State Army-North and the Shan State Army-South. Yet two of them, the Palaung State Liberation Front (PSLF) and the Kachin Independance Army (KIA), are still fighting against the Tatmadaw (Myanmar Armed Forces).

6 On 30 March 2011, the Union of Myanmar became the Republic of the Union of Myanmar.

7 With the support of the United Nations Population Fund (UNFPA) Myanmar plans to carry out a population census in 2014. Anxiety grows among the ethnic minority peoples about the outcome of the planned national census and its impacts on the issues of nationality and ethnic identity in National Registration Cards.

8 Increasing fighting between the Tatmadaw and the KIA since June 2011 has spilled over into northern Shan State where the KIA has some strongholds.

9 Despite a ceasefire agreement signed on 2 December 2011 between the SSA-S and the government, over fifty clashes have been reported by the non-state armed group which has warned the Tatmadaw to respect the terms of their ceasefire agreement or jeopardize the peace process.

10 Expression used by Mary P. Callahan (2007).

11 The situation has changed since June 2011 after the seventeen-year-old truce between the KIA and the Myanmar government ended.

12 In September 2003, China sent the People's Liberation Army (PLA) to crack down on Mongla's Chinese-invested casinos as Chinese high-ranking cadres lost millions of yuan in them.

13 In December 2010, Yangon Airways' licence was not renewed because it was associated with the Wa. Nevertheless, since October 2011, its flights have resumed after a ceasefire agreement deal was reached between the government and UWSA in September 2011.

14 In June 2011, fighting resulting from the dispute over areas close to Dapein dam, one of China numerous projects in Myanmar, broke out between the Myanmar forces and the Kachin Independence Army (KIA), Myanmar's second strongest ethnic armed group, and put an end to a seventeen-year long ceasefire.

15 On 6 November 2005, Myanmar moved its administrative capital to central Myanmar, Naypitaw replaced Rangoon.

16 Following the refusal of the BGF plan, the Tatmadaw had launched an offensive against the ceasefire Myanmar National Democratic Alliance Army (MNDAA) in the Kokang region in northern Shan State next to the Chinese border in August 2009, and won a rapid victory. In the wake of this attack, tensions were sparked off between the United Wa State Army (UWSA), MNDAA's ally, and Naypyitaw. On 8 November 2010, the very next day after the elections took place, fighting between the ceasefire Democratic Karen Buddhist Army (DKBA) Brigade Five that had rejected the BGF scheme, broke out in Myawaddy border town on the Thai-Myanmar border in Karen State. Finally, on 13 March 2011, the ceasefire Shan State Army-North (SSA-N) and Myanmar troops came to blows in southern and northern Shan State and clashes have been continuing to occur since.

References

AMEICC. *Study to formulate a Master Plan on Industrial Linkage between Thailand and Myanmar*. Bangkok: JODC, 2005.

Asia Development Bank. *Midterm Review of the Greater Mekong Subregion Strategic Framework (2002–2012)*. Manila: ADB, 2007.

———. "Transport and Trade Facilitation in the Greater Mekong Subregion: Time to Shift Gears". *Evaluation Study*. Manila: ADB, 2008.

Ballentine, Karen and Heiko Nitzschke. "Beyond Greed and Grievance: Policy Lessons from Studies in the Political Economy of Armed Conflict". New York: International Peace Academy, 2003.

Callahan, Mary. "Political Authority in Burma's Ethnic Minority States: Devolution, Occupation and Coexistence". *Policy Studies* No. 31. Washington, D.C.: East-West Center, 2007.

Collier, Paul and Anke Hoeffler. "On Economic Causes of Civil War". *Oxford Economic Papers* 50, no. 4 (1998).

Duffield, Mark. *Global Governance and the New Wars: The Merging of Development and Security.* London: Zed Books, 2001.

Guo, Xiaolin. "Towards Resolution: China in the Myanmar Issue". *Silk Road Paper*. Sweden: Central Asia-Caucasus Institute and Silk Road Studies Program, 2007.

Kramer, Tom. "Ethnic Conflict and Dilemmas for International Engagement". In *Trouble in the Triangle: Opium and Conflict in Burma*, edited by M. Jelsma, T. Kramer, and P. Vervest. Chiang Mai: Silkworm Books, 2005.

———. "Myanmar's Border Trade with China: Situation, Challenge and Prospects". In *Japan and Korea with the Mekong River Basin Countries*, edited by Kagami Mitsuhiro. *BRC Research Report*, no. 3. Bangkok: IDE-JETRO, 2010.

Kudo, Toshihiro. "Border Industry in Myanmar: Turning the Periphery into the Center of Growth". *IDE Discussion Paper*, no. 122. Chiba: Institute of Developing Economies, 2007.

Lintner, Bertil. *Burma in Revolt: Opium and Insurgency since 1948.* Boulder Colo.: Westview Press, 1994.

Mya, Than. "Myanmar's Cross-Border Economic Relations and Cooperation with the People's Republic of China and Thailand in the Greater Mekong Subregion". *Journal of GMS Studies* 2, no. 1 (2005).

Sciortino, Rosalia and Sureeporn Punpuing. *International Migration in Thailand.* Bangkok: International Organization for Migration, 2009.

Set Aung, Winston. "The Role of Informal Cross-border Trade in Myanmar". *Asia Paper*. Sweden: Institute for Security and Development Policy, 2009.

Shwe Gas Movement. *Corridor of Power: China's Trans-Burma oil and Gas Pipelines*. Chiang Mai: SGM, 2009.

Smith, Martin. "Burma's Ethnic Minorities: A Central or Peripheral Problem in the Regional Context?". In *Burma: The Challenge of Change in a Divided Society*, edited by P. Carey. London: Macmillan Press Ltd., 1997.

South, Ashley. *Ethnic Politics in Burma: States of Conflict*. London: Routledge, 2008.

Taillard, Christian. "Un exemple réussi de régionalisation transnationale en Asie orientale: les corridors de la Région du Grand Mékong" [A successful example of transnational regionalisation in East Asia: The corridors of the Greater Mekong Subregion]. *L'Espace Géographique*, 2009–1 (2009): 1–16.

———. "Border Trade and Economic Zones on the North-South Economic Corridor: Focusing on the Connecting Points between the Four Countries." *IDE Discussion Paper* no. 205 (2009).

Tsuneishi, Takao. "Development of Border Economic Zones in Thailand: Expansion of Border Trade and Formation of Border Economic Zones". *IDE Discussion Paper* no. 153 (2008).

Tun, Sai Aung. *History of the Shan State from Its Origins to 1962*. Chiang Mai: Silkworm Books, 2009.

Zaw, Myinmo and Toshihiro Kudo. "A Study on Economic Corridors and Industrial Zones, Ports and Metropolitan and Alternative Roads in Myanmar". In *Intra- and Inter-City Connectivity in the Mekong Region*, edited by Masami Ishisha. *BRC Research Report*, no. 6 (2011). Bangkok: IDE-JETRO, 2011.

9

SUMATRA TRANSNATIONAL PROSPECT BEYOND INDONESIAN INTEGRATION

Muriel Charras

The island of Sumatra is the Indonesian part of the Malacca Straits region, and forms one of its edges.[1] Its location on the route between the China Sea, the Indian Ocean and the Java Sea is exceptional, and has played a role in all the main periods of world history up to the time of Indonesian Independence. The building of this far-flung archipelago into a nation was marked by a fear of centrifugal forces; centralization of territorial management from the capital, Jakarta, and control by closing borders provided a solution to this problem. "Introverted" Indonesia developed over a fifty-five-year period, concentrating its investments on the central area, Java (Charras and Franck 2000, 2004), and turning Sumatra into an outlying area on the national scale. A major component of the Straits region due to its size (443,000 km^2), its natural resources and the production of its plantations, but also on account of its demographic weight (50 million inhabitants, representing at present 21 per cent of the population and 21 per cent of the surface area of Indonesia, a unique balance in the archipelago), Sumatra today is a long way behind Singapore and the peninsular part of the Malaysian Federation in terms of economic performance.[2]

This study presents the effects of the development policies since 1966, the beginning of the Soeharto regime, which led to this "sidelining", then examines whether the ten years' implementation of the regional autonomy law have finally enabled this large island to regain its place in the opening up and renewal of transnational relations on a regional scale. It deals with the central part of East Sumatra, which consists of three provinces — Riau, Jambi and South Sumatra (Sumsel) — facing the Malay peninsula, from Kuala Lumpur to Singapore, via Melaka and Johor Bahru. The coastline is bordered by a very wide swamp area and small islets encumbering the straits; it is the part closest to the opposite shore (see Map 9.1). It is also in this central part —

Map 9.1
Sumatra Island

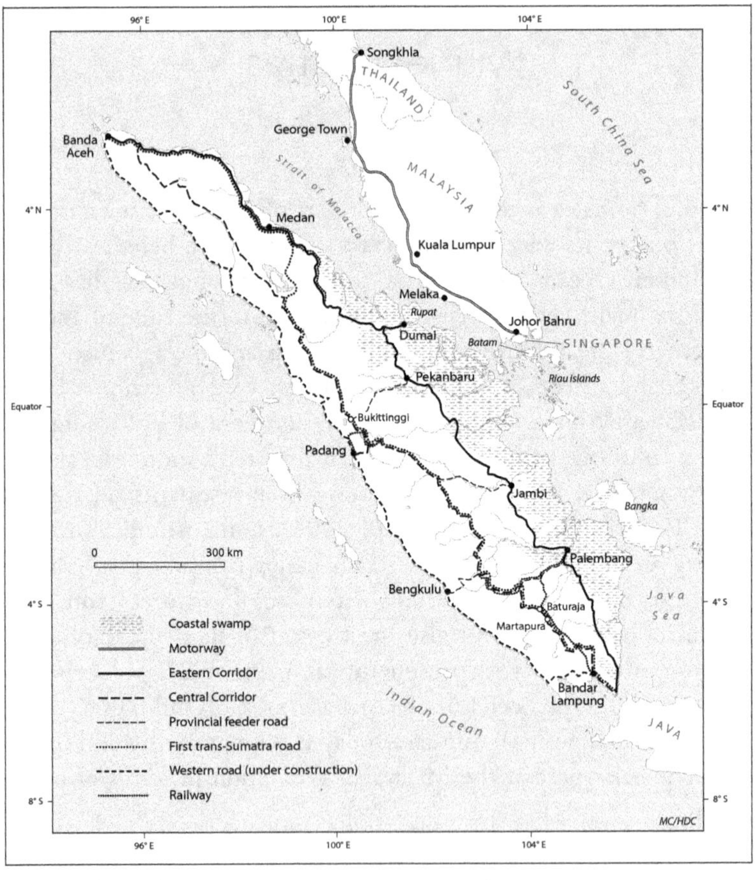

Source: The author.

rightly considered to be isolated and depressed in the regional study performed by the Japanese cooperation agency (JICA) in the 1980s — that recent development and spatial reshaping have been the greatest.[3] The province of West Sumatra, facing the Indian Ocean, is also associated with it, on account of the eastern tropism of the province (see Map 9.2). The Minangkabau were the first to colonize not only the eastern slopes of the Barisan mountains by following the Riau and Jambi rivers, but also, by crossing the Straits, the Malay peninsula, and this movement continues today in other forms.

Map 9.2
Sumatra Provinces

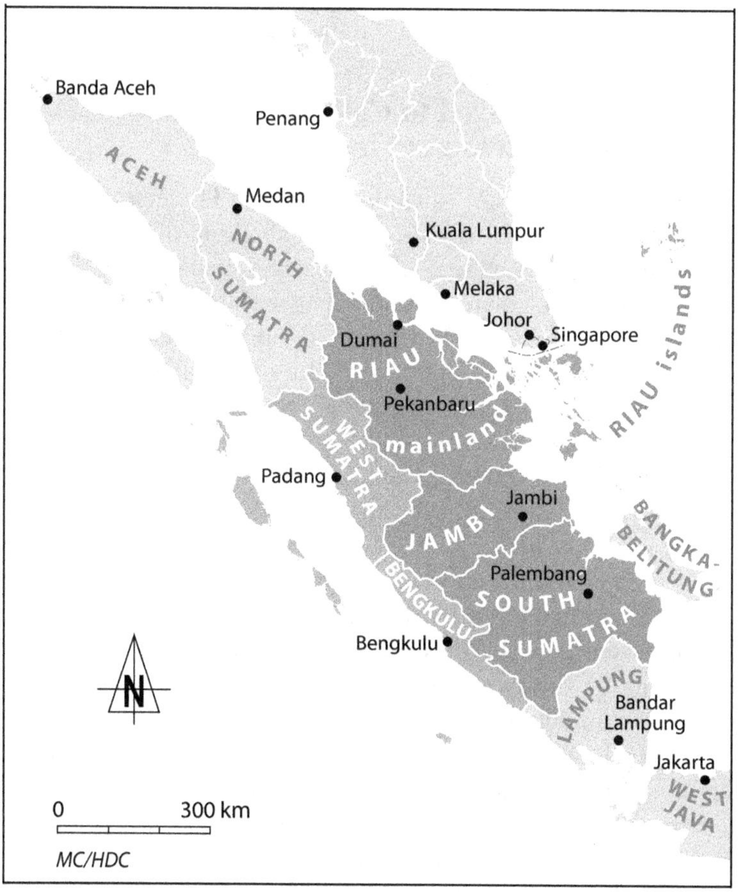

Source: The author.

This reshaping results from the creation of trans-island trunk roads (in the 1970s, 1990s, and at present), the SIJORI growth triangle (1990) and transnational economic zones (1993). Finally, after ten years' decentralization and regional autonomy (2000), it also results from strategies developed by the central administration and local elites in the new context of increasingly rapid opening up to globalization.

1. CHANGES IN CENTRAL SUMATRA UNDER THE NEW ORDER

Sumatra as a whole, and the central part in particular, has undergone many changes since the middle 1960s: strong demographic growth and disruption of ethnic composition, agricultural transition and reshaping of spatial organization.

Recent Settlement Linked to Colonization and Monopolization of Resources by the Centre

Two factors lie beneath the "sidelining" process. The first of these, as for all the outlying Indonesian islands, was the abolition of traditional powers after independence. In Sumatra, all the coastal sultanates and the networks linking them to the sultanates on the opposite shore were abolished. Then, in 1963, when the Malaysian Federation, grouping together all the British-ruled territories, was first being created, Indonesia dreamed of an overall structure (Malaysia-Philippines-Indonesia), and engaged in a confrontation policy (*Konfrontasi*) with Malaysia over its frontiers. Exports from Sumatra (vegetables, fruit and livestock), which supplied markets in Kuala Lumpur and Singapore, were suspended during the conflict and only resumed, on a small scale, in 2000.

Sumatra is the outlying island whose population has increased the most, in absolute figures, since independence. It is easily accessible from densely-populated Java, and has received large numbers of migrants since the beginning of the twentieth century due to spontaneous or State-organized agricultural colonization starting at its southern tip (Lampung), and the arrival of migrant contract workers in the privately-owned plantations of the northeast (Medan). Jambi and Riau, which are hard to reach, were the last provinces reached by these two pioneering fronts and remain today the least densely populated.

The completion of the roads at the beginning of the 1990s opened up access to land from the north, and accentuated the phenomenon of the pioneering front with the migration of Batak farmers, planters in search of new lands.[4] The population of mainland Riau increased from two to 3.75 million between 1980 and 2000, to which should be added the population of the Riau archipelago, still under its jurisdiction; one million inhabitants, including 426,000 on the island of Batam, a new economic centre opposite Singapore.

During the decade of major industrialization in Indonesia (1986–98), only Riau maintained strong demographic growth, whereas the other populations of Sumsel and Jambi went to look for work either in the Batam free trade area, or in the outer suburbs of Jakarta (see Table 9.1).[5]

In the last twenty years Indonesian agriculture has undergone structural changes with regard to land use, crops, the peasantry and rural life. Central Sumatra, apart from some rice-growing areas, was still dominated by forest in the 1980s, with a swidden cultivation system and small family plantations, mostly rubber trees, set up in the forest re-growth of previously cleared areas. Several factors were responsible for the reorientation of the rural population of Central Sumatra, including the policy of resettling migrants from densely populated islands (Java, Bali), settled by the government (transmigration) in the peneplains and along the marshy coastal strip, growing food crops in permanent fields. In fifty years, the central part of Sumatra has seen the arrival of one and a half million transmigrants

Tabel 9.1
Sumatra Demographic Evolution, 1971–2000

	Density Pop./km²			Population Growth Rate %		
Province	**1971**	**1990**	**2000**	**1971–80**	**1980–90**	**1990–2000**
Aceh	36	62	71	2.9	2.7	1.4
North Sumatra	94	145	164	2.6	2.1	1.3
West Sumatra	56	80	85	2.2	1.6	0.6
Riau	17	35	52	3.1	4.3	4.1
Jambi	22	45	54	4.0	3.4	1.8
South Sumatra	33	61	75	3.3	3.2	0.9
Bengkulu	25	56	74	4.4	4.4	2.9
Lampung	83	181	202	5.8	2.7	1.1

Source: National Statistics BPS.

Tabel 9.2
Transmigration in Central Sumatra, 1950–99

Province	Village Unit	Family	Population
Riau	265	113,418	491,598
Jambi	158	69,914	286,480
Sumsel	395	170,834	687,888
Total	818	354,166	1,465,966

Sources: National Statistics BPS.

(see Table 9.2). In their wake, spontaneous migrations have increased the flow and varied destinations, often strengthening the Sumatran rural economy (Charras and Pain 1993; Levang 1998).

At the beginning of the 1980s, the abolition of customary rules for a unified system of village organization enabled the State to dispose of forest land, including the rubber jungle. The authoritarian seizure of land for forestry concessions, given then to the same firms to create plantations: oil palms and fast-growing trees for wood pulp.

Since the 1990s, the State has pulled out of direct support to smallholders, including transmigrants, who find contracts directly with large plantations.[6] Oil palm plantations use the Nucleus Estates and Small-holders system, where the firm creates an estate and a refinery (*nucleus*), and surrounds itself with associated smallholders (*plasma*), transmigrants and local populations organized into cooperatives. It was in mainland Riau, the last province to be opened up, and still sparsely populated in 1990, that the movement was the most intense, with massive investment by regional conglomerates and planters from northern Sumatra, where space was becoming scarce. To oil-palm-growing areas should be added some 500,000 hectares of forest granted to three companies for wood pulp production, which is the main cause of deforestation, within and outside their concessions (see Table 9.3).[7]

In an article on the impact of roads on Riau, Freek Colombijn (2002) describes the process of spatial conquest and transformation as a "frontier society" where the actors succeed and replace each other: the oil companies (mainly Caltex) open the roads which are then used by a few migrants to nibble at the forest by starting small plantations of rubber trees; they are followed by forestry concessions, then large plantations; all these people are strangers to the region, tapping non-renewable natural resources without a qualm. However, Caltex should be recognized as

Tabel 9.3
Evolution of Palm Oil Plantation in Sumatra

	Planted Area (ha)			
Province	**1980**	**1990**	**2000**	**2005**
Aceh	22,500	90,530	218,493	261,101
North Sumatra	248,925	490,216	785,732	964,257
West Sumatra		35,604	229,575	324,332
Riau	2,078	240,181	815,646	1,340,036
Jambi		45,528	406,315	466,709
South Sumatra		61,939	557,849	532,365
Bengkulu		23,169	60,899	83,583
Lampung	12,226	14,960	97,445	163,589

Source: National Statistics BPS.

significant contributor "to schools, mosques, Universitas Riau (UNRI), the sports stadium and airport at Pekanbaru, as well as telephone lines, piped water, and electricity facilities" (Colombijn 2002, p. 748).

The ethnic composition of central Sumatra, a receiving land, has become increasingly varied. In rural areas, local groups, Melayu, are scarcely more numerous than the Javanese who represent some 30 per cent of the population, but these are now in the minority in comparison to migrants as a whole (Chinese, Bugis, Banjar, Batak[8] ...). In towns the Minangkabau predominate, except in Batam, where most labourers, but also most of the politico-economic elite, are Javanese. The proportion of the ethnic Chinese population in the towns and villages of the coastal region is also higher than in the rest of Indonesia (6 per cent in Batam).

Trans-island Routes: Increased Dependence on Java

The idea of trans-island routes in Indonesia was initiated in Sumatra, and the island has three, crossing it from the north (Aceh), at the head of the Straits, to the south (Lampung) on the Sunda Straits, and a stone's throw from Java. The first, upgraded in the 1970s from a former opening by the Dutch, goes through the centre, between the mountains and the piedmont; the second (1990s) goes through the lowlands on the eastern side, following the shore of the Straits; and the third, in the west, on the narrow coastal plain along the Indian Ocean.

The creation of communication routes (like any new type of transport or technical change) always gives rise everywhere to major restructuring of spatial organization; it leads to the emergence of new nodes, as well as the loss of activity in some central places, as illustrated by the situation in Palembang, the Sumsel provincial capital. Before the first trans-Sumatra was built in the 1970s, the whole of the Musi river basin (some 60,000 km^2) focused on Palembang, a river port eighty kilometres from the river mouth. The neighbouring river basin to the north (Batang Hari, Jambi), the high plateaus, and even the western coastal area (Bengkulu) depended on it, using the rivers or the roads built along their banks in the Dutch colonial period. Palembang was the port exporting products from an extensive hinterland (rubber, coffee, tea, rattan) and a dynamic trading centre.[9] At that time, as formerly, the flow of trade was oriented East-West, between the coast and the interior.

In the 1970s, the completion of the trans-island route following the piedmont reoriented the flow in a North-South direction. It enabled small towns to grow up as stopping places, at junctions between branches of the Musi river and the road crossing the interfluves (Lubbuk Lingau, Tebing Tinggi, Baturaja, Martapura). At the same time, it led to the decline of Palembang. In future, coffee, fruit and vegetables from the highlands and copra from the west coast would be taken directly to west Java, following the piedmont coal transported by the railway line built by the Dutch. In this way, most local exports vanished from Sumatran statistics. Because of this loss of activity, the river port of Palembang had to cut its maintenance expenses, especially the dredging of heavy alluvial silt, and the silting of the harbour justified the authorities' refusal to modernize it. All that was left for Palembang was the rubber and rattan production of its nearby hinterland, and oil from Jambi carried via a pipeline to the oldest refinery in the archipelago and the first fertilizer factory (*Pupuk Sriwijaya*).[10]

Before the building of the trans-island route, the dynamics of exchange flowing from the coast towards the interior were Sumatran (Kato 1997). The opening of the road brought about exchanges dominated by Java, both incoming (population and manufactured goods) and outgoing (natural resources and agricultural produce). The trans-Sumatra would mainly serve exports from the ports and processing industries in west Java.

The completion of the second trans-Sumatra, also oriented North-South, but following the east coast, did not change the overall purpose of this route, which was to reach Java. However, it has linked the provincial capitals of the central area: Pekanbaru, Jambi and Palembang, which are also all river ports, located between eighty and a hundred kilometres from the coast, at the point of contact between the peneplain and the swampland. The first of these is close to an access to the sea, Dumai, a former specialized transport harbour (oil), whose activities are now more varied (M. Franck in this work). Two effects are expected: firstly, a probable partial recapture of the flow of products from the interior, with a renewal of harbour activity and a rebound in urbanization; secondly, the linking of eastern Sumatra provincial capitals should give rise for the first time to an urban system on the island-wide scale.[11] The process will be slow, but probably inexorable, and we shall examine the first signs later on.

The western trans-Sumatra has been officially finished since 1994 but the sections crossing the three central provinces (Riau, Jambi and Sumsel) are particularly difficult to maintain, since alluvial or peat soils are poorly consolidated and often permanently saturated on the edge of swampland flooded daily by the tide, and the techniques used do not currently consider the very humid tropical climate zone of Sumatra. Also, these roads are used by lorries overloaded with timber for legal or illegal trade, during this period of intense deforestation. On this account, transport is organized, according to sections, on one or the other of the two roads — central or eastern — according to information provided by lorry drivers about the condition of the road network.

Economic Cooperation Zones: A Tightly-controlled Opening Up of Trade

Sumatra was also the first to establish economic cooperation zones in Indonesia. Since the beginning of the 1990s, starting with the island of Batam, located opposite Singapore, the eastern regions of Sumatra were supposed to be able to engage in exchanges with neighbouring countries on the other shore. Although the central powers had dreamed for several years of capturing part of the city state's trade, they were overtaken by the tenacity of the Singaporean private sector in search of space and labour. After lengthy equivocations, in 1990 the island became one of the components of the region's first Growth Triangle, SIJORI.[12] Batam became an industrial enclave, an export processing zone where the establishment of a genuine free trade zone was a lengthy process, and was controlled from the start by the central powers, bypassing local authorities. This was followed by the creation of two economic cooperation zones (1993), in the north of the Straits with Malaysia and the south of Thailand (IMT), and in the south with Malaysia and Singapore, which were supposed to prepare border regions for the opening up of trade in the ASEAN region. These zones were very quickly widened to the whole island, removing all meaning from the natural interface of the coastal area, which would have deserved specific development plans for border regions. Apart from a few annual meetings, under the aegis of the Asia Development Bank (ADB), there is, in fact, very little freedom of exchanges, which nearly all have to be agreed by Jakarta. Until 2000, very few tools encouraging trans-border cooperation were provided.

During the first years, while Malaysians and Singaporeans could go to Sumatra to conduct their business, Sumatran residents, except those from Batam, had to pay a tax of about a hundred euros each time they left their national territory, the same as any other Indonesian citizens. This exit tax was revised several times, and sometimes even completely waived for all Sumatran residents, without ever being clearly stated, thus hampering movements and disadvantaging Sumatran trading initiatives in neighbouring countries.[13] In fact, these economic cooperation zones contributed very little to the island's development, apart from Sino-Malaysian or Singaporean investments, mostly in forestry concessions and large plantations, after establishing contracts with the Sino-Indonesian conglomerates gravitating around the central power. Only Batam and the small islands surrounding it benefited from this trans-border cooperation, an opening up of trade "controlled" by Jakarta, to suit the needs of Singapore, using a novel model that would be difficult to reproduce in Indonesia (Phelps 2004). The dynamism of Batam, developed as an enclave, had no spatial impact on mainland Sumatra, on the contrary, it diverted investments from other areas of central Sumatra, especially Medan. The small islands surrounding Batam (e.g., Bintan, Karimun) were gradually engulfed by this model, which included the appropriation of land for enclaved development — tourist sites (resorts and golf courses), agricultural or industrial estates — from which the local population was excluded, and where staff were contracted migrants.

Thus, during Soeharto's New Order, central Sumatra was a reserve of land for resettling migrants from overpopulated islands, and setting up new plantations. This extraction economy was implemented without any significant local investment, but supplied processing industries concentrated in Java. In 1999, 75 per cent of the pulp industry's capacity is situated in Sumatra, 53 of the nation's 65 paper mills are located in Java, accounting for over 90 per cent of the industry's total production capacity (Barr 2001, p. 28). As examples, we may cite the manufacture of furniture (wood and rattan), rubber and palm-oil processing. The trans-Sumatran roads accentuated the overexploitation of natural resources and hyper-centralization in favour of Java. Economic cooperation zones existed only on paper, with the single exception of the strictly controlled opening up of Batam.

2. TEN YEARS OF REGIONAL AUTONOMY: A NEW CONTEXT, NEW OPPORTUNITIES?

The region entered the twenty-first century with two changes which should be capable of reorienting and boosting the development of the eastern side of the main island, and improve the region's integration in the Straits region. First of all, in 2000, after the fall of Soeharto, Indonesia underwent a major transformation, with the passing of drastic regional autonomy laws (Charras 2005; Booth 2011). Then, in 2006, under the aegis of ADB, there appeared the concept of "development corridor", aimed at structuring and integrating Southeast Asia as a whole, via a network planned on the scale of the Malacca straits, interconnected by three crossings of the straits. Local governments and populations should therefore be able, on the one hand, to free themselves from the directives of central government, and on the other, take advantage of the regionalization programmes of international organizations. In this new context, the eastern side of Sumatra should be equipped to reconnect with its geographical advantages and its transnational heritage (Tagliacozzo 2005). Given the scale of structural changes and the suddenness of their application, there is no doubt that, ten years on, the region is still in a state of transition.

Regional Autonomy: Reality and Limits

The regional autonomy laws changed Indonesia from a hyper-centralized territorial administration (based in Jakarta and regional capitals) to fragmented administration based on departments and towns with the status of autonomous communities. The effects of the law are considerable and transform the geography of the archipelago. The system of regional government was totally revised in 1999. The thread of territorial administration remains the same; the country is divided into provinces, themselves divided into departments (*kabupaten*) and cities (*kota*), which are made up of village communities (*desa*) or urban districts (*kelurahan*), coordinated in turn by the functional unit of the canton (*kecamatan*).

From the point of view of administrative organization, the major change is the abolition of the hierarchical relationship between the province and its territorial units (departments and cities).[14] The governor, formerly nominated by the central authority,[15] is no longer the agent of central government, as he has been elected by direct suffrage since 2004, as are departmental prefects and town mayors. The financial resources of the province have been considerably reduced: the allowance from central government, which used to be entirely at its disposal, is now distributed directly to the autonomous units, and the province only manages 10 per cent of it. The premises of the Sumsel provincial administration were in an advanced state of disrepair in 2008 (paintwork, office equipment, outdated computer equipment) and the civil servants unoccupied and uncommunicative.

Since the law providing departments with extra financial resources was passed, new autonomous territorial units have been created, especially outside Java and Bali, in places where departments were excessively large.[16] In ten years (1999–2009), Sumatra has grown from eight to ten provinces, from 52 departments to 116, and from 21 towns to 34. The average surface area of Sumatran departments has decreased from 8,800 to 3,200 km^2, which is still three times greater than in Java, and the average population per department decreased from 559,000 to 331,000 in 2009, i.e., 3.5 times less than in Java (Booth 2011, p. 36) (see Table 9.4). This is the major difficulty for development policies in the archipelago with, on the one hand, extreme concentration of population and economic activity on the islands of Java and Bali, and, on the other, a lack of real access to social infrastructures and little investment to generate economic activities in the other islands (Charras 1993).

The central part of Sumatra is concerned for more than one reason with the reshaping of its administrative grid with the gain of two new provinces (the two archipelagos on the eastern side[17]), 27 departments and eight towns, and the process of refining the grid continues.[18] Most of these creations are justified by a desire to bring infrastructures as close as possible to those under their jurisdiction.[19] These new units have often been created by communities which wish to stand out from their neighbours, on the basis of settlement catchment areas, based on former units of the colonial administration, which was unable to deploy its staff throughout this widespread archipelago, and therefore set up indirect government in conjunction with traditional local powers in the outlying islands.[20] This fragmentation leads to the deployment of urban development, since each department has to set up its capital, often a little rural town which will be enriched with new administrative functions and new social and economic infrastructures, thus increasing the supply of public services, and responsible for a major boom in the building sector in recent years.[21]

Tabel 9.4
Evolution of Sumatra Territorial Administration Grid, 1996–2005

	1996			2005		
Province	**Kabupaten**	**Average Area km²**	**Cities**	**Kabupaten**	**Average Area km²**	**Cities**
Aceh	8	6,924	2	17	3,258	4
North Sumatra	11	6,435	6	18	3,969	7
West Sumatra	8	6,222	6	12	3,519	7
Riau mainland	5	18,912	1	9	9,592	2
*Riau Islands**			1	4	1,994	2
Jambi	5	8,960	1	9	5,344	1
South Sumatra	8	12,961	1	10	9,719	4
*Bangka-Belitung**			1	6	2,695	1
Bengkulu	3	7,056	1	8	2,474	1
Lampung	6	5,551	1	8	4,721	2
Total	**54**	**9,032**	**21**	**101**	**4,829**	**31**

Note: * New provinces separated from mainland, Riau and Sumsel.
Sources: Charras (2005), Booth (2011).

Problems of shared responsibilities between the State and local units

The State is no longer responsible for development of the local level. Development planning must first be drawn up by cities and departments, then integrated into the province development plan. This radical and precipitate change has disorganized services concerned with development on both a national and a local level.[22] Although autonomous communities know what they want for their development, they do not have the necessary skills to implement their actions (e.g., planning, managing data). In fact, autonomy has led to a transfer of civil servants; managers, mostly well-educated Javanese, have returned to Jakarta to work in government agencies on a national level, or to their native regions. Also, databases are either non-existent or inaccurate; data were formerly collected and processed at a central level, via the province. In 2008, for example, civil servants in the city of Palembang still did not have a list of local firms, and had to comb the town to find them and collect data (status, jobs, financial reports) so they could raise taxes. The problem is the same for data concerning private plantations set up before 1999, whose contracts, surface areas and production are unknown to departmental authorities. Taxes and royalties were paid to the central government which originally provided the permits, with the department collecting only land tax. After concerted pressure on central government by the departments, a solution should be found as for dividends from the mining industry, which are still managed by the central powers, but now redistributed: to the department where the mine is located (6 per cent of the oil production and 12 per cent of the gaz), to the province (3 per cent and 6 per cent), and to be shared among other departments in the province lacking these resources (6 and 12 per cent).[23]

However, the State remains responsible for network infrastructures: inter-provincial roads, ports and airports, energy, communications ... Similarly, major investments must be approved by the National Agency of Investment (BKPN), especially foreign ones. Many projects have been subjected to feasibility studies by foreign investors but have not been approved by Jakarta, which fixes national priorities. A good example is the Sumsel seaport project (Tanjung Api-Api), to replace or to complement Palembang harbour whose possibilities for development are limited. Regional autonomy revived hopes[24] of restoring Palembang as a gateway for export, but it was not until 2010 that the creation of the harbour was given a place among priority projects. The province and the units concerned (Banyuasin and Palembang) anticipated the decision, and, using their own budget, started investing as an incentive package for foreign investors; in 2008, 80 per cent of land was freed, 70 kilometres of the access road and a coaling quay were built. Today, India, which imports coal, has proposed developing some of the harbour infrastructures, an industrial zone and the railway line to the mines in the interior.[25] Singapore is looking at the possibility of developing a container zone, possibly preparing to add a new feeder port to the international hub.[26]

Development Corridors: A Contrasted Perception

In its presentation of the development project for the Malacca straits region (IMT) in 2007, ADB presented a map of the corridor network, including:

the Singapore-Bangkok route, which relies on the high-quality motorway in the west of the Malay peninsula, the trans-Sumatran route from Aceh to Palembang, termed "Sumatra Eastern Economic Corridor", and two corridors crossing the straits between Penang and Medan and Melaka and Dumai, which would eventually be completed by a crossing further north, providing a link to southern Thailand. Since then, the keyword of meetings in the IMT economic zone has been "connectivity"; a vision and a motto which are not shared by everybody.

The reticence of the central government to open up Sumatran territory via its eastern side facing the Straits can be analysed by looking at the Indonesian participants' perception of economic corridors. The trans-Sumatran routes were planned, as we have seen, before the establishment of trans-border cooperation (IMT and IMS) and are seen by decision-makers in national planning only as national routes extending those crossing Java from West to East.

At no time was the importance of the three crossings of the Malacca straits, discussed untiringly at each IMT meeting, mentioned in Jakarta as a trump card for the development of the largest island. The term "development corridor" was not even mentioned until 2011, in the 2011–15 Master Plan (MP3EI).[27] However, since 2008, the "Sumatra Eastern Economic Corridor", has been mentioned locally in the development plans of certain departments crossed by the corridor (including Banyuasin, Palembang and Jambi). In 2009, a prospective study predicted that links across the Malacca straits would benefit the development of Sumatran regions as a whole (both coastal and interior): "The economic effects on the E&E industry in Medan and Dumai are outstanding ..." although "... surprisingly, Jakarta receives virtually no economic benefits" (ERIA 2009, p. 94). This interesting simulation was not adopted in the latest Master Plan, and if connectivity exists, it remains on a national scale. There is no doubt the last sentence ruffled the feelings of the authorities in Jakarta, which explains the reflection of a civil servant that there would be no connectivity with the exterior as long as Sumatra was not completely attached to Java, which appears to be the aim of the projected bridge across the Sunda straits which, according to investment plans, should swallow up about 80 per cent of future State-funded infrastructures in Sumatra (MP3EI, p. 72). However, on the ground, the vehicle loading quay for the ferry in Dumai was ready in 2008, but there were still no ferries in 2011. On the other shore, the state of Melaka is stalling; it does not want to build a quay in the city centre, in close proximity to a district classified World Heritage Site in 2008, but proposes using several piers on its coast already in use by the customs, immigration and quarantine services.[28]

The delay illustrates the difficulties encountered in setting up institutional software processed in both countries. On one side, Malaysia fears an influx of Indonesian migrants, and the arrival on the roads of vehicles without valid insurance; on the other, Indonesia is afraid that new or second-hand manufactured goods will flood Sumatra at the expense of those coming from Java, especially second-hand cars. An alternative to the ferries would be a 35-km-long bridge linking the two shores, starting on the islet of Rupat, envisaged by Riau province and the state of Melaka.[29] One thing is certain: although local governments are informed rapidly of major projects concerning their development, national agencies are in no hurry to remove obstacles.

Although the State is responsible for the maintenance of the eastern trans-Sumatra, it must be admitted that this national route does not operate in its entirety on a permanent basis, delaying coastal development, especially between Palembang, Jambi and Pekanbaru. Based on personal experience, the road between the first two towns was in very bad condition between 2002 and 2004. However, in 2008, this section was in perfect condition whereas sections to the north (Jambi-Riau) and south (Lampung to Palembang) were potholed, forcing north-south traffic to use both trans-island routes. Consequently, townships along the road between Palembang and Jambi[30] were unable to benefit from exchanges other than between these two regional capitals. Only the building of modern service stations anticipated an improvement in traffic; there were still only about a dozen restaurants, and no sign of any roadside hotels, either new or renovated. Interviews conducted on this section showed a certain satisfaction regarding the improvement of communications (more traffic, quick access to two large cities, less chaos and dust…), but also considerable doubt concerning the permanence of the situation; this anxiety was justified in view of the state of the roads observed three years later.[31] In 2008, firms in Jambi were not yet considering buying building materials from Palembang; they were still getting them from Jakarta.[32] According to an official in Jakarta, north-south traffic on the two trans-Sumatran roads is not yet sufficient to justify long-term investment in either of them, and it is by using alternative routes, paid for by local governments, that the north-south flow of traffic is maintained.

3. REAPPEARANCE OF TRANSNATIONAL CONNECTIONS

Since time immemorial, Sumatran populations have traded with those on the opposite shore. The establishment of frontiers in 1824, then the building of nations structured around their central zones, have not completely eliminated these cross-straits exchanges, and Sumatra's transnational tropism via its east coast continues to exist in other forms (Tagliacozzo 2007).

Transnational Coastal Areas Isolated from Mainland Infrastructures

Trans-straits relations exist for the coastal settlements of Sumatra as a whole, but they are essential for those separated from the land corridor by the swampy area which becomes wider from Riau and to Sumsel provinces. Apart from the Dumai-Pekanbaru road, opened by Caltex in the 1950s, other cross-routes are few and far-between, and often impassable.[33] The only alternative, therefore, is to get supplies from the Malay peninsula. These exchanges are performed by two types of actors. The network of businessmen (ship-owners and traders), mostly Sino-Malays and Sino-Indonesians, difficult to study but certainly dominant, use their fleet of wooden cargo vessels. Goods leaving Sumatra include wood (logs, scaffolding and charcoal), fish and agricultural produce (vegetables, livestock, palm oil), sand and other natural resources; from the Malay peninsula come consumer goods: food, medicines, building materials and manufactured goods, both new and second-hand. At the same time, without developing real commercial networks, Indonesian fishermen come in their frail fishing boats to sell fresh or dried fish[34] and return laden with manufactured goods, mostly second-hand, bought from shops located close to the piers. This is the principle of barter trade, requiring no shipping documents, and conducted without having to change any money.

Since 2004, with the easing of exchange rules in the economic cooperation zone, official barter trade piers with customs and immigration services have multiplied along the Malaysian coast in the federal ports (Penang and Klang), but also in small ports. On the Sumatran side, boats moor directly alongside the wooden quays of the coastal township's market where a customs agent keeps an eye on them. On both sides of the straits there is a multitude of places, inlets and bays, where illegal cargoes can be brought at night: human traffic, protected animal species, *kretek* cigarettes or subsidized products. This flow, substantial but non-quantified,[35] is very fluid and adaptable. Crossings or routes are changed, the sea enables boats to head for the best landing place according to the products, the best commercial contacts, and the compliance of the customs and immigration, which alternates with periods of increased vigilance.

At the end of 2008, faced with the strong development of this type of trade in new and second-hand goods — which now flooded not only the stalls of coastal townships but also the shopping centres of towns which were supposed to get their supplies from Java — the Indonesian government decided to reduce the number of import-export ports for electronic goods, textiles, foods and dairy products, retaining only Medan/Belawan in Sumatra, meaning that any goods entering at other points were considered to be smuggled. In Dumai, which had greatly benefited from the liberalization of trade, this decision sparked the anger of the local political elite, who threatened to block exports of oil and palm oil; Dumai was then added to the list of authorized ports.[36]

This border area may be considered to have remained typically transnational due to the flows crossing it and the actors working in it, whatever may be the central powers' wish to control it, only the status of trade classified as official or smuggling has changed. The straits is seen as a socio-economic space where the currency is the Malaysian Ringgit, ethnic Chinese communities are linked, Indonesian patients cross it to receive treatment, and young people go over to continue their studies.

Relations between the two national Melayu communities are complex and changing, and worthy of study in the present context.[37] In 2008, Sumatran Melayu were most frequently classed by Malaysian Melayu as Indonesian migrants performing 3-D jobs (dirty, difficult, dangerous), who for the past few years have been harassed and mistreated by the Malaysian government and society in general, whether or not their papers are in order.[38] Yet the majority of migrants are not Sumatrans, who do not traditionally do this type of job. Although symbolic exchanges are more frequent since decentralization and are organized by the Melayu aristocracy or political elite on both sides of the straits (Sakai 2009), Sumatrans lack trading partners among the Melayu on the opposite shore. In 2008, the students and the few Sumatran traders encountered in Melaka appreciated life in a modern city, but suffered from the ostracism of their own kind, and preferred the company or business of other communities.

With the ethnic Chinese community, only the Minangkabau can be considered to be transnational actors. Based on ancestral settlement processes (Andaya 2008, pp. 82–107), exchanges between the two sides have been constant and have become more varied. They are sufficient today to justify ten weekly flights, and include socio-cultural (education,[39] art, literature, religious studies, tourist visits), and economic exchanges in a great variety of activities (export of foodstuffs and handcrafted religious items from Sumatra). There is a strong Minangkabau presence on the Malay peninsula in the Muslim intellectual elite, the hotel and restaurant business and the textile trade, and even in industry. It is said that this community is the only one capable of rivalling the ethnic Chinese in the economic sector, as long as it is able to improve the development of its networks (Heryanto 2011).

The economic differential, Indonesia's closed-door policies, and the suspicions of disloyalty attributed to border populations[40] have all contributed to the passivity of actors on the Indonesian side, and the coastal areas are overshadowed by the Malay peninsula, and considered as an extension of its trading area.

Local Initiatives: Island Integration and Development of Regional Capitals

Regional autonomy has given impetus to initiatives foreshadowing future changes, even though the slowness of the processes involved has led some to doubt. Since 2001, all the governors of Sumatran provinces have met regularly to coordinate their views, their data and their efforts to improve the integration of development in the island (Charras 2005). Although there have been numerous projects, only a minimal number have so far been carried out, since governors do not have either the financial resources or the authority to carry with them their departments still focused on their own development, and not yet convinced of the merits of inter-community action, however necessary this may be.[41] The creation of a Sumatra Promotion Center in Batam is gradually imposing a Sumatran commerce label. Since 2000, thanks to pressure from governors, and supported by the IMT-GT, Padang, Palembang and Pekanbaru airports have been allowed to run direct flights to other countries in the straits (see Table 9.5).

Autonomy, economic cooperation zones and development corridors, linked to an opening-up towards the exterior have led to the development of regional capitals. Pekanbaru was the first to benefit from this process, as the administrative capital of the Riau archipelago, and then thanks to its own resources (oil and gas, forests and large plantations) and pivotal position in the central part of Sumatra, providing the link between the Minangkabau area on the west coast and the Malay peninsula (cf. M. Franck in this volume).

A similar process of internationalization is under way in Palembang, speeded up by the organization of the 2004 National Games with the creation of sports facilities, but also the modernization of the airport, and the building

Table 9.5
Direct International Flights In/Out Central Sumatra, June 2011[42]

		Weekly Number Flight	Companies
Padang	Kuala Lumpur	10	Air Asia (Malaysia)
Pekanbaru	Kuala Lumpur	7	Air Asia
Pekanbaru	Melaka	5	Riau airline
Palembang	Kuala Lumpur	3	Air Asia
Pekanbaru	Singapore	3	Silkair (Singapore)
Palembang	Singapore	3	Silkair

Note: Only stable links are retained as others open by low-cost companies are often cancelled.
Source: From internet search, September 2011.

of the first international class hotel (Novotel) and a hypermarket (Carrefour). In 2008, Palembang won the Nation's Cleanest Major City Award. In 2010, the new harbour finally featured in the list of national priorities. The city and the province have increased the number of national, or even international events organized,[43] and hotel infrastructures have multiplied. In 2011, Palembang will be the joint main host of the 26th Southeast Asian Games with Jakarta.[44] After twice unsuccessfully approaching the developer Ciputra, the city's mayor was able to persuade him in 2008 to invest in a 200-hectare complex twelve kilometres from the city centre, Citra Grand City,[45] which, apart from four standard housing areas, will include shopping centres, a water park and a golf course. Apparently, house buyers come not only from the city, but also from the rest of the province and from neighbouring provinces — Bengkulu and Jambi — and even include Jakartanese originally from south Sumatra who now have confidence in the city's future. The SEA Games and Ciputra's investments have announced the start of major projects such as the development of the Musi river waterfront in the city's historic centre. Palembang's new motto is "Goes to international city by 2013; a free medical, educational and wifi city". The reason behind this new development of Palembang is mainly that, for the first time since regional autonomy, the city was able to resume exchanges with its historical partner, Singapore, via a thrice-weekly airline connection.

CONCLUSION: THE FUTURE CHALLENGE — CONNECTING NATIONAL INTEGRATION AND TRANSNATIONAL LINKS

Until 1999, the centre of Sumatra was a region of natural resources on a national scale, a reserve for agricultural colonization from Java, and for large privately-owned plantations since 1986, preceding or concomitant with the exploitation of forest resources carried out under the care of the politico-economic elite of central government. The region provided a major part of exports of natural resources and plantation products, which as soon as they were processed, left for Java via trans-island roads. In spite of the existence of economic cooperation zones, the means to play the transnational game were limited to the Batam archipelago which polarized industrial investments and flows of migrants to the detriment of the development of major Sumatran cities — Medan, Palembang and Padang — which stagnated. Only Pekanbaru, the provincial capital of Batam, derived some indirect benefit from its activity. The only international gate on the east coast was Medan (airport and Belawan seaport), apart from a few specialized ports such as Dumai.

Research conducted in 2008, completed by the analysis of the latest Master Plan, shows that changes in central Sumatran space have eluded the analysis of the central planning authorities for whom the organization of Sumatra is still divided into three parts: a northern part (Aceh and North Sumatra provinces) organized around and from Medan, a southern part (Bengkulu, South Sumatra and Lampung) strong attracted to Java, and a depressed central part (Riau, Jambi and West Sumatra, with the latter revolving round two poles, Medan in the north and Java in the south), without taking into account its eastern tropism.[46] The 2011–2025 Master Plan confirms the island's function as a reserve of natural resources (plantations, energy) for the development of Java, by giving exclusive priority to the connectivity of national routes.

The centre's perception of the coastal and mainland border zones has remained frozen through fear of centrifugal forces; these are the areas where dividends escape the central authorities, and the loyalty to the Indonesian state of local populations, who have cultural affinities on both sides, is also questioned. The external coasts of the national territory may be considered as maritime frontiers (Tagliacozzo 2005; 2007) , and although some of them face the immensity of the Indian and Pacific Oceans, in other areas, such as East Sumatra, the frontier represents a significant reality since neighbouring countries are relatively close; the inland seas contribute to regional bonding, although this situation has not yet been considered as an advantage to fit into the network of sea-borne exchanges, and thus promote local and regional development.[47] For the planners, in the Malacca straits region, only Batam is considered as a frontier region; an "almost terrestrial" frontier, a mere twenty-five to thirty-minute trip away on one of the numerous ferries making the crossing to Singapore or the south of the Malay peninsula.

At the time when the great privately-owned plantations were being developed at the end of the nineteenth century, advantages and constraints between the north-east of Sumatra and the north-west of the Malay peninsula were fairly similar. During the Japanese occupation, and up to 1943, Sumatra and the peninsula were both under the same territorial command. Since then, enormous progress has been made in the development of the west coast of the peninsula, its harbour system (Nathalie Fau in this volume), and its road and urban network — it is true that this is the central zone of the Federation, with a nearly complete land route to the regional hub of Singapore — but what a contrast with the under-development of Sumatra, the delay in the spatial integration of the island, and the lack of openness towards the exterior! The Sumatran paradox is the perfect example of the introversion and hyper-centralization of Indonesian regional development policies.

Sumatra has been at the intersection of several exogenous spatial dynamics. The most active of these today is dominated by the Indonesian capital, which induces a centripetal flow within national limits (see Map 9.3). It came into existence thanks to the first trans-Sumatra passing through the centre, which has not favoured the spatial integration of the large island. The other exogenous dynamic is dominated by Singapore and the west coast of the Malay peninsula, which, because of transnational heritage, is now giving rise to transversal flows, still slight due to the reticence of the Indonesian state. The second trans-Sumatra road, running along the east coast is also, similarly, seen as a route leading to Java, but, by linking provincial capitals, it can reactivate areas formerly turned towards transversal dynamics, by taking advantage of the degree of autonomy given by decentralization.

Beyond crossings of the straits, we should also see the East-West crossing of the meridian strips of the island's relief by relying on feeder links created after the central trans-Sumatran,[48] thus disenclosing the west coast on the Indian Ocean which would not be served by the completion of the third trans-Sumatra alone. The resilience of East-West relations provides hope for regional integration in the straits area including the whole of the island of Sumatra via active trinomials: Penang/Medan-Belawan/Sibolga, Melaka/Pekanbaru-Dumai/Padang, Johor/Jambi/Kerinci, Singapore/Palembang/Bengkulu, as predicted in the ERIA study. To achieve this, Indonesia will have to break with its deep-rooted centralizing heritage, and at a local level, departments and towns will have to understand the advantages of inter-communality under the guidance of their provinces.

Map 9.3
Sumatra: Integration and Connectivity Process

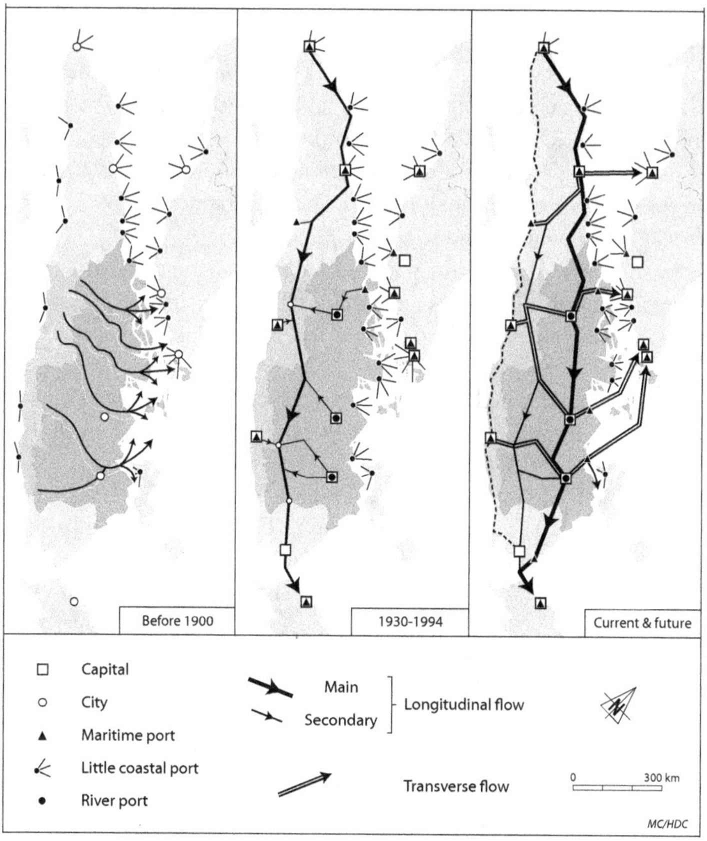

Source: The author.

Notes

1 The island, oblong in shape, is oriented North-West/South-East, with the Barisan mountain range running from North to South, structuring the relief in meridian strips with a fairly steep western slope and an eastern slope descending gently to lowlands on the Sunda rock shelf, where the rivers are often navigable until the piedmont. The island's soils are rich in natural resources (gold and silver, coal, oil and gas), and, in this very humid part of Indonesia, forestry has gradually given way more and more to perennial crops, first grown within abandoned swidden fields, then in large plantations (pepper, rubber, coffee, tobacco, tea, sugarcane, oil palm).

2 GDP per capita in 2008: Indonesia around US$2,000, Malaysia US$7,000 and Singapore US$35,000.

3 The analysis is based on many years of observation, mostly in Sumsel province, updated by a two-month mission in 2008, more specifically oriented towards the theme of the Transiter programme, financed by the ANR (French Agency for Research) in Sumsel, Jambi and Riau provinces and in the state of Melaka in Malaysia. The study is greatly indebted to the problems exposed in 2003 by Nathalie Fau in her unpublished thesis on the northern part of Sumatra (Aceh, North Sumatra, Riau and West Sumatra).

4 Riau is the last Sumatran reserve for transmigration, receiving transmigrants from Aceh, relocated for safety reasons, and the grandchildren of transmigrants from Lampung, which is now densely populated.

5 The west of the province is a special case: the Minangkabau, traditionally migrants, are settling in other regions of Sumatra, in the Malay peninsula, Singapore and Jakarta. The province has always had a negative migratory balance.

6 Many studies are devoted to the extension of plantation areas and the situation of the peasant community (McCarthy 2010; Potter and Badcock 2001; Levang 2003).

7 Sinar Mas (-PT Indah Kiat Pulp & Paper in Riau and PT Lontar Papyrus in Jambi) and the Raja Garuda Mas Group (PT Riau Andalan Pulp & Paper in Riau). In 1999, Riau mainland had 2 functioning pulp mills, 15 plywood mills, 3 chip mills, 27 moulding factories and 345 legal sawmills (Potter and Badcock 2001, p. 10). Concessions for the pulp industry covered 270,000 hectares in Jambi and 300,000 hectares in Sumsel developed by Tanjung Enim Lesatari from the Barito Pacific group, a holding company owned by Soeharto's eldest daughter.

8 The first three groups have long been settled along the coastal strip. The Chinese have engaged in the manufacture and sale of charcoal, gambier and other products in Singapore since the nineteenth century. The Banjar (South Kalimantan) and the Bugis (South Sulawesi) were an example of spontaneous agricultural migration, and exploit the coastal swampland (rice, coconuts), the Batak arrived more recently with the opening of the link road to Medan.

9 Touwen (2001, pp. 73–75) classifies Palembang among the places whose dynamism was qualified as local "indigenous performers" at the end of the Dutch colonial period. Goods were exported from the port, some directly to Singapore or Europe, and some to Java. See also Oki (1986) and Colombijn (2005) for the transportation history in the middle part of Sumatra.

10 The refinery and the fertilizer factory are national firms, whose products are transported by sea in special ships with very little impact on the rest of the activity.

11 Up to now, we have mainly seen an "archipelagic" urban. The one in the north is well organized, starting with Medan, and including Banda Aceh and a series of medium-sized towns (Fau 2003). Elsewhere, each province had its own urban system, organized from the capital on the coast towards the interior, without any real connection to the urban systems of neighbouring provinces.

12 Acronym for the three components of the Singapore, Johor and Riau triangle for Indonesia. When it was created, the island of Batam was part of the Riau archipelago department (KEPRI) of Riau province, until 2004, when the new province was created.

13 Today this exit tax has been abolished for all Indonesians who have paid their taxes, including Sumatrans.

14 The law was adjusted in 2004 and gives the provinces a certain coordinating power, but its effects have not yet been felt on account of the reticence of departments and towns, which are determined to keep their recently acquired autonomous status, and see it as a threat of re-centralization.

15 The governors were often Javanese (in 1984, in Jambi, Riau, Bengkulu and West Sumatra) and/or military (in 1984 in Riau and Sumsel) (Cribb <http://www.indonesianhistory.info/map/dwifungsi84.html>).

16 In total, for the whole of Indonesia, the number of second-level territorial units (town and department) increased from 290 (61+229) in 1995 to 495 (98+397) in 2009 (Booth 2011, p. 32).

17 These two creations can be explained, among other things, by their specific spatial profiles which were not taken into account by capitals situated on the mainland developed as a priority (Sakai 2002).

18 After a moratorium on new creations to organize the 2009 elections and parliamentary representation (national and regional), there is a national waiting list of over 155 departments.

19 The size and borders of French departments were determined in 1799 by the time taken by an administrator on horseback to travel from his head office to those under his jurisdiction and return in one day; the same applied to the population's access to services.

20 A return to cultural basics which worries many observers.

21 As an example, in Sumsel province, the former department of Ogan Komering Ulu (OKU), whose prefecture was Baturaja, has been divided into three; one part has retained the old prefecture, the other two departments have each created a new prefecture (Muara Dua and Martapura).

22 Since 2008, there have been a number of preparatory studies for reforms in different sectors, from national to local level (ports, transport, energy), made necessary by the regional autonomy laws, under the guidance of national development aid agencies.

23 The three provinces of central Sumatra are concerned by oil and gas dividends. They supply gas to industries and residential customers in Batam and Singapore via a network of undersea pipelines (Duri-Jabung-Grissik block). Jambi now requests that at least 15 per cent of its production be used locally.

24 The site, situated on the Musi delta, had been studied by the Dutch shortly before the Second World War, but no action was taken. To go further in the analysis of the strategic positions in the long run of the ports in Greater Southeast Asia, see Tagliacozzo (2007).

25 "Sumsel Realisasikan Proyek Tanjung Api-Api" [Sumsel resumes its project TAAJ] mediaindonesia.com, 26 January 2011.

26 Palembang exports to Singapore were twice as great (US$450 million) in 2006 as those from Medan, but mostly concerned transhipment for a third country.

27 *Masterplan for Acceleration and Expansion of Indonesia Economic Development 2011–2025* (Mp3ei), Ministry for Economic Affairs, Jakarta, 2011. Available at <http://www.ekon.go.id/media/filemanager/2011/05/27/p/d/pdf_mp3ei.pdf>.

28 There are plans for a barter trade quay at the mouth of the Sungei Linggi which is on the border with the state of Negeri Sembilan and another at Tanjung Kling/Bruas which has a better draught and is already used by cruise ships.

29 A project which would be in strong competition with the Ro-Ro crossing, an estimated six hours, compared with an hour-and-a-half by passenger ferry and less than an hour by air from Pekanbaru to Melaka.

30 In the Sumsel section, the stopover towns of Pangkalan Balai (which has grown since it became the prefecture of a new department), Betung (at the intersection with a road heading into the interior, with a renovated bus station), Sungai Lilin, Sri Gunung and Bayung Lincir (three townships which could benefit from the trunk road). From the provincial border with Jambi the capital can be reached in less than half an hour, without any real towns in between, but with long-established settlements along the road.

31 On the section in good condition in October 2008, the article entitled "Stuck in Congestion and Crumbled Roads in Jalintim", *Kompas* (Indonesia), 19 April 2011, points out that "Damages have also occurred on the roads from Pangkalan Belai to Betung and a little bit in Sungai Lilin"; and, according to a survey carried out among lorry drivers "Conditions in Sumatera's eastern highway … in the region of Riau, are indeed very lamentable. Once entering the territory of Riau Province from the direction of Jambi Province, there are only two choices of roads: bad or very bad".

32 Or maybe directly from the other side, from Malaysia; this possibility was not mentioned in interviews.

33 In Riau, a road leads up to the departure point for the island of Bengkalis, and another one under construction will go to Kuala Enok. In Jambi, a road goes to Kuala Tungkal port for the passenger ferry to Batam, and another is being built to Muara Sebak, which is set to become the seaport for the province, using infrastructures developed by PetroChina which exploits the oilfields.

34 Sometimes other products, such as young coconut fronds or coconuts for Muslim festivals. The fishermen wait for customers near their boats, without using middlemen who buy their cargo and sell it on.

35 Malaysia and Singapore are in favour of these exchanges, which are to their advantage, but do not publish import/export statistics for trade with Indonesia.

36 A parliamentary candidate told us "I was raised on milk from Malaysia, and I expect my grandchildren to be as well, not on cartons of Indomilk" (November 2008).

[37] To understand the historic building of identities as well as the settlement process in the Malacca straits region, see the work by Leonard Andaya (2008). The journal Bijdragen tot de Taal-, Land-en Volkenkunde published a special issue on the notion of Malayness in 1997 (pp. 153–54). Minako Sakai (2009) gives a lucid analysis of the revival of Melayu identity since decentralization in the four cities of central Sumatra. Note also the articles by Lian Kwen Fee (2001) and Nathalie Fau (2000, 2009), which analyse identity across nations.

[38] Suryadi (2005) analyses the history of relations between the two communities (pp. 140–44) from the Soeharto period until the first years of Reform. His survey done in Riau in 2002 emphasizes the fact that peninsular Malays despise and reject the poor, underdeveloped Melayu Riau.

[39] After the most recent earthquake in Padang, Malaysia immediately chartered a plane to bring home the 287 Malaysian students. A common Internet teaching project is being studied.

[40] Especially because Riau province threatened to leave the republic in 2002. Taking into account the extent to which land has been appropriated in the province, protests against centralization in Java have been among the strongest in the country, after East Timor, Aceh and Papua.

[41] Today, for example, almost all coastal departments aim at building a port with international status.

[42] Only regular links have been counted, others have been opened by low-cost companies but are often cancelled.

[43] These include events linked to the Pan-Malay community, which launched the first conference on Malay and Islamic World (2001) in Palembang, which claims to be the historic ancestor of Malay political entities. This movement has gained all the other provinces of Sumatra (Sakai 2009 and M. Franck in this volume).

[44] A total of 296 events will be held in Palembang (compared with 266 in Java), including the opening and closing ceremonies of the games.

[45] "Eddy Santana: Ciputra Tak Menduga Potensi Palembang" [Eddy Santana: Ciputra did not believe in Palembang's potential], "Penjualan Rumah di Citra Grand City Palembang Melesat" [House sales in Grand City Palembang explode], *Kompas* (Indonesia), 11 February 2011.

[46] Whereas, in 2003, the western province planned and built an improvement of the tricky area on the cross route in the direction of Pekanbaru in order to remove a major obstacle to the daily traffic flow of some 6,000 vehicles, an enormous project, almost finished today, using its recurrent budget. "Kelok Sembilan Selesai 80 persen" [The nine hairpin bends are 80 per cent finished], *Padang Ekspress* (Indonesia), 18 June 2011. Available at <http://padang-today.com/index?php?mod=berita&today=detil&id=29285> (accessed July 2011).

[47] For the history of the frontiers, see Eric Tagliacozzo (2005 and 2007). Indonesia has only two international ports on its periphery: Medan and Bitung (North

Sulawesi), the other three are situated on the inland Java sea (Jakarta, Surabaya and Makassar).

48 Pekanbaru-Bukittingi-Padang; Jambi-Sarolangun-Kerinci; Palembang-Lubuk-Linggau-Bengkulu.

References

Andaya, Leonard Y. *Leaves of the Same Tree: Trade and Ethnicity in the Straits of Melaka*. Honolulu: University of Hawaii Press, 2008.

Asia Development Bank. *IMT-GT, Building a Dynamic Future; A Road Map for Development 2007–2011*. IMT-GT Secretariat and Asian Development Bank, 2007.

Barr, Christopher. *Banking on Sustainability: Structural Adjustment and Forestry Reform in Post-Suharto Indonesia*. Washington, D.C.: Center for International Forestry Research (CIFOR) and WWF Macroeconomic Program Office, 2001.

Booth, Anne. "Splitting, Splitting and Splitting Again: A Brief History of the Development of Regional Government in Indonesia since Independence". *Bijdragen tot de Tard-, Land- en Volkankunde* (BKI) 167, no. 1 (2011): 31–59.

Charras, Muriel. "L'Indonésie extérieure en peine de perspectives" [Outer Indonesia left in need of prospects]. *Archipel* 46 (1993): 173–90.

———. "The Reshaping of the Indonesian Archipelago after 50 Years of Regional Imbalance". In *Regional Autonomy in a Multi-Cultural Indonesia*, edited by M. Erb, C. Faucher and P. Sulistiyanto. London & New York: Routledge Curzon, 2005.

Charras, Muriel and Manuelle Franck. "Quarante ans d'introversion de l'Indonésie; l'éclipse de toute une région. L'ASE insulaire dans les recompositions spatiales» [Forty years of Indonesian introversion: the fading of a whole region. Spatial reconfigurations of Maritime Southeast Asia]. In *Indonésie contemporaine, Hommage à Jacques Leclerc* [Contemporary Indonesia, a tribute to Jacques Leclerc], edited by F. Cayrac-Blanchard, G. Defert and F. Durand. Paris: L'Harmattan, 2000.

———. "Les configurations et recompositions spatiales dans l'Asie du Sud-Est insulaire contemporaine" [The Configuration and Spatial Recomposition in Contemporary Insular Southeast Asia]. In *Intégrations régionales en Asie orientale* [Regional integration in Oriental Asia]. Paris: Les Indes Savantes, 2004.

Charras, Muriel and Marc Pain, eds. *Spontaneous Settlements in Indonesia: Agricultural pioneers in Southern Sumatra*. Paris: ORSTOM, 1993.

Colombijn, Freek. "A Wild West Frontier on Sumatras East Coast: The Pekanbaru-Dumai Road". In *On the Road: The Social Impact of New Roads in Southeast Asia. Bijdragen tot de Taal-, Land- en Volkenkunde* (BKI), Leiden, 158, no. 4 (2002): 743–68.

———. "A Moving History of Middle Sumatra, 1600–1870". *Modern Asian Studies* 39, no. 1 (2005): 1–38.

Cribb, Robert. *Digital Atlas of Indonesian History*. Nordic Institute of Asian Studies (NIAS). Available at <http://www.indonesianhistory.info/>.

ERIA, Economic Institute for ASEAN and East Asia. *Geographical Simulation Analysis for Logistic Enhancement*. Research Project Report, nos. 7–2 (March 2009). Available at <http://www.eria.org/research/y2009-no7-2.html> (accessed July 2011).

Fau, Nathalie. "D'une rive à l'autre: les migrations entre l'Indonésie et la Malaisie dans le détroit de Malacca" [From one shore to the other: migration between Indonesia and Malaysia in the Malacca Straits]. *Revue Européenne des Migrations Internationales* 16, no. 1 (2000): 151–83.

———. *Le nord de Sumatra : une périphérie indonésienne sur le détroit de Malacca. Un espace partagé entre intégration régionale et recompositions transnationales* [Northern Sumatra: an Indonesian periphery on the Malacca straits. A shared space between regional integration and transnational reconstruction]. Ph.D. dissertation, Department of Geography, Paris 10, Nanterre University, 2003.

———. "Être malais de part et d'autre du détroit de Malacca: approche transfrontalière" [To be Malay on the two shores of the Malaka Straits: Transborder approach. *Aséanie* (June 2009): 107–44.

Herayanto, Hery. "The overview of Cultural Background of Minangkabau and Chinese Customers in Padang West Sumatra". *International Journal of Lean Thinking* 2, no. 1 (June 2011).

Kato, Tsuyoshi. "The Localization of Kuantan in Indonesia: From Minangkabau Frontier to a Riau Administrative District". In *Bijdragen tot de Taal-, Land- en Volkenkunde* (BKI), Leiden, 153, no. 4 (1997): 737–63.

Levang, Patrice. *La terre d'en face: La transmigration en Indonésie* [The Land Beyond: Transmigration in Indonesia]. Paris: IRD, 1998. [Translation: *Ayoh ke Tanah Seberang, Transmigrasi di Indonesia*. Jakarta: KPG Gramedia, 2003.]

Lian Kwen Fee. "The Construction of Malay Identity Across Nations Malaysia, Singapore, and Indonesia". In *Bijdragen tot de Taal-, Land- en Volkenkunde* (BKI), Leiden, 157, no. 4 (2001): 861–79.

McCarthy, John. "Shifting Resource Entitlements and Governance Reform during the Agrarian Transition in Sumatran, Indonesia". *Journal of Legal Pluralism* 55 (2007): 95–121.

———. "Oil Palm: Boom or Ruin for RI Farmers?". *Jakarta Post*, 19 November 2010.

Oki, Akira. "The River Trade in Central and South Sumatra in the 19th Century". In *Environment, Agriculture and Society in the Malay World*, edited by T. Kato, M. Lutfi and N. Maeda. Kyoto: Center for Southeast Asian Studies, Kyoto University, 1986.

Phelps, N.A. "Archetype for an Archipelago? Batam as Anti-model and Model of Industrialization in Reformasi Indonesia". *Progress in Development Studies*, no. 4 (2004): 206–29.

Potter, Lesley and Simon Badcock. *The Effects of Indonesia's Decentralisation on Forests and Estate Crops in Riau Province*. Jakarta, Indonesia: CIFOR, 2001.

Sakai, Minako. "Resisting the mainland: The formation of the province of the Bangka- Belitung (Babel)". In *Autonomy and Disintegration in Indonesia*, edited by Damien Kingsbury and Harry Aveling. London: Routledge Curzon, 2002.

———. "Creating a New Center in the Periphery of Indonesia". In *The Politics of the Periphery in Indonesia: Social and Geographical Aspects*, edited by Minako Sakai, G. Banks and J.H. Walker. Singapore: NUS Press, 2009.

Suryadi. "Identity, Media and the Margins: Radio in Pekanbaru, Riau (Indonesia)". *Journal of Southeast Asian Studies*, vol. 36, no. 1 (2005): 131–51.

Tagliacozzo, Eric. "An Urban Ocean: Notes on the Historical Evolution of Coastal Cities in Greater". *Journal of Urban History* 33 (2007): 911–93.

———. *Secret Trades, Porous Borders: Smuggling and States Along a Southeast Asian Frontier, 1865–1915*. New Haven: Yale University Press, 2005.

10

DRY PORTS POLICY AND THE ECONOMIC INTEGRATION PROCESS ON THE WESTERN CORRIDOR OF PENINSULAR MALAYSIA

Nazery Khalid

INTRODUCTION: MALAYSIA, A MARITIME NATION AT THE HEART OF A VIBRANT ECONOMIC REGION

Malaysia's economy, one of the brightest shining stars in the developing world, has undergone rapid transformation from being a commodities-based economy in the 1970s to a manufacturing-driven, export-oriented one today. A major producer of electronics and electrical goods, Malaysia was ranked the 8th largest trading nation in the world in 2009 by the World Trade Organization (WTO).[1] It recorded a total trade value of RM1.169 trillion in 2010 and its external trade balance was at a surplus of RM110.1 billion.[2]

As a nation surrounded by seas, it is not surprising that Malaysia depends heavily on seaborne transport to facilitate its trade. An estimated 95 per cent of the nation's trade is carried in whole or in part by maritime means.[3] Hence, the importance of its seaports and other supporting maritime trade infrastructures to the nation's economic well-being cannot be overstated.

Given this reliance, it should not come as a surprise that Malaysia was ranked 23rd in the list of the world's top maritime nations based on its registered dead weight merchant shipping tonnage.[4] The country contributed 10.25 million DWT[5] or 0.8 per cent of the world total merchant shipping tonnage as of 1 January 2010 and 0.8 per cent to the world total merchandise trade in terms of value.[6]

Underscoring Malaysia's clout as a maritime nation, it boasts two of the world's top twenty busiest container ports, namely Port of Tanjung Pelepas (PTP), the nation's transshipment hub, and Port Klang, the national load centre. Port Klang handled 8.87 million TEU[7] while PTP handled 6.53 million TEU in 2010,[8] and were ranked 14th and 18th, respectively, in the list of top twenty busiest container ports by throughput handled.[9] Johor Port, located in the southern tip of Peninsular Malaysia, is the world's largest palm oil export terminal, while Bintulu Port in the state of Sarawak on Borneo Island is the world's largest LNG export terminal. The ports of Kemaman and Kuantan on the east coast of Peninsular Malaysia serve the petrochemical and oil and gas industries there that feature major international names such as BP, Shell, Dow Chemical, BASF, Toray and Idemitsu, and the national oil company, PETRONAS.

These ports, servicing major economic corridors in Peninsular Malaysia created to boost economic growth and generate value-adding economic activities, are known for world-class infrastructures and services offered at competitive rates. Malaysia can also be proud of its national carrier, MISC, which has the world's largest LNG fleet and is a major global name in energy transportation. In addition, several Malaysian shipyards are able to build offshore support vessels, passenger vessels and leisure crafts for export markets.

1. TRANSPORT SYSTEM AND DRY PORTS IN PENINSULAR MALAYSIA

Transport and Spatial Layout of Peninsular Malaysia

The network structures, population distribution and economic activities, hence transport infrastructure development in Peninsular Malaysia, are influenced by its topography. This explains why the major ports, airports and highways in the Peninsula are located mainly along its western corridor which hosts most of the Peninsula's population and key economic centres. Transport development in the Peninsula is largely determined by site and suitability of location based on population distribution and economic activities. Their development reflects linkages and interaction among the population, infrastructures and economic activities.

Peninsular Malaysia is bordered by the Straits of Malacca, one of the world's busiest straits used for international navigation. By virtue of its location linking the West and East trade lanes that are critical to global trade, this feature lends the sea-lane enormous strategic importance to the littoral states and the international community. Over 76,000 vessel movements were recorded in the Straits in 2010;[10] and an estimated 30 per cent of world trade and 50 per cent of world energy are transported through it annually.[11]

To the East Coast is the South China Sea, a vast semi-enclosed sea that separates the Peninsula and the other half of the country, formerly unofficially referred to as East Malaysia, which contains the vast, resources-rich but under-populated states of Sabah and Sarawak on the island of Borneo.

A mountain range, Banjaran Titiawangsa, of meridian direction separates the two coasts from the peninsula. These two coasts feature some of Malaysia's most developed transport systems and cities. The Western coast has a multimodal corridor of transport organized around three major population and economic areas: namely the northern area around the state of Penang, the central area around Kuala Lumpur and the southern area around the city of Johor Bahru, Peninsular Malaysia's southernmost city bordering Singapore.

The Eastern coast is also served by a road system skirting the coast but the area is sparsely populated and includes only a few economic centres of importance. These areas are the oil and gas cluster along the coast of the state of Terengganu, the petrochemical cluster in Gebeng in the state of Pahang and near its capital Kuantan, and the bustling and retail-driven city of Kota Baru, the capital of the state of Kelantan.

Cost is also a crucial factor in determining what transport infrastructure is built, and where, when and how it is built. In terms of the agglomeration of transport development in the Peninsula, more development takes place in key economic zones, namely the Klang Valley. The specialization of activities also affects developments in the Peninsula. For example, development on the East Coast of the Peninsula is driven by the oil, gas and petrochemical industries. To cater to these industries, infrastructures such as ports, tank farms, refineries and plants are built.

Just as important as the factors mentioned above is the historical influence on transport development in Peninsular Malaysia. For example, the British built the nation's first railway line from Taiping to Port Weld to facilitate the shipment of tin ore, mined mainly in the Kinta Valley region, to the United Kingdom through Port Klang, the country's major gateway port.

This port, built to facilitate the export of raw materials and commodities from the country, has left a legacy that can be seen until today, in terms of its roles and layout.

Inland Ports in Peninsular Malaysia

In Peninsular Malaysia, inland cargo terminals, also known as inland ports or "dry ports",[12] play a significant role in providing access for manufacturers and producers in the hinterland to seaports which act as gateways to the nation's trade. They were built to promote economic development and encourage the growth of industrial areas in the hinterlands to complement the role of seaports.

The transit points provided by inland ports enable the exporters and importers in the nation's hinterlands to connect with the nearest transshipment seaports. They also provide linkages between shippers/exporters and end-users of products/commodities, hence increasing the attraction and feasibility of investing in industrial areas.

The roles of inland ports include: providing transit points/land bridges for cargos en route to/from seaports, providing linkages between shippers/exporters and end-users of products/commodities, acting as conduit between production/industrial centres and transport/distribution gateways, helping promote modal shift of cargo, intermodal connectivity and integration of transport nodes and helping to boost socio-economic development in the hinterlands they serve.

In the Eastern part of Malaysia on the island of Borneo, which includes the less developed states of Sabah and Sarawak, inland ports cater to the international trade between Malaysia and Indonesia along the latter's Kalimantan border. The inland port of Tebedu in the state of Sarawak not only facilitates the movement of goods across the borders but also people, hence contributing to the promotion of stronger bilateral ties between Malaysia and Indonesia.

No two inland ports in Peninsular Malaysia are alike, but they share some commonalities which include: being located in populous, or reasonably populous, areas and close to industrial/economic zones; being well-connected with the national highway and railway grids, and major seaports and airports; having mixed ownership by state agencies and private companies, including foreign companies (see Map 12.1 in Farhat contribution).

All the inland ports in Peninsular Malaysia are designated landing points for exports and imports, and are recognized by Royal Malaysian

Customs as seaports. At these ports, manifest documents can be processed, hence affording shippers cost and time savings without having to do so at the major seaports. In this manner, the inland ports play a complementary role to the nations' seaports.

Most of these inland ports feed cargos to Port Klang, Peninsular Malaysia's and the nation's largest seaport. It is also the national load centre, a designation accorded to it by the Malaysian Government in 1993 to create a critical mass of cargos and spur the development of a maritime cluster in Malaysia. The port is also well connected by roads and rail and not far from the major airports. Most importantly, it lies along the coast of the Straits of Malacca, one of the busiest shipping lanes in the world.

The choice of localization of the inland ports hinges upon their location, the surrounding infrastructures and the economic activities in the respective hinterlands they serve. For example, Prai Inland Container Depot lies within the Prai Industrial Estate which hosts many multinational electrical and electronics companies, and is in close proximity to Penang Port. Ipoh Cargo Terminal is built to tap cargos from both the northern part of Peninsular Malaysia and the Central and Southern corridors, and is also situated at a railway-highway intersection. Nilai Inland Port enjoys the advantage of being located near the border of Selangor, Malaysia's most industrialized state, and Negeri Sembilan which is a fast-developing state.

Given the load centring policy at Port Klang and the consideration of the surroundings in deciding the locations of inland ports in Peninsular Malaysia, it should not come as a surprise that there are no inland ports on the East Coast. This side of the Peninsula is sparsely populated, less developed, does not feature many industrial activities that can generate containerized cargos, and does not enjoy the more advanced trade infrastructures of the western part of the Peninsula.

It is interesting to observe the multiple effects that some of these inland ports generate. In the case of Prai Inland Container Depot, it was built in 1984 around the same time that the state of Penang, that hosts the inland port, rose as an industrial powerhouse. There is no doubt that the presence of the port, which provides a link between the industrial cluster in the Prai area and Penang Port, contributed to the growth of the industries and the area as a whole. The inland port had the same effect on the Free Industrial Zone in Bayan Lepas in the southeastern part of Penang island, the state's main industrial hub. In the case of Sungei Way Inland Clearance Depot, which was built in 1973, its development was almost simultaneous with the beginning of the industrialization of Malaysia's economy, especially the Klang Valley area. As the process grew rapidly, other inland ports mushroomed further south, such as in the states of Negeri Sembilan (Nilai Inland Port) and Johor (Batu Pahat Container Terminal and Segamat Inalnd Port), thanks to the domino effect of economic growth in the western part of Peninsular Malaysia.

Among the key activities carried out at the inland ports in Peninsular Malaysia are: storage of cargos in shed and open storage facilities, receiving, sorting and delivery of conventional/bulk cargos, packing and unpacking of containers transiting between borders, warehousing facilities for loading, unloading, packing and unpacking of cargos, packing and unpacking of containers and on-dock depot services including maintenance and repair of containers.

Inland ports are more than just facilities where cargos are loaded and unloaded. If planned, operated and managed well, they can play multi-dimensional roles beyond economics. From an environmental standpoint, inland ports play a significant role in reducing reliance on road transport. Well-planned inland ports can promote sustainable development by encouraging a modal shift of cargo transportation from road to rail which is less polluting than its road counterpart. When operated under the principles of sustainable development, they can contribute to efforts to reduce carbon footprints in the trade transport sector. With Malaysia committed to reduce its greenhouse gas emissions by up to 40 per cent by 2020 (from 2005) levels, its inland ports are expected to play a greater role in promoting green cargo transportation and help meet the national target for emission reduction.

2. TYPOLOGY OF INLAND PORTS IN PENINSULAR MALAYSIA

The key features of the inland ports in Peninsular Malaysia are discussed in the following section according to three categories: inland ports having national function, regional functions and bordering a neighbouring country.

Inland Ports Having a National Function

Inland ports in this category are the largest such ports in Peninsular Malaysia in terms of capacity and volumes handled. They are designated by the Royal Malaysian Customs as authorized landing places under Section 65E of the Customs Act 1967, hence giving them the status of "seaports". This designation allows shippers to declare their manifest forms at the inland ports instead of Port Klang, the National Load Centre port. The savings incurred from doing so are a pull factor for shippers to feed their cargos through the inland ports. By virtue of their proximity to the major seaports in Peninsular Malaysia, these ports play a national function in distributing cargos to and from the respective hinterlands they serve.

Sungei Way Inland Clearance Depot is Malaysia's first inland port. Built in 1973 when containerization came to Malaysia, specifically Port Klang, the depot is located within Klang Valley, Malaysia's most populated, economically dynamic and well-developed region. It is within a few minutes' drive of Kuala Lumpur, the nation's capital and Putrajaya, the Federal Government's administrative centre. It is also well connected to the nation's largest airport and seaport, Kuala Lumpur International Airport (KLIA) and Port Klang and the North-South Expressway (known in Malaysia as PLUS Highway).

The hinterland that Sungei Way Inland Clearance Depot serves features the Malaysian headquarters and factories of many multi-national corporations involved in a variety of industries including manufacturing, automobiles, electrical and electronics and fast-moving consumer goods. It facilitates break bulk and Customs inspection for exporters in Klang Valley en route to their markets within Malaysia and abroad. These services contribute to better speed and efficiency of the movements of cargos along the supply chains of producers in Klang Valley and beyond that use the terminal. The depot has been suffering from a drop of business with the proliferation of bonded warehouses — estimated to be around 20 — in the Port Klang area. The services offered by these bonded warehouses have reduced the need for shippers and importers to use the depot. However, since being taken over by Complete Logistics in 2007, an integrated logistics service provider, business at the depot has improved. The provision of integrated logistics services there has enabled the depot to secure business from big clients.

Nilai Inland Port provides road, rail and air links and logistics services to industries in the central region of Peninsular Malaysia. It is located within the Nilai-Senawang industrial area and the city of Seremban in the state of Negeri Semibilan. It is a joint venture between the private sector and the Negeri Sembilan state government. Nilai Inland Port's proximity to Kuala Lumpur and the highly industrialized area of Klang Valley, with good sea, air, road and rail connections, puts it in a strategic position to tap into the cargos of shippers south and north of the terminal. It is recognized by Customs as a seaport and boasts a range of services that can be found at seaports, including transportation, logistics, container handling and freight forwarding services. Among the facilities at the port, beside a Customs office, are container yards, a bonded warehouse, a container freight station, container handling equipment and a rail spur.

Ipoh Cargo Terminal is located in Ipoh, the capital city of the state of Perak (i.e. the contribution of Hamel Farhat in this section). The state, which means "silver", was once the world's centre for tin mining which was one of the principal economic activities in Malaysia when it was a commodities-driven economy. The terminal plays a vital function as a centre acting as a gateway for outbound cargos produced by businesses and industries in Kinta Valley in Perak, and also as a focal point for the distribution of imported goods and commodities to the economically vibrant and populous region. Its location not far from the PLUS Highway and seaports in Peninsular Malaysia allows the terminal to tap into the Klang Valley market to its south and the industrial areas to its north. In addition, there is a daily train service between the terminal and Port Klang, with a capacity of 70 TEU and 35 bcf of wagons.

Ipoh Cargo Terminal offers a range of services including container handling, bonded and non-bonded warehouses, cargo distribution and storage. It is also a designated authorized landing place by Customs, hence shippers and importers can pay their Customs duty and get their goods cleared there instead of at the ports of entry/exit. The terminal faces keen competition from hauliers offering very competitive rates to meet the Just-In-Time requirements of shippers and importers. In reaction to this, the terminal has introduced a service called Pit Stop Operations in rail service which has improved the timing and schedules of its services.

Prai Inland Container Depot is located within the industrialized area of northern Peninsular Malaysia of Seberang Prai in the state of Penang. It offers container handling facilities. The establishment of a Free Commercial Zone in North Butterworth Container Terminal in Penang Port had caused an adverse effect to business at the depot. Operators of warehouses at the container terminal enjoy cost and time savings from being spared from having to load on and load off their containers, which is necessary at the depot. Space limitation at Prai Inland Container Depot has prevented it from growing and realizing its full potential.

Inland Ports Having Regional Functions

The second category of inland ports in Peninsular Malaysia is inland ports which are smaller in capacity compared to the first category. Although these ports are also designated by Customs as landing places and enjoy the status of seaports, they serve only the hinterlands within which they are located.

Batu Pahat Container Terminal is located near the town of Batu Pahat in the southern state of Johor within what used to be a vibrant textile industry region. It is located not far from the Straits of Malacca and the North-South Highway (PLUS Highway), Peninsular Malaysia's main interstate highway system. This gives it a unique advantage as a "coastal inland port". It serves the fast-growing and fast-developing north-central Johorean towns of Batu Pahat, Muar and Kluang. The terminal has a barge feeder trans-shipment terminal which it jointly developed with Hong Kong-based Overseas Orient Container Line (OOCL), the world's 11th largest container shipping line in terms of fleet capacity (2010 ranking). The barge feeder terminal links the Pasir Gudang industrial zone (further south), which hosts various industries such as palm oil, petrochemical and manufacturing, to Port Klang. The investment is the first by a major liner player in a Malaysian inland port.

Since Batu Pahat Container Terminal is a Customs-authorized port, it offers shippers cost and time savings to get their export consignment document, known as K2 Form in Malaysia, cleared there in advance instead of at Port Klang. Being a designated seaport, shippers not only within the hinterland but elsewhere in Peninsular Malaysia can feed their outward-bound containers and inward-bound empty containers through the terminal. The innovative inter-modal service provided by the barge feeder trans-shipment terminal allows shippers from the surrounding hinterland to barge their containers not only to Port Klang but other ports including Pasir Gudang in southern Johore and even the neighbouring Singapore Port, to be shipped by large 'mother vessels'. The barge feeder trans-shipment terminal not only helps shippers in the surrounding area to feed to these ports but also to reposition empty containers from Port Klang to them and further south.

Segamat Inland Port is located in the northern part of the state of Johor. It functions as a collection and distribution centre for business and economic centres in the South-Central region of Peninsular Malaysia. It offers feeder rail links between this region and Port Klang. The inland port is a joint venture between the private sector, the Johore state government, Keretapi Tanah Melayu (the national railway company) and Port Klang Authority (the authority body for Port Klang). Among the services offered by Segamat Inland Port are cargo handling, consolidation and storage; port services; documentation; shipping; repositioning of empty containers and freight insurance. It is also an authorized landing place, hence offers Customs clearance services as well.

Built in 1998 to serve Port Klang, Segamat Inland Port has not met with the success imagined by Port Klang Authority. Currently handling low volumes of cargos, the inland port has not been able to compete with hauliers and is also moving boxes to the southern ports of Johor Port and Port of Tanjung Pelepas. This has raised questions on the choice of location, viability and long-term prospect of the inland port. Segamat Inland Port wound up its operations in 2007 and is now used on an ad-hoc basis only.

Inland Ports Bordering a Neighbouring Country

The third, but certainly by no means least important category of inland ports in Peninsular Malaysia is inland ports located near the international border. There is only one such port in Peninsular Malaysia, namely Padang Besar. However, to expand on the discussion, this section has included another such port in the state of Sarawak in Borneo, Tebedu Inland Port.

The Padang Besar inland container depot bordering the state of Perlis which borders Thailand was established by Multimodal Freight Sdn. Bhd., a subsidiary of Keretapi Tanah Melayu, the national railway line. The depot handles mainly laden containers from fast-developing Southern Thailand which are then transported by rail to Penang Port, a good 180 km away. These boxes are transported by road to the depot and then loaded on board trains to be taken to Penang Port, and vice versa. There are twelve daily train services between the inland depot and Penang Port, with a capacity of 8,000 TEUs a day. The depot was funded by the Federal government when it was set up. It is the only inland port in Malaysia offering only road-rail services, unlike others which offer a wide range of services.

Tebedu Inland Port is situated at the border of the Malaysian state of Sarawak and the Indonesian province of Kalimantan in Borneo Island. Acting as a trans-shipment point, it caters to trade, mainly commodities and consumer goods and items, between the two areas and also facilitates the movement of people. In this manner, the inland port plays a role in strengthening bilateral ties between Malaysia and Indonesia. The inland port is operated by SM Inland Port Sdn. Bhd. which was appointed by Sarawak State Government through Kuching Port Authority, as gazetted in the Sarawak Government Gazette in 2004. It features cargo and container yards to facilitate cross-border trade, warehouses, storage yards and facilities for packing and unpacking cargos.

Tebedu Inland Port sits on a land crossing between the Tebedu area on the Malaysian side with Entikong on the Indonesian side. The inland port acts as a gateway for producers from Kalimantan to send their goods by road to Kuching Port in Sarawak, and enables traders to bring in goods via Sarawak. It is worth mentioning here that on the Kalimantan side, modern trade infrastructures are lacking.

3. PENINSULAR MALAYSIAN INLAND PORTS STRATEGY

Issues Faced by Inland Ports

The Peninsular Malaysian inland port experience is not unlike those in other developing countries, where infrastructure developments are patchy, capital is not always available and the hinterlands are not as well developed as major population and economic centres. Despite the growing importance of the inland ports, some are faced with issues and shortcomings which prevent them from handling a bigger trade volume.

The most glaring weakness in the development of inland ports in Malaysia is the lack of a national policy on inland port development. Without it, development of inland ports in the country is haphazard. The failure of inland ports like Segamat Inland Port underscores the need for a cohesive, long-term plan to develop these facilities to ensure they are built at the right location and can serve the needs of their target markets. There is also the problem of poor intermodal connectivity and poor interface with transport corridors that prevent them from operating at optimal capacity and with maximum efficiency to serve the hinterlands that host them. Other issues include limited capacity, limited service offerings, small market share, poor linkage to international distribution networks, and strong competition for business and investment.

As the development of trade facilitating infrastructures in Malaysia is largely determined by site and suitability of location — which takes into account factors such as population distribution and economic activities, among others — the development reflects linkages and interaction among the population, infrastructures and economic activities. Cost is also a crucial factor in determining what transport infrastructure is built where and when, and how it is built, and this applies to inland ports as well.

To understand the disparity in the development of trade infrastructures in Malaysia, it is necessary to develop an appreciation of the agglomeration of transport development in the country. Naturally, more development takes place in key economic zones in the country such as the Klang Valley that hosts the capital city, Kuala Lumpur; the Southern corridor where Port of Tanjung Pelepas and Johor Port are located; and the northern state of Penang that hosts Penang Port and the cluster of electronic plants. Specialization of activities also affects the development of trade infrastructures, for example the oil and gas corridor in the East Coast of Peninsular Malaysia and off the coast of the states of Sabah and Sarawak where deepwater exploration and production activities are carried out. There is also historical influence; ports like Port Klang and Penang Port were built to facilitate the export of raw materials/commodities to colonial countries, leaving a legacy that affects connectivity with their hinterlands to this day.

New dynamics created by several developments and trends will have a bearing on the way inland ports are developed and managed. These developments include:

1. Increase in regionalism/integration to boost economic and trade growth
2. Move from border control to post-clearance and audit-based controls
3. Better inter-agency link-ups in facilitating trade
4. Establishment of Free Trade Areas (FTAs)
5. Proliferation of free industrial/commercial/trade zones
6. Lowering of import duties hence trade cost
7. Alignment of trade practices in developing countries to international trade standards
8. Growing focus on addressing common economic and trade challenges in the face of global recession

In the case of inland ports bordering neighbouring countries, continuous efforts should be made to ensure they are capable of handling bigger volumes of trade more efficiently and at competitive cost. Should the neighbouring countries decide to establish inland ports of their own to cater to growing cross-border trades, Malaysia's inland ports could find themselves losing business if they fail to upgrade the facilities and improve their services.

Stepping Up to the Plate

As Malaysia's trade and commerce expands, it is envisioned that the nation's inland ports will play a more visible and active role in facilitating the nation's trade and enabling goods to be distributed between producers and industries in the hinterlands and the seaports. Despite their low-key features and lack of visibility and glamour compared to the major seaports, the role of inland ports in the economy cannot be underestimated. By virtue of their location in the hinterlands, they can create a virtuous cycle in the economy by linking businesses and consumers to suppliers and by promoting investment and development (see Figure 10.1).

Figure 10.1
Virtuous Cycle Created by Inland Ports

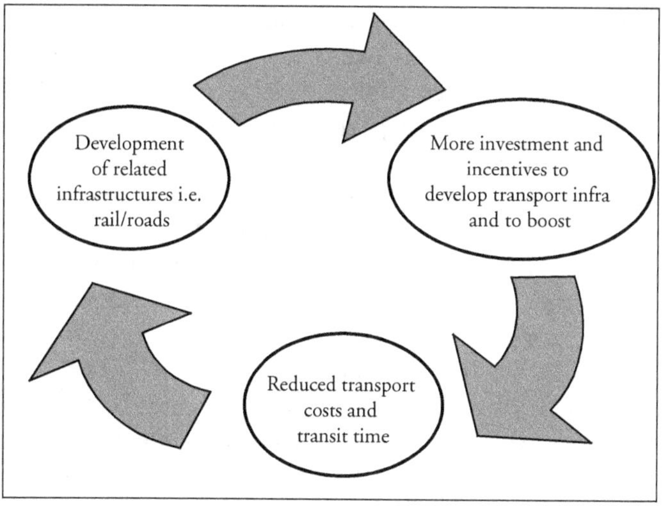

Source: ESCAP.

As the trade landscape changes and new dynamics are unleashed from economic integration, liberalization and globalization, and as Malaysia's export-oriented economy grows, its hinterlands will depend more on its dry ports to facilitate business, trade and economic growth and enhance integration among the transport modes. This is crucial not only for the development of Malaysia's hinterlands but also to foster international trade, which is vital to the nation's economic growth. Given this, the inland ports in Peninsular Malaysia must be developed and organized efficiently to enable them to handle bigger trade volumes. Failure to provide adequate facilities and good services will prevent these ports from growing and realizing their full potential. It can also result in the loss of business of these ports to foreign competition.

A concerted effort is needed to enable inland ports in Peninsular Malaysia to play a more significant role in facilitating the nation's growing trade. This goes beyond focusing on the physical development of inland ports alone. Equal attention should be given to developing supporting infrastructures and the trade supply chain which are crucial to the success of inland ports, and to improve management of the supply chain connected to the inland ports.

To this end, the Government, inland port owners and operators, and players along the trade supply chain must work together to boost intermodal connectivity to cater to more trade and improve cargo handling capability in the inland ports, and to enhance integration across the supply chain. No efforts should also be spared to attract investment to bridge the development gaps among economic regions in the country. The so-called "last-mile connectivity" of seaports must also be improved to attract companies and shipping lines. Last but not least, there must be a national plan in place to develop inland ports in a cohesive and orderly manner.

To match the bullish growth prospect of Malaysia's trade, its inland ports, especially in Peninsular Malaysia where the economic engine of the country lies, must improve their competitiveness and capacity to handle more business. As highlighted, many factors will exert their influence on Malaysia's inland ports' development and operations; hence, inland ports must keep up to speed with the emerging trends and dynamics that affect their business. However, the changing business, investment and trade dynamics will also give rise to many opportunities that inland ports can capitalize on.

Growing intra-Asian trade and trans-shipment business arising from regional economic integration, presents a lucrative segment that Malaysian inland ports can tap into. In this context, the inland ports can act as a conduit of international trade by way of providing "land bridge" connection between Malaysia and its neighbouring countries. A strong push by the Government to provide support and incentives to boost the development of inland ports in Malaysia will not hurt the growth prospect of these "unsung heroes" in playing a more significant role in enhancing Malaysia's trade growth.

The following actions are needed to enhance inland ports' economic contribution to Peninsular Malaysia hence the nation: boost intermodal connection and enhance the efficiency and cost-effectiveness of intermodal services to cater to more trade and improve handling capability, boost efficiency of the logistics network to support regional integration across the supply chain, attract investment to bridge the development gaps among economic regions in the country, improve the "last-mile connectivity" of seaports, have in place a national plan to institutionalize development of inland ports, increase marketing and promotional efforts to convince shippers and shipping lines that the inland ports are viable places of exports and create niche markets and service differentiation among the inland ports.

CONCLUSION: INLAND PORTS AND REGIONAL INTEGRATION

The outlook for Malaysia's container trade should remain bullish in the near future, barring unforeseen circumstances. Intra-SEAN trade is growing, the trend of containerization of cargos is unmistakable, and Malaysia's economy, which is trade-driven, looks set to continue its dependence on trade to power its growth.

Amid these dynamics, inland ports in Peninsular Malaysia are poised to play a more active role and assume greater prominence in facilitating the nation's trade. They must prepare to be more involved in supplementing the roles of major ports and handle greater volumes of trade. To do so, they need to invest in capacity, equipment, systems and human capital to improve their efficiency, productivity and competitiveness to handle more business. At the same time, the state and Federal governments must also help in terms of building infrastructures and providing incentives to boost the development and growth of inland ports in Malaysia.

Many internal and external factors will exert their influence on Malaysia's inland ports' development and operations. They will be subjected to intense competition from one another so they must endeavor to offer a variety of services at competitive rates. But amid the challenging operating environment, there are plenty of opportunities that the inland ports can reap from. The generation of economic activities from the economic corridors they serve, growing intra-Asian trade and increasing regional integration will generate demand for the service of inland ports.

The Malaysian Government has spelled out an ambitious target for Malaysia to attain global competitiveness and to become a fully-developed nation by 2020. To this end, it has identified twelve National Key Economic Areas that will generate high-income activities for the country to boost its income per capita. While the services sector has been identified to power the engine of growth to realize this ambition, trade is expected to continue to play just as important a role. In its push to become a fully developed country, Malaysia will need all of its trade transport components to run at full throttle. Its inland ports will also be expected to stand up and be counted, and play a bigger role to increase the country's trade volumes and put Malaysia in the league of globally competitive trading nations.

It would be interesting to compare the development of the Malaysian transportation system and the transportation systems of other South-east Asian nations as the region aspires to attain full-scale, European

Union-like political and economic integration through the ASEAN Economic Community. With the region rapidly modernizing and approaching regional economic integration, there will be a high degree of convergence between the transport systems.

The maritime sector is no exception; there are already various measures initiated through regional platforms such as the Senior Transport Officials Meeting to integrate the region's maritime transport systems, for example the ASEAN Single Shipping Market. With integration and liberalization going full-scale in the region, the spotlight will be focused on Malaysian inland ports to find a niche within a fully integrated regional economy. They might have to redefine their roles and strategies, and even reconfigure their infrastructures, to survive the onslaught of open competition in a borderless economic region. They might even have to reconfigure their business to play more than just a regional and national function and move up the value chain to service the lucrative integrated regional market of 550 million people in Southeast Asia and assume a more transnational function.

In the final analysis, inland ports cannot be developed on a piecemeal basis. The planning and development of these infrastructures must take into account many factors such as location of major ports, business activities, population centres/consumer markets and raw materials; state and Federal government development strategies; existence of economic corridors; manpower availability/requirement; investment/financing and many others elements. To this end, Malaysia would do well to set up a policy to develop its dry ports, or at the very least to embed their development into a National Port Policy that Malaysia does not yet possess. The emergence of various economic corridors and economic clusters in Malaysia calls upon a more nuanced approach to development of trade distribution infrastructures like ports. Without a long-term cohesive plan to develop the inland ports, they will not be able to realize their full potential as a facilitator of trade and distributor of goods and materials to the nation's hinterlands.

Notes

1 WTO, *WTO Annual Report 2010.*

2 MATRADE, "Malaysia's Trade Performance in 2010", available at <http://www.matrade.gov.my/en/malaysia-exporters-section/122-trade-performance-2010/1105-malaysias-trade-performance-in-2010> (accessed 11 April 2011).

3 Malaysian Shipowners Association, *Malaysia Maritime Handbook 2007–2008* (2010): 29.

4 UNCTAD, *Review of Maritime Transport 2010*, 2010, p. 152.

5 Deadweight tonnage.

6 UNCTAD (2010), p. 70.

7 Twenty foot equivalent unit, the standard unit for container boxes carried by container ships.

8 Ministry of Transport Malaysia port throughput statistics, 2010.

9 Containerisation International.

10 Marine Department Malaysia statistics.

11 S. Bateman, C.Z. Raymond, and J. Ho, "Safety and Security in the Malacca and Singapore Straits: An Agenda for Action", Institute of Defense and Strategic Studies, Policy Paper, May 2006.

12 According to a definition by UNCTAD, a "dry port" is an inland terminal which provides services for the handling and temporary storage of containers, general and/or bulk cargos that enters or leaves the terminal by any mode of transport such as road, railways, inland waterways or airports. A dry port may feature full Customs-related services and other related services such as essential inspections for cargo export and import.

References

Bateman, S., C.Z. Raymond, and J. Ho. "Safety and Security in the Malacca and Singapore Straits: An Agenda for Action". Institute of Defense and Strategic Studies, Policy Paper, May 2006.

Malaysian Shipowners Association. *Malaysia Maritime Handbook 2007–2008* (2010).

MATRADE. "Malaysia's Trade Performance in 2010", 2011. Available at <http://www.matrade.gov.my/en/malaysia-exporters-section/122-trade-performance-2010/1105-malaysias-trade-performance-in-2010> (accessed 11 April 2011).

UNCTAD. *Review of Maritime Transport 2010* (2010).

WTO. *WTO Annual Report 2010* (2010).

PART III

NEW NODES OF ECONOMIC CORRIDORS: Urban Pairs and Twin Border Cities

11

TWIN CITIES AND URBAN PAIRS, A NEW LEVEL IN URBAN HIERARCHIES STRUCTURING TRANSNATIONAL CORRIDORS? A Case Study of the Pekanbaru-Dumai Urban Pair

Manuelle Franck

Specialists in transport geography define a corridor as an area of particular intensity in terms of flow and of spatial, economic, and demographic accumulation, generally supported by transportation infrastructures, on an axis between two poles (Debris and Comtois 2010; Rodrigue et al. 2006). According to this definition, Asia is structured by transnational continental and maritime corridors which are hierarchized according to the intensity of their traffic and trade patterns, be they local, regional, or international (Rimmer 1999). These corridors, whether planned or the result of existing flows, are structured by nodes whose ranks are linked to the ranks of the corridors and whose dynamics are closely tied to international exchanges along its length. These nodes function as logistic and communications hubs or gateways into territories and participate in articulating the space of flows (Castells 1996) and the space of places (Albrechts and Coppens 2003). The head nodes of the corridors are usually large first or second rank metropolises. Between these poles of command, the corridors are structured by a hierarchical series of hubs: by inland province capital cities and by smaller border towns along the continental corridors or, in maritime corridors, by smaller ports.

In the area studied, the strongest corridors, which are commanded by some of the most powerful global cities,[1] are the maritime corridors (Malacca Straits and the corridor that runs along the Asia's Pacific coast). Asia also possesses continental corridors, especially the planned corridors of the GMS, a dense grid of transport corridors that connects southern China to Southeast Asia and the eastern and western coasts of the Indochinese peninsula (see Christian Taillard's article in this volume).

The dominant spatial organization is north-south both at the transnational level as it does in continental Southeast Asia at the national level in terms of spatial constructs. The dominant corridors, whether continental or maritime, are north-south oriented (or north-west-south-east in the case of the Malacca Straits) due to the intensity of the maritime traffic in one of globalization's dominant poles or, for continental corridors, due to the intensity of flows with China. The transverse corridors, which feed the main meridian ones, are emerging secondary corridors even though they are sometimes based on traditional spatial organizations, as in the GMS where some portions of the corridors redraw ancient caravan linkages or in Sumatra where transversal rivers were the main transport modes.

The maritime corridors are structured by cities located at the entrance and along the corridors, the competition between ports as much as the intensity of flows allowing the development of several commanding cities. The corridor that runs along the coastline of Asia Pacific is commanded by several global cities. Singapore, Hong Kong, and Tokyo, which have structured the corridor since the nineteenth century, share their nodal functions with Seoul and Pusan across the Sea of Japan, Taipei and Kaoshiung across the Taiwan straits.[2] In the Malacca Straits, Singapore, located at the Straits' East Asian entrance, controls the Straits. Kuala Lumpur claims a portion of the command functions over port and air flows in the Straits, placing it in a competitive relationship with Singapore. The other extremity of the Straits is controlled by Penang and Medan, two cities in the midst of a fully engaged process of metropolization.[3]

The cities that command the continental corridors are less numerous and of lower rank in world urban hierarchies than in maritime corridors and are only situated at the extremities of the corridors. Kunming to the north and Rangoon, Bangkok, and Ho Chi Minh City to the south are the head nodes of the North-South corridors of mainland Southeast Asia. Except in the case of the planned "Capital cities transverse corridor" of Ho Chi Minh

City-Bangkok-Rangoon, the poles of command of the transverse corridors are mostly second rank regional cities.

Transnational corridors are key vectors to channel flows. The combined regional, national and transnational dynamics are responsible for the more rapid development of existing cities or promote the emergence of new types of cities along these corridors, through the addition of activities or management functions of flows. The development of second-tier cities initiates this process of densification of the urban framework. At the lower end of the urban hierarchies, it is also responsible for the development of smaller cities. The New Economic Geography concludes by stressing a direct link between agglomeration forces and growth (Fujita et al. 1999 or for a summary Fujita and Thisse 2008). However, in direct contrast with this concept, Hardoy and Satterthwaite (1986) and Moriconi (2010) postulate that the dynamics of small cities might also occur far from large metropolises thanks to new opportunities offered by transnational dynamics.

Studying the transformation of the lowest ranked cities (in urban hierarchies) is another approach to analysing the impact of transnational dynamics. However, accounting for their transformations requires studying them not as individual entities, which in many cases would ignore much of the process; but rather as functionally complementary or competing doublets situated across borders or within the same national territory. Multiple parameters need to be examined: their location (central or peripheral) within national space, their place in national urban hierarchies, their relationship with former command centres, and for border cities, the type of borders involved (land, river, or sea). This chapter will try to attempt a typology of these new nodes linked to the regionalizing process of globalization.

To explore how urban duos are formed through combined regional, national, and transnational dynamics, this chapter aims at studying the emergence of the paired cities of Pekanbaru-Dumai, which are located in the province of Riau, in the centre of Sumatra Island, a location where there is a marked discontinuity in the island's urban hierarchy. Resulting from the combined action of Indonesian national planning and transnational dynamics, the dynamics of these cities, which work together due to functional complementarity despite possible competition, are linked to their position at the crossroads of a dominant north-south axis and a reactivated transverse axis that articulates both national and transnational processes. These pair cities represent a new urban scale in maritime Southeast Asia, consisting of two main hierarchized poles undergoing rapid transformations and operating as a network thanks to improved infrastructures supporting the creation of corridors.

1. URBAN PAIRS AND TWIN CITIES, TOWARDS A TYPOLOGY OF NEW URBAN CONFIGURATIONS ON CORRIDORS

It is necessary to introduce a distinction in terminology between twin and pair cities. The distance between cities, and, where appropriate, from the border, is a key differentiator. In this chapter, the term "twin city" indicates border cities, located along land or river borders, whose geographic proximity is a key factor in their integrated operations. The term "urban pairs" or "pair cities" will encompass cities located at greater distances than "twin cities"; but which are often linked by hierarchical, complementary, and sometimes competitive relationships. These "pairs" either exchange particularly important flows between each other or are poised to do so due to their position within corridors. This concerns the relationship between both regional capitals and border or port towns within the same national context as well as between cities facing each other across the Straits of Malacca (see Map 11.1).

Twin Cities in Border Areas: An Integrated Functioning?

The border town and its tandem functions with its urban counterpart on the other side of the border is an evident first form of these new types of hubs. This holds true whether this functioning is planned within the context of national planning strategies or transnational cooperation strategies, or whether it is the result of spontaneous economic or cultural ties resulting from historic legacies or mere geographic proximity. Eventually, these twin cities will likely form new, transboundary, urban configurations as a result of important investments in transport infrastructure: bridges, roads, urban bypasses, as well as the intervention of private actors responsible for internationalizing local economies. The diversity of situations is obvious. In addition to how cities are situated within their respective national territories and their respective national urban hierarchy as previously stated, we must also consider their distance from their respective borders and the inclusion or exclusion of border areas in transnational cooperation programmes such as the GMS corridors or growth triangles.

The type of border regions is a first step to refining the typology of border doublets. When the border separates peripheral regions, the twin cities are small border towns, in relation to their national scale, with comparable sizes and economic weight. They tend to occupy lower ranks in their respective national urban hierarchies, as shown by the urban twins that are

Map 11.1
Cities and Corridors in Southeast Asia

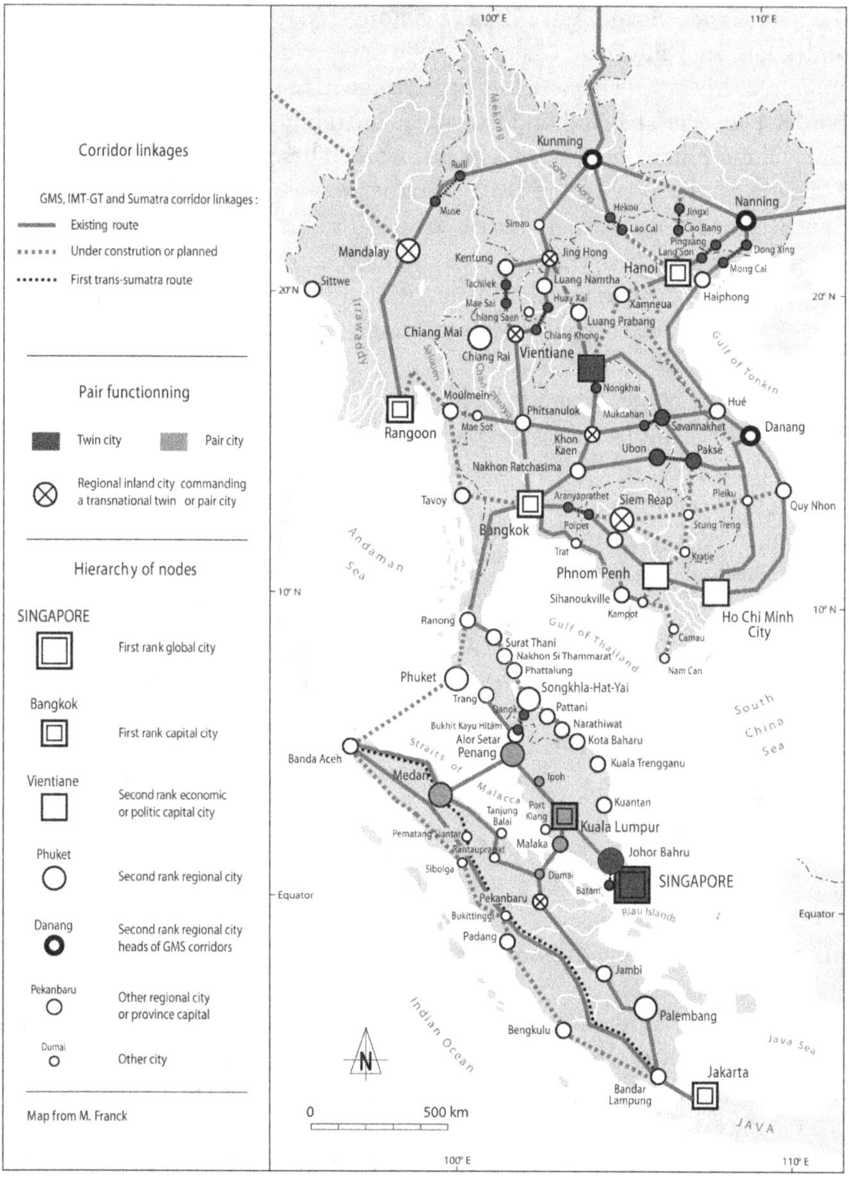

Sources: C. Taillard (2009); N. Fau (2003); Asian Development Bank.

strung along China's southern border: such as Ruili-Muse between Yunnan and Burma, Hekou-Lào Cai between Guangxi and Vietnam (studied in this work), or on the upper Mekong like Mae Sai-Tachilek between Thailand and Burma, and Chiang Khong-Huay Xai between Thailand and Laos. In Peninsular Malaysia, twin border cities are linked with the Indonesia-Malaysia-Thailand Growth Triangle (IMT-GT)[4] as well as with the border between Thailand and Malaysia, including that of Danok (Sadao)/ Bukit Kayu Hitam, also studied in this work. These twin cities on the Thai-Malaysia border present the same patterns as those observed in continental Southeast Asia. Most of these twin border cities could be considered as integrated border peripheries, a useful concept to describe these towns that weigh little in their respective urban national hierarchies, but which benefit greatly from transnational dynamics despite their remoteness from more important cities.

When the border separates highly asymmetrical territories — a peripheral region at the national level on one side compared to a central zone at the national level on the other side as is the case along part of the Mekong border between the northeast of Thailand and Laos, the cities involved are no longer very small towns, but medium-sized cities, such as Savannakhet facing Mukdahan, a twin studied in this work, or more openly asymmetrical twins such as Vientiane, a national capital, and Nongkai, a provincial capital. Note also that Thailand's national economic dominance over Laos is often inverted when it comes to their asymmetrical twin cities: the Laotian component is largely dominant in the twin pairings of Vientiane-Nongkai and Savannakhet-Mukdahan.

In all cases, twin cities whose borders are defined by a river are not as closely linked, for water offers greater opportunities for exchanges between hubs that are not necessarily located opposite each other. Bridges, important crossing facilities that maintain terrestrial continuity and concentrate flows, are the necessary condition for a more integrated functioning of cities on either side of border rivers. Political commitment to facilitating trade is also needed.

These twin border cities are often the result of planning at the national level and within the framework of transnational cooperation programmes. They benefit from bilateral agreements that accompany the implementation of the GMS corridors or growth triangles. The cities in this context often benefit from public investment in transport infrastructure, industrial and mixed-use free trade zones, and the facilitation of border crossings. The difficulty then is to encourage the private sector to take over public investment in the economic sector with long term returns. The private sector accompanies and even anticipates the boom effect linked to the construction of these new infrastructures, but it tends to focus on those services that are most directly profitable.

National and Transnational Urban Pairs

Another case is represented by urban pairs located on both sides of the Malacca Straits. The Straits, regardless of the sections considered, feature high differential levels of development between both sides that generate asymmetries. The Malaysian shores compose the centre of peninsular Malaysia whereas their Indonesian counterpart is located at a distant periphery from the nation's core. Given the importance of the Malacca Straits in global circulation, its shores are punctuated by large cities, some are even ranked as global cities. However, due to their asymmetrical development, there are more cities, with higher rankings, on the Malaysian side than on the Indonesian shore. A series of various sized ports relay these cities and constitute the specificity of the hubs of this maritime corridor. Indeed, the maritime patterns of the Malacca Straits corridor requires careful consideration of its ports and their hierarchy, as suggested by Nathalie Fau in her presentation of Straits whose width and international status render the corridor's architecture even more complex.

Even more than river borders, maritime fluidity limits the reality of paired operations of ports on both shores in favour of polycentric functioning of the ports in the Malacca Straits, despite the crushing weight of Singapore. Within the context of IMT-GT corridors, the Asian Development Bank (ADB) projects of corridors across the Malacca Straits do map out possible pair cities such as Medan and Penang, or Dumai and Melaka. However, relationships are in fact multidirectional and changing, depending on the volume and nature of the flows as well as on the final destination of the transported goods.

Transnational dynamics are also responsible for the development of pair cities at the national level, strongly hierarchical structures that add territorial depth to the impact of border dynamics, as territories are integrated into a world system. Maritime Southeast Asian ports and their pairs, either dry ports or inland cities, are the very urban border figures that are most affected by transnational dynamics. The case of Ipoh, studied in this book, shows how this city, located mid-way between Penang and Kuala Lumpur, has regained strength thanks to its dry port function after a period of crisis due to the decline of its previous base economic sectors: mining and industry. Today, it is Southeast Asia's main dry port, thanks to its complementarity with costal ports. The functional complementarities between coastal ports controlled by towns further inland also allow the development of paired operations. This type of functioning is even more closely tied when the port was created to serve the city, as in the case of Kuala Lumpur and Port Klang. The Malaysian capital developed during the nineteenth century as a mining town and not a port town.

In mainland Southeast Asia, such pairs can be found between border cities and regional capitals located further inland, on the same transnational corridor and benefiting from the resumption of economic relations since the 1990s between Yunnan and mainland Southeast Asia. This is the case for Chiang Rai, which plays an important role in the functional management of border towns such as Mae Sai and Chiang Kong. In Burma, Kengtung controls the border town of Tachilek. The current dynamic of Mandalay is strongly associated in Burma with the activities of the Ruili-Muse twins and with the resumption of economic relations with the Yunnan along north-south communication axes.

These processes have reinforced key provincial capitals, whose functions have diversified beyond simple administrative functions, particularly in the GMS where infrastructure investments are very important as are incentives provided by national governments to attract businesses to these peripheral areas (Lainé 2011). Chiang Rai's Special Economic Zone (SEZ), Mukdahan's industrial park, and especially Khon Kaen's dry port infrastructures (regrouping containers) should over time promote the industrial dimension of these provincial capitals located within the GMS corridors.

National urban pairs, and especially urban pairs connecting regional cities and ports which are specific of maritime configurations, have been identified as one of the urban categories that are emerging from the ongoing globalization process and the transnational dynamics it generates. The Pekanbaru-Dumai case allows us to elaborate on this specific configuration.

2. THE CHANGE OF SCALE AND RHYTHM OF PEKANBARU'S DEVELOPMENT

The Foundations of Pekanbaru's Recent Economic Take-off

Pekanbaru, founded in the eighteenth century according to official historiography, is situated a hundred kilometres from the mouth of the Siak River and numbered 900,000 inhabitants in 2010. It is the capital of the

Riau province of Sumatra.[5] This sparsely populated province, located along the Straits of Malacca, grew rapidly in the 1950s with the discovery of oilfields and the development of plantation crops (see Muriel Charras in this volume). Riau's economic structure was modified with the exploitation of oilfields (Duri, Minas), roughly sixty kilometres north of Pekanbaru, by the American company, Caltex (California Texas Oil Company), whose current operator, Chevron, is also one of its subsidiaries. Caltex built the deep water export port of Dumai in an inlet sheltered by the island of Rupat, across from the Malaysian city of Melaka. This port is connected to the oilfields by 150 kilometres of pipelines and a small road, which in 1958 was extended from Duri to Pekanbaru, home to Caltex' headquarters and where a portion of its staff resides.

The Pekanbaru-Dumai road opened up hectares of forest to industrial use and thus allowed other activities to develop: forestry, plantation crops, and near Pekanbaru, business activities tied to the presence of storage companies and to the arrival of many migrants. The arrival of these workers, whether spontaneous or through Indonesian transmigration programmes (especially starting in the late 1970s), triggered conflicts over land control and between ethnic groups. In the 1970s, Pekanbaru was the capital of a "frontier" province (Colombijn 2002), leading a rapidly developing territory with growing population growth rates that had undergone profound economic and social upheavals over the previous half-century. The province's population has multiplied by five since 1971, mainly through immigration, and the city of Pekanbaru has multiplied by six.

In 2010, Dumai already had a population of 253,000 inhabitants, despite its short history as an oil port built in the 1950s.[6] Dumai's port activity met the needs of the oil industry but not the harbour requirements of Pekanbaru, whose river port remained active for quite some time. Up until the 1990s, Dumai oriented its activity towards exports and therefore towards the Straits of Malacca. It had few ties with other cities on Sumatra's east coast (Fau 2003), despite the existence of Pekanbaru-Dumai road.

Despite all this, the necessary conditions currently exist for these two cities to perform as a pair by playing on functional complementarities. One of the conditions was the improvement of road infrastructures that places Dumai and Pekanbaru on the eastern longitudinal axis dominating Sumatra's road system. It also necessitated the reactivation of a transversal axis from Padang to Pekanbaru, allowing Pekanbaru to recover its pre-colonial era functions as a link between the western highlands, the coast, and the Straits of Malacca.

Pekanbaru: From a Node on a Transversal River Axis to a Pole on Sumatra's North-South Corridor

Nathalie Fau (2006) has analysed the changes in the spatial organization of the island of Sumatra since the colonial period; starting with a west to east pattern and ending with today's north-south organization. Rivers were the traditional means of transportation and thus highland products from the West were carried to river ports located a hundred kilometres from its mouth, to then be ferried out towards the Straits of Malacca.[7] Freek Colombijn (2005) accurately traces land then river transportation of goods and the spatial organization of the eastern part of central Sumatra, with its hierarchy of merchandise collection centres, located at the contact points between the river and its major tributaries. This "dendritic system" allowed a single river port, located at the contact point between peneplain and marshy plain — a point where most tributaries flowed into the main watercourse but where the delta distributaries have not yet taken shape — to control the traffic of its river basin (see also Manguin 2000, p. 163). This is the case of Pekanbaru on the Siak River, closely linked to the Minangkabau highlands and the Straits of Malacca.

The direction of flows, corresponding to the transversal orientation of the rivers linking the Barisan highlands to the Straits of Malacca, was changed during the colonial era. The development of coffee plantations in the Minangkabau highlands starting in the 1800s was accompanied by the Dutch construction of roads leading from the highland production areas to the west coast. This was accomplished in order to favour the port of Padang, which became the third largest port of the East Indies in the mid-nineteenth century after Jakarta and Surabaya, and diverted part of the traffic heading towards the Straits of Malacca (Colombijn 2005, p. 22). Pekanbaru continued to ensure contact with the Straits of Malacca. In 1878, the city was connected by a regular steamboat route to Singapore, where it exported timber cut along the river banks, but Pekanbaru no longer received supplies from its traditional hinterland because a large part of the highland flows escaped its control.

The construction of the first trans-Sumatra route by the Dutch contributed to marginalizing Pekanbaru even further (Fau 2003). Favouring the Padang and Medan poles, the latter developed during the last third of the nineteenth century at the heart of a rich belt of plantations on volcanic soils, the Dutch sought to link these two poles directly to Java and therefore built a trans-Sumatran road from Medan to the southern end of Sumatra. This road cut across the highlands and turned its back on a large part of Eastern Sumatra, bypassing the East Coast river ports, including Pekanbaru.

It was not until the 1950s, with the development of oil production and the construction of the Pekanbaru-Duri-Dumai road by the Caltex oil company, that activity resumed in the eastern part of central Sumatra. Pekanbaru, thanks to the Duri-Pekanbaru portion of the road and its extension to the port of Dumai, would begin to feature on a new eastern north-south axis and thus recover a portion of the activity of a new hinterland, no longer located in the highlands of Barisan, but along the eastern coast. This process really took off with the construction of the trans-Sumatran eastern route in the 1990s.

For Pekanbaru, its position on the north-south axis is all the more important in that the post-independence period, because of increasing centralization of the country's governance, did not change the dominance of a north-south orientation of flows initiated during the colonial era. The north-south orientation of the trans-Sumatran route, which allows Sumatran flows to pour into Jakarta and perpetuates the breaking off of transversal relationships, corresponds to the centralizing efforts of the independent government. This is especially so in that until the late 1980s, Jakarta's centralization led to the closing of the country's borders, the breakdown of direct links between the Sumatran and Malaysian shores of the Straits in favour of transit through Jakarta. This process transformed Sumatra into an Indonesian peripheral zone, even though it lies along international Straits through which large flows of international trade have always transited (see Charras 1993 and in this volume)

Pekanbaru: New Policies that Encourage Renewed Activities on the Eastern Seaboard and in Relation with the Straits

The shift of economic policies in Indonesia has liberalized several areas of national economy and opened the country up even further to foreign capital as part of a policy of promoting non-oil and gas exports, initiated in the late 1980s. It has also triggered the loosening of Jakarta's stranglehold on international flows. Both the establishment of growth triangles in the 1990s, whose successive expansion ultimately included all the island of Sumatra, and the establishment of AFTA (ASEAN Free trade area) in 2002, have offered various opportunities throughout the country to deal directly with foreign countries. Riau was thus free to do business with the other side of the Straits. However, it is mainly the loosening of centralization due to the fall of the New Order regime in 1998 along with the decentralization process that began in the 2000s, and the economic opening, that has allowed the resumption of direct ties between Riau and the Straits' other shore (Charras in this volume).

Decentralization has endowed Pekanbaru, like other departments, with new decision-making powers — despite the ongoing importance of central decision-making, which remains responsible for issuing permits for major investment projects — and new financial capabilities while the province's expansion provided new revenue. Riau is one of the richest provinces in Indonesia, in terms of gross provincial product. Since decentralization measures were enacted, its wealth also comes from its share of oil revenues. The increase in provincial and departmental revenues has helped fund urban planning projects to improve the city, visible in the construction of new, iconic office buildings, especially in the newly created district capitals,[8] as well as attracting domestic and foreign investors. This is even truer in that the administration of Riau province, whose governor actively supports business and foreign investments, strongly encourages cross-Straits exchanges.

Since the separation in 2002 of Riau with the island portion of its territory in favour of the creation of the new province of Kepri (*Kepulauan Riau*, Riau Islands),[9] where Singaporean investments in the industrial sector are concentrated, the province of Riau, thus redefined yet maintaining Pekanbaru as its capital, currently directs international flows almost exclusively to Malaysia as it has lost most of its business ties with Singapore. This reorientation has forced Riau to refocus its activities on developing mineral and land resources, especially by attracting major Malaysian investments to its plantations. It has also confirmed Pekanbaru's role as the main administrative and business pole of Central Sumatra. Riau thus stands as a province whose development is closely linked to transnational dynamics, taking advantage of cultural ties based on its *Melayu* identity to strengthen its relations with the Malaysian shores of the Straits.[10]

3. PEKANBARU, A REGIONAL CITY FOR CENTRAL SUMATRA OPENING ONTO THE STRAITS OF MALACCA

Pekanbaru: At the Crossroads of North-South and East-West Axes Structuring Central Sumatra

Confirming the New Order's nation-building focus on north-south organization, three north-south trans-Sumatran road projects were developed in the west, the centre, and the east of the island. The construction of the eastern trans-Sumatran began in the 1990s, starting in the north with Banda Aceh and heading south to Bandar Lampung. It travels through the major cities on the eastern seaboard (Medan, Pekanbaru, Jambi, and Palembang) and relies on pre-existing sections including notably the Pekanbaru-Duri-Dumai road[11] to which were added sections such as those from Pekanbaru towards Rantau Prapat to the north and towards Rengat to the south.[12]

The eastern trans-Sumatran route was built in sections of uneven quality with insufficient investments to maintain the whole (Charras in this volume). The Pekanbaru-Dumai section is so poorly built that it takes up to six hours to travel the 150 kilometres. This seriously hampers relations between the two cities and limits investment. The reasons behind this include the lack of State interest; for quite some time, the Sumatran road was not an administrative priority.

The ambitious new strategy outlined in the 2011–2025 *Master Plan for the Acceleration and Expansion of Indonesian Economic Development* (MP3EI) holds the promise of new possibilities of mobilizing necessary capital, with its focus on developing national economic corridors with clearly identified economic specializations that will concentrate public and private investments. Launched by Indonesia's President in 2010, this strategy aims to enable Indonesia to join the most advanced countries by 2025 (Republic of Indonesia, Coordinating Ministry for Economic Affairs, 2011). The Master Plan complements the regional and national planning documents (RTRWN, RTRW), and falls within the broad guidelines for Indonesia's mid-term (2010–14) and long term (2005–25) planning (RPJMN and RPJPN). These corridors will be composed of growth poles formed by the development of industrial clusters and special economic zones, which are linked to existing poles by connective infrastructures that will focus investments (see Map 11.2).

In Sumatra, as with road projects in the 1990s, the strategy aims primarily for national integration, without of course ignoring the need to plan for international gateways in the context of Indonesia's increasing integration into global networks. Thus, the flagship project will build a bridge linking Java to Sumatra across the Sunda Straits and develop the steel industry in South Sumatra, to complement Cilegon's facilities in western Java. The programme also recognizes the strategic aspects of better integrating eastern Sumatra and recommends developing North-South road and rail links in this region (MP3EI, p. 73). At the level of the Riau province master plan (Riau RTRW 2007–2026), the Pekanbaru-Dumai section already represents a strategic national road that could become a toll road.

Map 11.2
Sumatra Economic Corridor Projects

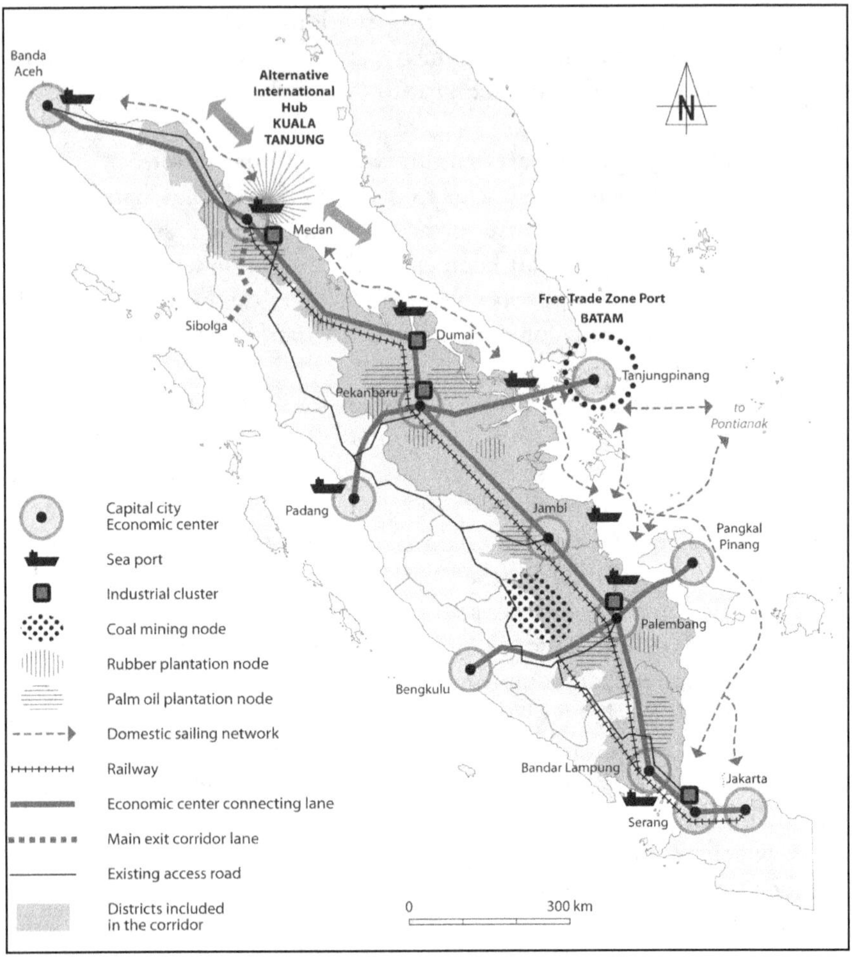

Sources: Republic of Indonesia, Coordinating Ministry for Economic Affairs (2011). Master plan for Acceleration and Expansion of Indonesia Economic Development 2011–2025, p. 51.

Finally, the MP3EI programme situates Pekanbaru at the intersection of North-South and East-West axes, recognizing the need to renew relations between Padang and Minangkabau highlands (Bukittinggi region) with Pekanbaru and the Straits of Malacca, especially through investments in communication infrastructures (see Maps 11.2 and 11.3).

Map 11.3
The Pair City Pekanbaru-Dumai at the Crossroad of a Meridian and Transverse Axis

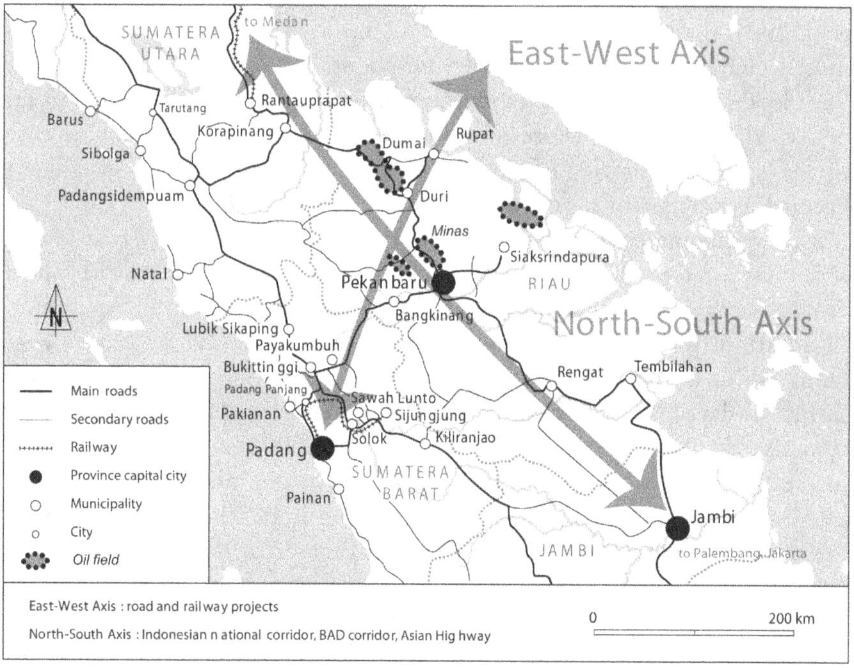

Sources: N. Fau (2003); various maps.

Seen from Padang, this means recognizing the need to redirect the west Sumatran flow towards the Straits of Malacca. Since the 2000s, relations have begun to develop with Malaysia and Singapore. Apart from a handful of Malaysian investments in the industrial area of Padang (Fau 2003), Malaysians and Singaporeans are increasingly interested in the region of Padang, the highlands around Bukittinggi, and also the western islands of Nias and Mentawai as tourist destinations, due to accessibility by air from Padang or by road from Pekanbaru. Direct flights by low-cost carriers (daily Air Tiger flights between Padang-Singapore and Air Asia flights between Kuala Lumpur and Padang) have further increased West Sumatra's reputation as a tourist destination. Available flights have also increased medical tourism from Padang to Kuala Lumpur. Padang exports cement, shipping it across the Malacca Straits. A recent survey of West Sumatran small and medium enterprises shows that exports to Malaysia and the

IMT-GT have increased, notably handicrafts, embroidery, and traditional textiles, products from the sea and coal (Syahrial Syarif 2004). The Padang-Pekanbaru section of the route, along with the Medan-Pekanbaru portion, is already considered Sumatra's densest route for passenger traffic (Lubis et al. 2005, p. 50), while freight traffic intensifies,[13] road transport being the most widely used mode of transport for goods.

Goods flow either directly to countries on the other side of the Strait, or pass through Riau, whose relations with Padang are important because of the complementarity of the two provinces. West Sumatra has again become a supplier of agricultural products from the highlands, especially market gardening (vegetables, fruits) and freshwater fish for the East Coast. It has also become the most sought-after and popular tourist destination in central Sumatra, especially for the middle class and Pekanbaru's expatriates. In addition, plantation farming is moving inward in Riau, towards west Sumatra due to the scarcity of land along the North-South route (Lermoyer 2011). Today, Padang province's growing interest in reorienting its flows towards Pekanbaru and the Malacca Straits (Fau 2003) and its reintegration in the Straits' transnational dynamic is visible in the initiatives undertaken by West Sumatra Province. This is especially true in relation to improving the transport network between Padang and Pekanbaru, as existing roads are already congested due to the resumption of flows.

Since 2000, the province of West Sumatra has already invested in the reconfiguration of roads and the construction of bridges to span the bends in mountain roads. It now plans to double the road from Padang to Payakumbuh to improve transport infrastructures for tourism and consolidate links with the eastern seaboard, as many products headed for Malaysia are transported along the Pekanbaru road. The province also plans to financially support the State project of resurrecting the Padang-Payakumbuh-Pekanbaru railway.

As for Riau, the 2007–2026 master plan (RTRW Propinsi Riau) provides, in addition to the construction of the Padang-Pekanbaru railway, for the construction of a highway from Pekanbaru to Bukittinggi.

Pekanbaru and the New Grid of Transnational Corridors Crossing the Straits

The route of the Sumatran corridor of the MP3EI project cuts across the path of two transnational corridors seeking to promote transnational integration at the level of the Straits and of Asia as a whole. The eastern trans-Sumatran route is one of Indonesia's sections of the Asian highway network, a proposed network of 141,000 km of standardized roads throughout 32 Asian countries. The project is the result of a 2003 intergovernmental agreement, whose implementation is supported by the United Nations through its Economic and Social Commission for Asia and the Pacific (ESCAP). Its north-south path has been incorporated into the economic corridors of the IMT-GT,[14] which are the focus of investments for infrastructures and supporting facilities financed within this framework. On the main North-South corridor linking Aceh to Palembang, three transverse corridors branch out, connecting Sumatra to the Malay Peninsula at three different points: from Banda Aceh to Phuket; from Medan to Penang with an axis extending to Songkhla on the east coast of the Peninsula; and from Dumai to Melaka (see Map 11.1).

By drawing three corridors across the Malacca Straits, the Asian Development Bank, responsible for the very concept of economic integration through corridors and author of the IMT-GT corridor routes, has transposed into a marine context the model of continental corridors, especially from the Greater Mekong Region (see Taillard in this volume). It has also planned for ports to function in pairs, thanks to improved connective infrastructures: bridges, fast passenger ferries, or Ro-Ro ships. Though the vision is reductive in terms of the actual origins and destinations of maritime flows across the Straits (see Fau in this volume), the proposal for the paired ports of Dumai-Melaka reflects the reality of the local exchanges of men and goods that has resumed between these two ports.

However, the ADB corridor does not have a transversal extension to Padang, even though all the provinces of Sumatra, including Padang, are part of the IMT-GT. This contrasts with the proposed extension of the Medan-Penang corridor across the Straits, through the Malaysian Peninsula to Songhkla on the eastern coast of southern Thailand. The first corridor routes are based on pre-existing infrastructures and links that focus on the projects; other corridors may be added later. The absence of a lateral route in Sumatra is, however, a sign that integration is conceived first as a transboundary issue rather than a transnational one with deeper territorial impact. It also illustrates the mediocrity of transversal relations in Sumatra in the 2000s as compared with north-south flows.

4. THE FUNCTIONAL COMPLEMENTARITY OF PEKANBARU-DUMAI, A NEW URBAN SCALE?

In the master plan for the province of Riau (RTRW 2007–2026), Pekanbaru and Dumai appear as its two national centers (*Pusat Kegiatan Nasional*).

Dumai exercises industrial functions and possesses an international hub port, whereas Pekanbaru exercises tertiary and administrative functions as the regional emergent metropolis of Central Sumatra. The socio-economic development of the province of Riau, overlooking the Straits of Malacca, physically close to the other shore, and closely linked to transnational dynamics; the mobilization of infrastructure investments along the north-south axis; and the reactivation of the transverse axis adding territorial depth to transnational dynamics, all combine to contribute to their functioning as pair cities.

The functional complementarity between the two cities, located 150 km apart, is most evident in trans-Straits transports. Port functions are dissociated between the two cities: Dumai, located on the Straits, is Central Sumatra's main export port, while the river port of Pekanbaru, located on the lower Siak, is more specialized in imports towards the regional capital. In the ranking of the public company Pelindo 1, which manages the ports of Dumai and Pekanbaru, Dumai is classified as a first-class port whereas Pekanbaru is rated second class.

Complementarity in Transport Functions: An Export Port Versus a Regional Port and an Air Hub

Dumai is primarily an export port, 80 per cent of the province's export value transits through its ports (BPS Riau Province Statistics Yearbook 2008). The city has a general port and ports specialized in palm oil[15] and hydrocarbons, which are the province's two main exports. Certain ports are managed by the public company Pelindo and others by Chevron. Dumai is also the most important ferry terminal to Malaysia from central Sumatra in terms of the number of weekly trips; as well as being the scene of all forms of trafficking with the other shore: smuggling, migrants and drugs. Dumai could become Riau's main export port for the province's products to the other side of the Straits. It could even play this role for all of central Sumatra as it is capable of draining goods from West Sumatra, the southern part of North Sumatra, and the northern most part of Jambi.

The MP3EI Master Plan draws a transverse corridor through central Sumatra, connecting Padang to Pekanbaru and then Batam, but not Dumai, suggesting that Batam's port could take over the export flows of West Sumatra. However, Batam's utterly outward orientation that maintains few ties with the hinterland of Riau, with its free zone status, and operating more like an enclave than a gateway to Indonesia, renders this hypothesis implausible. Dumai could better play this role for the central region of Sumatra if, besides the improvement of Sumatra's roads, connections were facilitated with the Straits' other shore. Two projects, under way since the 1990s, tend to limit the impact of transshipment constraints between both shores. The first is a project for a bridge between Dumai, Rupat, and Melaka, supported by Riau province and the State of Melaka. While this is not an Indonesian state priority, it could receive funding as part of the IMT-GT because Dumai-Melaka is one of the transversal corridors of the growth triangle. Second, there is the long overdue establishment of a shuttle ferry (Ro-Ro), whose infrastructure, a 110 metre pier and a terminal, have been completed on the Indonesian Dumai side; but not yet on the Malaysian shore.

Pekanbaru's harbour functions complement those of Dumai.[16] Half of the province's import value transits through the ports of Pekanbaru (BPS Riau, Riau in Figures 2008), specialized in container transport. The old port, inside the city, is declining due to its limited accessibility.[17] The port of Pekanbaru is currently the Port of Perawang, located downstream on the Siak River.[18] Pekanbaru does not handle large vessels, which cannot navigate up the Siak. This has become even truer since 2007 when the construction of a bridge in Siak Sri Indrapura, only 23 metre high, permanently prevented the passage of large ships. Imported goods are transported by tug boats or on barges loaded with containers that navigate up the river and are loaded in Kuala Lumpur or in Singapore,[19] more rarely after transshipment in the port of Dumai or Bengkalis. The expansion planned by Pelindo I in the port of Perawang (dock extensions, expansion of the container storage area, adding two crawler cranes to accelerate unloading) attests to the importance of this traffic that consists mainly of imports transported by container. As long as the road from Pekanbaru and Dumai is not redesigned, Pekanbaru port is also an alternative to Dumai port for exporting products from West Sumatra and Riau. In 2008, Perawang was the province's third port in terms of export value (which represented, however, only 6 per cent of the province's export values, compared to Dumai's 80 per cent). However, Pekanbaru's port activity will always be limited by the physical constraints inherent in river ports.

Nevertheless, Pekanbaru is Riau's air hub for service to the other side of the Straits, an important function at a moment when low-cost air transportation is competing with ferry transportation. Pekanbaru is connected to major cities in Sumatra and Jakarta as well as Kuala Lumpur, Melaka, and Singapore, by scheduled airlines (Garuda) and low-cost airlines such as Tiger Airways, Lion Air, Batavia Air, and Air Asia. Riau province also established the Air Riau airline to link up to Dumai, a cost-effective option as long as the highway connecting the two cities is not completed. It also flies to major cities in Sumatra as well as Jakarta. In January 2011, it opened a

Pekanbaru-Melaka line, which could be extended to Bangkok. The airport's expansion is 80 per cent complete.

Complementarity in Economic Functions: Industry Versus Services

The two cities are also complementary in terms of activity sectors, secondary for one and tertiary for the other. These specializations will no doubt be strengthened with each new project. Dumai has a confirmed industrial vocation. Its location along the border, as well as its activities, have designated it as a National Strategic Center of Activity in mid-term national planning (RPJMN 2010–2014), with the status of a Special Economic Zone with port infrastructures and direct links to Sumatran corridors. Beyond storage activities and refineries for crude oil (Pertamina and Chevron) and palm oil (Wilmar facilities), Dumai's five industrial zones[20] will ultimately host export industries, particularly post-extraction palm oil facilities. Of these five zones, the most dynamic is Pelintung (Kawasan Industri Dumai or KID), inaugurated in 2010, with an export platform capable of handling up to three tankers. One of the country's largest agrofuel plants has already been built there. Though barriers to the installation of these industries have been lifted, notably poor roads (palm fruit must be processed quickly) and difficulties obtaining raw material at competitive prices.[21] Taking advantage of the proximity of the Straits of Malacca, the central government's project also provides for the creation of a shipyard that should ultimately take over Java's facilities.

Pekanbaru is not lacking in industry; but its focus is more on the processing industry and consumer goods (food industry, wood products etc.). An industrial zone is under development at Tenayan on the city's eastern ring road; however, national planning has not included any major projects in Pekanbaru in these sectors of activity. Pekanbaru concentrates high-level tertiary activities and administrative functions at the provincial and municipal levels. Riau's largest oil producer, Chevron (PT Chevron Pacific Indonesia, or CPI), has established its headquarters in the city; Pertamina, the state oil company, has done the same for its Riau headquarters. The city is also home to a large number of companies working in conjunction with oil or plantation industries, particularly in transportation and equipment supply. The city's thirty-three public and private banks manage the highest assets in the province, ten times higher than in Dumai (Bank Indonesia, Regional Economic and Financial Statistics). Pekanbaru also houses the consulates of Singapore and Malaysia.

Paired Operations in the Making

This paired operating system between Pekanbaru and Dumai suggests a hierarchical relationship between a centre of command, Pekanbaru, and a centre for production and maritime interface, Dumai. This hierarchical relationship is also shown by the higher level of equipment in Pekanbaru as compared to Dumai, and its ongoing metropolization process.

Pekanbaru's level of equipment is unparalleled in Riau's other cities. The number of international level hotels in the city attests to its role as a business destination.[22] Pekanbaru is also a university town, with a public university and several private universities (see Map 11.4). Hospital care has recently expanded with the construction of a large public hospital completed in 2007 and two private general hospitals. Since 2009, a mass public transport system has been brought into service with the Trans-metro Pekanbaru, an air-conditioned scheduled bus system serving three routes, to be expanded to six in 2011, a very rare service in medium-sized Indonesian cities. The city bears physical evidence of the start of a metropolization process, driven by the internationalization of the province's economy and amplified since the 2000s by the financial capacities of decentralization. The city centre has been reconfigured and different business centres have sprung up with the construction of large north-south arteries. New buildings have sprouted along these avenues, emblems of the city such as the main library or the province's new headquarters. Currently, luxury estates are opening up along new axes, especially around the Sudirman Square, on the road to the airport. In addition to large hotels, several modern malls have been built, including one by Ciputra, one of Indonesia's largest developers. In Asian cities, the crossing of rivers by the built environment and development of river banks are some of the signs of ongoing metropolization processes and account for a change in urban scale. In Pekanbaru, the construction of several bridges on the Siak and the continuation of the ring road on the other side of the river allowed the expansion of the urban built environment towards the Chevron industrial area of Rumbai. Besides the Chevron headquarters, huge facilities, such as hospitals, universities or the large stadium that will host most of the events of the 28th National Sports Week in 2012, create new polarities there. The development of the riverbank was initiated near the first bridge crossing the Siak and should continue with the installation of a water front city including the old port; a project supported by the Province, but still in need of investors. In comparison, Dumai's internationalization through its transit, export and industry functions, does not produce similar markers of metropolization and its urban landscape has hardly changed.

Map 11.4
The City of Pekanbaru

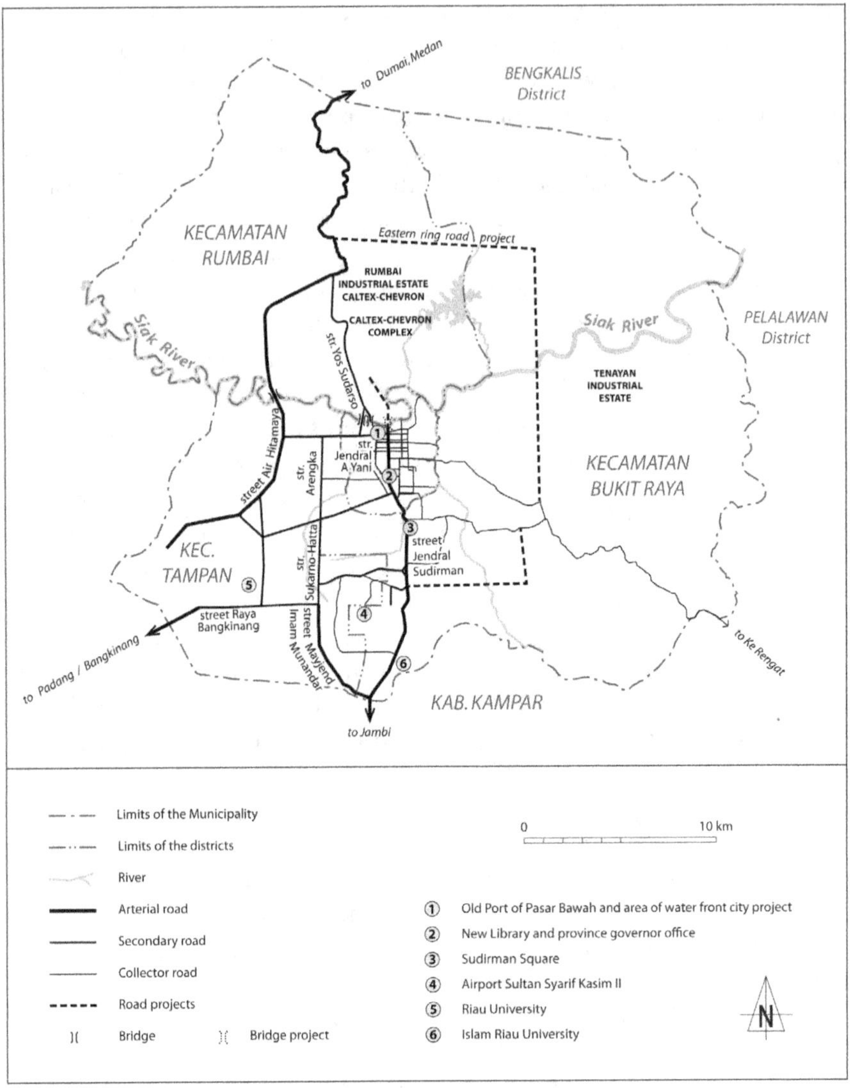

Sources: Various maps from BAPPEDA Pekanbaru.

This complementarity between Pekanbaru and Dumai, however, must be organized as equivalent powers granted to both cities by decentralization could trigger competition and undermine existing synergies. For example, the Municipality of Dumai is already planning to expand its airport runway to accommodate the Boeing 737s of international airlines and open a direct line to Batam, thus bypassing Pekanbaru and creating new alliances between those two port cities. The possibility to function as paired cities is also highly dependent on the rapid execution of the redesigning of the road and rail transport system between the two cities, which alone would allow an efficient connection between the two cities and with the Straits maritime flows.

CONCLUSION

The fact that cities are closely and deeply affected by globalization is not a new concept. Cities function as hubs, articulating material and immaterial flows through their infrastructure and services, as well as handling the presence of transnational actors. Are global flows responsible for urban development or, conversely, is it investing in infrastructures, particularly transport and services, that renders global flows possible, all the while generating specific urban configurations? Globalization impacts cities at all hierarchical levels. This influence is visible in their capacity to command and control flows: the network of large cities is made denser by second-tier cities and by a series of towns with lesser command capabilities which are nonetheless developing connections with global flows. The bottom tier of the urban hierarchy of cities affected by globalization appears highly complex and is rarely studied.

In Asia today, there exists a context of generalized trade liberalization coupled with an ideology, driven notably by the ADB but widely used elsewhere, that focuses investments on corridors to optimize returns, to promote transnational integration, and in some cases to open up peripheral territories. This general context promotes the emergence of new types of nodes along these corridors, which operate as twins or are paired with other urban constructs. Along the land borders of mainland Southeast Asia and the Malay Peninsula, these hubs are mainly positioned on the periphery of national territories, on both sides of the border, at the corridors' national crossing points. They are additional nodes within urban hierarchies. These border cities often pair with higher-ranking, inland cities that handle command functions in administrative, logistic or economic sectors. In maritime Southeast Asia, urban pairs emerge between ports and regional inland cities or dry ports.

There is a marked imbalance in the urban and port hierarchies between both sides of the Straits of Malacca, at Straits as well as at Sumatra island level with a discontinuity of these hierarchies in central Sumatra. As Sumatra is structured by Medan to the north as the island's main international port and Palembang to the south providing access to Jakarta, the emergence of Pekanbaru and its port connection on the coast in Dumai will likely reduce this discontinuity, a function that could not have been taken over by one of these cities alone.

Notes

1 See for instance Taylor et al. (2010) and the GAWC's renowned roster which ranks cities based on their connectivity through advanced producer services, thus measuring their degree of integration into the world system; <www.lboro.ac.uk/gawc>.

2 See the results from "New regional organization in Oriental Asia" (NORAO vol. 1 and vol. 2) research programme, which focuses on new regional patterns that emerge within the regionalization of the world system (Pelletier vol. 1 and Taillard vol. 2, 2004*a*) which show the permanence of the Asia-Pacific maritime axis and of the structural hubs of this Asian maritime corridor (Tertrais 2004; Taillard 2004*a*). See also Gipouloux (2011).

3 See Franck et al. (2012) for a study of metropolization processes in Southeast Asia second tier cities.

4 The IMT-GT growth triangle is a transnational cooperation area, supported by the Asian Development Bank. The institution was created in 1993 to encourage economic ties between regions belonging to three national groups. It covers the southern provinces of Thailand; the Malaysian Peninsula; and includes the whole island of Sumatra after the progressive addition of all its provinces.

5 Riau province was created in 1957, when Sumatra's three provinces were subdivided. These provinces were originally created with the establishment of the unitary Republic of Indonesia in 1950. The subdivision resulted in the resumption for the most part of the limits of the former colonial residencies, in order to better account for regional specificities and improve administrative control. The province of Central Sumatra was divided between western Sumatra, Riau, and Jambi. The capital of the province of Riau was, for two years, Tanjung Pinang, on the island of Bintan, seat of the former sultan of Johor Riau in the eighteenth century and then the Dutch Residency. It had always been very oriented towards Singapore. The capital moved to Pekanbaru in 1959, when Caltex discovered new oil deposits and decided to finance the construction of infrastructures for the city such as roads, universities, and hospitals.

6 Seventy per cent of Indonesian crude oil, a major part of Indonesia's economy, was exported from Dumai in the 1970s and 1980s, thanks to the province's abundant deposits (Colombijn 2002).

7 The physical structure of the island of Sumatra is composed of a series of longitudinal, physiographic groupings with, from east to west, a marshy plain opening out to the Straits of Malacca and a relatively barren peneplain, and the Barisan mountains whose valleys concentrated populations. There is also a narrow plain overlooking the Indian Ocean. Sumatra is drained by a series of rivers with transversal orientations that begin in the Barisan mountains and flow in vast loops towards the east before flowing into the Straits of Malacca and the short rivers to the west.

8 Decentralization has notably resulted in the creation of new administrative entities. For example the district of Bengkalis, divided into three districts (Bengkalis, Rokan Hilir, and Siak) and one municipality (Dumai), has helped small cities accede to the rank of administrative centre. The small port town of Tanjung Siapi-Api has thus become an administrative centre with huge administrative buildings marking the entrance to the city.

9 The province of Kepulauan Riau consists notably of the islands of Batam and Bintan, whose development as an industrial and resort extension of Singapore has been entirely planned and managed by the central government since 1990.

10 For example, cultural associations, such as *Dunia Melayu Dunia Islam* (Malay World, Islamic World) based in Melaka, bring together representatives from Sumatra and the Peninsula. There are a number of student exchanges between Islamic universities on both shores.

11 Until 2006, planning documents mentioned the 50 km Duri-Dumai section as a feeder of the eastern trans-Sumatran route, which was given the status of national route. Since, it has become a part of the trans-Sumatran route.

12 Though it has paved the way for logging and the expansion of plantation crops along the route throughout the central-eastern zone of Sumatra, it also risks creating new inequalities with departments that are far from the route.

13 See the Ministry of Transports (2008). This study compares the result of the origin-destination survey of 2001 and 2006.

14 See Ruth Banomyong's article in this volume and the IMT-GT website <http://www.imtgt.org/Documents/Books/roadmap-development.pdf>.

15 The special port for crude palm oil export in Dumai is also operated by Pelindo 1. It is the largest port handling exports of crude palm oil and derivatives in Indonesia. This port is equipped with various facilities for loading and unloading goods, especially crude palm oil.

16 The traffic from Dumai and Pekanbaru does not of course exhaust the totality of Riau's maritime flows with the Malaysian peninsula. Several large companies, established along the Siak, notably handling wood pulp, own their own docks and export their production directly from their own docks to the other shore of the Straits or by shipping their production to ports located at Siak's mouth, such as Sungai Paknin. Furthermore, small ports dot the coast, exchanging copra — sometimes even as far as Singapore — and latex for basic

consumer goods across the Straits.

17 Especially since the municipality has forbidden trucks to circulate on the roads to the port.

18 The city's old port, Pasar Bawah, and the Perawang port are managed by the Pekanbaru branch of the public operators of Pelindo I ports.

19 Several Singaporean and Malaysian navigation companies stop at Pekanbaru. A regular barge service was established in 2003 between Northport, the container terminal of the Kuala Lumpur port, and Perawang.

20 Pelintung, Lubuk Gaung, Dockyard, Bukit Kapur, and Bukit Timah. The Special Economic Zone covers a surface of roughly 1,000 ha and benefits from fiscal incentives.

21 Four Indonesian exporters of crude palm oil hold the near-monopoly of the market and impose high prices.

22 There is one 5 stars hotel and six 4 stars hotels, all of international standing. The majority belong to large Indonesian real estate development companies.

References

Albrechts, Louis and Tom Coppens. "Megacorridors: Striking a Balance between the Space of Flows and the Space of Places". *Journal of Transport Geography* 11 (2003): 215–24.

Castells, Manuel. *The Information Age: Economy, Society and Culture, Volume I: The Rise of the Network Society*. Oxford: Blackwell Publishers, 1996.

Charras, Muriel. "L'Indonésie extérieure en peine de perspectives" [Outer Indonesia Left in Need of Prospects]. *Archipel 46* (1993): 173–90.

Colombijn, Freek. "A Wild West Frontier on Sumatras East Coast: The Pekanbaru-Dumai Road". *BKI* 158, no. 4 (2002): 743–68.

———. "A Moving History of Middle Sumatra, 1600–1870". *Modern Asian Studies* 39 (2005): 1–38.

Debris, Jean and Claude Comtois. "Une relecture du concept de corridors de transport: illustration comparée Europe/Amérique du nord" [A re-reading of the concept of transport corridors: a comparison between Europe and North America]. *Les Cahiers Scientifiques du Transport* 58 (November 2010): 127–44.

Fau, Nathalie. *Le nord de Sumatra: une périphérie indonésienne sur le détroit de Malacca. Un espace partagé entre intégration régionale et recompositions transnationales* [Northern Sumatra: an Indonesian periphery on the Malacca Straits. A shared space between regional integration and transnational reconstruction]. Ph.D. dissertation. Department of Geography, Paris 10 — Nanterre University, 2003.

———. "La région nord de Sumatra: une périphérie indonésienne sur le détroit de Malacca" [The North Region of Sumatra: A Peripheral Area on the Strait of Malacca]. *Mosella* 28 (2006): 3–4.

Franck, Manuelle, Charles Goldblum and Christian Taillard. *Territoires de l'urbain en Asie du Sud-Est, métropolisations en mode mineur* [Urban territories of Southeast Asia and minor metropolisations]. Paris: CNRS Alpha Editions, 2012.

Fujita, Masahisa, Paul Krugman, and Anthony Venables. *The Spatial Economy: Cities, Regions and International Trade*. MIT Press, 1999.

Fujita, Masahisa and Jacques-François Thisse. "New Economic Geography: An Appraisal on the Occasion of Paul Krugman's 2008 Nobel Prize in Economics". *CEPR Discussion Paper* #7063, December 2008.

Gipouloux, François. *The Asian Mediterranean: Port Cities and Trading Networks in China, Japan and Southeast Asia, 13th–21st century*. Edward Elgard Publishing, 2011.

Hardoy, Jorge E. and David Satterthwaite, eds. *Small and Intermediate Urban Centers: Their Role in Regional and National Development Within the Third World*. IIED, 1986.

Lainé, Elsa-Xuan. "Les villes frontalières thaïlandaises dans la régionalisation: planification et production de nouvelles formes urbaines" [Thai border towns in regionalization process: spatial planning and production of new urban forms]. *Mousson*, no. 17 (2011): 51–75.

Lermoyer, Martin. *Organisation et développement du secteur de l'huile de palme dans la province de Riau en Indonésie depuis 1997* [Organisation and development of palm oil sector in Riau province since 1997]. Master degree paper. University Paris Diderot, 2011.

Lubis Harun Al-Rasyid, Sjafruddin Ade, Isnaeni Muhamad, and Dharmowijoyo Dimas. "Multimodal Transport in Indonesia: Recent Profile and Strategy Development". *Proceedings of the Eastern Asia Society for Transportation Studies* 5 (2005): 46–64. Available at <http://www.easts.info/on-line/proceedings_05/46.pdf> (accessed June 2011).

Manguin, Pierre-Yves. "Les cités-Etats de l'Asie du Sud-Est côtière. De l'ancienneté et de la permanence des formes urbaines" [City States of Coastal Southeast Asia: Some Permanent and Ancient Urban Forms]. *BEFEO*, no. 87 (2000): 151–82.

Ministry of Transports (Indonesia). *Perencanaan technis physical distribusi angkutan barang strategis di pulau Sumatra* [Physical planning of the distribution of strategic good transports in Sumatra]. 2008.

Moriconi, François, Eric Denis and Kamala Marius-Gnanou. "'Repenser la géographie économique': Les arrangements du rapport de la Banque Mondiale avec les sciences géographiques urbaines" ["Reshaping Economic Geography": The World Bank Report Arrangements with Urban Geographical Sciences]. Cybergeo: European Journal of Geography, online 20 May 2010. Available at <http://cybergeo.revues.org/23144> (accessed 25 January 2013).

MP3EI. *Master Plan for Acceleration and Expansion of Indonesia Economic Development 2011–2025*. Republic of Indonesia, Coordinating Ministry for Economic Affairs, 2011, p. 207.

Pelletier, Philippe, ed. *Identités territoriales en Asie orientale* [Territorial Identities in Oriental Asia]. NORAO vol. 1. Paris: Les Indes Savantes, 2004.

Rimmer, Peter. "The Asia-Pacific Rim's Transport and Telecommunications Systems: Spatial Structure and Corporate Control Since the Mid-1980s". *Geojournal* 48, no. 1 (1999): 43–65.

Rodrigue, Jean-Paul, Claude Comtois and Brian Slack. *The Geography of Transport Systems*. Routledge, 2006.

Syahrial Syarif. "Characteristics of Small-scale Enterprises in West Sumatra". Universitas Andalas, Padang. *Asian Academy of Management Journal* 9, no. 2 (July 2004): 87–95.

Taillard, Christian. "Un Exemple Réussi de Régionalisation Transnationale en Asie Orientale: les Corridors de la Région du Grand Mékong" [A Successful Example of Transnational Regionalisation in Oriental Asia]. *L'Espace Géographique*, no. 1 (2009): 1–16.

Taillard, Christian, ed. "Les Nouvelles Intégrations Régionales en Asie Orientale" [New Regional Integrations in Oriental Asia]. In *Intégrations Régionales en Asie Orientale*. NORAO vol. 2. Paris: Les Indes Savantes, 2004*a*.

———. *Intégrations Régionales en Asie Orientale* [New Regional Organisations in Oriental Asia: Regional Integration in Oriental Asia]. NORAO vol. 2. Paris: Les Indes Savantes, 2004*b*.

Taylor, Peter, et al. *Global Urban Analysis: A Survey of Cities in Globalization*. Routledge, 2010.

Tertrais, Hugues. "L'axe de Croissance de l'Asie-Pacifique" [The Growth Axis of Pacific Asia]. In *Nouvelles Organisations Régionales en Asie Orientale: Intégrations Régionales en Asie Orientale* [New Regional Organisations in Oriental Asia: Regional Integration in Oriental Asia], edited by Christian Taillard. Paris: Les Indes Savantes, 2004.

12

THE RE-EMERGENCE OF IPOH CITY
Toward a New Urban Pair with Kuala Lumpur

Amel Farhat

The implementation of transportation infrastructures has played a key role in the economic development of Malaysia by supporting the rise of urban and industrial centres. The extraction of natural resources, notably tin and rubber, has encouraged the establishment of roads and railways from the nineteenth century. Production areas, while witnessing an increase of their transport network, are linked to the major international gateways such as Klang, Penang and Singapore (Fau 2003). The States on the west coast of Malaysia — Penang, Perak, Selangor, Negeri Sembilan, Melaka and Johor — have thus seen an early development rarely known across Southeast Asia (World Bank 1955). Consecutive governments of Malaysia have conducted a development policy focused on the implementation of clusters along a corridor from North, Bukit Kayu Hitam and Padang Besar, to South, Johor Bahru, which is neighbouring Singapore (Fau 2003). The nodes of the corridor, by having a range of multimodal transports based on rail, highways, airports and ports, concentrate flows of investments, goods and labour. The connectivity based on a modern transportation network, does indeed allow the emergence of nodal cities, but also an integration of margins (see Map 12.1).

Map 12.1
Transportation Network of Malaysia Peninsula

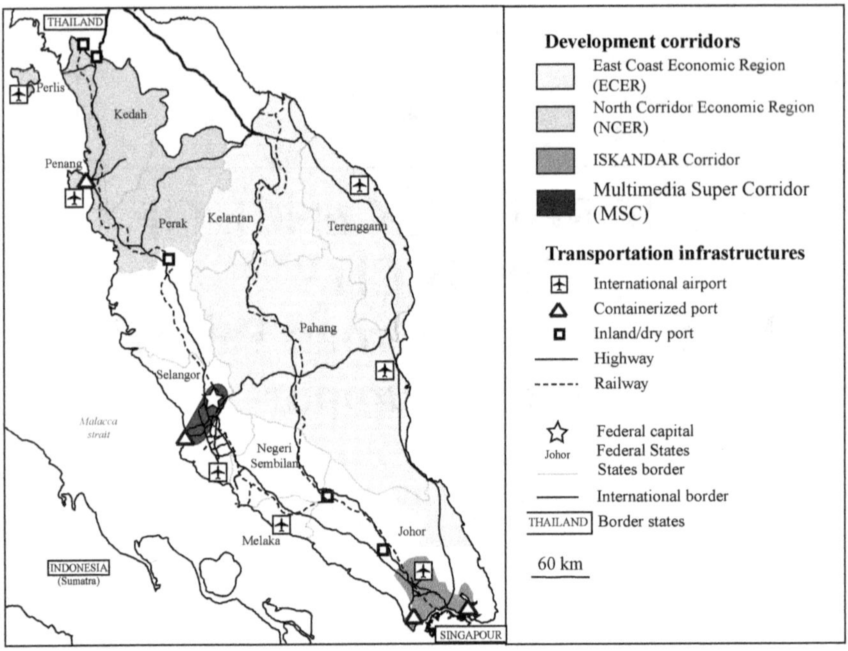

Source: Ministry of Transport, Malaysia.

The first-rank cities of Malaysia, Kuala Lumpur and Penang, cover an international scope but are also facing congestion issues which can to a certain extent slow down their growth. To tackle the congestion matter, the use of storage areas and better interconnection with the hinterland has been implemented. The result of this spatial development is an increased complexity of the national and regional urban network and hierarchy (Goldblum and Franck 2007). This spatial reorganization leads to a redistribution of growth through the region and stimulates a great nodal region. The logistic model of a dry port is therefore used as a main tool which by its hinterland location and its land transportation infrastructures allows a terminal to act as supporting port (Walter and Poist 2003). Although it is not a landlocked country, Malaysia benefits from a dry port to relay its congested coastline (i.e. Naziri Khalid's contribution). By absorbing the excess of flows from the littoral, the dry port acts as a structural interface influencing the emergence of a second-rank city (Fujita and Mori 1996; Haynes and Button 2001;

Rodrigue et al. 2010). The implementation of such an infrastructure was made in the late 1980s in Ipoh, the old and declining industrial capital city of Perak. By establishing a dry port, it is not only a project of decongestion of Selangor and Penang which is planned, but also the redevelopment of a whole urban and industrial space of peninsular Malaysia. This node, nestling between two main hubs of national and regional importance, not only acts as an articulation, but is also an emerging "pair-city" relying on the gateways of the North/South corridor (cf. Manuelle Franck's contribution). It is therefore important to observe the impacts of Ipoh dry port on urban and regional scales. In this case, we examine how it may redefine the economic composition and include the city in the main national and transnational corridor.

The connection between Ipoh and the main poles of Malaysia has been conditioned by the richness in natural resources of the Kinta Valley. However, the artificial increase in tin prices combined with the decrease in its worldwide demand has caused a phase of economic crisis, influencing the integration and the marginalization of the city within the development policies. Long forgotten, Ipoh has again become a major logistical hub by handling container flow through its dry port. It is therefore no longer on the map for its natural resources but because of its infrastructural asset, its links with the main international gateways and its inclusion within the corridor.

1. THE STRATEGIC LOCATION OF KINTA VALLEY

The strategic location of Ipoh between two major cities, combined with the richness of natural resources at the city's disposal, has eased its integration within the North-South corridor. The role of the British in influencing the early development of the transportation network along the western part of the peninsula to serve their need for moving extracted tin and rubber, cannot be overlooked. As a direct result of this network, the Kinta valley gained importance as a node on the Penang/Ipoh/Kuala Lumpur axis.

At the Halfway-point between Two Economic Nodes of Malaysia

Ipoh, the capital city of Perak, leans at the confluence of the Pinji and Pari rivers, within the Kinta valley (see Map 12.2). This territory is mainly shaped by mountainous landscapes, made from limestone and granite hillocks from northwest to southeast and from the Keledang Mountain, extended from north to west. This topography is however not a geographic difficulty since it abounds with natural resources, such as tin, limestone and some marble.

Map 12.2
Geographical Setting of Kinta District

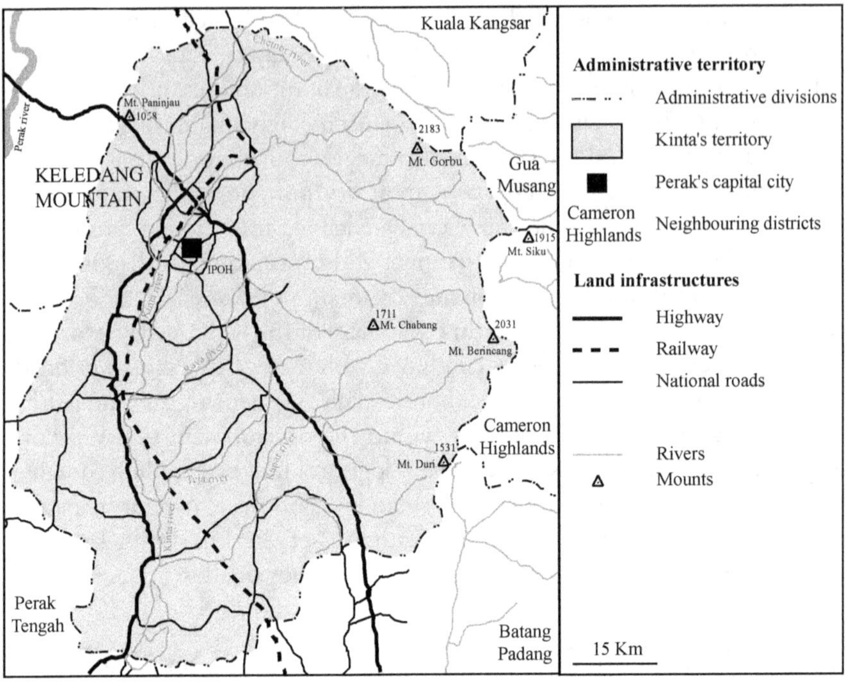

Source: From the map "Daerah oleh direktorat permetaan Negara, Malaysia, 1992".

At first, the river system, notably Perak and Kinta, allowed the Perak State to transport its production and to develop interconnections with the western coast of Malaysia. The land transportation implemented from the nineteenth century led to the building of a North-South axis linking Perak to Selangor and Penang, through Ipoh.

Nowadays Ipoh represents an effective junction within a multimodal corridor from Penang to Kuala Lumpur. Ipoh is an unavoidable passage for land-based transportation which gives this territory significance on a national scale. As a result of its location close to economic and decisional centres, 150 km from Penang and 200 km from Kuala Lumpur, this node has developed into a major connection which became an asset within the national transport policy. As the capital of Perak, Ipoh concentrates many administrative offices allowing a direct link between the national policy and a local application.

The hinterland situation did not allow the development of port infrastructures and made Ipoh economically and spatially dependant on the connections with the international gateways of Kuala Lumpur and Penang. Therefore, even if Kinta was a centre of mineral resources, it could not have been a first rank pole and became strongly linked with Klang Port and Butterworth Port. However, its economic importance with regard to mineral and industrial production has enabled a rapid development of transports, providing a first-rate spatial integration within Peninsular Malaysia. Thus, although Ipoh is not included at the top of the urban hierarchy, the connections built with Kuala Lumpur and Penang through its integration within the transportation systems has made it an important node of the North-South corridor.

Ipoh, a City Inserted into a Multimodal Transport System

Ipoh is a nodal city structuring the west coast of Peninsular Malaysia. Its location has allowed the development of transport infrastructures rapidly connected to the political and economic centres. Although the historical transportation was only possible through the Kinta River, rail and road have rapidly developed as the main transport system.

First, the North-South highway, surrounding Ipoh, appears to make the road network optimal within the *daerah* Kinta. Moreover, the city centre is directly accessible through a dense and well-developed road network. However, this road infrastructure should not only be studied within Ipoh boundaries, but must take into consideration the Kinta valley's conurbation. As an important city of Perak, both politically and economically, Ipoh is located at the centre of a metropolitan area. Indeed, the dense mesh of roads alongside the highway have an integrated approach across the urban area and have contributed greatly to the transition from one city to another without any real differentiation of urban landscape. Within this urban network, the most dynamic cities located on the outskirts of Ipoh are directly connected by expressways which are the direct result of a development strategy. Although Ipoh is the main hub of Perak, key industries and free trade zones are indeed organized along the urban belt surrounding the North-South highway.

Secondly, the route of the railway follows the path of the expressway and is soon to become a major axis of Peninsular Malaysia. By linking Kuala Lumpur to Penang through Ipoh, its modernization allows an optimization of the dry port's relay and logistics functions. The railway station, having been established during the colonial period, remains an important landmark and infrastructure located at the heart of the historical city centre. The national railway company of Malaysia, KTMB (Keretapi Tanah Melayu Berhad), has heavily invested in the upgrading of the railway station in order to rationalize and distinguish the flow of passengers and goods. While its colonial architecture and its central location provide a tourist attraction, local authorities have also used it as a dominant logistic tool. The construction of the dry port has been paired with the already existing railway station, avoiding the construction of an additional station on the periphery of Ipoh. The location of a logistic platform within the city centre constitutes multiple advantages. On one hand, the site is perfectly integrated into the dynamics of input and output flows near the highway. On the other hand, the dry port is directly accessible to the professional transport companies based in Ipoh without requiring a shift to the periphery.

Both road and rail infrastructures represent the basis of this multimodal logistical hub linking major seaports of Malaysia to Ipoh. The modernization of the railway network for the Kuala Lumpur-Ipoh and the Ipoh-Penang sections are of major importance and contribute to strengthen the position of Ipoh as a node of the peninsula.

A Situation Enhanced by Natural Resources Assets

The natural resources of Kinta valley are mainly based on tin and limestone mines. Although the extraction of raw materials was carried out before the colonial period, it was really the British administration from 1878 and the application of modern techniques that made Perak a fast and early developing pole of Malaysia. The establishment of an organized policy for export led to modern extraction techniques and to the urban and transport planning of this area.

The landscape of Kinta is made of a multitude of limestone hillocks that can be seen along the highway. The hilly setting has resulted in an industrial network scattered around Ipoh, making it one of the most important extraction sites in Malaysia. Indeed, the construction manufacturing is likely to need limestone since it represents the main component of cement (80 per cent) and grained mortar (75 per cent). The impact of mineral resources on the development of an urban core and the implementation of a logistic system has been crucial. During the nineteenth century, the "tin rush" marks the beginning of a spatial organization focused on its extraction, refining and transport. The extent of Ipoh tin fields was unmatched and constituted the world's most important mines. This abundance allowed the extraction to focus on the upper layers of soil, representing thus a substantial financial manna that did not require heavy investments in building deep mines. The first facilities were made in 1898 when a rail route was built in order to connect Ipoh mines to the Kinta River to transport the tin to Selangor and Penang (Leinbach 1975; Ho 2009). This kind of infrastructure, which was usually located along the coastline, was made for the first time on a hinterland territory, indicating the significance of Kinta.

Mine exploitation has represented a true economic boom for Kinta and Perak State, allowing the development of an early industrial foundation. The actual land planning has been shaped from this situation, strengthening the strategic location of Ipoh which became highly connected within the Malaysian peninsula.

2. THE INDUSTRIAL INHERITANCE: AN EARLY NODE OF MALAYSIA

The economic history of Kinta consists of phases of growth and crises. By these fluctuations, the industry, as a structuring element, has led to many socioeconomic disruptions. The evolution of the demography and its influences on the economic structures has been highly dependent on the mining sector. The great crisis period is correlated with the decline in tin use which led to a significant deterioration of the urban and industrial setting. However, the redevelopment of Ipoh as a nodal city and the re-emergence of Kinta on a national scale have been carried out through the areas of logistics and transportation systems. This rediscovery of Ipoh has allowed an optimal integration of the Kinta Valley with the main international gateways along with urban renovation, industrial rebirth, and the emergence of the tertiary sector.

A Socioeconomic Structure Based on Natural Resources

Ipoh was originally a Malay *kampung*, surrounded by a tropical forest populated by Orang Asli. The discovery of tin led to an attractiveness of the site, promoting a high level of immigration, mostly composed of Chinese and marginally of Indians (Hirschman 1986). More complex methods imported by the Chinese and the British have allowed an increase and a significant enhancement of tin production. This demographic change has also resulted in a strong decrease in the Malay ratio to date.

Among the miners, the colonial authorities established that 90 to 95 per cent of them were Chinese (Hirschman 1986; Ho 2009). This immigration was certainly widely promoted by the economic attractiveness of mineral resources, but also by the colonial administration's policy (Dodd 1969; Chin 2000). Ipoh illustrates the case of racial classification established during the colonization period. Indeed, during this period the composition of the population in Malaysia was modified and restructured (Dodd 1969; Ho 2009). Beyond racial assumptions, a strong spatial segregation took place through a strict ethnic distribution of economic activities, imposing a single ethnicity for each company and association. The British thus favoured the absence of horizontal links between communities, causing a rigorous ethnic division of the Malaysian population until the decolonization period (Chin 2000).

Demographic growth has naturally shaped the formation of an important urban pole. Ipoh, as a tin mining zone, has attracted foreign labour which allowed the establishment of a commercial network and significant enrichment. Thus, from 8,900 inhabitants in 1879, 58,600 in 1892, the Kinta district's population is nowadays over 700,000 (Ho 2009). During the early nineteenth century, about two-thirds of the population were tin miners, while the rest were involved in commercial activities and limestone quarries. From a traditional Malay kampung, it gradually became a modern colonial city. The importance of the population size has allowed the emergence of commercial sectors, including large trading companies within Ipoh's growing city centre. The predominance of the Chinese community led to the establishment of social patterns that influenced the economic activities of the city. Mainly based on gaming businesses (casinos), banks, (Hong Kong and Shanghai Bank), hawkers and restaurants, and leisure activities (theatre, opera, shadow puppetry) the economy was flourishing throughout the nineteenth and up to the mid-twentieth century (Ho 2009). The increasing number of banks reflects the financial flow generated from Ipoh, and as a result, this golden age led to the transfer of political powers, from Taiping to Ipoh, which became the capital city of Perak State in 1927. The colonial period remains a lasting mark on the urban space, highlighting the old city centre and providing abundant historical landmarks.

A City Specialized in the Industrial Sector: Risks and Declines

Ipoh industries were based on two elements: the transformation of limestone and the processing of tin ready for export. However, the lack of diversification in the secondary sector contributed to a significant exposure to the risks of economic fluctuations.

Ipoh rapidly became one of the leading industrial centres in Malaysia between the nineteenth and the mid-twentieth century. The establishment of companies such as Lafarge helped to develop extraction and industrial processing settlements. However, the importance of the secondary sector was mainly due to the abundance of tin. Mining industries were concentrated in the immediate vicinity of Ipoh, with direct access to the railway line. The fields were unevenly distributed between communities, with a near-monopoly of British companies. The leadership of a few companies in the tin industry, not only across Malaysia, but also worldwide, led to a corrupt and risky price control policy.

The prices of tin underwent an important fluctuation after the First World War. In order to avoid massive economic loss, private companies and governments of Great Britain and The Netherlands decided to stabilize prices during the Bandoeng Pool in 1921. By trying to regulate the collapse of tin prices, private companies started to virtually increase the production value. The holder of a near worldwide monopoly, the Anglo-Oriental Mining Corporation, through its economic choices, implemented a risky financial system by directly buying its own production in order to ensure the price increase. Indeed, the policy of the owner of a quarter of the world's tin supply resulted in an unstable economic situation for Kinta.

With the reduction of tin use, these kinds of financial decisions have been disastrous for tin-dependent cities. The introduction of plastic, including PVC, caused the reduction of demand for tin, a non-renewable resource. The advantages of plastic, regarding its weight, but also its affordability and modernity, rapidly dethroned the use of tin. The desire to replace tin with plastic in the worldwide markets from the second half of the twentieth century onwards has been the trigger of a latent crisis period. As a result of the mono-industrialization of Kinta, Ipoh's secondary sector and most of its economic activities declined rapidly. In such an atmosphere, only the limestone industry has managed to continue, digging further and further from the city. However, this sector was not such a structural element as tin, and could not prevent the downfall of the local economy.

Until the 1980s, Ipoh was an important nodal city of the urban hierarchy of Malaysia. However, its downgrading was not followed by a proper integration of Kinta in the national development policy, allowing the gap to increase with Selangor, Penang, and more recently Johor. On the contrary, the hinterland location of Ipoh, left without any major natural resources, led to an isolation factor after its rapid decline between 1980 and 2000.

3. THE KEY ROLE OF IPOH CARGO TERMINAL WITHIN THE DEVELOPMENT STRATEGY

Despite the economic downturn, Ipoh managed to overcome this crisis and became one of the leading points of Malaysia's development plan. Mainly because of the infrastructure heritage and its central location in the peninsula, the city is rapidly gaining back its importance. Based on a federal decision, the construction of a dry port began in 1989, at the core of the industrial crisis. This kind of infrastructure is implemented in order to link back and renew the industrial importance of Ipoh within the country. As the only well developed city located in the hinterland, it was also the occasion for Malaysia to convey the importance and its concern for a modern and efficient logistic system. Moreover, the Ipoh Cargo Terminal (ICT), makes it possible to restore the value of the comparative advantages of specialized workforce in the industrial sector, close to both Kuala Lumpur and Penang's expensive labour. The ICT is a valuable tool for Ipoh, providing a possibility for the reindustrialization and reintegration of the whole of Perak in the peninsula.

A Strategic Location and Organization for the Ipoh Cargo Terminal

The ICT was an ambitious project in 1989 and was directly integrated into the logic of redistribution of flow from the overcrowded ports of Klang and Butterworth, allowing Perak to implement new advanced port infrastructures. Apart from Lumut port, which is home to a small and non-containerized flow of goods, Perak cannot accommodate large tonnages. Moreover, the use of a third port on the western coast of Peninsular Malaysia would not have been of a proven relevance. The choice of a dry port implementation in Ipoh was therefore the result of geographical setting, but also of its already existing logistic facilities. The proximity and the connection with two of the three most important containerized ports of Malaysia gave ICT a chance to host a significant flow of goods. The construction of such logistic infrastructure has been pioneering in the context of a non-landlocked emerging State.

The administrative area of customs administration and storage is situated immediately to the East of the railway station, downtown Ipoh, offering an ideal location as it enables prompt input and output of containers from the ICT, joining a few metres further the KTMB network. This direct connection is also coupled with a fast link with the highway, located a hundred

Table 12.1
Shareholder Distribution of Ipoh Cargo Terminal in 2009

Shareholder	Share Capital
KTMB	30%
Perak Development Authority	25%
Port Klang Authority	15%
Penang Port Authority	15%
Johor Port Authority	15%

Source: Ipoh Cargo Terminal, 2009.

metres from the dry port. The location and the logistics skills of ICT have allowed it to attract three-quarters of container flows handled by KTMB, becoming one of the most important hinterland logistic platforms of Southeast Asia (see Table 12.1).

ICT is a semi-private company, owned by public and semi-public capital, placed under the authority of the Ministry of Transport, resulting in a total control of stocks by the federal State. Among them, the national company of Malaysian railways, the only semi-public company, owns more than one quarter of the shares which reflects the importance of transit flow through rail networks. Furthermore, the second largest shareholder, the Perak Development Authority, plays a dual role: on one hand, it stands as the public relation office with private companies for Perak; and on the other hand, it relays the national state policy made in Selangor. Finally, the three major Malaysian ports are also equal investors, highlighting the relay function of logistic activities. Although these ports possess the same number of shares, their relationship with ICT is nevertheless not the same in terms of flow.

Since Johor developed its own dry port in Nilai, ICT, which is over 500 km north, does not represent an efficient inland logistic platform for southern Malaysia and therefore does not provide logistic services. However, its links with Port Klang and northern Malaysia are historically strong and the storage and transport services are a useful tool that contributes to the unloading of containers from Selangor and Penang.

Roles of Ipoh Cargo Terminal as a Dry Port

The ICT is arranged so that all elements are visible and easily accessible. This organization helps to simplify the business procedures and flows encountered in the logistic platform. The Ipoh dry port has a range of services designed to manage the flow of containers and goods. Companies using the facilities pay for a service that includes a full administration and supervision of each container, including the insurance of goods. ICT infrastructure allows for fluid trade, especially by combining its site with the railway station as well as taking care of tax collection. This site does indeed have a Malaysian customs office located right in front of the container yard and financial and information procedures can be easily done with the ICT offices which provide a traceable informatics system. Finally, the dry port also hosts logistics and transportation companies, acting as a professional link between the ICT and the industrial sector.

The organization of the site is also related to a process of urban development. The Perak State government, responsible for the promotion of the economic development of Ipoh, relies heavily on this logistic asset. This infrastructure certainly represents a comparative advantage since its prices are more competitive than those of Selangor and Penang container yards.

Ipoh Cargo Terminal Links with Seaports: An Opening to Global Flows

The container yards are able to contain six rows, filling up to 800 and 1,440 TEUs. Furthermore, the recently renovated warehouses, located a few metres away, are able to host a mass of goods within 41,000 sq. ft. Given the limited capacity of the yard, warehouses compose the main tool and show the optimal management of input and output flows. The management of goods is, indeed, directly conducted to make available space for both local businesses, but also the important needs of seaports.

The use of the Information and Technology (IT) sector enabled ICT to have a real-time contact with the leading ports of Malaysia. It is thus a structured management system that works in order to anticipate the movements of input and output flows. By understanding the needs which arise in port management, ICT contributes to the congestion regulation of the international gateways of the Malaysian west coast.

The ICT's infrastructure is a strategic logistic platform attracting a large flow of goods; however, the downward trend reflects the cessation of operation with Penang Port from 2008 (see Table 12.2). As a Malaysian economic pivot, the federal government, historically led by the UMNO (United Malays National Organization) and the *Barisan Nasional* (National Front) coalition, has always kept an eye on Penang State's elections and was seriously concerned by the triumph of the opposing political force, the *Pakatan Rakyat*

Table 12.2
The ICT Container Flows between 2001 and 2009 (TEU)

Year	Full	Empty	Revenue
2001	34,408	23,044	–
2002	27,991	19,261	–
2003	26,575	18,079	–
2004	27,764	18,489	278,341,528
2005	25,296	17,811	264,563,573
2006	26,518	18,073	281,736,539
2007	26,710	17,562	289,843,169
2008	26,284	19,603	334,026,809
2009	19,972	15,150	–

Source: Ipoh Cargo Terminal, 2010.

(People's Alliance) led by the Democratic Action Party (DAP). As what can be seen as retaliation or a challenge for the new ruling party of Penang State, the federal power influenced a partial redirection of Penang Port's activities to Klang Port, and among them its strategic link with ICT.

4. IPOH'S URBAN RENEWAL SINCE 2000

The industrial crisis that has affected Ipoh has enabled to highlight the risk of an economy dependent on a limited number of resources. The Malaysian development policy, often focused on Selangor, has been renewed in order to smooth the diffusion of national growth to other parts of the country, including Ipoh. The diversification of activities in secondary and tertiary sectors thus represents the desire for a local economic restoration. Perak, which used to be only focused on limited industrial activities, is currently strengthening its services sector.

The Urban Development of Ipoh

Since the beginning of the 2000s, the economic integration policy of Malaysia is at a turning point. Since 2003, the Badawi government has been redeploying its national interests in order to produce growth, not only in the first-rank cities, but also in its peripheries. The development of Ipoh and its outskirts is then included within a national plan in order to produce a new industrial and tertiary cluster (see Map 12.3).

Map 12.3
Ipoh's Old Town: A Tourism Plan based on Urban Renovation

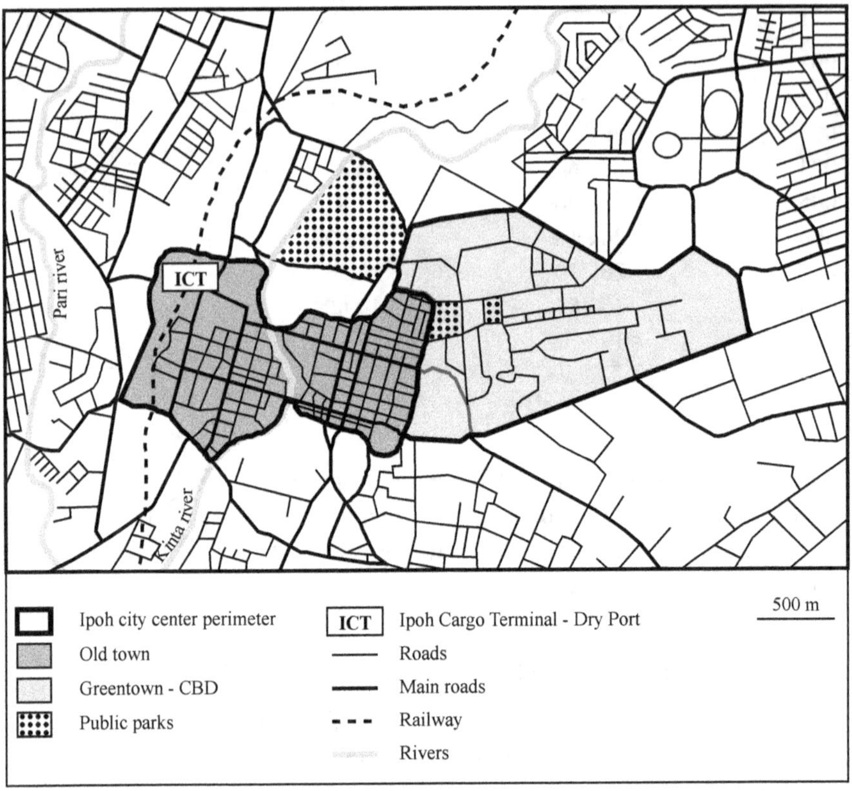

Source: Google Earth. Amel Farhat (2011).

Ipoh's Old Town: A Tourism Plan Based on Urban Renovation

Although Ipoh has developed modern facilities and a CBD, it is without a doubt a city that highlighted its Chinese and British colonial heritage. As a result, a double centrality has been developed. The colonial part is located to the west of the city centre and is organized around the train station, the ICT, the municipality and the Perak government offices. The presence of these three elements illustrates the importance of this area which is not only touristic but it is also of local use.

The old city is certainly the main tourist area of Ipoh. Far from being an abandoned space, this area is undergoing an urban rehabilitation, especially in the area surrounding the *Concubine Street*. Indeed, the municipality's development policy follows an Asian trend, highlighting a cultural awareness by avoiding the destruction of its historical heritage. However, conservation is perfectly understood as a tourist asset, following the example of Penang's old city project. By its location, Ipoh does indeed have the possibility of attracting visitors going from Kuala Lumpur to Penang. Tourist activities are both a success and a failure; while it is a popular leisure area for Malaysians and Singaporeans, this might not be the case for other nationals. It is therefore a city of domestic tourism, hardly going beyond its national borders. Far from being only a tourist place, the old town of Ipoh is also concentrating higher administration offices, such as the magistrate's offices, the municipality and the State government buildings.

Greentown: An Indicator of Economic and Urban Modernization

Greentown, the new district of Ipoh, is located immediately on the east of the old town. As the name suggests, small parks and urban gardens are among the main characteristics of the area. Since its completion in 2009, this new centre concentrates urbanite economic activities around a main building known as the "Business Center" and the newly established companies are largely from the tertiary sector (banking, insurance, start-up, private schools etc.). Greentown has become the new recreation centre for residents of Ipoh's conurbation. Its two malls and numerous cafes and restaurants make it a great spot for modern consumers. However, far from being a copy of Kuala Lumpur City Centre, the economic centre is largely inspired by Chinese architecture, with low buildings. This example also illustrates a modern urban development based on airy streets, gardens and inspiration from the local culture.

This dual centrality is complementary and does not produce a competitive situation. The creation of a modern district along with a real will to conserve the old city was until recently not part of a process commonly found in developing countries. This unique urban rehabilitation and renovation is directly linked to its economic development. Indeed, Ipoh, with its many logistical advantages, is a destination that has rediscovered its attractiveness for business implementation and the relocation of secondary and tertiary sectors.

The Evolution of Ipoh's Urban Frame

Map 12.4 allows the residential and industrial development of Ipoh from 1985 to 2000 to be studied. Firstly, the residential settlements are clearly growing, especially between the 1995 and the 2000 period. The implementation of new residential areas makes dense outskirts for Ipoh. Former districts were rebuilt and new areas are highlighted within the development plan. The immediate periphery of Ipoh is the site of middle and upper middle class neighbourhoods. Meru is the perfect area that sets an example of Ipoh metropolitan expansion. Exclusively residential and tertiary, this part of the city is organized around the "Meru Valley" gated community, accommodating wealthy Chinese families, western expatriates, retired Japanese and Middle Eastern visitors. On the contrary, it is significant to note that the residential parts of Ipoh city centre have been impoverished. However, along with the rehabilitation of the colonial part and the implementation of Greentown, these areas are soon to undergo a renovation according to the urban planning unit. Recent residential projects drawn up after 2000 are few and mainly located in the outer periphery. These new spaces, mainly consisting of modern condominiums, present similarities with the property market of Penang or Kuala Lumpur.

Secondly, the industrial areas are experiencing a strong growth, starting from 1995. These new facilities do not constitute the only clusters, but complement the existence of a former industrial park. On one hand, the expansion of the industrial sector is mainly composed of heavy and processing industries in the south-west, while on the other hand the north focuses on high technology. The tertiary sector was established mainly between 2005 and 2009. Within the region of Ipoh, there are three main poles aimed at services. The modern city of Ipoh, Greentown, is presented as the new Central Business District of Perak. In addition, the Simpang Pulai Free Zone can complement and support a pre-existing heavy and chemical industry. Finally, the main tertiary pole in the region of Ipoh is located in Meru, a site exclusively focused on research and development and new technologies. The implementation of a gated community in the North of this area reflects the need for a new type of housing suited to the tastes of expatriates and local educated workforce.

Under the 9th Malaysia Plan (2006–10) the federated States of Malaysia were part of a strategy based on the distribution of economic growth through seven development corridors across the country. The official structure and the institutional framework are used as an implementation tool promoting investment and interstate development planning. Hence the NCER includes, since 2007, the States of Perlis, Kedah, Penang and northern Perak in order to coordinate and plan economic policies adapted to north-western Malaysia. The Kinta valley, as a new local centre, managed to obtain a derogation in order to be included within the Northern Corridor Economic Region (NCER) and became a free zone providing a multimodal option of transportation and research infrastructures (implementation of Universiti Teknologi Petronas' department). This development of the periphery led the urban structure into forming a conurbation centred on Ipoh and supported by an institutional framework promoting local development under the NCER (see Map 12.4).

Map 12.4
Ipoh Development Region Layout from 1985 to 2000

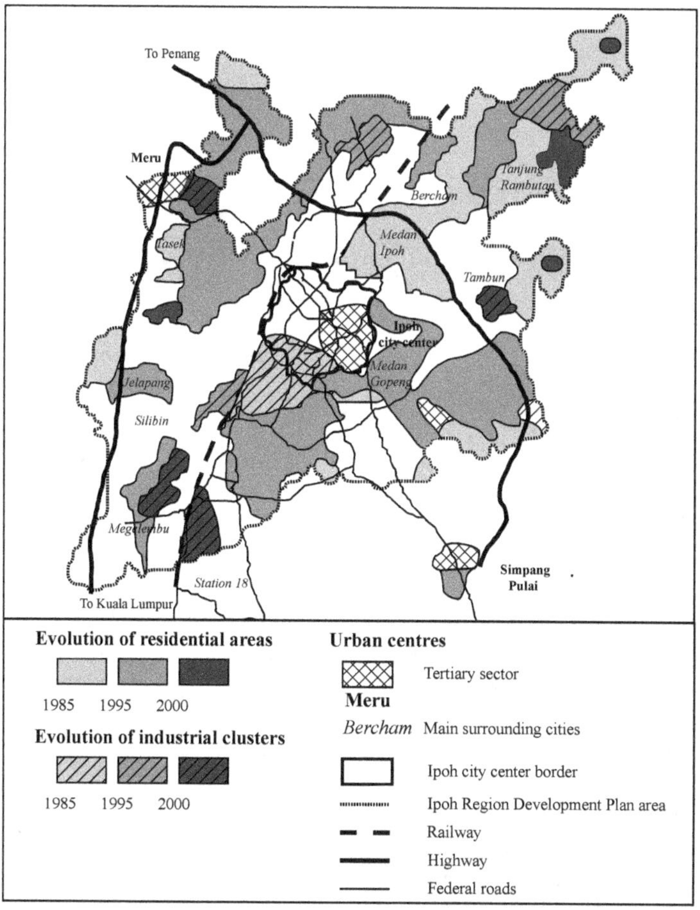

Source: Ipoh Structure Plan 1986, 1995; Amel Farhat (June 2010).

The Emergence of Economic Corridors in Kinta Valley Development Region

The local development policies of Ipoh are structuring specific spatial axes of expansion. Economic growth is thus not only due to the city centre, but also incorporates all the peripheries, forming a true metropolitan system within Kinta. Although the borderlines between Ipoh and the neighbouring cities are not well marked, it is not difficult to distinguish the economic activities of each part by observing the landscape.

The formation of many corridors between Ipoh and its periphery tends to strengthen the city on a national scale while structuring the regional level. While Ipoh is a tertiary centre, the periphery focuses on new settlements of the industrial sector. The types of activities are distributed between the heavy industries and high technology (see Map 12.5).

Map 12.5
The Regional Organization of Kinta Valley Development Region

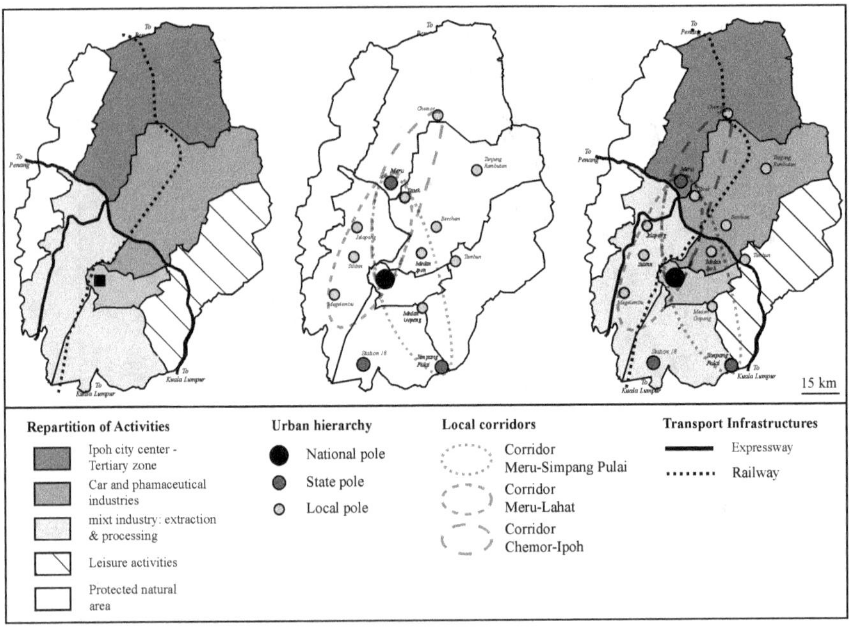

Source: Ipoh Structure Plan; Amel Farhat (June 2010).

Heavy industries are based on two main sectors. The extraction of limestone has never ceased to make quarries and processing plants marking out Simpang Pulai area, in the southeast part of Kinta. Moreover, the automotive industry is a major part of the heavy industry located in Bercham and Taman Tasek Ipoh. These production clusters are combined with research and development infrastructures of companies like Toyota and Kia. High technology is the other industrial sector and is mainly concentrated in the northwest and west of Ipoh. This type of activity is based on electronics, cosmetics, pharmaceuticals and information. These industries and the implementation of two clusters, one in Simpang Pulai specialized in heavy industry and the other in Chemor-Meru, focused on high technologies, were based on a highly skilled workforce whose wages are far more affordable than in the other urban centres of Malaysia.

Three major corridors structure the Kinta valley development region: Meru/Simpang Pulai, Meru/Lahat Ipoh and Ipoh/Chemor. The ICT is a key infrastructure strongly related to the industrial and technological development of these corridors. The dry port acts as the central logistical hub for companies, allowing the supply and diffusion of manufactured goods. Each of these axes uses the infrastructures of this dry port in order to export their production to Kuala Lumpur and Penang. According to the ICT management, in 2009 approximately 40 per cent of the goods produced in Kinta used the dry port facilities in order to export from Perak. The convergence of these corridors highlights a fourth axis of growth between Ipoh and Meru. This area constitutes the heart of Kinta's conurbation and industry. It reveals a slight shift of the regional centre of gravity from Ipoh to Meru, whose importance is growing at a regional level. Indeed, these two sites are places of concentration of the tertiary sector in which Kinta is trying to specialize.

These important industrial and tertiary centres gave a forward motion to many local centres sprinkled along Kinta's corridors. The integration of peripheries with a regional development policy was partially encouraged by the presence of a logistic platform of a national importance. However, in order to maximize the economic capacity of the dry port and its integration capabilities, indeed, the dry port, although easily accessible and equipped with storage capacities, is still an insufficient logistical infrastructure. Industrial growth has allowed an increase in the production of manufactured goods with high added value but has also dramatically increased the need for storage and goods management. The ICT has therefore been renovated in the first part of 2011 in order to adjust to the rising logistic demand of Kinta. However, it is important to notice that any spatial expansion of the ICT site is not an option due to a lack of surrounding urban space.

The layout of Kinta's clusters reveals a double integration related to the main centre of Ipoh and is established by economic, industrial, logistical and transportation means. The project of implementing associated branches of ICT within some clusters shows indeed the need to enhance and diversify the logistic offer of a region in growth.

CONCLUSION

The establishment of dry port as a logistic relay has also been accompanied by a relocation process of a part of industrial and commercial activities. Although the headquarters of most large companies remain in the main metropolis, Ipoh attracts large functions of industrial systems. Thus, the implementation of secondary and tertiary sectors does not only consider the production phase, but also incorporates Research and Development activities. The current specialization of Ipoh allows it to develop a leading local industry which takes advantage of the infrastructural facilities of a dry port and hence optimizes access to seaports for export.

Furthermore, the local corridors of Ipoh metropolitan region, stretching from Meru/Simpang Pulai, Meru/Lahat Ipoh and Ipoh/Chemor are following the meridian orientation centred on Ipoh. The development of industrial and logistic infrastructures of the Kinta valley is indeed following the lines drawn by the North-South Expressway and the KTMB railway network, the spines of the west coast economic corridor. However, due to Ipoh's location at the centre of Peninsular Malaysia located between two main ports, its direct links with other countries appear to be limited and seem rather to use the coastal ports as intermediaries. Firstly, its connections with Singapore are obstructed by the strategic location of Johor Bahru in the southernmost part of the peninsula and its links with Nilai dry port. Secondly, its connections with Thailand face the same dilemma because of Penang's infrastructures and its links with Bukit Kayu Hitam inland container terminal (Abdul Rahim Muszafarshah Mohd Mustafa and Amel Farhat's contribution). However, the construction of the electrified double-track railway following the North-South corridor and linking the Thai border (Padang Besar) to Ipoh may allow the question of the connection of Kinta to the northern part of the Peninsula and the direct consequences for ICT growth to be reconsidered.

Logistics and transportation infrastructures of Kinta have followed the revitalization of a pre-existent colonial axis not only used to channel goods to main ports, but also built as a link halfway between two historical growth poles of Malaysia, Kuala Lumpur and Penang. As a multimodal link, Ipoh reinforced its nodal role within the peninsula by implementing a dry port directly linked by railway to these cities. Hence the logistical and industrial assets facilitated its integration within a main economic corridor stretching from Klang valley to northern Penang and allowed Ipoh to re-emerge as a growth pole turned towards export through its linkage with international gateways. However, the political and economic tropism exerted by Kuala Lumpur has questioned the central position of Kinta metropolitan region within the west coast economic corridor. As shown by the ceasing of relations between ICT and Penang Port, it seems that the local economy is turned toward the federal capital and Klang Valley in terms of exports acting as an "outskirt of Selangor" and not any longer as a strong interstate link. As an infrastructure of growing importance, the dry port not only strengthens Ipoh as a new second-rank pole, nestled between Malaysia's first-rank cities and gateways, but also along the Southeast Asian corridor stretching from Singapore to Bangkok. While the geographic location and infrastructure assets serve its linkage within the urban hierarchy, it is counteracted by the political pressure of the federal capital exerted on Penang port's development. The complementarities between Kuala Lumpur and Ipoh are therefore strengthening at the cost of the Kuala Lumpur-Ipoh-Penang tripolar dynamic. Ipoh is consequently forming a duo with Kuala Lumpur, looking towards the hinterland but highly dependent on the maritime dynamism of the Malacca Straits. It can be assumed that a political will influencing the isolation of Penang has radically changed the articulation role of Ipoh, from a relay within the corridor, into a "pair-city".

As a main manufacturing and transit nation, Malaysia demonstrates its capabilities in the transportation system sector and its understanding of a need for multi-level polarity of nodes within such an extended axis. The permanence and the centrality of the Malaysian section of the corridor are partially due to the relaying role established by the spatial integration of second-rank cities, such as Ipoh. However, the deepening of the "pair-city" role of Ipoh by the federal State is not only influencing an economic growth located in the hinterland, but also strengthening the macrocephalous nature of Kuala Lumpur in Malaysia and its regional weight within the Malacca Straits and Southeast Asia.

References

Chin, Christine. "The State of the 'State' in Globalization: Social Order and Economic Restructuring in Malaysia". *Third World Quarterly* 21, no. 6 (2000): 1035–57.

Dodd, J.W. "The Colonial Economy, 1967: The Case of Malaysia". *Asian Survey* 9, no. 6 (1969): 438–46.

Fau, Nathalie. "Le rôle des infrastructures dans l'organisation spatiale du détroit de Malacca (Indonésie, Malaisie, Singapour)" [Infrastructures Role in Malacca Strait's Spatial Organization]. *Cahier du GEMDEV*, no. 29 (2003): 163–83.

Fujita, Masahisa and Tomoya Mori. "The Role of Ports in the Making of Major Cities: Self-agglomeration and Hub Effect". *Journal of Development Economics* 49 (1996): 93–120.

Goldblum, Charles and Manuelle Franck. "Les villes aux marges de la métropolisation en Asie du Sud-est" [Cities on the Margins of Metropolisation in Southeast Asia]. *Espace géographique* 36, no. 3 (2007): 229–36.

Haynes, K. and K.J. Button. "Transport System and Economic Development". In *Handbook of Transport Systems and Traffic Control*, vol. 3, edited by Kenneth J. Button and David A. Hensher. Elsevier, 2001.

Hirschman, Charles. "The Making of the Race in Colonial Malaya: Political Economy and Racial Ideology". *Sociological Forum* 1, no. 2 (1986): 330–61.

Ho, Tak Ming. *Ipoh when Tin Was King*. Ipoh: Perak Academy, 2009.

Jabatab Perancangdan Desa Semenanjung, Pengarah jabatan perancang bandar dan desa Perak darul ridzuan, *Draf rancangan struktur negeri Perak 2001–2020* [Peninsular Malaysian Planning Unit, Perak State, *Perak structure plan 2001–2020*], 1996.

Leinbach, Thomas R. "Transportation and the Development of Malaya". *Annals of the Association of American Geographers* 65, no. 2 (1975): 270–82.

Majlis Perbandaran Ipoh. *Draft Ipoh Structure Plan* [Ipoh Municipality, *Ipoh Structure Plan*]. Ipoh, 1986.

Rodrigue, Jean-Paul et al. "Functions and Actors of Inland Ports: European and North American Dynamics". *Journal of Transport Geography* 18, no. 4 (2010): 519–29.

Walter, Clyde K. and Richard F. Poist. "Desired Attributes of an Inland Port: Shipper vs. Carriers Perspectives". *Transportation Journal* 45, no. 5 (2003): 42–56.

World Bank. *The Economic Development of Malaya, Report of a Mission Organized by the International Bank for Reconstruction and Development*. Baltimore: John Hopkins Press, 1955.

13

DANOK-BUKIT KAYU HITAM, TWIN BORDER TOWNS ON THE THAILAND-MALAYSIA'S BORDER

Abdul Rahim Anuar, Muszafarshah Mohd Mustafa and Amel Farhat

Cross-border twin cities have a particular spatial organization in the sense that geographic proximity is an intermingling factor of urban areas and economic interdependence allowing a transnational development. Their transnational setting, i.e. through a border, covers a multiscalar approach on a provincial scale (within a State) on the one hand, and a regional scale on the other, combining at least two States. This kind of urban layout therefore leads to a spatial reorganization networking a new regional territory which responds to the economic guidelines of partner States (Raffestin 1974; Taillard 2010). Infrastructure networking, administrative harmonization, respective comparative advantages and complementarities have made these urban areas a layout that may support an economic dynamism specific to transnational regions (Maneepong et al. 2004; Taillard 2010; Goldblum et al. 2007). The border is thus no longer exclusively a component of separation, but may also turn into a spatial bridge between two bordered provinces (Arnold 2010).

Among the twin towns bordering Thailand and Malaysia,[1] the Danok-Bukit Kayu Hitam pair may have the greatest potential to be developed into a major growth area. Although the urban and economic layouts fall into the category of a twin city, it has not however been officially stated by the Malaysian and Thai governments so far. It is thus by no means as institutionalized as those at the southern boundary of China, which are discussed in more detail by Sébastien Colin in this book. This study aims to investigate how the twin cities of Danok and Bukit Kayu Hitam may become a peripheral growth centre for both Thailand and Malaysia, thus promoting sustainable development in the region.

This study is organized according to approaches on different scales (see Map 13.1). The following section will provide a local scale by studying the district location and socio-economic backgrounds of Sadao and Kubang Pasu and the twin cities in question. The second part will cover a comparative analysis of the twin cities to determine whether their status may imply identical dynamics; we shall also touch on the institutional framework of the aforementioned cities. Finally, the regional scale will be examined through the links existing between the studied area and the urban poles of northern

Map 13.1
Interscalar Modelization of the Malaysian-Thailand Spatial Organization

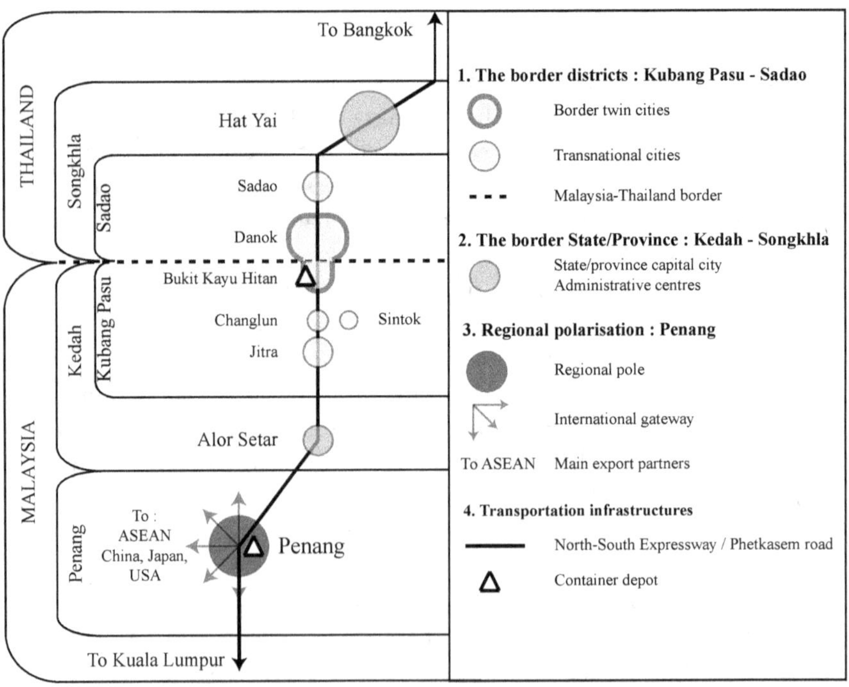

Source: Abdul Rahim Anuar, Muszafarshah Mohd Mustafa, Amel Farhat.

Malaysia and southern Thailand. This approach will highlight the influence of Penang, an important Southeast Asian metropolis, over the Malaysia-Thailand border.

It is worthwhile mentioning that this article was first presented in the ANR Transiter/Southeast Asia Regional Workshop Vientiane Transnational Dynamics and Territorial Redefinitions in Southeast Asia: the Greater Mekong Subregion and Malacca Straits economic corridors workshop held in Vientiane on December 2010.

1. AN OVERVIEW OF THE STUDIED TRANS-BORDER SPATIAL ORGANIZATION

In order to control the Sintok jungle territory, known as a communist hiding haven, Malaysia and Thailand agreed to open the Danok-Bukit Kayu Hitam border checkpoint in 1967. From this date, the implementation of civil and military administrations was quickly followed by the economic development of the two districts facing the border: on the one hand Sadao on the Thai side, and on the other hand Kubang Pasu on the Malaysian side.

The Border District Scale

Sadao district, located in the south of Songkhla province, covers an area of 860 sq. km and numbers 115,109 inhabitants, mainly Buddhist (60 per cent) and Muslim (39 per cent). Its territory goes along the Malaysian border for 73 km and includes two customs and immigration checkpoints, Padang Besar and Danok. The geographical setting shows a predominance of plains and mountains and the fertility of its soil has led to activities mainly turned towards the primary sector. Agricultural land indeed covers 646 sq. km and is based on rubber trees (95 per cent), but also food and palm oil. Moreover, manufacturing is also an important sector of activity and is essentially dependent on rubber and rubber wood processing. Lastly, retailing and services are not significant all over Sadao district, but are concentrated in strategic locations, such as Sadao town, and the border towns of Padang Besar and Danok. The transportation infrastructure of Sadao district is connected with two highways, No. 4, also known as the Phetkasem, linking Bangkok to Southern Thailand, and No. 4054, linking Sadao to the border checkpoint of Padang Besar. These infrastructures are not obstructed by the border situation but show a continuity of transportation systems since they are linked with the North-South Expressway, spanning the west coast of Peninsular Malaysia as far as Singapore (see Map 13.1).

Among its nine sub-districts, the district of Sadao includes two poles of local influence, the city of Sadao with 18,099 inhabitants and Danok (also known as Dan Sadao or Samnak Kham) with 10,654. Both cities are 15 km apart and are located along the Phetkasem highway, the most structural infrastructure of Songkhla. The spatial distribution of activities allows us to establish a specialization of each urban pole. Sadao town is an important administrative centre hosting the main police offices, social and health services, whereas Danok is specialized in retailing, transportation companies, and entertainment activities.

Kubang Pasu is the largest district of Kedah State, covering about 950 sq. km and numbering 213,000 inhabitants in 2010 (see Map 13.2). With more than 80 per cent Malay population, Islam is not only the main religion, but it can be seen as a great influence on local lifestyle. Only halal food and beverages can be purchased from the local shops and restaurants. The geographical setting is dominated by plains and a few limestone hillocks, and dotted by tropical forest. The primary sector occupies a

Map 13.2
The Kubang Pasu District

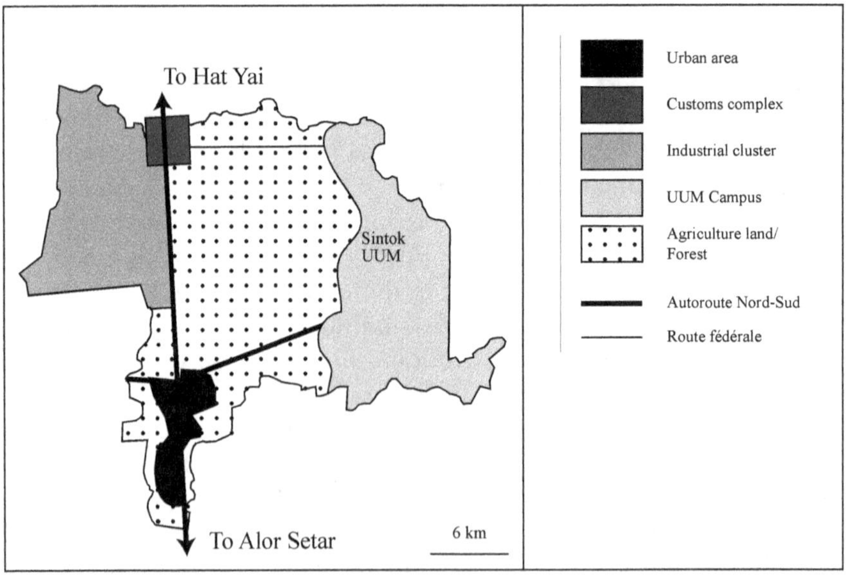

Source: MAJLIS DAERAH KUBANG PASU, 2000. Draf rancangan tempatan Bukit Kayu Hitam-Changloon-Sintok 2000–10. Amel Farhat.

Table 13.1
Distribution of Activities in Kubang Pasu, 1991–2005

Economic Sectors	1991	1995	2000	2005
Primary	82,7	78,6	73,2	68,3
Secondary	5,5	9,1	13,7	17,6
Tertiary	11,8	12,3	13,1	14
Total	**100**	**100**	**100**	**100**

Source: Rancangan Struktur Majlis Daerah Kubang Pasu, Kedah Darul Aman, 1997. Draf Racangan Tempatan Daerah Kubang Pasu 2020, 2010.

large part of the land, as in Sadao, and consists of rubber trees and *padi* fields (rice crops). Although agriculture is an important part of the local economy, it is decreasing in favour of the secondary and tertiary sectors, as can be seen in Table 13.1. The Kubang Pasu district is indeed undergoing an important development policy turned towards processing and high technology clusters. The development programme is coordinated by the North Corridor Economic Region — *Utara Korridor* — (NCER), an institutional framework controlled by the federal State including the north-western Malaysian States of Perlis, Kedah, Penang and Perak.

The opening of the Danok-Bukit Kayu Hitam checkpoint allowed Kubang Pasu to become one of the busiest crossing spots between Malaysia and Thailand. The North-South Expressway starting in Kubang Pasu acts as a substantial vehicle for development according to two criteria: on the one hand it allows Malaysia to convey an efficient transportation system all the way to Thailand by being linked to the Phetkasem; but on the other hand it also makes Kedah more easily accessible to the other major States of Malaysia, such as Penang and Selangor. The combination of the highway and a border location has therefore become a major asset that has led to the implementation of industries and businesses turned toward logistics and trade.

The spatial organization of Kubang Pasu reveals an urban hierarchy based on the polarity of Jitra-Changloon, Sintok and Bukit Kayu Hitam (Majlis Daerah Kubang Pasu 2009). Each of these urban poles is specialized in specific activities and they have recently been renovated. As the third most important city in Kedah, Jitra is the biggest and most populated urban centre of Kubang Pasu with 20,000 inhabitants and accommodates administrative offices, services and a halal agro-industry hub. The urban development projects of Kubang Pasu are indeed located in the south part of Jitra, notably in the township of Bandar Darulaman, hosting higher education institutes[2] and high standard accommodations, and are expected to receive up to 50,000 newcomers. The city of Changloon, with 1,500 inhabitants, is not a demographic pole but is rather a commercial and leisure city coupled with Jitra, and since it is frequently visited by students it has therefore become greatly dependent on the campus of Sintok. Sintok is the remains of a tropical forest and as an academic city hosts the vast campus of Northern Malaysia University and its 14,000 resident students coming from all over Malaysia and some foreign Muslim countries (mostly from Indonesia and Middle East). Finally, Bukit Kayu Hitam is an industrial and logistic cluster bordering Thailand and sharing an urban continuum with its twin city Danok. It is mainly a passing-through point leading to Danok and therefore does not represent a demographic centre, numbering 14,900 inhabitants.

While Sadao district's territory is organized in a locally concentrated bipolar structure, Kubang Pasu has developed a multipolar structure geographically distributed over large spaces, as illustrated by Sintok and the upcoming expansion of Jitra. However, it can be noted that both sides of the border have established a specialization for each of their poles: they are characterized as a demographic pole, a leisure and commercial centre, an administrative concentration or an industrial, logistic and academic cluster.

Danok and Bukit Kayu Hitam: The Cross-border Twin Cities

Facing the Malaysian border city of Bukit Kayu Hitam and numbering 10,654 inhabitants, Danok, from a quiet town in the 1990s, grew into one of the busiest gateways into Thailand. The city centre of Danok is organized along the Phetkasem road and is directly accessible immediately after the Thai immigration checkpoint. Unlike the quiet cities of Kubang Pasu, Danok offers a vibrant nightlife for Thai but mostly Malaysian and Singaporean visitors taking advantage of their higher purchasing power. By taking advantage of its border location this city became the second most popular touristic spot of Songkhla after Hat Yai, receiving thousands of weekend tourists every month. Danok is indeed known for its food stands, restaurants and shopping opportunities, but is mainly a well-known spot for the inexpensive sex industry. These businesses shaped the layout of the city centre, dotted mainly with restaurants, hotels, clubs and brothels. The growing residential area is therefore located along the roads adjacent to the Phetkasem.

Bukit Kayu Hitam is Kedah's main border town on the Malaysia-Thailand border. It marks the beginning of the North-South Expressway and Federal Route 1, the longest Malaysian road, which runs until the Singaporean border in Johor Bahru. Bukit Kayu Hitam covers 36 sq. km and numbered about 14,900 inhabitants in 2010, living in specific gated communities and mainly civil servants and military officers working on the border, whereas the other workers settled in Changloon or Jitra.

While Bukit Kayu Hitam is a rising residential centre, its importance is in its border location and its infrastructural assets. Not only is it the beginning of the North-South Expressway, but it also hosts a unimodal inland port allowing the handling and customs clearance of a large number of containers. Such infrastructures led to the implementation of numerous logistic companies and wholesalers within the trade centre area, which is located in the north-eastern part of the city, allowing the private sector to benefit from the Free Commercial Zone status. The companies in this location consist of small and medium enterprises (SMEs), offering logistical services such as freight, forwarding and warehousing. Moreover, the city is home to an industrial cluster that is of major significance. Manufacturing activities are concentrated in the Bukit Kayu Hitam Industrial Estate, in the southwestern part of the city, which has been developed by the federal government.

Among these activities there are a few which are of international interest. These include the composite production and assembly plants of Boeing 777 aircraft and the Kota Perdana project, the output point of a pipeline for fossil fuel. Such initiatives put Bukit Kayu Hitam on the map as a national and regional industrial and logistic cluster.

As a political partition of a national territory, a border is a specific geographic setting whose status can vary from simple closure to free openness. The case of Malaysia and Thailand allows us to illustrate how the border can present both situations: it is a component of separation in the economic, legislative and geographical landscape; but it is also a place characterized by the abundance of its traffic. By analysing the urban dynamics of Danok and Bukit Kayu Hitam, we are showing here how the border urban layout takes advantage of dual complementarities which are not only strictly located on both sides of the border, but also advance deeply within the hinterland of Malaysia and Thailand. Danok and Bukit Kayu Hitam do not share the same economic interests. While Danok is oriented towards entertainment and leisure industries, Bukit Kayu Hitam is turned towards industrial activities. By facing each other and not competing economically, these two twin cities may be able to develop complementarities and grow alongside each other.

2. URBAN DYNAMISM AND TRANSNATIONAL SETTING OF DANOK AND BUKIT KAYU HITAM

Although Danok and Bukit Kayu Hitam are characterized as twin cities, this does not necessarily imply similarities. Not only do these cities lack a comparable economical structure but the physical structure of the urban layout is far from being identical. Such dissimilarities do not, however, question their twin city status due to the strong interlacement and complementarities linking Danok and Bukit Kayu Hitam.

A Dissymmetric Urban Continuum

The twin cities of Danok and Bukit Kayu Hitam are not based on the same layout and did not develop the same urban landscape (see Map 13.3). Although the Malaysian side is growing fast in terms of residential areas, urbanism is mainly based on a kampung style settlement resulting in a patchwork of neighbourhoods. Unlike the industrial and trade areas, housing is not built directly either along the highway or next to the border. The Malaysian side of the border therefore does not resemble an urban setting at first sight, but it gives a strong vibe for a working centre with an industrial cluster.

On the contrary, the other side of the border is far from the quiet lifestyle of Bukit Kayu Hitam and presents all the attributes of a typical Thai city. First the geographic landscape, made out of sounds, smells, commercial and residential areas draw a picture of a crowded and lively city. Passing the immigration checkpoint, you can immediately see where you are. This urban layout setting shows, finally, the different development models applied in the case of these two cities: whereas Bukit Kayu Hitam is part of a government-led or "sponsored border-town" agenda, Danok is more under the influence of a market-led development of Sadao or a "non-sponsored border-town" (Maneepong et al. 2004). These two facts have led to an uneven development of Bukit Kayu Hitam and Danok. The political border is therefore not an invisible component but it contributes to shape a dissymmetric urban continuum and uncoordinated development policies which are immediately obvious when crossing.

Comparative Advantages and Complementarities

These border cities are identified as new economic bases linking regional cities to stimulate decentralization to peripheral areas. The divergence

Map 13.3
Danok and Sadao Urban Layout

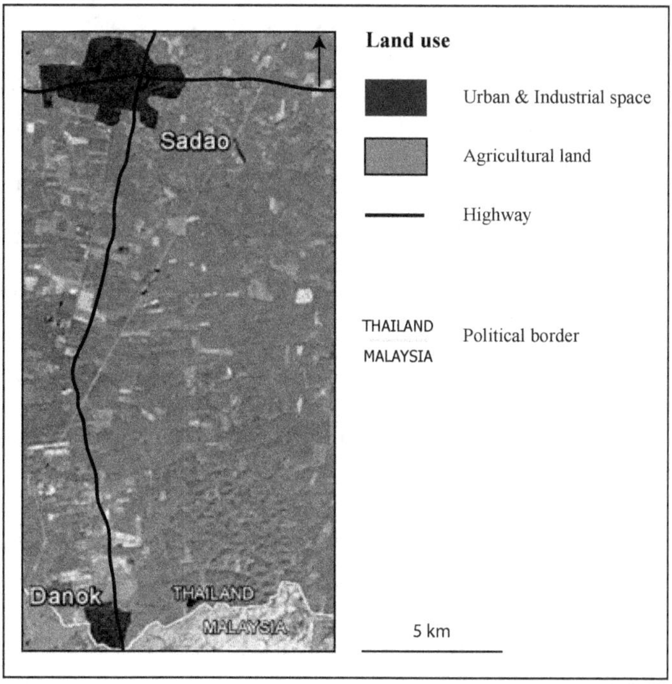

Source: Amel Farhat.

between Danok and Bukit Kayu Hitam twin cities is not however a factor of separation, but constitutes the key to strong complementarities intertwining the economies (Tsuneishi 2008).

Firstly, Danok developed its commercial and leisure activities mostly for neighbouring Malaysia, and these facilities are greatly appreciated by Malaysian visitors. This results in time-dependent immigration peaks during the weekend from Malaysia towards Thailand.

Secondly, the Thai labour force of the Sadao district is dependent on both the industrial Estate of Bukit Kayu Hitam and the rubber plantations of Kubang Pasu. As a result, the Malaysian side enjoys a large, cheap workforce, sometimes unprotected by labour law due to illegal issues concerning work permits. This labour dependence results in Thai males commuting across the border, which becomes congested in early morning and evening.

Thirdly, the industries of Malaysia are highly dependent on Thai products, both in the agro-industry business and in manufacturing. The import of raw materials is undergoing transformations within an agro-industrial process, whereas imported electrical parts are assembled in Malaysian plants. As a result, Malaysia exports predominantly value-added goods, such as electrical appliances (51 per cent), while Thailand exports raw materials, mainly rubber (31 per cent) through the Danok Bukit Kayu Hitam border (Royal Customs Department Malaysia 2009). This economic process is therefore based on an international industrialization of the region resulting in a value-adding procedure from Malaysia and based on cheap Thai materials.

Finally, most trade activities between Thailand and Malaysia are conducted by road and rail transport services. While Bukit Kayu Hitam dominates the road transport service with more than thirty forwarding agents, Padang Besar mainly handles rail transport. Based on the 2007 Malaysia-Thailand bilateral trade data, the value of total trade that passed through Bukit Kayu Hitam was worth RM11 billion and accounted for approximately 65 per cent of the total bilateral trade values by road transport. The logistic assets of Bukit Kayu Hitam, notably its inland port facilities and a direct access to the North-South Expressway, are a structural element allowing its hinterland and Southern Thailand to export goods efficiently to each other's countries.

Such a situation, even if it involves interdependency, is largely to the advantage of Kubang Pasu, regarding its local and regional economic development. The relationship between Danok and Bukit Kayu Hitam as twin cities is economically established; however, their institutional status is still an issue that needs to be figured out by Thai and Malaysian authorities.

The Institutional Framework of the Cross-border Twin Cities

Both border cities present great potential to grow and complement each other even more, but are constrained by their existing national development policies. Even though some administrative measures have been taken in order to ease the border zone human and goods traffic, the lack of institutionalization restrains coordinated development opportunities.

Institutional entities have been created in order to coordinate the economic development of transnational territories. Thus, Malaysia and Thailand based their attempt to foster joint development on two complementary scales: on the one hand a regional scale covering the entire bordering territory; on the other hand a local level considering exclusively Kedah and Songkhla.

In 2004 the committee on Joint Development Strategy for Border Areas (JDS) was established in order to coordinate the development plan and policies by creating an institutional bridge over the border for the mutual benefit of both countries. The area of coverage includes the border States of

Malaysia — Perlis, Kedah, Perak and Kelantan — and Thailand — Satun, Songkhla, Yala, Narathiwat and Pattani (Malaysia-Thailand Committee on JDS for Border Areas 2004). Although it focuses on a local scale, its influence includes a regional level by addressing a range of fields related to infrastructure and cross-border transportation linkages, human resources, tourism, culture and civic relations, trade and investment, primary sector upgrading, fiscal and financial development, energy, and disaster prevention and management. Moreover, the Mal-Thai Border Commission focuses on a local scale by being responsible for a coordinative development between the province of Songkhla and the State of Kedah. This transnational cooperation led to the facilitation of cross-border flow for goods and people by implementing an administrative framework. As a result, any border citizen is entitled to a free pass, allowing them to avoid the use of passports and easing the procedures for the authorities. Moreover, handling of cargo and containers has been harmonized by creating a single *modus operandi* and a single bilingual form for both Malaysian and Thai customs. These measures have been taken in order to remedy the congestion issues faced at the border.

The establishment of such cooperation bodies is however not fully efficient and does not involve proper coordinated transnational development planning. The launch of a special border economic zone which encompasses the two border towns would enhance the border towns' development and their mutual comparative advantages.

An Institutionalization Beyond Malaysia and Thailand Alone

The location of Danok and Bukit Kayu Hitam twin cities at the border is an even more strategic asset with the ongoing transnationalization development promoted under the Indonesia-Malaysia-Thailand Growth Triangle[3] (IMT-GT) and the ASEAN. By being on the route of the ASEAN highway these two cities also provide an inland connection to the crisscrossing transport and urban corridors of the Greater Mekong Subregion (ADB 1998, 2002, 2007*b*, 2009). Therefore, as a key area located in both the centre of the IMT-GT and at the periphery of the ongoing intra-ASEAN corridors, Danok and Bukit Kayu Hitam can become a stronger pair acting as a link between the subregions of Southeast Asia. The current improvement of internationalization of the ASEAN borders within the Free Trade Areas and Agreements (AFTA and FTAs) allows the intra-ASEAN trade to be strengthened by reducing customs taxes and easing the way across the border (ADB 1998, 2007*a*).

As indicated by its status, the ASEAN cannot, however, act on a supranational scale and cannot therefore counteract some national practices, although sovereign, incompatible with international trade. Protectionism is indeed carried out via measures obstructing the ease of cross-border flow. This blocking process involves, for instance, a compulsory Malaysian and Thai insurance for heavy trucks instead of issuing a bi-national special pass. By not being able to access to a bi-national insurance scheme, the transported goods must proceed to a change of gauge in order to cross the border, resulting in almost permanent heavy congestion. In order to obtain bi-national insurance one must apply for it directly in Kuala Lumpur. Since only four logistics companies[4] possess the precious cross-border pass it shows a monopoly of a few powerful companies over the cross-border trade as well as the complexities of obtaining such authorization. The cross-border trade is not only obstructed by physical means but also by institutional ones, as shown by the strong protectionism of Malaysia over rice and fruit imported from Thailand.

These "one step forward two steps backwards" measures show however the limits of the border as an economic bridge fully open between Malaysia and Thailand. The twin border cities of Danok and Bukit Kayu Hitam may be able to capitalize on the intra-ASEAN trade enhancement with more willingness on the national level. By reducing border taxes and easing the international trade border along the ASEAN infrastructure corridor the transnational territories can gain a broader influence at a provincial scale. Therefore, as one can observe along the other Asian borders, twin cities do not only cover a local or even a transnational scale, but are involved on a regional level, allowing them to benefit from economic and institutional support from both the States and international cooperation (Taillard 2010).

3. THE HINTERLAND LINKAGE OF DANOK — BUKIT KAYU HITAM WITH NORTHERN MALAYSIA AND SOUTHERN THAILAND

As integrated margins of Thailand and Malaysia, and as border trade and transnationalization centres, Danok and Bukit Kayu Hitam are strongly linked to their hinterland. In this section, the hinterland is studied according to two scales and notions: on the one hand the political roles of Kedah and Songkhla's capital cities, Alor Setar and Hat Yai; and on the other hand the economic and logistic influences of Penang, acting as an international gateway for northern Malaysia and southern Thailand.

The State/Provincial Capitals of Kedah and Songkhla: A Political Influence

Located respectively about 60 and 45 km from the border, Hat Yai and Alor Setar are subjected to more or less the same comparative observations as Danok and Bukit Kayu Hitam. Demographic data are sufficient to illustrate the difference between the Great Hat Yai Metropolitan Area, numbering 800,000 inhabitants, and Alor Setar and its suburbs, home to 300,000. Whereas the Thai city is well known as an attractive commercial and tourist spot, on the contrary, the capital of Kedah is a quiet but growing government centre, notably through its outskirts with the new Anak Bukit administrative city.

Even if this urban opposition confirms the comparative situation of the twin border cities, it does not, however, influence the political weight of one city over another. As marginal cities located far from Bangkok and Kuala Lumpur, they fulfil an important role in the application of national policy on a provincial and State level. The cities of Danok and Bukit Kayu Hitam are thus under the political relay power of Hat Yai and Alor Setar and are well connected through the Phetkasem and the North-South highway. As administrative centres, both cities work on the implementation of the national policy regarding the twin border cities. Local public policy is an issue that is not strictly covering the concerned twin cities but the provincial/State capitals, whereas the real decisional power is located in the hands of the national capitals. Both Hat Yai and Alor Setar host offices related to national development planning and to border institutions, such as the Mal-Thai border commission.

A Border under the Influence of a Malaysian Metropolis: Penang

Although Penang seems relatively far from the studied area, since it is located about 145 km from the border, it represents, however, an influential national and regional pole. The links between the Thai border and Penang show how a second-rank metropolis of Southeast Asia maintains a centre/periphery spatial organization. By capitalizing on its gateway position and its strong transportation infrastructure assets, Penang constitutes the main exit point within the IMT-GT's territory, exercising a strong influence over the export flow of southern Thailand.

After Selangor, Penang and Johor are the most prominent growth poles of Malaysia. Mainly developed on research and development and on manufacturing and high technology clusters, Penang is the main economic centre of the north. As such, the State has implemented efficient multimodal transportation infrastructures, including the third most important seaport (Penang Port) and the second biggest international airport of the peninsula (Bayan Lepas), container depots, railway and highway. Its impact on Kedah, Perlis and Perak is extremely strong and the city constitutes one of their main economic and urban poles. As an international gateway linked mainly to Asia through the Malacca Straits, the city maintains a centre-periphery relationship over the States included within the IMT-GT framework. As the main R&D pole it attracts a skilled workforce from Malaysia and neighbouring countries. Moreover, its importance also plays an institutional role by hosting important bureaus and international conferences related to the transnationalization of Malaysia and the Malacca Straits.

The Main Exit Point of Southern Thailand

Efficient logistics and transportation systems are indeed a strong element within the development process of Penang and contribute to making it the main exit point for manufactured goods of the IMT-GT territory. Among Malaysian ports, Penang is indeed a strategic infrastructure platform region due to its strong connection with the hinterland. Therefore, while Bukit Kayu Hitam represents a main entry point to Malaysia, Penang links the country on an international level, allowing southern Thailand to rely on modern logistic chain management and infrastructures.

The lack of efficiency and modernity of the whole of southern Thailand, including Hat Yai, leads most of the export goods to pass through Penang Port. As a result, Hat Yai has developed a dense mesh of logistic services used by Thailand, such as forwarding companies, specialized in the import and export of containers (full and empty) from Penang Port. Accessible by both rail and road, northern Malaysia is therefore considered as a main territory of Thai export policy. Not only does it represent a great opportunity for Penang to maintain its growth rate into a regional scale metropolis, but it also allows the Malaysian side of the border to host a greater inflow through the inland ports and container depots of Bukit Kayu Hitam. The Malaysian choice to implement such an infrastructure at the border was probably based on the logistic and infrastructure deficiency of southern Thailand.

On a regional scale, both economically and logistically, the border seems to be more dependent on Malaysian assets, notably its industrial clusters and its modern transportation system. Kedah State is thus enhancing its logistic and transport infrastructures in order to strengthen its role in the cross border trade.

Not only have the northern territories of Malaysia developed a polarity that reinforces Penang, but the southern part of Thailand also relies on it more than on its own regional pole, Hat Yai. The same mechanism can be observed across the Malacca Straits with Sumatra relying on both Selangor and Penang in order to access an international gateway. The influence of Penang is therefore obviously crossing the borders and it has achieved the status of a main city within the IMT-GT framework.

CONCLUSION

In terms of traffic, Malaysia is one of the largest hubs of Southeast Asia, just behind the leadership of Singapore. Its transportation infrastructure, located in the land and maritime gateways, has allowed the uptake of most of the Thai and Malacca Straits' flow and a stronger foothold in Asian global trade. The economic links between Malaysia and Thailand are strengthened by the shared border and contribute to redefine marginal territories located far from main decisional and urban poles. The twin border cities have thus gained their interest by being included in a transnational process covering Southeast Asia and by acting as urban, industrial and logistic links between two partner countries.

Urban and economic complementarities are indeed highlighted by the case of Kubang Pasu-Sadao districts. However, the spatial organization on either side of the border developed different patterns and penetrates the hinterland in different degrees (see Map 13.1). While the "border effect" is noticeable only in the Thai Songkhla province, on the Malaysian side its influence goes far beyond Kedah State and can be noted as far away as Penang. The island infrastructures are indeed tools of international junction and act as a gateway linking the studied area with a broader region. This difference of depth resulted in a strong imbalance met by cross border relations between the two partners and is mainly due to two elements. Firstly, these two States demonstrate a different spatial organization, with on the one hand Malaysia, which is based on tripolar development (Selangor, Penang, Johor Bahru); whereas on the other, Thailand, as a macrocephalous State, has favoured an extremely high concentration of activities and wealth within the Bangkok metropolitan region, leading to a strong structural disequilibrium of the country particularly noticeable from the south of the Kra isthmus (Webster 2006). Secondly, Songkhla province, as a Thai peripheral territory, is turned towards northern Malaysia by being economically integrated but also by presenting a high degree of dependency on Kedah and Penang's infrastructures and industrial clusters.

Furthermore, as seen by the study of Bukit Kayu Hitam and Danok twin cities, transnationalization between Malaysia and Thailand is nevertheless not a completed process and has to handle constraints based on protectionism of national authorities in order to become an integrated cross-border space. Although it is currently not an institutionalized transnational area, the recent ASEAN policy promoting regional trade has put border territories in the spotlight. The Free Trade Agreements, indeed, mark a genuine regional trade by removing or reducing border tax and hence it results in a significant new strategy of border territories by counterbalancing the protectionism of the states.

Notes

1 The cross-border twin city list is exhaustive between Thailand and Malaysia: Padang Besar (Thailand)/Padang Besar (Perlis), Wang Prachan/Wang Kelian (Perlis), Danok/Bukit Kayu Hitam (Kedah), Ban Prakob/Durian Burong (Kedah), Betong/Pengkalan Hulu (Perak), Ban Buketa/Bukit Bunga (Kelantan), Golok/Rantau Panjang (Kelantan) and Tak Bai/Pengkalan Kubor (Kelantan).

2 The Institut Perguruan Darulaman (IPDA) [Darulaman Teachers' Institute], Politeknik Muadzam Shah (POLIMAS) [Muadzam Shah Polytechnic], Institut Latihan Perindustrian (ILP) [Industrial Training Institute], Institut Aminuddin Baki (IAB) [Aminuddin Baki's Institute] and the Kolej Komuniti MARA [MARA Community College] are the academic institutions newly implemented in Bandar Darulaman.

3 The IMT-GT is a transnational institution established in 1993 which includes some provinces of Indonesia (Aceh, North Sumatra, West Sumatra and Riau), northern Malaysia (Perlis, Kedah and Penang) and southern Thailand (Nakhon Si Thammarat, Narathiwat, Pattani, Phatthalung, Satun, Songkhla, Trang and Yala).

4 MISC Haulage Services Sdn. Bhd.; Konsortium Logistk Sdn. Bhd.; E.A.E. Freight & Forwarding Sdn. Bhd. and Bintang Bulk Mover Sdn. Bhd.

References

Arnold, Dennis. "Administration, Border Zones and Spatial Practices in the Mekong Subregion". Ph.D. Dissertation, University of North Carolina, 2010.

Asian Development Bank. *Economic Cooperation in the Greater Mekong Subregion: An Overview.* Manila: Asian Development Bank, 1998.

———. *Building on Success: A Strategic Framework for the Next Ten Years of the Greater Mekong Subregion Economic Cooperation Program.* Manila: Asian Development Bank, 2002.

———. *Mid-term Review of the Greater Mekong Subregion Strategic Framework 2002–2012.* Manila: Asian Development Bank, 2007*a*.

———. *IMT-GT: Building a Dynamic Future: A Roadmap for Development 2007–11.* Manila: Asian Development Bank, 2007*b*.

———. "Sharing Growth and Prosperity: A Strategy and Action Plan for the Southern Economic Corridor", 2009. Available at <http://www.adb.org/GMS/Economic-Corridors/SEC-SAP-draft.pdf> (accessed 29 November 2010).

Goldblum, Charles and Franck Manuelle. "Les villes aux marges de la métropolisation en Asie du Sud-est" [Cities on the Margins of Metropolisation in Southeast Asia]. *Espace géographique* 36, no. 3 (2007): 229–36.

Kerajaan Negeri Kedah. *Draf rancangan struktur negeri Kedah 2002–2020* [Kedah State Structure Plan Draft 2002–2020]. Alor Setar: Department of Town and Country Planning, 2008.

Majlis Daerah Kubang Pasu. *Draf rancangan tempatan Bukit Kayu Hitam-Changloon-Sintok 2000–2010* [Local Plan Draft Bukit Kayu Hitam-Changloon-Sintok 2000–2010]. Jitra: Kubang Pasu District Council and Country Planning Department, 2000.

———. *Draf rancangan tempatan daerah Kubang Pasu: Laporan Rancangan Tempatan 2002–2015* [District of Kubang Pasu Local Plan Draft: Local Plan 2002–2015]. Jitra: Kubang Pasu District Council and Town and Country Planning Department, 2002.

———. *Draf rancangan tempatan daerah Kubang Pasu 2020* [District of Kubang Pasu local plan draft 2020]. Jitra: Kubang Pasu District Council and Town and Country Planning Department, 2009.

Malaysia-Thailand Committee on JDS for Border Areas. "Joint Development Strategic Action Plan". Unpublished report. Kuala Lumpur: Ministry of Foreign Affairs of Malaysia, 2004.

Maneepong, Chuthatip and Chung Tong Wu. "Comparative Borderland Development in Thailand". *ASEAN Economic Bulletin* 21, no. 3 (2004): 133–66.

Raffestin, Claude. "Eléments pour une problématique des régions frontalières" [Questioning Border Regions]. *L'espace géographique,* no. 3 (1974): 12–18.

Taillard, Christian. "Corridor Linkages in the Greater Mekong Subregion: New Implications for Peninsular and Regional Powers". In *New Dynamics Between China and Japan in Asia: How to Build the Future from the Past,* edited by Guy Faure. Singapore: World Scientific, 2010.

Tsuneishi, Takao. "Development of Border Economic Zones in Thailand: Expansion of Border Trade and Formation of Border Economic Zones". *IDE* Discussion Paper No. 153. Tokyo: IDE, 2008.

Webster, Douglas. *Supporting Sustainable Development in Thailand: A Geographic Clusters Approach.* Washington: World Bank, 2006.

14

MUKDAHAN AND SAVANNAKHET, INTERNATIONALIZATION PROCESS OF TWIN MEKONG BORDER CITIES ON THE EAST-WEST ECONOMIC CORRIDOR

Elsa-Xuan Lainé

This chapter characterizes and evaluates the internationalization process on going for twin cities located across the Thai-Lao border and on the Greater Mekong Subregion (GMS) corridors. Two corridors implemented within the Asian Development Bank (ADB) GMS framework at the beginning of the 1990s have a great importance for Thailand and Lao PDR, i.e. the North South Economic Corridor (NSEC) linking Bangkok to Kunming *via* Lao PDR or *via* Myanmar's Shan State, and the East West Economic Corridor (EWEC) linking Moulmein port in Myanmar to Danang port in Vietnam *via* Thailand and Savannakhet Province in Lao PDR (cf. Taillard, Map 2.2 in this volume).

The "twin cities" can be defined according to multiple criteria. The first characteristic used for this article is the fact that the selected cities are border towns, i.e. located on an international border (the Mekong River between Thailand and Lao PDR) and with some functions directly related to the presence of this border. Furthermore, the other criterion used is a geographic one, implying the existence of two urban centres facing each other on each side of this international border. If they can be linked by an institutional agreement ("twin cities" or "sister cities" agreement), this institutional aspect is an optional requirement, compared to the interactions (formal and informal) existing between the two cities. Finally, the expression "twin cities" should be nuanced as it implies a resemblance or symmetry between the two urban centres, which is limited in reality, in terms of population, urban landscape or economic activities.

This chapter will try to discuss this theoretical frame by studying an example of these so-called "twin cities", Mukdahan and Savannakhet, focusing on their characteristics, functions and interactions.

Since the end of the 1980s, development of border regions has been supported by national governments as well as international institutions such as the ADB to promote decentralization and enhance cross-border trade and economic cooperation with neighbouring countries (Maneepong 2010, p. 1). In the case of Thailand, border towns have been identified as economic gateways since the 7th National Plan, established for the 1992–96 period (NESDB 1999). For Lao PDR, the largest border towns have been developed under the ADB Secondary Towns project at the end of the 1990s.

The implementation of ADB corridors strategy within the GMS framework in the 1990s is likely to give more functions to cities located near the borders and on these corridors (border cities as well as inland cities).[1] As stressed by ADB in different reports, corridor towns, due to their potential to attract private sector investments, can transform transport corridors into economic ones.[2] On the local scale, they can also have an economic importance in their respective hinterlands (ADB 2010, p. 3).

Internationalization describes the process by which a city can be integrated not only into its hinterland and national environment, but also into broader networks and flows of capital, people, goods or even cultural products. Although internationalized or global cities have been much studied (for example Sassen 1991), this process is still ongoing for most of the Southeast Asian second or third rank cities, and the regional economic cooperation schemes can be seen as enhancing internationalization, at least on the regional scale. Regarding internationalization, special attention will be paid in this chapter to two dimensions of this process: first the integration into international flows (trade and tourism flows due to improved transport infrastructures; capital flows from foreign direct investments and assistance); secondly the presence of international stakeholders in addition to the local and national ones. Focus will be made on the spatial consequences of this internationalization.

The cities of Mukdahan in Thailand and Savannakhet (known also as Kaysone Phomvihane District)[3] will be used as case studies to illustrate this topic. They are located on a GMS corridor (East-West Economic Corridor or EWEC) and have the particularity of being river cities, developed along the Mekong riverside that constitutes an international border. Mukdahan is linked by the corridor to Khon Kaen, which emerged during the last decade as the capital of the Northeastern region, whereas Savannakhet appears as a node on the national scale between the first-decade ADB corridor EWEC and national road 13, the main road axis of the country and a corridor of the second decade. Targeted in the GMS framework as border cities located on a corridor and promoted as centres for economic growth by Thai and Lao governments, these two cities have indeed been the place of many transformations, including soft policies for easing border crossing, industrial promotion or new cross-border infrastructures.

The main issue is to evaluate to what extent the internationalization process observed on the city scale and originating from various actors, i.e. public as well as private, is likely to change these cities' position on the subregional scale. To investigate these issues, the following set of research questions has been elaborated. First, what are the main factors and actors of change within the GMS framework for these cities? Secondly, to what extent are the selected towns facing an internationalization process? And more important, how is this dynamic impacting these cities, on the city scale, between the two "sister cities", and on the national and regional scale?

The first part of this chapter will briefly present the two cities to underline their characteristics as border towns and discuss the concept of "twin cities". The second part deals with the internationalization process observed in the two cities that differs in terms of rhythm and level, and tries to identify factors and actors of change. The manifestations of this internationalization process will also be described in each city's organization. The third part investigates the implications of this process on different scales, in other words to what extent these cities are being upgraded in the subregional (GMS) urban frame.

1. MUKDAHAN AND SAVANNAKHET IN THEIR NATIONAL AND PROVINCIAL CONTEXTS

Mukdahan and Savannakhet are two cities of the EWEC, and share this first characteristic of being cities located in a GMS corridor.

National and Subregional Scale: Mukdahan and Savannakhet as Nodes of the East West Economic Corridor (EWEC) Targeted by National Plans

The 1,450-kilometer-long EWEC corridor has been one of ADB's priority projects since the implementation of the programme in 1992. Linking the Andaman Sea in the Indian Ocean to the South China Sea through two ports (Danang port in Viet Nam and Moulmein port in Myanmar), the EWEC crosses Thailand's territory without including any of the biggest Thai cities, and Laos' territory through the province of Savannakhet (Taillard Map 2.2). ADB is providing a definition of corridor towns. They can be defined according to their economic functions: they are "towns which produce basic goods, which transform products into higher value products and towns which provide national and international access to these products" (ADB 2009, p. 4).

The location of Mukdahan and Savannakhet on the corridor and in their national territories should be underlined to characterize these cities. Savannakhet has a strategic location, at the crossroads between the EWEC connecting to the Thai and Vietnamese road networks, and the backbone axis of Lao PDR, i.e. the Mekong River and road 13, linking to Vientiane and to the Southern Corridor in Cambodia. The two roads do not cross in the city centre, but in Seno, located about a dozen kilometers northeast of Kaysone Phomvihane city. Mukdahan, with a more peripheral location on the national scale, was initially planned to be connected with Udon Thani, a centre of the Thai northeastern region and the main access to Lao PDR (NESDB 1992, p. 123). With the implementation of the EWEC, Mukdahan is nowadays more connected to the Lao side but also, *via* Kalasin, to Khon Kaen, which has emerged as the main node and center of the Northeastern Region.

According to ADB typology of corridor towns, different types of these have been identified: commercial nodes/border nodes/gateway nodes (access points to external markets) and interchange nodes (intersection points).[4] As far as the EWEC is concerned, the two sea ports (Danang and

Moulmein) are considered as gateway nodes, whereas Mukdahan and Savannakhet are border nodes and commercial nodes. Savannakhet is furthermore recognized for its location as an interchange node, as well as Phitsanulok and Khon Kaen on the Thai portion of the corridor. Apart from different strategic location in their national contexts, both cities are integrated in national plans, as their potential as economic centres has been acknowledged in the context of regional economic cooperation.

Thailand national policies started targeting border areas and towns at the beginning of the 1990s. The Regional Cities Development Policy had been initiated twenty years before to reduce economic and population disparities between Bangkok and the other cities. The 7th Plan is the first one to mention explicitly the border towns as potential centres for economic growth. Mukdahan is there referred to as a "border trade point" to be linked with the regional centre of Udon Thani (NESDB 1992, p. 123) — before the implementation of the EWEC — whereas the expression "economic gateways" is used for Mukdahan in the 9th Plan (NESDB 2003, p. 56). In order to give border towns the opportunity of playing this economic role, the Thai Government has been investing in urban services in these cities, such as water supply, road improvement or waste management, and ADB was asked for technical assistance in 1997 (Gibb 1999, p. 1).

The Socio-economic Development Plan of Lao PDR (1996–2000) has been also putting emphasis on promoting greater regional socio-economic development through investment in provincial infrastructures in a context of increasing provincial urbanization. At the end of the 1990s, the Lao government requested ADB's assistance in developing urban infrastructures and services as well as the management of these services in the four largest secondary towns of the country, namely Luang Phrabang, Thakhek, Savannakhet and Pakse. Located on the Mekong riverbanks, these cities are secondary towns as well as border towns (with the exception of Luang Phrabang), even if they are not explicitly referred to by their border location. An ADB loan of US$27 million was approved in June 1997 (ADB 1997, p. 2).

The implementation of regional economic cooperation through the GMS Flagship was concurrent with an improvement of urban services in border towns located in the corridors for Thailand and Lao PDR; at the end of the 1990s, Savannakhet as well as Mukdahan are included in these government strategies.

Provincial Scale: Two Provincial Capitals Differently Integrated in the National Territory

Mukdahan and Savannakhet are also provincial capitals. Table 14.1 presents some of the urban and provincial basic indicators in order to underline the differences between these two provinces and to replace them in their national contexts.

According to the national census, the population of Mukdahan is 44,000 people, and 84,000 for Savannakhet (population located inside the urban boundaries), making Mukdahan a small city in the Thai urban hierarchy, and Savannakhet the second largest city of Lao PDR after Vientiane, the capital city.[5] On the provincial scale, Savannakhet was the most populated province in the country in 2007, with more than 800,000 inhabitants, whereas Mukdahan province, created in 1982 and numbering slightly more than 300,000 inhabitants, is one of the least populated on the national scale. Given the population difference between the two countries and the land coverage of both provinces, Mukdahan province's density is nevertheless higher than that of Savannakhet (272.2 inhab/km^2 compared to 39.4). As far as urban population is concerned, the level of urbanization is low for both provinces, around 20 per cent (22.4 per cent for Savannakhet province and 16,8 per cent for Mukdahan in 2005). However, if this rate is slightly lower than the national average of Lao PDR (27 per cent according to the 2005 Population Census), it is even lower for Thailand (32.2 per cent in 2007, NSO).[6]

From the economic point of view, the Gross Provincial Product per capita (GPP/capita) is also higher in Mukdahan province (US$1,245

Table 14.1
Basic Indicators for Savannakhet and Mukdahan Provinces, 2007

	Savannakhet Province	Mukdahan Province
Land coverage (km^2)	21,774	1,235
Capital population (Urban Planning Boundaries)	74,000	44,000
Total provincial population	859,000	336,107
GPP/inhab (US$, 2007)	720	1,245
Contribution to the GNP (%, 2008)	12	0,78

Source: NSO (2008), LNSC, Savannakhet Province's Department of Planning and Investment.

compared to US$824, NSO 2008). However, the contribution of both provinces to national wealth, i.e. Gross Domestic Product (GDP) differs drastically: whereas Savannakhet province contributes to 12 per cent of Lao GDP, Mukdahan province represents less than 1 per cent. Furthermore, the agricultural sector contributes to 24 per cent of GPP for Mukdahan province, compared to 54 per cent for Savannakhet (although recent data reveal a rapid growth of the industrial sector and a decline of the agricultural one). To sum up, it should be underlined that demographic weight, structure of the GPP, as well as contribution to national wealth, are completely different for the two provinces.

City Scale: Two Cities Facing Each Other and Developed Along the Mekong River Banks

Located almost face-to-face on an alluvial plain separated by the Mekong River, these two cities share a common history and cultural features. They used to be part of the same territory: first the Lao Lane Xang Kingdom and then the Kingdom of Siam in the eighteenth century; in 1893, Siam ceded the western banks of the Mekong to French Indochina, and the border was set on the Mekong River. The interactions between the two cities are furthermore eased by an ethno-linguistic proximity, their inhabitants share, for example, a similar language and religious festivals. A common point between these cities is that they are organized by the river and developed from it.

Mukdahan was established in 1767 on the confluence of the Huai Muk River and the Mekong River; although Savannakhet was created in the final years of the nineteenth century, the precise choice of the site remains unclear (Labarthe 1969, p. 71). Savannakhet was developed under French colonization as an administrative centre. The grid pattern of the streets and French colonial buildings in the old city centre are characteristic of this period of urbanization. In 1930, the city ranked third in terms of population after Vientiane and Luang Prabang (Labarthe 1969, p. 73). The main facilities, administrative buildings, hospitals, markets and temples were on the river banks or in the streets running parallel to the river, even if some of them have been displaced recently. Mukdahan was also organized along the river, where the Indochina market, administrative buildings, main temples and the ferry pier are located (see Map 14.1). Some old wooden houses can still be found in this area. Before the completion of the bridge, the border crossing was made by ferry; this transportation is still in use, but is limited to Thai and Lao nationals only. The piers were centrally located

Map 14.1
Mukdahan and Savannakhet Historical Centres: An Organization Along the Mekong River

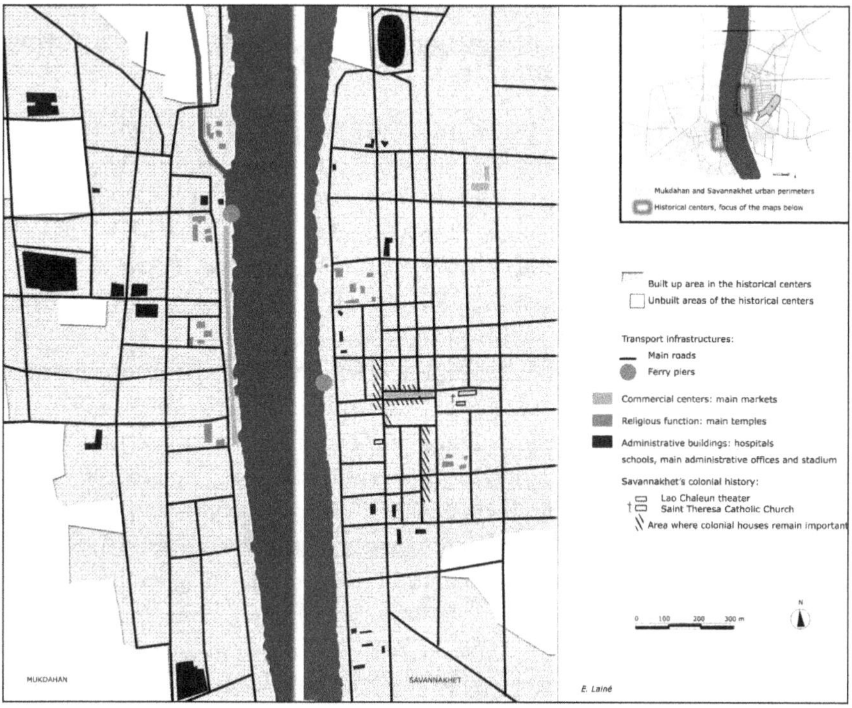

Source: Labarthe (1969), Information from Savannakhet Tourism Department and field trips.

for both cities' historical centres. Until the beginning of the 2000s, most of the cities' development was linear, along the main axis of urbanization represented by the Mekong River. The construction of Mukdahan Tower (a building 65 metres high with an observation deck and used as a museum) in 1996 illustrates also the extension of the city towards the south and along the same axis.

2. INTERNATIONALIZATION PROCESS ONGOING IN BOTH CITIES: ACTORS AND FACTORS OF CHANGE

With the complete re-opening of borders at the end of the Cold War, both cities initiated a process of internationalization in the 1990s. However, it remained low and the key element identified as a turning point in this process is the construction of the Second Mekong International Bridge (SMIB)[7] and the cross-border infrastructures related to the bridge.

A Low Level of Internationalization Observed for Both Cities before the SMIB Completion

If we take the example of Savannakhet, it appears that before 2006 urban transformations were low, and the city shows only few signs of internationalization. Official Development Assistance (ODA) providers and foreign private investments are the main actors of these transformations. The low level of foreign direct investments (FDI) can be explained by the fact that the investment climate was liberalized in Lao PDR in the late 1980s and the FDI encouraged — with the so-called FDI law — in 1994.

The first ODA project targeting the city of Savannakhet was the water supply improvement financed in 1971 through a grant from the French government. Three main construction projects shown in Map 14.2 had a later impact on city organization and functional zoning. The first one is the displacement of the fresh products market from the city centre to the north of the urban perimeter in 1998. The new market, know as *Talat Savanxay*, was built with private Singaporean funds and creates a new commercial centre in the north of the city (see Map 14.2). As a consequence, land prices in this area exceeded those of the old city centre in 2003 (Land Management Authority interview 2009). The second project that changed the city organization is the administrative centre created by the ADB under the Secondary Towns project in 1998; most of the administrative buildings, spread along the river, have been relocated in a newly built neighborhood on the east side of the city. The last project also has a specific function: it is the construction of the stadium in 2005 in the north of the city to host the 7th National Games. This construction cost US$800,000 and was financed by Oxiana Ltd., an Australian-based company involved in the mining industry in Sepon, as a gift to the province (Oxiana Ltd. 2005, p. 13).[8] These three projects show a first stage of internationalization by foreign actors and funds, and created new functional centres in the city of Savannakhet, as presented in Map 14.2.

In the case of Mukdahan, although the city grew at a rapid pace during the 1990s (see Map 14.3), the level of internationalization appears even lower before 2006. Most of the small-scale industries and services are held by Thai nationals, and the most remarkable example of foreign-owned company impacting the city development is the supermarket Tesco

Map 14.2
Stages of Development and Spatial Organization in Savannakhet

Source: Data compiled by E. Lainé (2011).

Lotus, a Thai-British joint venture constructed in the early 2000s, located in the north-west of the city near the bus station and creating a commercial centre on the urban scale (see Map 14.3).

The Main Factor of Change: Infrastructure Improvement within the Gms Framework

The transport infrastructures built within the GMS framework, especially the Second Mekong International Bridge (SMIB) led to an acceleration of the internationalization process in both cities, even if it seems to be higher for Savannakhet. The SMIB construction started in 2004 and

Map 14.3
Stages of Development and Spatial Organization in Mukdahan

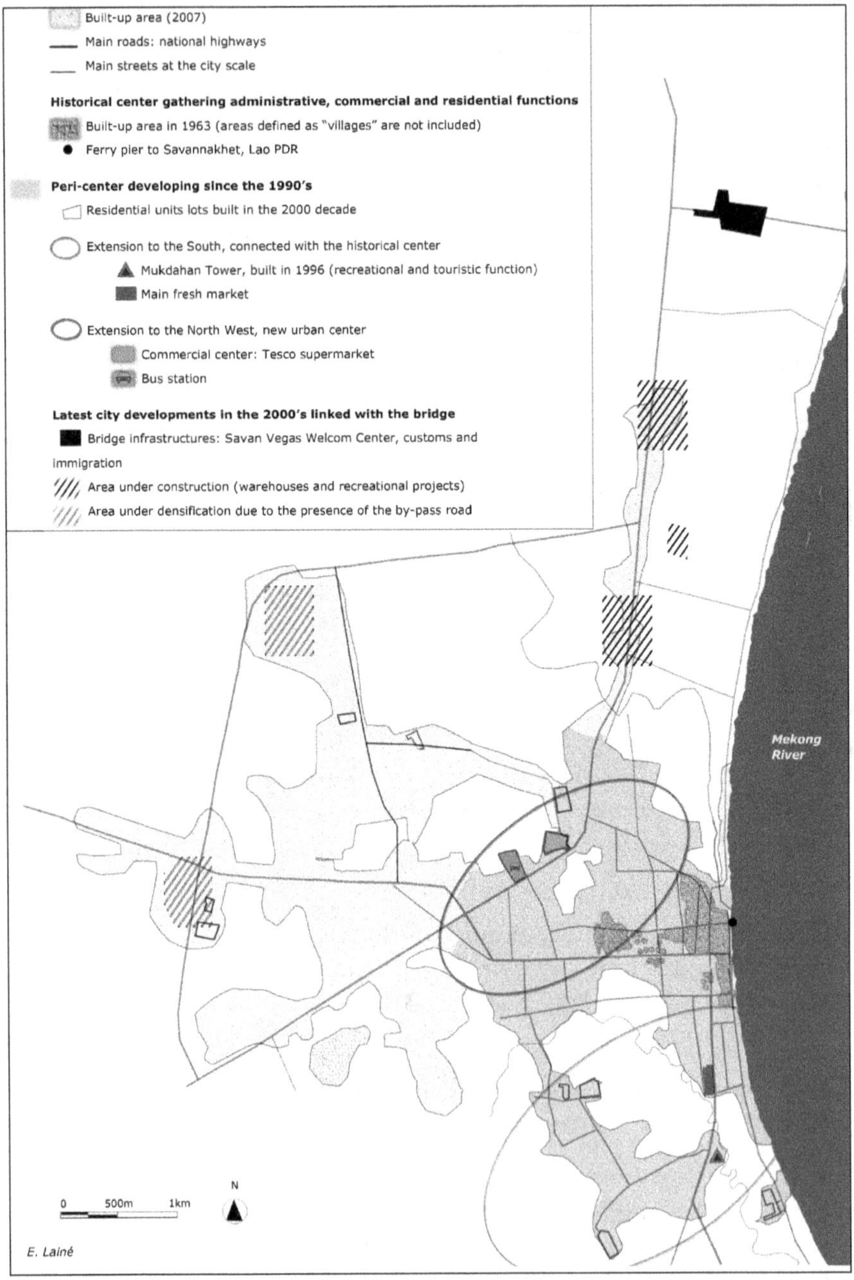

Source: Data compiled by E. Lainé (2011).

was completed in December 2006 for a public opening in January 2007. Financed through a JICA loan, this bridge had several expected outcomes on different scales: integrating both cities in a regional transport network, i.e. the EWEC, strengthening the link between Thailand and Lao PDR via Mukdahan and Savannakhet, and increasing competitiveness for both cities on national and regional scales. Other infrastructures have been implemented with the SMIB construction: customs houses, immigration buildings and an international bus system, linking Mukdahan bus station to the one in Savannakhet (January 2007). Another unprecedented project formulated in the early 2000s and still not implemented is the shared use of Savannakhet airport by both Thai and Lao users. In both cases, the origins of the loans point to an internationalization process.

Two main projects decided by national governments testify to the ADB and national strategy of promoting industrial development and boosting cross-border trade in the border region: Savan-Seno Special Economic Zone (SSEZ) in Savannakhet province, and Mukdahan Distribution Center and Industrial Estate.

Established by Prime Ministerial decree in 2003, the SSEZ has shown great dynamism since the completion of the bridge. The initial plan established two sites, yet this plan has been recently modified to four: Savan City (A), Logistic Park (B), Savan Park (C) and a resettlement site (D). All these sites are located in the north of the urban area, near the bridge (site A), along route 9 (sites C and D), except site B located in Seno (see Map 14.2). Joint ventures between the Lao government and foreign companies are in charge of developing this new commercial, industrial and logistic hub.[9] The target group of investors identified is labour-intensive industries, such as manufacturing of shoes, garments, toys and assembly of component parts (SavanPark, <www.savanpark.com>). In 2009, the Lao government approved eleven investment projects: four Japanese companies, three Thai, two Malaysian, one from the United States and one Lao company.[10]

On the other side of the border, the creation of an industrial estate in Mukdahan working in collaboration with SSEZ has been delayed. The project has been under discussion for many years, but contrary to SSEZ, the Thai government is counting on the private sector alone to develop it. The Industrial Estate Authority of Thailand (IEAT) performed a feasibility study in 2006, and plans include a logistic centre of 8 ha and an industrial estate of 80 ha near the bridge location (Tsuneishi 2008, p. 24).

Actors of Change: Low Level of Tourism and Importance of FDI and Foreign Entrepreneurs

Although these two cities received massive planning and investments from their national governments as well as ODA for financing infrastructures, other actors are taking an active part in the internationalization process, i.e. private actors.

International tourists can generally be seen as major actors of internationalization. Yet, in these cases, data presented in Table 14.2 shows evidence that they cannot be identified as relevant driving actors for Mukdahan and Savannakhet internationalization. The level of tourists remains low in Mukdahan province, compared for example to Ubon Ratchathani province, which also has an international crossing post to Lao PDR (to the province of Champassak), and the share of international tourists is marginal. For Savannakhet province, although the number of tourists is higher than in Mukdahan, as well as the share of international tourists among them, it should be underlined that the share of Thai tourists is really high (and can be partially explained by the presence of the casino). Champassak province in the southern part of Lao PDR seems to be more internationalized by tourism, as shown in Table 14.2, due to the presence of its tourist attractions such as the Vat Phou heritage site or the Bolaven Plateau.

On the other hand, the main actors identified to characterize and explain the ongoing internationalization process in Savannakhet are the

Table 14.2
Number of Tourists and Share of International Tourists by Province, 2007

	Province	Total Tourist Arrivals	Share of International Tourists, %
Thailand	Mukdahan	286,678	1,7
	Ubon Ratchathani	1,081,861	3,0
Lao PDR	Savannakhet	661,929	25,2 (+46,6 Thai)
	Champassak	220,214	46,9 (+36,8 Thai)

Source: Development of Planning and Investment, Savannakhet Province.

foreign investors and entrepreneurs. Table 14.3 shows that the number of joint ventures and foreign companies is only a little higher in Savannakhet than in Champassak. However, the number of these companies located in the urban district is much higher for Savannakhet than for Pakse, and the origin of capital is more diverse and internationalized due to the presence of Chinese companies. Furthermore, the field trip showed that most of the new joint ventures and foreign companies in Savannakhet province are not only located in Kaysone Phomvihane district, but more especially in the bridge area.

However, more interesting than a quantitative approach, a qualitative one can show how the FDI and presence of foreign investors can impact the urban landscape. The Savan Vegas Casino provides a good example for studying this dynamic, as it has an impact on both Mukdahan and Savannakhet and the relationship between the two cities. The project of Laos' second casino was launched in 2006, although the construction of the building only started in 2007, and was opened to the public in April 2009. The shareholders are Sanum Investments, from Macao (60 per cent), Lao government (20 per cent) and Lao private funds (20 per cent).[11] The casino complex, located in the north of the bridge area (B. Nongduen on Map 14.2), has a big impact on the urban landscape, as it gathers on a 50-hectare land concession a 180-room hotel, one floor of which is dedicated to the casino, a 300-seat restaurant and other smaller services. On a broader scale, this casino has also an impact

Table 14.3
Comparison of FDI in Champassak and Savannakhet Provinces, Situation in June 2009

	Champassak	**Savannakhet**
Total number of foreign companies and joint-ventures	81	83
Number of foreign companies and joint-ventures located in the main urban district	Pakse district: 21	Kaysone Phomvihane district: 49
Number of investing countries	16	17
Main investors	Thailand, Vietnam	China, Thailand, Vietnam

Source: From the Department of Planning and Investment, Savannakhet and Champassak provinces.

on the relations with the city of Mukdahan, as most of the customers are Thai, especially from the neighbouring provinces; a shuttle departing every hour from Mukdahan (in the bridge area and from the bus station) provides a transfer to the Casino, which represents an alternative way of crossing the border. Two welcome offices have opened in the above-mentioned areas to guide customers. The company also provides transfer from Vientiane, Pakse, and cities of central Vietnam (to the Vietnamese border on a daily basis, but also to Hue and Quang Tri three times a week and to Danang and Hanoi twice a week).

3. FROM BORDER TOWNS TO CORRIDOR HUBS: A PROMOTION OF THESE CITIES ON THE GMS SCALE?

ADB and national strategies turn Savannakhet and Mukdahan cities into a hub at the subregional scale. Do their location on the corridor and their programmed functions have an effect on the cities' position at national and GMS level? Because ADB and national strategies, as well as projects, are currently being implemented, not all the conclusions can be stated yet and this chapter aims to give only elements of response.

City Scale: Mukdahan and Savannakhet are Both Facing a New Dynamism with the Emergence of a Polycentric Pattern

The internationalization process going on with the upgrading of the cities' services and infrastructures has had an impact on the urban landscape of both cities in recent years. Savannakhet as well as Mukdahan are indeed facing unprecedented dynamism and expansion, formal — or planned — as well as unplanned. Three main dynamics are impacting the urban built-up area: modernization and requalification of old city centres; extension of the urban perimeter and production of new centres.

As shown in Map 14.1, the presence of administrative and commercial services as well as the ferry pier as a way to cross the border established the historical centres along the Mekong River as the central area. Mukdahan seems more affected by the modernization of the historical city located on the Mekong riverbanks: the road near the river and the Indochina market were rebuilt and extended at the beginning of the 2000s; on the other hand, Savannakhet's old city centre stays mainly unchanged, apart from the upgrading of the road network. The question of preserving the architectural heritage of the historic quarter seems highly accurate, although not clearly handled.[12]

However, urban growth and the construction of the bridge are inducing new patterns of urbanization and the emergence of a polycentric structure at the city level. Both cities have been facing a linear urban sprawl, i.e. along the main axis of transport.

The dominant axis of urbanization is East-West for Mukdahan; the renovation of the road between Khon Kaen and Mukdahan *via* Kalasin and the growing importance of Khon Kaen in the region certainly created an extension of the urban perimeter in Mukdahan and a densification of the area along this road. This axis of urbanization is so far predominant compared to Roi Et road (to the south-west, linking to Khon Kaen too but without being part of the EWEC), but also, to the road to the bridge (north-south axis). As a consequence, the main access to the city is by the Khon Kaen road (see Map 14.4). The new urban centre presented in Map 14.3 with the bus station and the commercial area is likely to get reinforced by its strategic location at the crossroads between the road to Khon Kaen, the historical centre and the road to the bridge. On the other hand, the residential extension in the south of the city (along the road to Kemmarat, which is not an important urban centre on the Northeast region scale) or that along the road to Roi Et will probably not be developed so much in the future.

In the case of Savannakhet, the emergence of new centralities appears simpler, following an axis to the north. If both cities are extending in the direction of the new bridge with peripheral areas in the north of cities facing a densification of the urban frame, Savannakhet has been more affected by this dynamic, probably also because the infrastructures linked with the bridge are more important (see Map 14.2). The first step of urbanization for Savannakhet was the creation of a new centre with the construction of the new market and the bus station; the second step is currently a north polarization with the construction of the bridge. The emergence of these new centralities raises unprecedented challenges for the local authorities, such as the connection between the old and the new urban centres, as pointed out in Savannakhet by the Urban Development Agency (UDAA), or the increasing population in these new centres.

Furthermore, both cities are affected by a change in the road network hierarchy, particularly with the growing importance of by-pass roads linking the new bridge to the main roads on the provincial scale (highway to Khon Kaen for Mukdahan and road to Seno for Savannakhet, linking to Vientiane by road 13 as well as to Vietnam through road 9) without entering the towns. Three consequences can be identified from this dynamic: a densification of the areas located along these roads, the emergence of a main access to the cities dominating the historical entries (see Map 14.4) and a risk of a "tunnel

Map 14.4
Savannakhet and Mukdahan New Patterns of Urbanization

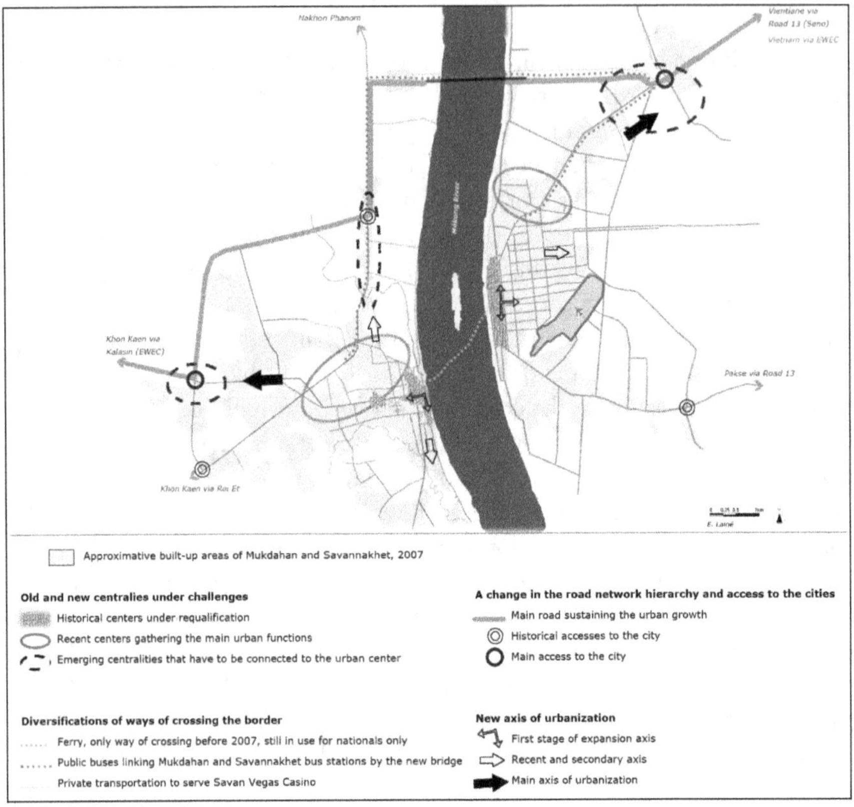

Source: Data compiled by E. Lainé (2011).

effect", i.e. the infrastructures linked with the bridge and the new transport axis will be used to cross the border without benefiting the cities that will be circumvented.

An Increasing Connectivity between the Two Cities: The Example of Cross Border Trade

Mukdahan and Savannakhet are declared "twin cities" under the ACMECS cooperation programme. This institutional agreement that includes, for example, contract farming programmes between the two provinces, seems to have little direct impact on the relationship between the two cities.

During the interviews, local authorities did indeed acknowledge a low level of cooperation with their so-called "twin". However, due to the recently upgraded transport infrastructures as well as historical interactions, flows of capital, people and goods are increasing between the two cities. On each side of the border, the new centres related to the bridge are connected by physical infrastructures used by different types of actors: nationals and tourists using private vehicles or public buses linking the two bus stations to cross the bridge or customers of the Savan Vegas benefiting from minivans from the Welcome center in Mukdahan to the Casino in Savannakhet (see Map 14.4).

This increased connectivity can also be seen through the level of cross border trade, for example. The value (and volume) of cross border trade has expanded since the bridge's completion, as shown in Figure 14.1. Furthermore, the Mukdahan-Savannakhet customs post concentrates an increasing part of the total volume of border trade between the two countries, showing that the corridor location can be strategic in terms of trade flow control. Mukdahan was in 2008 the second post in terms of value of border trade between Thailand and Lao PDR after Nong Khai, accounting for nearly 35 per cent of the total border trade between the two countries (see Figure 14.1). In 2008, copper from Savannakhet province mines accounted for 87 per cent of total exports from Lao PDR to Thailand. On the Thai side, petroleum and building equipment represented together 22 per

Figure 14.1
Thai-Lao Border Trade at Mukdahan Customs Post, 2006–08

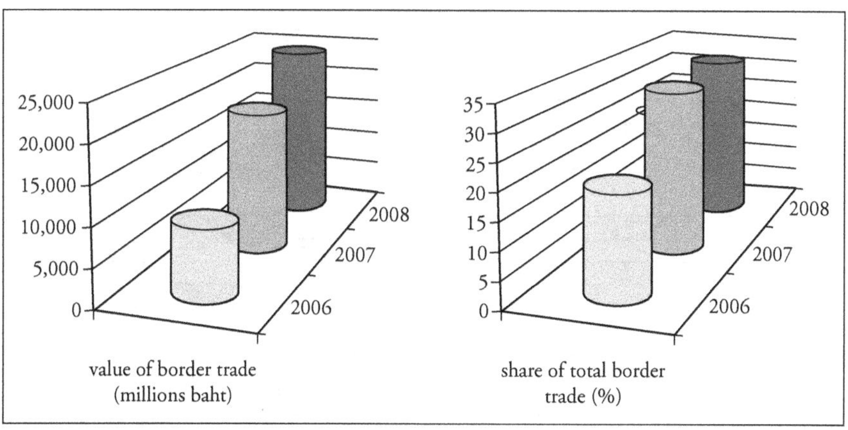

Source: Mukdahan Customs House (2009).

cent of total exports (ADB 2009, p. 46). According to the same report, total transit trade is marginal, less than 5 per cent of the total value of trade at the Mukdahan-Savannakhet checkpoint. However, these data do not take into account informal trade that is considered as important.

National and Regional Scale: Are These Twin Cities Benefiting from Their Corridor Location?

If ADB and national strategies aim to transform border towns into regional hubs or nodes, evidence shows an upgrading in the level of infrastructures in Mukdahan and Savannakhet. Apart from this aspect, does it imply that these cities are benefiting from their location on the corridor, giving rise to a promotion from small towns to secondary cities playing a role at the GMS level? Even if the completion of these infrastructures is too recent to draw conclusions, some indicators suggest not only a greater internationalization as shown previously, but also a potential to gain more weight in the sub-regional urban hierarchy. However, this potential seems for the moment more important for Savannakhet than for Mukdahan.

In terms of foreign direct investment, the construction of the bridge had a clear impact on the number of foreign companies and joint ventures (see Figure 14.2). The amount and number of FDI has been rising significantly

Figure 14.2
Number of Foreign Companies and Joint Ventures, Savannakhet Province

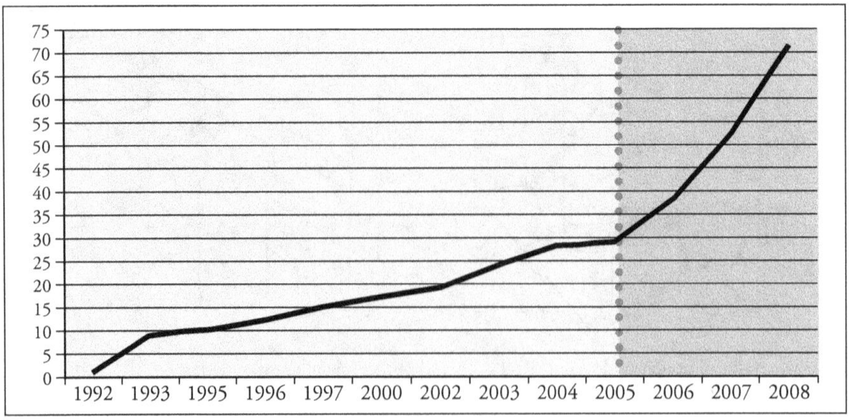

Source: Mukdahan Customs House (2009).

Table 14.4
Number of Projects Approved by BOI in Selected Provinces of Northeastern Thailand

	Numbers of Projects	Including Numbers of JV and FDI
Mukdahan 2005–10	3	2
Ubon 2005–10	22	5
Khon Kaen 2005–7	33	12
Udon Thani 2005–7	14	5

Source: From Board of Investment (2010).

since 2005, illustrating the city's greater competitiveness. This dynamic has not affected Mukdahan to the same extent: as shown in Table 14.4, the number of projects approved by the Board of Investment (BOI) since 2005 is very low, and a small part of them are from foreign capital. Comparison with other cities in Thailand Northeast Region, inland as well as border towns, states that fewer projects have been approved in Mukdahan than in these cities, particularly Khon Kaen. One can thus make the assumption that the location on the GMS corridor and the upgrading of infrastructures are not sufficient in this case to attract these flows of capital that mainly target the capital of the Northeastern region.

CONCLUSION

For the last decade, due to their corridor location, Mukadahan and Savannakhet have benefited from large projects involving different actors: central governments, international agencies through ADB as well as international and national private investors. These actors are impacting the cities' structure that is changing into a polycentric pattern with new centres of activities, and creating a greater internationalization. The changing scale of transport infrastructures through the GMS flagship sustains this dynamic, and is partially relayed in Savannakhet by foreign investments. The changes are not that important in Mukdahan, that benefits for the moment less from its corridor location, as for example the FDI are still concentrated in Khon Kaen.

However, unprecedented challenges are rising at various levels for both cities: at the city level, how to connect the new central areas with the old ones, and how to preserve the historical value of the old city centres? On the national scale, Mukdahan is still a third-rank city with a deficit in competitiveness compared to other urban centres in northeast Thailand, especially the regional city of Khon Kaen, whereas Savannakhet could have the potential to be promoted to an industrial centre, capital of central Lao PDR and the city linking Thailand to central Vietnam.

The arrival of private investors for developing Mukdahan industrial estate as well as a greater dialogue among stakeholders seem to be prerequisites for gaining more benefit and monitoring the eventual negative side effects of the twin cities' corridor location. Both cities can be promoted both at the GMS level and in their national contexts, with the effective implementation of planned cross-border projects, such as the development of a cross-border economic zone and the creation of the shared-use airport of Savannakhet for both Lao and Thai residents that has been proposed since 2004.

Notes

1 Illustrating this importance, the ADB is currently funding a project on corridor towns focusing on 10 cities, 5 border towns and 5 inland towns in Cambodia, Lao PDR and Vietnam (ADB 2010).

2 See the contribution of Ruth Banomyiong in this book.

3 The name of Savannakhet town or Khantabouly has been changed into Kaysone Phomvihane District in 2005, from the name of Lao PDR's former President, who was born there. However, this article will employ the former appellation of Savannakhet, which is still mainly used among the population and in international reports.

4 ADB, "The Strategic Role of Corridor Towns in the Development of GMS Economic Corridors", Background Paper for 5th GMS Development Dialogue, 20 November 2009.

5 Including peri-urban areas, Savannakhet city population is estimated at 120,000 inhabitants, which is more than the population of Luang Phrabang (Nolintha 2011).

6 The urban population in Thailand encompasses population living in urban areas, called "municipal areas". Using this methodology, the urban population is 19.9 per cent for the Northeast region of Thailand in 2007, which is less urbanized than the national average. Yet, NESDB points out that a wider definition that covers both urban and peri-urban areas will raise the national rate to 39.2 per cent (NESDB 2008, p. 12). For Lao PDR, urban population is living in urban villages, and five criteria are used to define the latest: "The village must lie in the municipal vicinity where the district or provincial authority is located, there are more than 600 residents or more than 100 households / There is a road for motor vehicles to get access to the village / The majority of

households in the village are electrified / There is a tap water supply in service to the majority of households / There is a market in the village" (LNSC).

7 The First Mekong International Bridge linking Thailand and Lao PDR was built in 1994 between Nong Khai and Thanaleng. A third and a forth one are currently under construction, linking Nakhon Phanom and Thakhek, and Chiang Khong and Houexay respectively.

8 This company changed its name to "OZ Minerals" in 2008 and since 2009 the majority of the company has been acquired by a Chinese one (China Minmeals). Available at <http://en.wikipedia.org/wiki/OZ_Minerals>.

9 Site A (Savan City) is developed by Savan City Co. Ltd., a joint venture between the Lao government (30 per cent) and a Thai company; the agreement was signed in June 2007. Site C (Savan Park) is developed by Savan Pacifica Development Co. Ltd., a joint venture between the Lao government and Pacifica Streams, a Malaysian company; the agreement was signed in February 2008. Both companies opened a website to describe their activities: <www.savancity.org> and <www.savanpark.com>.

10 Interview with the Savan Seno Special Economic Zone Authority.

11 Sanum Investments is a Macau-based company investing heavily in Macau, China, Cambodia, Thailand, and Laos. Slot machines have also been constructed by the company at the Dansavan-Lao Bao checkpoint. Interview with the Vice-President of Advertising and Public Relations of the Savan Vegas, 2009, and <http://www.sanuminvestments.com/home.html>.

12 Some projects related to tourism are currently implemented in Savannakhet Province, including a Tourism Service Center along road 9 supported by the JICA and ADB; none of them are directly targeting Kaysone Phomvihane district though, except the renovation of the Dinosaur Exhibition Hall and the publication of a brochure on a walking tour of Savannakhet historic quarter by the JICA (interview with the Department of Savannakhet Provincial Tourism 2009).

References

Arnold, Dennis. *Administration, Border Zones and Spatial Practices in the Mekong Subregion*. Ph.D. dissertation, University of North Carolina, Chapel Hill, 2010.

Asian Development Bank. *Report and Recommendation of the President to the Board of Directors on a Proposed Loan to the Lao PDR for the Secondary Towns Urban Development Project*. ADB: Manila, 1997.

———. "The Strategic Role of Corridor Towns in the Development of GMS Economic Corridors". Background Paper for 5th GMS Development Dialogue, 20 November 2009.

———. "Preparing the Corridor Towns Development Project". Technical Assistance Report, 2010.

———. "East West Economic Corridor: Strategy and Action Plan", 2009. Available at <www.adb.org/GMS/Economic-Corridors/EWEC-SAP.pdf> (accessed December 2010).

Board of Investment, Thailand. List of projects approvals, 2011. Available at <www.boi.go.th>.

Gibb Ltd. and Pal Consultants Co. Ltd. *Border Towns Urban Development Project: Final Report*. Asian Development Bank and National Economic and Social Development Board, Bangkok, 1999.

Krainara, Choen. *Cross-Border Trade and Commerce in Thailand: Policy Implications for establishing Special Border Economic Zones*. Master Thesis, Bangkok: Asian Institute of Technology, 2008.

Labarthe, Christian. *Quelques aspects du développement des villes du Laos* [Considerations on Laotian Cities' Development]. University of Bordeaux, 1969.

Maneepong, Chuthatip. *Dynamics of Cross Border Development in the Mekong Sub-region: A Case Study of Thailand*. Saarbrücken: LAP Lambert Academic Publishing, 2010.

National Statistical Office. *Provincial Statistical Report: 2008*. Mukdahan Province, Mukdahan, 2008.

Nolintha, Vanthana. "Cities, SEZs and Connectivity in Major Provinces of Laos". In *Intra- et Inter- City Connectivity in the Mekong Region*, edited by M. Ishida. *BRC Research Report*, no. 6. Bangkok: IDE-JETRO, 2011.

Office of the National Economic Social Development Board (NESB). 7th National Economic Social Development Plan, Bangkok, 1992.

———. 9th National Economic Social Development Plan, Bangkok, 2003.

Oxiana Ltd. "Sustainability Report Summary". Melbourne: Oxiana Ltd., 2005.

Sassen, Saskia. *The Global City: New York, London, Tokyo*. Princeton University Press, 1991.

Taillard, Christian. "Un exemple réussi de régionalisation transnationale en Asie orientale: les corridors de la Région du Grand Mékong" [A Transnational Integration Example: The Continental Grid of Economic Corridors in the Greater Mekong Subregion]. In *L'Espace Géographique*, 2009–1. Belin Paris, 2009, pp. 1–16.

Tsuneishi, Takao. "Development of Border Economic Zones in Thailand: Expansion of Border Trade and Formation of Border Economic Zones". IDE Discussion Paper no. 153. Institute of Developing Economies, JETRO, 2008.

15

PRIVATE COMMITMENT Marital Alliance in the Establishment of Business Networks at Hekou-Lào Cai, Twin Sino-Vietnamese Border Cities

Caroline Grillot

Decision-making policies, and economic development plans make up the changing framework in which people are moving and organizing their life in the emerging corridors of Southeast Asia. This chapter aims to bring the local and human dimension back into the picture by focusing on the Chinese actors in the building up of these zones, on their business practices and their personal involvement apart from financial investments. It provides a perspective on the social changes both affecting the implementation of economic corridors and resulting from them, since they bring together eclectic categories of population. Most of those, by various means, are trying to build or extend some economic networks to benefit from the implementation of these regional projects. I propose to draw our attention to one community who live scattered in and around the Chinese border town of Hekou, in Yunnan province: the community of Sino-Vietnamese couples.[1] Many Hekou observers consider this community as an important element in the development of economic collaboration between China and Vietnam. It may symbolize the nature of the "possible" and the existing "challenge" of the reshaped space of pre-existing interactions.

Indeed, these mixed marriages are less the consequence of romance than pragmatic and conjectural alliances between individuals needing to develop their businesses and to improve their material life. As this is the case in many alliances, it is not considered to be a negative factor, but a personal and patriotic way of penetrating further into a land where local guidance is necessary in order to establish long-term collaboration despite obstacles. Besides, on a more personal level the individual commitment that marriage entails for each spouse needs to accommodate local discourses that interfere with the official intention to "work together toward the harmonization of relations between Chinese and Vietnamese communities" at large (as stated in such widespread street slogans). The various prejudices and discrimination that these couples embody remain a sensitive issue that needs to be addressed to understand the importance as well as the limits of their role in economic development.

But first of all, in order to understand how this particular group of people emerged and why it deserves closer attention, we need to briefly outline the principal characteristics of the population of Hekou.

THE COMPLEXITY AND DIFFERENTIAL EXPERIENCES OF BORDER COMMUNITIES OF HEKOU

Nowadays, Hekou has acquired a regional and national strategic position since it is situated on the road that links Kunming to Haiphong, via Hanoi. It is located right at the intersection where the Red River meets its affluent, the Nanxi River, and faces Lào Cai in the northwest of Vietnam. Although Hekou used to represent no more than just a passage between China and Vietnam, its importance has changed in the course of the twentieth century. From a small port mainly populated by Yao villagers on its highlands and by Yunnanese Han on its lowlands, it became a sort of refuge for outcasts of contemporary history. During the Chinese revolution, Republican Army soldiers established themselves in this village of petty traders and began the first noteworthy migration to the area. Later, after the establishment of the People's Republic of China in 1949, some new settlers arrived. Indeed, one of the main goals of foreign policy was then to protect China's frontiers, and large groups of young people from Yunnan and Guangxi were sent to these remote places along the border with Vietnam. During the Cultural Revolution and its "*xiaxiang* [down to the countryside]" political movement (1966–76), more people from all over the country settled in the area. Although this migration mainly took place during the twentieth century, these different layers of residents now constitute the "local" people of Hekou. This majority of Han people and a minority of urbanized ethnic group members still make for a certain human diversity.

But in 1978–79, before and during the armed conflict between Vietnam and China, sudden migrations again changed the local demography pattern in a different way. Following a patriotic call from the Vietnamese government, the Vietnamese people who were living as long-time residents in Hekou, as well as in the whole province of Yunnan, fled back to Vietnam. Whereas on the other side, the Hoa (the Chinese ethnic group in Vietnam) were "encouraged", then forced to resettle in their "homeland" China. Many of the Hoa ended up as refugees whose status remains partly unresolved thirty years after the conflict. They also account for a significant proportion of Hekou's inhabitants, especially in its suburb Nanxi. Others obtained Chinese nationality and decided to stay near Vietnam to maintain family links. Today, living together in peace despite remaining resentments, both communities play a crucial role in establishing communication between Hekou's traders and their Vietnamese business partners since their language skills make them the ideal translators.

When the border gate of Hekou officially reopened after 1991, and following the official call for the development of Hekou as the soon-to-become "Border Economic Cooperation Zone" (created in 1992), new waves of immigrants arrived. Initially coming from the surrounding cities of Yunnan and Guangxi, entrepreneurs started to visit and then to settle in the city to establish their new, promising border trade business. The acceleration of these new settlements coincided with the first decade of the Great Mekong Subregion's economic corridor that emphasizes the Kunming-Hanoi trading route. Quickly, as soon as the news of business opportunities spread among traders' networks, more of them came from Zhejiang, Guangdong, Sichuan, Hunan, and even further north to engage in border trade. Some brought along their families while others came alone.

In the meantime, simplified procedures to enter China[2] also allowed a significant migration of Vietnamese workers to Hekou through various sectors. Prostitution, tourism, transport, construction, and small trade brought thousands of men and women into town on a regular basis or permanently. It is together with the growing flux of population crossing the border, legally or not, that the human trafficking issue[3] started to attract attention. Numerous cases of trafficked women were officially registered while the tracks of thousands of others were lost. Local people even mention a peak period in the mid-1990s. In the context of the informal economy, along with prostitution and all sorts of smuggling, human trafficking has been recognized as a rampant social issue in the area but nothing very efficient has yet prevented the spread of this social phenomenon. On the structural level, the one-child policy, rural-urban migration, increase of bride-price in rural areas as well as women's emancipation and social discrimination stands have created a deficit in marriageable women in many Chinese rural areas (Grillot 2010). As a consequence, conventional explanations emphasize the demographic issue to describe increasing numbers of Chinese bachelors in search of a bride among the overpopulated regions of China and the neighbouring countries. Vietnam represents, for various cultural and economic reasons, one of these pools, and some men do not hesitate to employ all available means to find their other "half of the sky". In a context that witnesses rampant prostitution, arranged marriages as well as conventional cross-border romances in Hekou, the newly arrived Chinese traders face various ways to satisfy — when needed — their own intimate ambition if it happens to serve their business as well.

Nowadays, Hekou's population is composed of a mix of people coming from the various continuous waves of migration that have transformed local human settlement throughout the last century. All of them have different perspectives on and experiences with their Vietnamese neighbours and they generally maintain pragmatic relationships. However, one can observe an extensive mobility of the population due to the instability of the economic situation. Hekou remains a turntable for migration and its current development policy is further accentuating this tendency, but the human landscape keeps rotating, especially among the recent wave of traders and workers.

With a clearer picture of the complexity of settlement in Hekou, we may now examine how the current waves of migration provide a particular, specific social space for the population involved in the economic development of the region. Here I will focus on the matrimonial exchanges that occur in such a human setting and the implications they may have in the promotion of local cross-border trade.

VIETNAMESE WOMEN AS CRUCIAL LINKS IN THE ESTABLISHMENT OF BORDER TRADE NETWORK

In the contemporary period, cross-border families have appeared in a more obvious way in urban settings like Hekou. What used to be associated with intra-ethnic marital arrangements occurring regardless of the actual citizenship of the spouses has now turned out to be an illustration of the constant interaction between individuals from both sides of the border. However, one of the most noticeable differences between marital exchanges in the past and those in contemporary times is the individuals they concern: Vietnamese women migrate to China to marry while Chinese women who marry Vietnamese men are very rare. Furthermore, a situation which did not raise much concern a few decades ago now attracts scrutiny.

The circumstances in which these men and women from different backgrounds and with different expectations meet, and the types of relationships they establish, suggest some explanations. From the perspective of Chinese men, this could happen as follows: the first encounters between Chinese men and Vietnamese women can actually occur in different locations of the city of Hekou, and the contrasting impressions that these encounters leave on the newcomers tend to shape an ambiguous experience of the relationship that may result from these initial meetings.

When new cross-border "traders to be" arrive in Hekou, they usually already have some acquaintances in the city. In such a migration space dedicated to economic development, most of them establish their businesses according to their familiarity with two sorts of spheres: the sphere of expertise or the business sector, and the sphere of *laoxiang* [fellows from the same city or province]. Hence, besides the necessary guidance into specific aspect/rules of trade that these acquaintances provide them with, newcomers are also quickly introduced to the local social context through their peers. But as in most places in China, business practices often imply a ritual visit to local entertainment places with business partners. In the case of Hekou, this includes restaurant meals, Karaoke sessions, visits to the border casino and maybe an express trip to Vietnam. This may mean a simple one-day visit to Lào Cai, a slightly longer trip to the local tourist attraction Sapa, and even a visit to the capital, Hanoi. However, most newcomers soon hear about the most infamous attraction of Hekou, the Vietnamese market, which they tend to experiment at some point during their stay. In this place that clearly represents the presence of the Vietnamese population in town, newcomers can try Vietnamese food, buy exotic products from their neighbours (most of the shops are managed by Vietnamese women) and try a massage offered by a young Vietnamese woman. Generally, the ultimate entertainment activity consists of sex services provided by a Vietnamese prostitute, whether on the spot (in one of the market's numerous brothels) or in a nearby guesthouse. Vietnamese prostitutes represent an entertainment activity that has become a significant tool for local business associates to create familiarity with the Other and to build a partnership.

Indeed, these women also offer a seductive representation and a promising human perspective of Vietnam.

Whereas the newcomers enjoy these encounters with their new neighbours, they also come to observe in a more specific way how their peers organize their business in collaboration with their Vietnamese partners. For instance, consumer goods wholesalers, a large majority of Chinese traders in Hekou, use Vietnamese carriers to transport their goods to Vietnam, whether by legal means or by smuggling. The majority of these carriers are Vietnamese women whose ability, autonomy and efficiency are very much appreciated. Later, during their initial business trips in Vietnam, Chinese traders can enjoy the services of local guides. In this field, Vietnamese women also dominate the market. Although this position of guides often gives these women the opportunity to know future employers or extend their customer network, it also puts them in a delicate position since they may be solicited for other sorts of services or become the target of flirting. Still in this field of services, these same traders often use the services of Vietnamese women (many of them are of Hoa origin) to translate any negotiations, discussions and transactions with their Vietnamese clients. These women can directly work in Chinese shops or they can be employed on a regular basis for business meetings and business trips on either side of the border.

The role of go-betweens that they often play in addition to their language ability helps to overcome cultural misunderstanding and to establish efficient cooperation and trusting relationships between Vietnamese and Chinese people. Being the privileged actors of direct interactions between trade associates makes these often young, unmarried women — consciously or not — become representative of the Otherness and they eventually turn into potential private partners when they are available (single). Hence, the longer newcomers spend in Hekou, the more they notice the different levels of relationships between local Chinese men and Vietnamese women. They can be associates, lovers, both, and eventually spouses. They may even live together and raise a child. Together, they create an inevitable extension of Hekou's population: cross-border families.

SOCIAL NETWORK AS A BRIDGE

Although not a new phenomenon, Sino-Vietnamese cross-border social networks nowadays have reached an unprecedented degree through the variety of the individuals they concern. They create an active social space much larger than its historical precedent. Chinese migrants in Hekou come from various provinces of China, sometimes from as far off as the northeastern area of the country where there was previously no particular connection with Vietnam. It was only when some of these migrants started to come back home with the accounts of their various experiences and the substantial results of success in the business field that they encouraged others to follow their example, facilitating geographically extended economic cooperation. They transformed it from a simple deal between a local Hekou trader and his Vietnamese partner, to a long chain of associates ranging from — for instance — a factory owner in far-away Zhejiang, to a Chinese bamboo bed-mat shop in Hanoi, transiting through a wholesaler in Hekou.

But many business networks stem from personal involvement in relationships such as friendship or intimate affairs. Thus, these Sino-Vietnamese alliances are the result of a double dynamic: a rather natural one and a more pragmatic one. On the one hand, business exchanges facilitate human encounters and mix marriageable men and women, whereas on the other hand, marrying a spouse from the other side of the border is encouraged since it provides broader access to an already existing network of kin, friends, business or administrative partners. Hence, a diversification of potential partners adds to the ongoing extension of the network scale. But it is precisely this double movement of people and the motivations or consequences that such connections imply that make these couples suspect. In the local discourse, it is perceived that many couples are actually business partners using the cover of a private relationship to extend their potential area of influence.

As a matter of fact, these men do benefit from a sort of compensation by accessing the Vietnamese internal market through their marital alliances. Family members, acquaintances, business relationships or connections among local authorities represent a highly valuable advantage for those who wish to participate in the Vietnamese economy. In exchange, these Chinese men tend to provide their partner with a better life since the expected standard of living in China is higher than in Vietnam. Hence, although it might not be the motive for such alliances, the economical pragmatism they are associated with is often perceived by the observers as a deliberate strategy to extend and settle one's network across the border in a more reliable, if not sustainable, way.

But pragmatic purposes associated with marriage are not necessarily considered as negative when they only concerns Chinese people. Another factor, directly linked with Vietnamese women, prevents this cross-border marriage phenomenon from becoming more popular and improving the efficiency of local border trade relationships despite their apparent convenience. As noticed earlier, all the different groups of Hekou's inhabitants have experienced a certain degree of interaction and engagement with their Vietnamese neighbours, which can be elaborated by a complex family background, a war experience, or business collaboration. Therefore, a whole range of contradictory feelings can emerge from these relationships: resentment, distrust, enthusiasm, patriotism, condescension, compassion, etc. So when others talk about a person, his/her family and community, that is to say when their privacy is concerned, the impact of these feelings becomes significant and disturbing.

Furthermore, since the opening of the frontier and the reactivation of trading activities at the border, human exchanges have allowed cross-border marriages to reappear in the social landscape. Although they have always existed at the local level, especially among ethnic groups who share common space despite the existence of a formal borderline that goes through their cultural territories, they are also practised among lowland populations, through village-based interaction, that is to say through historical family and social networks.

However, over these last two decades, other forms of marital exchanges have also attracted attention. They are usually referred to by the term "human trafficking" and imply the deception of Vietnamese women followed by a coercive marital arrangement with a Chinese man via a go-between. This phenomenon is actually an extension of the human trade and slavery tendency that has existed in this region for centuries up to the present time within China itself, among Chinese and Vietnamese communities along the border and further north in China. Thousands of cases have been locally and officially reported during the last two decades. Yet, accurate data are scarce and they probably underestimate a phenomenon that usually goes unnoticed, partly because of the various forms it can take at various stages of its process, and partly due to the agency of the individuals involved all along their migration and the transaction they were the subject of. Nevertheless, accounts of such marriages under coercion have spread in the local communities, and people have spent time discussing and analysing these facts. However, local opinion on these marital exchanges interestingly differs from the current victimization discourse often heard at the macro level. Indeed, for observers or those who are directly affected by these local "dramas", Vietnamese women may not necessarily be described as victims and Chinese men are not always perceived as perpetrators of unfair actions.

THE AMBIVALENCE EMBODIED BY VIETNAMESE WOMEN

The specific nature of exchange between Chinese and Vietnamese populations at the border has a strong impact on how they perceive each other. Both countries shared relatively similar economic conditions before their reform eras, especially in the border areas that have been long neglected. The official introduction of a market economy and free trade started in 1979 in China and 1986 in Vietnam, even though significant changes really appeared from the beginning of the 1990s on both sides. Over the years, differences in the development of the neighbours' economy has made China a much more successful and active place to deal with for trade and it has thus attracted migrant workers and traders in search of improvement of their life and business. Hekou's inhabitants are very much aware of their position and of the opportunity their active economic environment represents for the Vietnamese. Thus, the former tend to interpret each of the latter's moves towards economic improvement with suspicion.

In Hekou, the overwhelming number of Vietnamese prostitutes also leads to assumptions about their real prospects: according to locals, these young women are voluntarily there, work three to four years, and earn enough money to return to Vietnam, build a house and get married. China only represents a way to earn fast and easy money that means a lot back home. But among those young, beautiful and "heartless" women, some use other means to reach their goal: they seduce Chinese men. To substantiate such allegations, numerous stories of Vietnamese women who once pretended love and care in order to control and take away their Chinese partner's belongings are continuously retold to those who would give them the benefit of the doubt. The confusion comes from the fact that in discourse, the association is easily made between young prostitutes, ambitious educated women and those truly engaged in a cross-border romance. In the eyes of Chinese men, especially those who come from Inner China and who are not familiar with local realities and the complexity of social interactions in Hekou, Vietnamese women embody the ambiguity of the exotic woman and have become the subject of unlimited fantasies, as do women from ethnic minority groups from Southern China. But these fantasies are often contradictory and blur the picture: from the inconspicuous dutiful and devoted housewife to the affectionate and seductive prostitute, it seems that Hekou's people always credit Vietnamese women with a more ambiguous and dangerous second nature. Hence, when a Chinese man chooses a Vietnamese woman as a life partner, he would expect some questions, comments and scrutiny from his acquaintances. Everyone has in mind many stories with unhappy endings.

Besides, although we can roughly distinguish two sorts of marriages in and around Hekou (those based on coercion and those based on mutual agreement or romance), there exists quite a clear tendency in popular discourse to assimilate all of them into a blurred category of suspect people. The actual image and reputation of the people directly involved in these marriages are much more ambiguous than expected and it is often confusing to distinguish who did what, for which purpose and according to whom. It would be a mistake to underestimate these accounts of cross-border marriages on the basis that they may only be the results of gossip and misinformation. For instance, it is disturbing to find out that a large part of Hekou's population, especially those who have spent at least the last two decades in and around Hekou, do not have a compassionate attitude towards the Vietnamese women who have experienced or who still experience a forced marriage with Chinese men, usually in rural areas. On the contrary, those Chinese men who are living with their willing Vietnamese partners are often reminded of the former cases of their fellows who have been cheated by foreign women. So even though Vietnamese women who have been forced into marriage do not have much in common with those who have deliberately made the choice of a Chinese husband, they often fall into the same category of women endowed with a strong sense of tactics. One must keep in mind that Hekou represents a turntable at the provincial level for economic migration. Those Chinese and Vietnamese who pass by or settle for short or long periods in the city are usually motivated by pragmatic reasons, and they struggle to improve their economic situation.

An opinion widely shared and believed among Hekou communities states that when a Chinese man chooses a Vietnamese woman as a life partner, it is the sign of a failure and becomes a stigma. Whether the facts confirm the rumours or not, the man will be suspected of bearing all sorts of social stigmas or be assumed to have adopted non-normative social behaviours. These include being a divorcee, having a family back home and enjoying a local concubine, being unable to find a Chinese wife, being too poor, too lazy, lacking ambition, hiding his past, or being an outcast. Even though choosing a Vietnamese wife is often considered as a relevant option for business strategy, it is also paradoxically understood as a sign of personal weakness. Indeed, especially among businessmen and traders, those who get involved in a relationship with Vietnamese women, especially if they officially marry them, are marrying down. This means that they are not choosing the most desirable partner in life since Vietnamese women are generally regarded as poor, uneducated and unreliable in the long term. Besides, in the case of undocumented marriage, men as well as women are suspected of being irresponsible, uncommitted or having a hidden past or situation that prevents one or both of the partners from registering their relationship. In China, particularly for semi-urbanized society, living as spouses without registering the relationship as a proper marriage is not an appropriate behaviour. The birth of a child complicates the picture.

Also, the precariousness of such personal circumstances, from which these couples may suffer, often affects how colleagues and business partners perceive them. Even if they are accepted and integrated into local society, when their marriage and children do have an official status, they remain in an outcast position in regard to normative conjugality. But besides the perceived weakness of these married Chinese men, maybe in contradiction with their praised business strategy, the core of the issue lies in Vietnamese women's reputation. No matter how familiar they get with Hekou's surroundings, no matter how good their Chinese language level and their ability to adapt to Chinese culture and society, no matter what their background is, and how convincingly they perform their aptitude to become "Chinese", Vietnamese wives are deemed as ambiguous characters who have proved in numerous cases how far they can go to fulfil their secret purposes.

Hence, Sino-Vietnamese marriages may become the sites of contradictory feelings shared among the population of Hekou. On one hand, they represent a perfect combination for those who wish to extend their professional network to the Vietnamese market through personal acquaintances, and avoid the long and troublesome process of establishing trustworthy relationships between perfect strangers. On the other hand, these couples may be considered as people who act in their own interest: if they can utilize their personal life for pragmatic purposes, this may imply that they would not hesitate to employ such means in their business negotiations. These representations are compounded by the general feelings of Hekou's traders towards their Vietnamese business partners: "unreliable", "untrustworthy", "cheats", and "lazy" are among the terms often heard when discussing the issues of implementing fair and sustainable economic partnerships. In fact, the mere existence of Sino-Vietnamese couples questions these reactions: what are the real expectations behind a marital arrangement, what makes it different from a business partnership, and what would prevent both the former and the latter from being a disappointment?

MARRIAGE AS A PRAGMATIC MEANS?

Nowadays, many cross-border marriages in Hekou are the result of a migration of Vietnamese women searching for work opportunities and personal emancipation combined with the new settlements of businessmen from inner provinces of China. Although Vietnamese wives of Chinese entrepreneurs remain inconspicuous or unidentified/unrecognized actors in general statistics and overviews of Hekou's cross-border economic development, they do play an active role in the establishment of traders' long-term relationships by reactivating and creating cross-border kin networks. By doing so, they may improve their social position although they rarely get rid of the ambiguity and social stigma associated with their presence in Hekou, and which those men who get too personally involved with them by mixing business and personal life consequently also carry.

Crossing a border, engaging with a partner who brings along an uncertain reputation can sometimes weaken some social and economic relationships. It requires both spouses involved to remain extremely cautious in the way they expose their private life, their activity and their purpose in life and the way they may be used and abused by third parties. But even though they acknowledge a feeling of discomfort, some Chinese men married to Vietnamese wives need to maintain a circle of friends and business relationships in order to remain public figures whose lives should not be regarded as different from those of the majority. In the end, the way in which cross-border alliances might affect as much as they may improve the social position of the involved individuals is unpredictable and might vary case by case. But the constant interaction and interplay between Vietnamese women and their Chinese partners with social networks of various levels on both sides of the border definitely make them key figures for understanding the changes in Hekou's border society and economic activities.

The number of cross-border marriages in Hekou is increasing, is changing in nature, and is visible enough to deserve close attention. These mixed couples count as significant actors of cross-border trade, but they also crystallize the present atmosphere between Chinese and Vietnamese communities at the border and question the harmony and friendly relationship among economic partners claimed and praised in official discourse. Official encouragements to reinforce cooperation on the economic level often — if not always — forget about human and cultural compatibility that may represent a significant challenge for sustainable cross-border trade. Vietnamese resentment toward Chinese investors and the general lack of trust of their Vietnamese partners by these same Chinese is often stressed. Marital exchanges, among many other possible means, could be a way to overcome these obstacles by bringing intimacy into partnerships. But these practices still have ambivalent results and show the limits of personal commitment with economic partners since the ambiguity embodied by Vietnamese women is believed to reflect a general misunderstanding between two communities with possible divergent practices and values.

These issues may be minimized when Hekou traders deal with the Vietnamese Hoa network since they share specifically Chinese values and practices. But when rules and practices differ, when the frequency of more ambivalent attitudes and uncertainties characterizes Chinese and Vietnamese business relationships, policies and guidelines become more difficult to implement. Stereotypes and suspicions about cross-border marriages whose partners become active members of the trading communities can be read as a form of intimate resistance to collaborating with a former enemy and a still uncertain economic partner.

Listening to social discourse provides a more ambiguous picture of Sino-Vietnamese cross-border marriages which challenges usual general beliefs on economic and demographic determinism in the understanding of the phenomenon but also the expectations and perceptions of the actors themselves. The evolution of both the scale and nature of economic exchanges as well as the varying profiles of mixed couples in recent history all have a significant impact on the reputation of current couples. Therefore, they often use strategies to respond to a widespread local discourse that gives a derogatory picture of their marriages. This means distinguishing oneself from "similar others" (i.e. the other Sino-Vietnamese couples) whose reputation is regularly being altered or is seen as suspicious. Although the general image of mixed couples is slowly improving, they still engage in an implicit negotiation of their relationship with the rest of their community: their neighbours, friends, and business partners. Still these strategies can be interpreted as performances meant to convince and they may nourish and even reinforce the existing discourse. In the public sphere for instance, Vietnamese women who are involved with Chinese traders tend to adopt a very virtuous behaviour, to master Chinese language to their best ability, to remain discreet and efficient in their work, and to offer a positive and healthy image of their couple. Chinese men emphasize the qualities of their Vietnamese partners or they explain how they find few difficulties adjusting to each others' habits. But in certain circumstances, being aware of rumours, they may also avoid mentioning being married to a Vietnamese wife to prevent unwelcome comments or assumptions. The way an individual plays a "foreign partner card" depends on how strategic these elements are in regard to business interactions. It may help to mention a Vietnamese wife when dealing with Vietnamese traders to enhance trust, but it may also be a better idea to leave her to one side when negotiating with local economic and political agents, to avoid questioning morality. In any case, men and women constantly need to navigate between the suspicion and the exoticism they embody to serve their own interests and sustain their network in both China and Vietnam. However, like any other traders, the success of these mixed couples depends primarily on their business skills, which, in addition to language and cultural understanding, facilitate common business practice. Thus, "seduction" strategies are elaborated — often tacitly — on a case-by-case basis, according to the need to keep trading networks alive and trustworthy, although findings show quite a defensive and protective attitude among couples, especially when one's inappropriate curiosity on their mixed marriage sounds like scrutiny. This back-and-forth movement of words and behaviours reflects the difficult social position of some individuals who struggle to gain recognition and trust within their social environment. In sum, for a Vietnamese woman or a Chinese man, getting married signifies crossing different sorts of boundaries, sometimes back and forth. These boundaries are geographical, legal, social, economic and even moral. Even though the cross-border marriage phenomenon occurs in a borderland setting, such a conjugal choice is still considered as problematic. For the individuals involved, this implies coping with the need to establish their social position and reputation, precisely when they believe they can benefit from such an alliance within the trading community, which counts on face and the reliability of connections.

Notes

1 This contribution is based on fieldwork partly conducted in Hekou and Lào Cai between 2006 and 2010 for two parallel projects: the publishing of a book on cross-border marriages (cf. bibliography) and a Ph.D. doctorate on the same topic currently prepared at Macquarie University (Australia).

2 For border residents, obtaining a "resident entry & exit permit [*bianmin tongxingzheng*]" allows unlimited border crossing and it replaces the classic combination of a passport and a visa to enter Vietnam.

3 The human trafficking paradigm has been criticized a lot in the last decade. I adopt it in this chapter for the sake of convenience since I have already addressed this issue elsewhere (Grillot 2010). Further research needs to be done on the actual impact (nature and significance) of these marriages on Vietnamese families who may benefit from their daughters' remittances.

References

Andaya, Barbara Watson. "From Temporary Wife to Prostitute: Sexuality and Economic Change in Early Modern Southeast Asia Southeast". *Journal of Women's History* 9, no. 4 (1998): 11–34.

Chan, Yuk Wah. "Trade and Tourism in Lào Cai, Vietnam: a Study of Vietnamese-Chinese Interaction and Borderland Development". Ph.D. dissertation, Department of Anthropology, Chinese University of Hong Kong, 2005.

Chau Thi Hai. "Trade Activities of the Hoa along the Sino-Vietnamese Border". In *Where China Meets Southeast Asia. Social & Cultural Change in the Border Regions,* edited by Grant Evans, Christopher Hutton and Kuah Khun Eng. New York: St-Martin's Press, Singapore: Institute of Southeast Asian Studies, 2000.

Grillot, Caroline. *Volées, Envolées, Convolées... Vendues, en fuite ou re-socialisés: les "fiancées" vietnamiennes en Chine* [Stolen, Vanished, Wedded ... Sold, Runaway or Re-socialized: the Vietnamese "Brides" in China]. Paris: Connaissances et Savoirs, 2010.

He, Zhixiong. "Migration and the Sex Industry in the Hekou-Lào Cai Border Region between Yunnan and Vietnam". In *Living on the Edges: Cross-Border Mobility and Sexual Exploitation in the Greater Southeast Asia Sub-Region*, edited by Darwin Muhadjir, Anna Marie Wattie, and Susi Eja Yuarsi. Yogyakarta: Center for Population and Policy Studies, Gadjah Mada University, 2003.

Loos, Tamara. "A History of Sex and the State in Southeast Asia: Class, Intimacy and Invisibility". *Citizenship Studies* 12, no. 1 (February 2008): 27–43.

Wang, Xiaodan. "Zhong Yue bianjing kuaguo hunyin dui Yunnan bianjiang shehui fazhan de yingxiang yu yingdui" [The Influences and Coping Strategies of Sino-Vietnamese Mixed Marriage to Social Development of Yunnan Frontier Regions]. *Chuxiong Shifan Xueyuan Xuebao* 25, no. 2 (2010): 96–99.

Zhang, Juan. "Border Opened Up: Everyday Business in a China-Vietnam Frontier". Ph.D. dissertation, Department of Anthropology, Macquarie University, 2011.

PART IV

IMPACTS OF ECONOMIC CORRIDORS ON LAOTIAN BORDER SOCIETIES

16

THERE IS MORE TO ROAD
Modernity, Memory and Economic Corridors in Hướng Hoá-Sepon Lao-Vietnamese Border Area

Vatthana Pholsena

Lao PDR lies in the centre of the Greater Mekong Subregion (GMS), an area which has received substantial investment from the Asian Development Bank (ADB) and other institutions with the aim of further integrating the GMS member states' economies. Economic corridors (i.e. regional transport infrastructure) are a key element of the ADB's strategy and Lao PDR is traversed by three of these: the Northern Economic Corridor (NEC, linking Northern Thailand with Southwest China via Northwest Laos), the North-East Economic Corridor (linking Northern Thailand with Northeast Vietnam via Vientiane and Northeast Laos), and the East-West Economic Corridor (EWEC, linking Eastern Burma with the port of Danang in Central Vietnam via Northeast Thailand and Savannakhet Province, Southern-Central Laos). In the eyes of the ADB and the Lao government, the planned road grid in the GMS will place Lao PDR at the centre of the Subregion's transport network and thus transport the country out of her geographic and economic isolation, one of the chief causes (in the view of political and economic decision-makers) of its structural vulnerability.

This chapter[1] focuses on a border area straddling the districts of Sepon and Hướng Hoá between southern Laos (Savannakhet Province) and central Vietnam (Quảng Trị Province) (see Map 16.1). Hướng Hoá-Sepon area

Map 16.1
East-West Economic Corridor (EWEC) under the GMS Framework

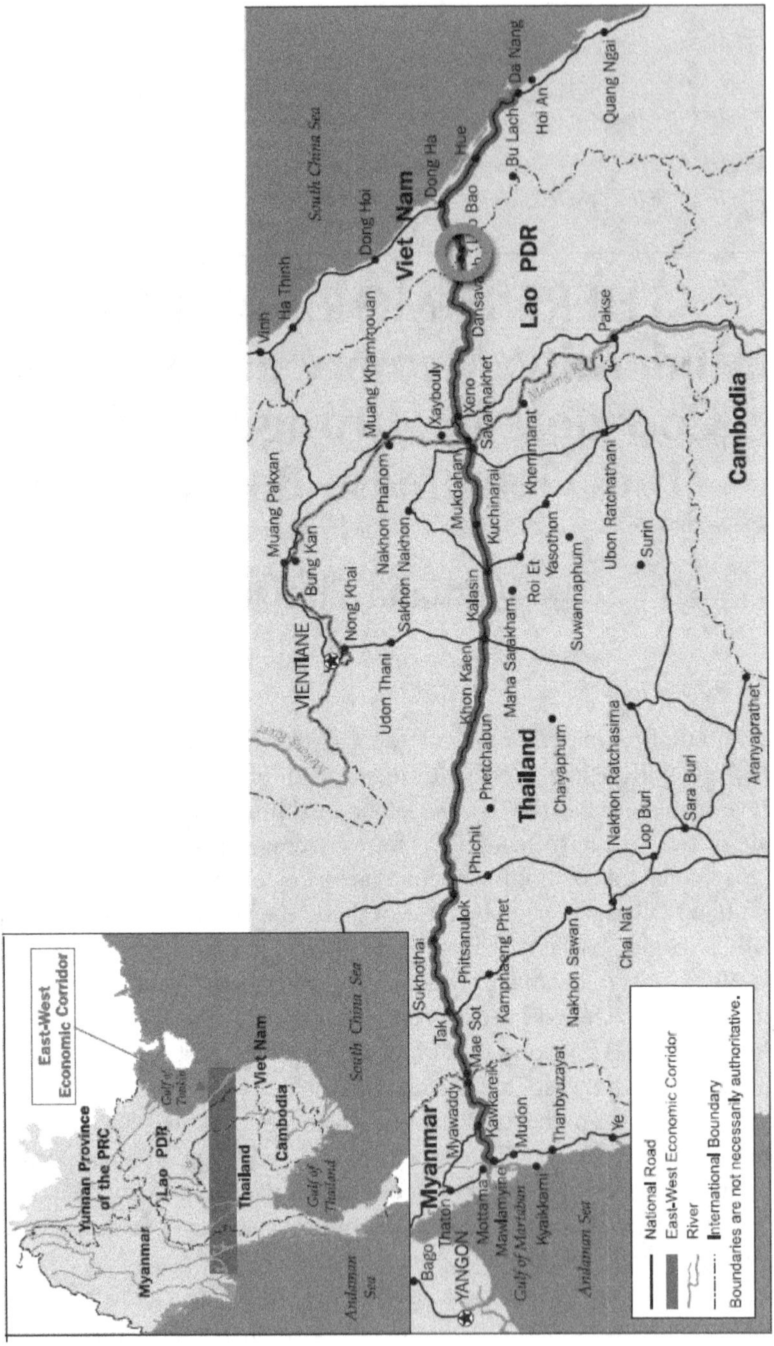

Source: Ruben Marmita.

is traversed by Road No. 9, which forms the backbone of the East-West Economic Corridor.[2] This border area includes two border towns: Lao Bảo on the Vietnamese side and Densavanh, its twin town on the Lao side. The populations living in this area comprise the Bru, an upland population, who belong to the Austro-Asiatic ethno-linguistic family;[3] the Phutai, a Tai-speaking population, who mainly live in the lowlands of Sepon district; and the Kinh, the dominant ethnic group in Vietnam, who reside in both the uplands and the lowlands of Hướng Hoá district.

Roads in developing countries have generated contrasting comments. Governments and international aid agencies generally stress the beneficial impacts of new or upgraded roads on local economies and communities. On the other hand, these positive effects are being frequently downplayed, if not downright refuted, by NGOs and academics (Colombijn 2002, pp. 595–98). As Nishizaki sums up the latter's views nicely: "At worst, roads are part of pretentious and dehumanizing social engineering projects undertaken by arrogant "high-modernist" elites bent on making society eligible (…). At best, roads are the tools that enable the state to promote economic growth or to enforce administrative and security control (…)." (Nishizaki 2008, p. 433). I wish to investigate the underside of grand schemes such as the Economic Corridors and explore — on a modest scale — the rural population's views of these roads that run nearby or through their homes. In other words, I inquire how roads are perceived from below. In the first part, I look specifically at two functions of Road 9: one that is economic — I examine the border towns that are located on this road and (briefly) assess the efficacy of this newly improved road in developing these areas' local economy and improving their residents' livelihoods — and another that is symbolic — I introduce postwar memories of the road being recalled by elderly members of Sepon's local communities. In the second part, I aim to show that roads constitute a strong symbol of 'modernity' to which some communities aspire.

1. FUNCTIONS OF A ROAD

Government Policies and Individual Agency

So far, the most visible changes that appear to be related to the improvement of the East-West Economic Corridor between Southern Laos and Central Vietnam have occurred on the Vietnamese side of the border. Indeed, the Vietnamese government has set up a commercial and industrial zone, the "Lao

Bảo Special Economic — Commercial Area" (*Khu Kinh tế Thương mại Đặc biệt Lao Bảo*), in Lao Bảo in 1998.[4] The state measures have aimed to attract domestic and foreign investments in order to develop industrial production,[5] as well as to encourage tourism, both domestic (mainly from Hanoi and Saigon) and foreign (mostly from Thailand, but also Laos) with the construction of the Lao Bảo market in the shape of a pseudo-Katu "traditional house".[6] The policies in Lao Bảo have been driven by two overall objectives: to boost the local economy, and at the same time, to further integrate these remote areas into the national economy, by attracting domestic and foreign companies and tourists into the border area. But the market appeared intermittently busy when we visited the area in 2007 and 2008. It was supposed to be a night market as well, but never actually opened (in evenings) due to the insufficient number of visitors, as the manager of the hotel where we stayed told us. Also, most of the companies that had invested in the Lao Bảo SECA were Vietnamese; only 8 out of 58 companies were foreign (Thai and Chinese) (2006–07 figures), which seemed to indicate that the Lao Bảo SECA might still be characterized by a weak regional integration into international capital flows. Nonetheless, new economic zones, independent businesses, emerging cash crops, such as coffee, cassava, and rubber, and a budding service sector appeared to have transformed this formerly war-ravaged landscape into a promising zone of industrial production, commerce and trade. In the last ten years, for example, the Lao Bảo special economic-commercial zone has attracted 41 investment projects with a total capital of more than 3.1 trillion Vietnamese đồng (US$187 million), contributing approximately 30 per cent of the provincial budget (Dương Vương Lợi 2010).

On the ground, the more favourable economic conditions in recent years seem to have benefitted some villagers who reside in the border area as well. In 2007, for example, approximately forty households from the village of Katang, which is located next to the border gate, were involved in various forms of trading between Lao PDR and Vietnam.[7] During our visit, we met a forty-four-year old female trader and an inhabitant from this village. Thương Thư moved to Lao Bảo in 1998 (she was born in Quảng Tri Province). She started cultivating upland rice and cassava, but life was then "difficult" ("we had nothing to eat but cassava and maize"). She decided to move to her present location in 2001 (she and her husband built their present house with the support of government funding). Thương Thư first started working as a porter at the border gate in that year. She got together with other women from Katang village and applied to the People's Committee and to the border station for the authorization to constitute a group of porters. In the beginning they had four carts and formed a group of forty porters. Traders from Lao Bảo would buy their goods in Lao PDR and hire them to carry their purchase back when they reached the immigration checkpoint. The market in Lao Bảo was not as large as it was now, and from 2001 to 2003, they could "earn good money" (as much as VND 150,000 a day). They eventually expanded from four to twenty-two carts and from forty to fifty porters.

But after the new Lao Bảo market was built and trading between Lao PDR and Vietnam increased, some traders from Lao Bảo shifted from carts to trucks to transport their goods. In consequence, their workload declined. Thương Thư eventually stopped in 2006 and decided to borrow some money to buy herself a truck and started her own trading business. She was now conducting a long-distance trading. It usually took her two days to buy goods in Lao PDR and another six days to sell them in Vietnam to traders in Lao Bảo. The goods she bought were diverse; they included coal, rice, fruit, as well as vegetables. She went around villages in various places in Savannakhet Province, such as Densavanh, Ban Dong, Phalane, and Seno (she could get by in Lao and "Lao Theung"[8] languages (her husband was Bru-Vân Kiều)). Roads were "very dangerous" a few years ago — she even sometimes feared for her life — but "now it's fine". Thương Thư had since bought a second truck.

Unlike Lao Bảo, Densavanh on the Lao side appeared to be a border town in relative decline from the point of view of some of its residents.[9] In 2006, we interviewed Nang Thon, a Lao national of Vietnamese origin, who owns a small shop. She was born in 1963 in the district of Nongbok, Khammouane province. Her parents came and settled in Laos in 1959 from Quảng Bình Province. The Lao state granted them Lao citizenship. She and her husband settled in Densavanh in 1988 (they had previously lived in Savannakhet town). The "border market" (*talat daen*) of Densavanh had been created the previous year (i.e. the year when the border demarcation was officially agreed and adopted between the Lao and Vietnamese authorities). The chief of their former village had encouraged them to leave. Profits were then higher at the border than in the western areas of Savannakhet. They were selling daily-use items and did not have to compete with companies. Nang Thon explained to us:

> We were buying in Savannakhet and selling to the Vietnamese. Business was good before because the road wasn't good, and cars used to be stuck for a while. People had to stop and there was no competition from

> Vietnam. But today the road is good and not as many people stop. Now the Vietnamese people use the phone and can order directly from Vietnam and the goods are transported on the river, whereas before the Vietnamese used to come and buy the goods themselves. They make their order [of Thai products] which they then sell in Vietnam. Shops and restaurants in Densavanh are owned by the Vietnamese, and Vietnamese people go and buy and eat there.

But business was not as flourishing as Ms Nang Thong thought for, at least, some Vietnamese traders in Densavanh. We met during the same visit Lê Van Phục, a thirty-five-year-old Vietnamese citizen from Hue, and the owner of a restaurant. He had been living in Densavanh for six years. "In Vietnam, there's no work and there are a lot of people", he explained. He lived in Densavanh with his twenty-nine-year-old wife and their two-year-old son. He sold bread and ice-cream for the first two years, and then opened the restaurant the following year. He bought his goods at the Lao Bảo market where they were cheaper. "But business has declined since two, three years, because the road is better and people don't stop as often as they used to. Yet, there's always at least one bus that stops by", he noted. Lê Van Phục had a brother who had been running a business in Vietnam for six years. He had already bought a house in Lao Bảo and did not plan to come and live in Laos. "In Densavanh, business is dominated by [Lao] companies. In Lao Bảo, there's more activity, it's more dynamic, so there are more business opportunities", Lê Van Phục summed up.

According to these residents, economic activities in Densavanh appeared to be declining as migrants and tourists did not stop to rest as often as they used to now that the road was in good condition. However, there existed an unofficial, underground, and dynamic commercial activity in the vicinity of Densavanh, i.e. smuggling. In other words, there existed a relatively thriving business on a parallel route on river — a water road — that was used to transport merchandise to Vietnam in order to avoid paying taxes on imported goods. The marketing director of a company based in Densavanh gave us some interesting information on these underground activities. The company opened its local office in the border town in 1988. The premises contained fridges, television sets, fans, and auto parts. Television sets, fridges and fans were assembled in Laos; all the parts came from Thailand and the finished products were exported to Vietnam. The company depended on orders from Vietnam: they could sell up to 10 fridges a day and 60 fridges a month, as well as approximately 100 TV sets and 100 fan units. After making their orders, Vietnamese labourers — in groups varying between 20–50 individuals — would come and collect the goods themselves at the shop. They did not come through the official border, however, because heavy importation taxes were imposed on some of these products. "Whereas the Vietnamese can buy the goods in Laos, their problem is to bring them over to Vietnam by their own means", explained the company's director. In consequence, Vietnamese porters would transport them by boat in order to avoid crossing the official land border and paying tax.[10] In this case, at least, the renovation of Road 9 has produced little impact on the trade relations as informal economic actors (i.e. Vietnamese traders) prefer using a different means of transportation (i.e. waterways) that is economically more advantageous.

In the next section, I would like to show another aspect of Road 9, unrelated to its (potential) economic impact, but which is no less significant in studying the social impacts of roads among local communities.[11]

Historical Memory of a Road

In Sepon District, Route 9 projects multiple images. Today, the road runs through open, flat lands filled with rice-fields and vegetation, encircled by mountain chains as one gets nearer the Vietnamese border, where physical remnants of war have all but disappeared (except for war commemorative sites in Phine and Sepon Districts on Route 9). Sepon villagers are no longer confronted with a landscape in ruins on Route 9; the road is no longer scattered with wartime debris — unexploded ordnance (UXO) and bomb casings, cracks and craters. But the absence of ruins does not equate to an absence of memory.

Because of its strategic alignment and economic potential, the completion of the upgrading of Route 9 in the first half of the 2000s was duly fêted in Lao national newspapers. In fact, the restoration was closer to a revival: at the end of the Vietnam War (also called the "American War") the road was in ruins, a war-ravaged landscape. Just a few years before turning into a conduit to "modernity" at the dawn of the twenty-first century, Route 9 was known among military planners and local inhabitants as the "death road". When one travels through Savannakhet Province's former conflict zones, it is hard to match the present-day scenery with this graphic description of the road in the aftermath of the fighting: "Route 9 was like a cemetery. The bodies of dead enemy soldiers, damaged tanks, cannons and military vehicles were scattered along the road. Unexploded land mines and bombs continued to create carnage long after the battles had ended" (Vietnam Museum of Ethnology 2009, p. 41).

As initially laid out, the covert route between North and South Vietnam that came to be called the Ho Chi Minh Trail (in reality, a maze of interlocking dirt roads that were gradually and partially upgraded to cobbled roads so that North Vietnamese trucks could move in an almost uninterrupted flow all year round), followed the eastern slope of the Trường Sơn mountain range, crossing Route 9 just south of the demilitarized zone in South Vietnam. But the route had to be changed early in 1960 after a stack of weapons left behind by North Vietnamese infiltrators on the Trail was discovered and reported to the ARVN (Army of the Republic of (South) Vietnam) (Prados 1998, p. 15). The 559th Transportation Group (the North Vietnamese logistics unit in charge of building the Trail and carrying supplies across the DMZ demarcation line) reacted to the consequent intensification of patrolling by shifting the main infiltration route in early 1961 to the western side of the Trường Sơn mountain range. This new route ran through the panhandle of southern Laos to the vicinity of Sepon on Route 9, a village that soon became a strategic centre and logistics base area for the North Vietnamese Army and one of the most important nodes on the transportation network of the relocated Trail. Sepon, one of the most important centres for the North Vietnamese's transportation network, was captured by the communists as early as 1961.

From 1964 onwards, Washington's first priority was to assist the military effort in the southern Republic of Vietnam. The U.S Air Force subsequently launched in December 1964 its first airstrikes against fixed targets and infiltration routes (particularly the Ho Chi Minh Trail) throughout Laos, which soon expanded in April 1965 to a day-and-night offensive air campaign in south-eastern Laos in an attempt to stop supplies flowing into South Vietnam from the North. As the U.S. bombing of the Trail grew in intensity, it led inevitably to a widening of the war into central and southern Laos.[12] Fighting on and around Route 9 reached a climax during the Operation Lam Son 719 launched by South Vietnamese armed forces, with logistical, aerial and artillery support from the United States, in early 1971. The limited offensive campaign aimed to attack westward from Khe Sanh (Quảng Trị Province) along Route 9, seize the storage depots and destroy the supplies stockpiled around Sepon, and withdraw back to South Vietnam. The campaign was a disaster for the Army of the Republic of Vietnam and led to much bloodshed on both sides (Nalty 2005, p. 151).

In mid-1961, the North Vietnamese forces captured the districts of Sepon, Phine, and Phalane in the east of Savannakhet Province and gained control of Route 9 in these areas. The road embodied on the ground the Cold War-induced Lao civil war: to the west of Phalane District on Route 9 lay the "U.S-controlled puppet/enemy zone" (*khet sattu*), while the eastern region formed the "liberated zone" (*khet potpoı/khet issala*), as villagers in Sepon recalled the divided territories. Key locations on Route 9 constituted the shifting front line. Members of families remained separated for the duration of the war, unable to travel either eastward or westward, living on either side of what may be called an 'internal border'. Some — such as soldiers and personnel of the Royal Lao Army and Government, as well as petty traders (most of whom were Vietnamese and Chinese migrants) — never returned, having fled upon the advance of the communist troops. To combatants and civilians alike, the space enveloping Route 9 figured as a fractured landscape, the road having lost its *raison d'être* — as a technology of rule and communication — and turned instead into something approaching the metaphor of an "iron curtain" (Pholsena 2010).

It took exactly thirty years for the road to revert to its status of "motor of modernity". Those inhabitants in Sepon old enough to remember had been keeping close track of the road's rehabilitation process.

> In 1983, Route 9 was a dirt road, in terrible condition, with holes and cracks. In some places, you even had to bend to go through; otherwise your face would get scratched by brushwood", recalled a villager who was in his late fifties. "It'd take 3 days and 3 nights to reach Savannakhet town from the [Lao-Vietnamese] border: one night in Sepon village, a second night in Setthamouak, and a third night in Seno or Donghen. In 1985, the Vietnamese [soldiers] came to help repairing the road, after that the travelling time came down from three days and nights to one day and night only. In 1994, it was possible to cover the distance in one day, from 6 am to 6 pm. In 2005, we had the road as it is today: 4 hours by bus or less than 2 hours by car."

The man's precise account resembles the discourse of government plans for economic development in that it portrays the renovation of the road as a crucial factor in the development of the district. In post-socialist Laos, one of the resonant symbols of "modernity" as opposed to "backwardness" is the existence of paved roads linking villagers to markets, and thus to economic "progress". Road construction is a top priority for state economic planners in developing countries as part of the effort to lift the rural population out of "backwardness" (both material and cultural) by integrating them with the modern, market-based economy (Flower 2004, p. 649). But unlike plans and reports authored by state agents and international aid agency experts insisting on the road as an example of modern advancement, villagers made clear that they also situated their narration of the restoration process in a historical timetable that followed their own recovery from the war.

> "After the war, the road was nothing like it is today. It was an area covered by forest and vegetation, it was a dangerous road. To those who weren't living here it is very hard to imagine that", a woman told me. "We never thought that we'd survive and live long enough to see Route 9 becoming a beautiful road", elderly villagers would often say.

In post-war Laos, roads embody state development projects that "deliver" modernization and civilization to remote rural communities; however, to villagers of Sepon, those ideologically loaded terms such as "progress" and "civilization" have not been simply passed down by outside authorities, but to a large extent have been shaped by villagers' very own wartime experiences. I argue that rural survivors saw the revival of the road as a metaphorical route that lifted them out of war-born rift and violence and back to a civilized state. To villagers in this impoverished rural area, the two-lane paved road both constitutes and represents economic development, but is also a place of darker times, of memories of violence and rift. Thus, there exists a salient contrast between today's physical landscape around Route 9 and the villagers' historical memory of the road.

2. ROAD TO "MODERNITY"

In the second part, through a case study in a remote sub-district in Sepon, I wish to show that the arrival of a road, along with other public development provisions (such as schools and electricity), should not be systematically analysed as a form of domination by the state of local populations, but should be considered, in some cases, as a result of a process of negotiation between the former and the latter.

Khet Lakô: A Government Policy

Lakô sub-district (an administrative unit now known as *khet* in Lao) situated in Sepon district a few kilometres from the border with Vietnam was a "focal site" (*khet jut sum*).[13] With the stated aim of improving the efficacy of rural development, the Lao government adopted a policy in 1994 that had as its central thrust the strategy of establishing "focal sites" (FS), i.e. clearly defined geographical areas on which development efforts were to be focused.[14] The slogan for the FS policy was *chatsan asib khongti*: *chatsan* means "to organize", *asib* "profession" and *khongti* "in one place". In other words, the objective was to promote sedentary farming settlements.

Khet Lakô was located 42 km to the north of Road No. 9. It was an uphill journey: the road was built during the Vietnam War as part of the Ho Chi Minh Trail and it took over four hours to cover the distance. At about half way, the road got narrower and started to climb steeply through the forests and reached Ban ("village") Laviền, the administrative centre of the *khet*. The low buildings that surrounded the large and empty field to the north of the village that would in the evening be filled with cows and buffaloes were the only concrete signs of the state's efforts to create infrastructure in that remote place. Next to the meeting hall stood the primary school, and facing them about 100 meters away was the visitors' lodge, with individual rooms and common toilets and showers. We learnt later that it used to be the health centre. One of the rooms was occupied by the border guard whose task was to register people on their way to the border with Vietnam, barely two kilometres away.

The local authorities began to invest in the focal site of Lakô in 1996. Aside from infrastructure, funding was directed towards the development of rice fields, animal husbandry, small coffee plantations and irrigation schemes, as well as agricultural training. Some villages were also resettled (*nyok nyai*) to the area in the years following the establishment of the *jut sum*, in order to bring them closer to this zone of so-called integrated development and as part of the Village Consolidation policy.[15] The focal site of Lakô was not typical, however. The government, through the Focal Site policy, aimed to encourage resettlement in lowland areas so as to move highlanders (most of whom are of ethnic minority origins) closer to markets, roads, facilities and services (e.g. hospitals and schools), rather than bringing these basic services to the upland areas.[16]

A different logic guided the establishment of *khet* Lakô; as late as 2006, officials had to journey on a poor road through forests and mountains for several hours before reaching the site, or more often, travelled through the district of Hướng Hoá to reach the area opposite the *khet* Lakô and then cross back, as road access to rural areas on the other side of the border was more comprehensive.[17] That hardly fitted the government's long-term plan to have most Lao citizens easily within their reach. And yet, the opening of the meeting hall, the school and the health centre was duly celebrated in the late 1990s in the presence of provincial and district authorities, as well as Vientiane representatives of the Lao Front for National Construction (LFNC), the main government mass organization.

The head of the *jut sum*, Mr Vang,[18] a Bru Tri, whom we had met in Sepon town a few days earlier, was there to greet us when we reached the focal zone of Lakô. He used to live in Vietnam and actually served in the People's Liberation Army during the Vietnam War. In fact, we quickly learnt that the territory where *khet* Lakô is situated used to belong to Vietnam until the early 1980s. The international border was officially drawn after strenuous and lengthy discussions between Lao and Vietnamese officials, state-building projects and security imperatives driving those tough negotiations (sentiments of socialist brotherhood being left aside on that occasion).[19] The latter were certainly complicated by the fact that as far as the border regions of Savannakhet Province[20] were concerned nation-states' cardinal principle of political sovereignty lost its significance during the war.

As a Lao district official euphemistically told us, the frontier area where *khet* Lakô is presently located was "temporarily left in the care (*fak*) of Vietnam [i.e. the Democratic Republic of Vietnam]" by the Lao communist leadership, for "it was impossible to reach those areas from the Lao side at that time". In reality, this "provisional" transfer of territorial "guardianship" resulted from the changing North Vietnamese military strategy in the late 1950s. In January 1959 during the Fifteenth Plenum of the Vietnamese Workers Party's Central Committee, the DRV leadership authorized the use of armed struggle in the south, to be headed by the National Liberation Front for South Vietnam (NLFSVN) (Qiang 2000, p. 81). In early 1961, after negotiations with the Pathet Lao, Hanoi decided to extend the Ho Chi Minh Trail into Laos (Prados 1998, p. 15).

Following the end of the war, Mr Vang decided to leave for Laos in 1979 (he even gave us the precise date of his departure, "on July 25th, 1979"). He regretted his decision these days, though: "I feel like an idiot now. My comrades who stayed [in Vietnam] enjoy their war pension, have a house, while I got nothing but my eyes to cry!" Mr. Vang applied three times (in 1993, 1994 and 2003) for a pension as a war veteran in Laos. He got so angry and bitter that he openly threatened the Lao authorities to return to Vietnam. "I forbid you to go back! We're going to make plans!', was the vigorous response, as he told us, thereby elicited from Bounyang Vorachit, then Governor of Savannakhet Province (and later the country's Prime Minister).

In truth, Mr Vang was not the only frustrated war veteran in the *khet* (during the village meeting in the evening, we realized that half, if not more, of the male farmers aged above fifty had been PLA soldiers). He was, on the other hand, a well-known figure at the district and provincial levels, as well as in Vientiane where he attended Party meetings as a member of the LFNC at the village level. Did Mr Vang's complaints persuade the authorities to choose Lakô as a focal site and to invest several thousand dollars there? His outburst was certainly persuasive enough to get a strong reaction from Bounyang Vorachit himself. In any case, Mr Vang was appointed as the deputy-head of the focal site from its inception.

State Territoriality Negotiated

More significantly, Mr Vang's grievances reflected a larger problem, i.e. the unequal pace of development between the Lao and Vietnamese border regions. In one of our conversations, the former head of the LFNC's Research Department on Ethnic Groups (who had visited *khet* Lakô several times) acknowledged that in the late 1980s and the early 1990s the upland population residing in these areas "threatened to cross over to Vietnam unless the [Lao] state offered them the facilities that they believed existed in Vietnam"; hence, the imperative of "developing" these border regions. In 2004 and 2005, the strategy failed to stop several households from villages in *khet* Lakô from leaving the country to settle in Vietnam.[21] The trend was alarming enough for district and provincial officials to travel to the area and to meet the population.

Khet Lakô was not merely another government-sponsored development site: it was a political issue that challenged the Lao state's capacity to retain those of its citizens who could choose the 'exit' option, or to use James Scott's terminology, exercise their 'capacity to flee' in search of a better livelihood (Scott 1998, pp. 52–53). Not unlike the Lauje hill people in the Tinombo-Tomini district of Central Sulawesi, who, as Tania Murray Li showed, aspire to a lifestyle endowed with modern infrastructure (Li 2001, p. 58),[22] villagers in *Khet* Lakô crossed the border in search of a better livelihood and, in doing so, they defied state territoriality in their own small way.

State territoriality was in effect negotiated in the focal zone of Lakô. The villagers maneuver in a context that Janet Sturgeon has analysed as the "always territorializing but never completely territorialized state" (Sturgeon 2004, p. 481). In other words, these villagers in the Khê Sanh-Sepon border zone are neither completely "free" (i.e. outside the reach of the state) nor "isolated" (i.e. excluded from decision-making processes). The Poverty Reduction Fund, financed by the World Bank and the Lao state (to a lesser extent), has supported several projects in the villages that composed *khet* Lakô (e.g. building of schools and health centres [*suksala*], installation of gravity-fed water systems [*nam lin*]) since the Fund's inception in Sepon district in 2003. In addition to his responsibilities as the focal site's head, Mr Vang was also the appointed liaison person (*phu phasaan ngaan*) between the Fund's Sepon office and the villagers in the *khet*. I would argue that the Vietnamese-born Lao war veteran of Bru Tri ethnic origins played an intermediary role between the upland villagers and the state, conveying to the latter messages underscoring the danger that a lack of state support to the local population could bring about "spontaneous" migration to Vietnam.

The Focal Site strategy was a national policy, and Mr Vang worked hard so that it would benefit the population of the *khet*. In his view, the solutions that would enable the state to "hold onto the people" were simple and could be summarized in three words: trade (Mr Vang mentioned a possible trade agreement with a Vietnamese company in Quảng Trị that was interested in investing in the plantation of Jatropha (*mak diaew*) in *khet* Lakô; he had applied, and was still waiting, for the province's authorization), road and electricity (*sengsavang* ["brightness"], as he explained in a distinctively exasperated voice; "*it's not so complicated, we just need to be connected to the grid in Vietnam two kilometers from here!*"). These are hill people who aspire to secure themselves a place within the state space as part of their survival strategy as mentioned in another context by Li.[23] Certainly, what the villagers were telling us about the sight of those lights that brightened up the mountain slopes on the other side of the border every night resembled the vision of a promised land where life appeared easier, brighter, and modern.

The prospects of having asphalt reaching the *khet* looked more promising than connecting Lakô with wire and pylon for the time being.[24] According to the Head of the Construction, Transport, Post and Communication (CTPC; since renamed Public Works and Transport) district office, works in order to improve the road to *jut sum* Lakô were under way, jointly funded by the World Bank's Poverty Reduction Fund, the state and the Asian Development Bank. The state strategy to control people and resources had in this case been inverted, from a policy of bringing the people down to a logic of reaching out (and "up") to them. The official explained to us: "*If we build roads that reach their villages, they will survive. But if we relocate them, they wouldn't know how to make ends meet. We must reach them, instead of relocating them to the lowlands.*" What is more, the CPTC office head added, "*the villagers are telling us: 'If we don't have roads, we are afraid that we will be displaced'.*" These are words indirectly reported that can be variously interpreted, but certainly, in the official's view, the provision of road access could keep people from fleeing in order to avoid resettlement in the lowlands. It is difficult to predict whether these efforts would prove sufficient to prevent farmers from migrating to Vietnam, despite development projects and Mr Vang's hard work.

CONCLUSION

This chapter examines different functions and representations of Road 9, a key component of the EWEC, among the local communities of the border area straddling the districts of Sepon and Hướng Hoá between southern Laos (Savannakhet Province) and central Vietnam (Quảng Trị Province). While for some residents the higher mobility the road offers has brought about some new economic opportunities (such as the expansion of cross-border trading in Katang village for example), it has, on the contrary, negatively impacted on the livelihood of some inhabitants in Densavanh. Paradoxically, the border town on the Lao side has suffered to some extent from the improvement of the road; and instead, one of the most dynamic cross-border movements across the Lao Bảo-Densavanh area appears to be the smuggling of imported goods on the waterways.

Residents of Sepon, one of the district towns on Route 9, welcome the potential for economic advancement the road represents. But, to older residents, the cause for celebrating the restoration of the road in the early 2000s went beyond the renovation of an important piece of national infrastructure: they metaphorically see the revival of the road as a route back to civilization. Civilization here is defined in fundamental terms, i.e. as a state of affairs counterpoised to the violence and savagery that residents experienced during the war (from bombing and land battles) as well as in its aftermath (from unexploded ordnance and "counter-revolutionary" attacks). Therefore, for local residents, the road's significance goes beyond the tangible benefits it brings in the form of economic growth and access to services.

Contrary to the commonly held view on state policies and ideology in Mainland Southeast Asia,[25] my analysis, through the case study in the second half of the chapter, has tried to show that an encounter between "the state" and "the people" does not necessarily always entail the asymmetrical imposition of the former's rules and apparatus on the latter's lives. To be sure, there are a number of related issues on the Lao-Vietnamese border on which insufficient information is currently available; for example, how many villages or households in the border peripheries have gone through similar experiences to those in the case studies (crossing over to another country, or negotiating with local authorities in order to improve their livelihoods); or how common is it to come across political "entrepreneurs" or "brokers" of ethnic minority origins, such as Mr Vang. Nevertheless, this case study compels us to critically revisit the conventional views on roads or any public development goods in developing countries. Some villagers, in some instances, even ask for stronger, not weaker, state intervention through the provision of public development services (such as roads, schools, electricity) so as to be better equipped to survive in an environment in which subsistence economy is being relentlessly subsumed by market-based livelihoods.

In other words, there is more to roads than social engineering projects, economic growth or security control. Impacts of economic corridors in the region in Hướng Hoá -Sepon area therefore must also be considered from a historical and symbolic perspective, which takes into account people's enduring memories and perceptions of their local environment. Thus, for some inhabitants in Sepon the renovation of Road 9 represents a victory over a painful past, while for other residents access to paved roads brings them closer to their particular image of modernity.

Notes

1 The fieldwork upon which this chapter is based was funded by a Regional Collaboration Grant from the SEASREP Foundation and by the ANR Transiter Program. The materials used in this chapter are drawn from research trips carried out between March 2006 and January 2008, amounting to a total of four months.

2 Route 9 is also one of Laos' most important national highways. The road runs for 324 km across Quang Tri (Central Vietnam) and Savannakhet Provinces to the Thai border. Route 9 is also commonly called Road No. 9 or Highway No. 9.

3 In Vietnam they are called Vân Kiều (or Bru-Vân Kiều), while in Laos, they are classified into two groups, the Makong (or Mangkong) and the Tri.

4 Lao Bao SECA encompasses Lao Bao and Khe Sanh towns, as well as the communes of Tan Thanh, Tan Long, Tan Lien, Tan Lap, and Tan Hop of Hướng Hoá district, covering an area of 15,804 ha and comprising a population of around 45,000 people.

5 These measures include: (1) Projects investing in Lao Bao SECA are exempted from corporate income tax for four years from their taxable incomes; then benefit of a 50 per cent reduction on their income tax the subsequent nine years; and thereafter, a tax rate of 10 per cent will apply. (2) Investment projects in Lao Bao SECA are exempted from land rent for the first eleven years (land where the infrastructure has been built) and benefit a rate of 30 per cent of land rent (applicable in Quang Tri's highland regions) from the twelfth year onwards.

6 For example, domestic and foreign tourists on entry into Lao Bao SECA are allowed to purchase duty-free imported goods with value of not exceeding VND 500,000 per person and per day to bring into domestic markets.

7 The village of Katang comprised 147 households (17 of which were Kinh, 127 Van Kieu, and 3 Paco) in 2007. Their livelihood activities mainly consist of agricultural works: upland rice (400 ha), bananas and cassava cultivation. Most households own fish ponds and cows.

8 'Lao Theung' in Lao popular parlance roughly refers to upland populations, generally of Austro-Asiatic ethno-linguistic origins.

9 The population of Densavanh amounted to 1,761 individuals, including 602 "*Lao Loum*" and 1159 "*Lao Theung*" (source: Government Office of Sepon District 2008).

10 Thanks to the cooperation of a local official, I was able to directly witness these smuggling activities that were taking place a few kilometres from Densavanh town.

11 Several anthropologists and historians have stressed the importance of symbolic meanings and socio-psychological dimensions of roads in studying the social impacts of roads among local communities (see Colombijn 2002; Flower 2004; Rosman 1996; Thomas 2002; Nishizaki 2008; Weber 1976).

12 From 1964 to 1973 American planes dropped 2,093,100 tons of ordnance in 580,344 bombing missions, which approximately equated to a bombing every eight minutes, twenty-four hours a day, for nine years. According to these statistics, Laos is the most heavily bombed country per capita in the world. Attacks on the Trail intensified from 1968 onwards; whereas 52,120 missions were carried out in 1967, this figure almost trebled in 1969, amounting to a total of 148,069 sorties flown in that year. The bombing finally halted in 1973. Bombs

still continue maiming and killing scores of people in today's Laos, however: at the end of the war, it was estimated that 78 million unexploded cluster bomblets had been left in the country's rural areas. All figures are quoted from the very informative article by Channapha Khamvongsa and Elaine Russell (2009). A recent survey by the National Regulatory Authority (NRA) for the UXO/Mine Action sector in Laos reveals that more than 50,000 people were hurt or killed by UXO (unexploded ordnances) between 1964 and 2008, *Vientiane Times*, 6 February 2010.

13 I first heard of Lakô subdistrict (*khet* in Lao) during a field research trip in 2005.

14 Though Baird and Shoemaker mention in their article that international aid donors also played a critical role (by funding most of the cost) in the implementation of the Focal Site Strategy (Baird and Shoemaker 2007, p. 875).

15 The Village Consolidation Policy combines smaller settlements by relocating people into larger permanent villages.

16 In many hilly areas, the *khet jut sum* does not necessarily have to be in a lowland area; indeed, in some areas, this is impossible because of the topography. In consequence, the *khet jut sum* would be the nearest road accessible point to many more remote villages in the area. However, *khet* Lakô was not such accessible point; it was the nearest location to the Lao-Vietnamese border.

17 In 2007, it took us one hour driving at a moderate pace on a motorbike from Khê Sanh town in Quảng Trị province to reach Hương Phúng just opposite the *khet* Lakô.

18 I have changed the individuals' names.

19 For more details on Lao-Vietnamese border agreements after 1975, see Gay Bernard (1995).

20 Today's districts of Vilabuly, Sepon, Phine and Nong.

21 Information on the number of villagers who voluntarily relocated from Lao PDR to Vietnam in 2004–05 was collected from district officials. It concerns 15 households from Ban Laviền, 7 households from Ban Kapaï, 2 households from Ban La-an, and 2 households from Ban Lakô. These 4 villagers belong to the focal site of Lakô.

22 Tania Murray Li argues, thus: "*The goal consistently expressed is for a closer relationship with government authorities and access to facilities currently available only on the coast. Hill folk reject resettlement on practical grounds: there is no land available on the coast, and they must live close to their swiddens to guard them from the wild pigs. Nevertheless, they would like to be on official maps, included in the desa [a unit of about two to five thousand people under a headman] population counts, and made the beneficiaries of 'development' hand-outs. (…) they too seek to enmesh themselves more firmly, and more advantageously, within the government administrative system.*" (Li 2001, p. 58).

[23] Li sensibly noted, "*populations excluded from official maps and invisible in the national census may be more deeply taken by the idea of 'the state' than savvy, urban skeptics; therefore, they devise strategies to position themselves closer to what they imagine to be the center*" (Li 2005, p. 385).

[24] Indeed, when we returned to the area in September 2008, the road had been paved for approximately two-third of its length.

[25] Several studies have argued that the "state of marginality" among societies inhabiting Mainland Southeast Asia's upland areas is being cultivated by government development and nation-building policies containing a "civilizing" agenda, that is, the compliance of upland "peripheral" peoples to the ruling majority's cultural and social norms — be it Thai, Kinh, Lao, Han, Burmese or Khmer — that ultimately aim to keep them under the state's tutelage (Duncan 2004; McCaskill and Kampe 1997; McKinnon and Vienne 1989).

References

Baird, Ian G. and Bruce Shoemaker. "Unsettling Experiences: Internal Resettlement and International Aid Agencies in Laos". *Development and Change* 38, no. 5 (2007): 865–88.

Colombijn, Freek. "Introduction: On the Road". *Bijdragen tot de Taal-, Land- en Volkenkunde* 158, no. 4 (2002): 595–617.

Duncan, Christopher R., ed. *Civilizing the Margins: Southeast Asian Government Policies for the Development of Minorities*. Ithaca: Cornell University Press, 2004.

Dương Vương Lợi. "80 công ty tìm hiểu đầu tư vào khu kinh tế Lao Bảo" [80 Companies Survey Investment in the Lao Bảo Economic Zone]. *VietnamPlus*, 10 March 2010. Availabe at <http://www.vietnamplus.vn/Home/80-cong-ty-tim-hieu-dau-tu-vao-khu-kinh-te-Lao-Bao/20113/81053.vnplus>.

Flower, John M. "A Road is Made: Roads, Temples, and Historical Memory in Ya'an County, Sichuan". *The Journal of Asian Studies* 63, no 3 (August 2004): 649–85.

Gay, Bernard. *La nouvelle frontière lao-vietnamienne. Les accords de 1977–1990* [*The Lao-Vietnamese New Border. The 1977–1990 Agreements*]. Paris: L'Harmattan, 1995.

Khamvongsa, Channapha and Elaine Russell. "Legacies of War". *Critical Asian Studies* 41, no. 2, (2009): 281–306.

Li, Tania Murray. "Relational Histories and the Production of Difference on Sulawesi's Upland Frontier". *The Journal of Asian Studies* 60, no. 1 (2001): 41–66.

———. "'Beyond 'the State' and Failed Schemes". *American Anthropologist* 107, no. 3 (2005): 383–94.

McCaskill, Don and Ken Kampe, eds. *Development or Domestication? Indigenous Peoples of Southeast Asia*. Chiang Mai, Thailand: Silkworm Books, 1997.

McKinnon, John and Bernard Vienne, eds. *Hill Tribes Today: Problems in Change*. Bangkok: White Lotus-ORSTOM, 1989.

Military History Institute of Vietnam. *Victory in Vietnam: The Official History of the People's Army of Vietnam, 1954–1975*. University Press of Kansas, 2002.

Nalty, Bernard C. *The War against Trucks: Aerial Interdiction in Southern Laos, 1968-1972*. Washington, D.C.: Air Force History and Museums Program United States Air Force, 2005.

Nishizaki, Yoshinori. "Suphanburi in the Fast Lane: Roads, Prestige, and Domination in Provincial Thailand". *The Journal of Asian Studies* 67, no. 2 (May 2008): 433–67.

Pholsena, Vatthana. "Life under Bombing in Southeastern Laos (1964–1973): Through the Accounts of Survivors in Sepon". In *European Journal of East Asian Studies* 9, no. 2 (2010): 267–90.

Prados, John. *The Blood Road: The Ho Chi Minh Trail and the Vietnam War*. New York: John Wiley & Sons, Inc., 1998.

Qiang, Zhai. *China and the Vietnam Wars, 1950–1975*. Chapel Hill: University of North Carolina Press, 2000.

Roseman, Sharon R. "How We Built the Road": The Politics of Memory in Rural Galicia". *American Ethnologist* 23, no. 4 (November 1996): 836–60.

Scott, James C. "Freedom and Freehold: Space, People and State Simplification in Southeast Asia". In *Asian Freedoms: The Idea of Freedom in East and Southeast Asia*, edited by D. Kelly and A. Reidpp. Singapore: Institute of Southeast Asian Studies, 1998.

Sturgeon, Janet C. "Border Practices, Boundaries, and the Control of Resource Access: A Case from China, Thailand and Burma". *Development and Change* 35, no. 3 (2004): 463–84.

Thomas, Philip. "The River, the Road, and the Rural-Urban Divide: A Postcolonial Moral Geography from Southeast Madagascar". *American Ethnologist* 29, no. 2 (May 2002): 366–91.

Vietnam Museum of Ethnology. *Highway no. 9: Opportunities and Challenges*. Hanoi, 2009.

Weber, Eugen. *Peasants into Frenchmen: The Modernization of Rural France, 1870–1914*. Stanford, California: Stanford University Press, 1976.

17

POPULATION'S MOBILITY IN NORTHERN LAOTIAN TRANSBORDER AREAS

Vanina Bouté

The aim of this chapter is to understand and analyse the effects of the opening of frontiers, the increase in cross-border exchanges and subsequently, the creation of economic corridors among local societies in Northern Laos. Until the last decade, a picture of Northern Laos would have been approximately the following: on the one hand, Tai populations (including the Lao majority) settling in the lowlands, growing rice, and on the other, highlanders practising shifting cultivation, husbandry and poppy growing. Roughly 75 per cent of the Laotian population of 5.5 million practised shifting cultivation. During the last ten years, some significant changes have raised questions concerning this situation. These changes are the result of the combination of several policies:

A first set of changes concerns the border areas and has its origins in regional political and economic transformations induced by the opening of borders. In the northern part of Laos, most border regions have been isolated for a long time since they are located in the outer regions of the country. Since the policy of Economic Corridor Development, these previously isolated regions are now mostly included in the new trade routes and there has been a particular increase in pressure on these peripheral areas. Due to the regional context mentioned elsewhere, the government considers border area trade as an important component of its foreign trade policy. Facing external demands for developing market-based agriculture, the Lao State has promulgated various laws (e.g. land-forest allocations), following the injunctions of the World Bank and the UNDP. These two organizations strongly encourage land allocation in order to develop a land market attractive to investors and to ensure a better use of agricultural land and environmental protection. At the same time, and as the contribution of Danielle Tan will show, the opening of borders has made possible the establishment of several foreign companies (mainly Chinese) which are, for the most part, dedicated to the agricultural sector (growing rubber trees, corn, tea, or buying NTFP). As a result, each province and district is witness to the burgeoning growth of foreign concessions owned by these companies. This rapid growth is also having a snowball effect and is leading to the appearance of Lao citizen-owned concessions — generally smaller in size.

The second kind of change was brought about by the national strategy related to agriculture which tends to produce similar results, e.g., the shift from rice growing for families' self-subsistence to cash-crop agriculture. From about 2000 to 2005, poppy growing was forbidden. In some areas of northern Laos, it strongly affected some highlanders who used to rely on it for their livelihood; during the same period, a new set of reforms was introduced regarding new land tenure, such as zoning and land allocation (e.g., the Land and Forest Allocation Programme). Finally, in 2005, the eradication of shifting cultivation was introduced and was expected to be completed by the year 2010 — but it is still a mode of production that concerns a third of the total population of Laos, including most of those living in highland areas.

All these changes have had several effects. In this chapter, I will focus on two processes, territorial change and social change. First, I will propose an overview of the changes noted within the territory, in particular an overview of the appearance and strengthening of demographic and territorial imbalances throughout the Northern provinces of Laos. Then I will focus on the social impacts introduced by these territorial changes, through the study of some resettlement areas for migrants, especially some urban villages which have attracted a lot of migrants over the last ten years. I will argue that the land and demographic pressure on these areas goes with an increasing social differenciation leading, paradoxically, to the control of land by the first inhabitants of the cities — civil servants and traders.

I will first add a few words about my fieldwork and my methodology. The study area is located in Northern Laos and is spread across three provinces: Phongsaly, Oudomxay and Luang Namtha. These three provinces all share a border with China (and also with Vietnam in the case of Phongsaly, but cross-border exchanges between Phongsaly and Vietnam are so slight that they are hardly worth mentioning). Within these three provinces, I particularly focused my survey on four towns chosen as relevant examples to illustrate the situation in Northern Laos regarding cross-border exchanges and development corridors:

1. Two provincial capitals, Phongsaly and Oudomxay cities. Oudomxay is located in the lowlands, at the crossroads between China, Thailand and other provinces in Northern Laos; Phongsaly is more or less enclosed by mountains. Despite their different size in terms of inhabitants and their very different locations, they do share some common features: both witnessed substantial demographic growth in the last twenty years, especially a peasant migration in the last ten years; and at present both are facing similar problems; growing social stratification, land pressure and ongoing land speculation.
2. Two small cities, the main town of Boun Tay district (Phongsaly province) and Nateuil, a small town in Namtha district, located 19 km from the Mohan border. Because of its strategic position, provincial authorities have planned to make Nateuil a main town (*müang noy*) in a few years' time. Located in the lowlands, the populations of both have significantly increased: in the first, because rubber can be planted and it is one of the rare locations in Phongsaly province where this is possible at present; in the second because of its location, at the junction between the R3A road from Luang Namtha (which is part of the North-South corridor) and National Road No. 1 to Oudomxay and Luang Prabang.

For the present study, several inquiries were conducted in different places, using both a qualitative method (interviews with institutional actors such as provincial and district authorities, village headmen, etc. and villagers) and a quantitative one: questionnaires were handed out to some fifty families in Phongsaly city and Oudomxay city, and to about thirty families in the small town of Boun Tay and the large village of Nateuil.

I will now go on the two topics previously mentioned — territorial change and social change.

1. A GROWING DEMOGRAPHIC IMBALANCE IN NORTHERN LAOS

Development planning and policies are some of the main causes of population displacement in Lao PDR (Goudineau 1997, Vandergeest 2003). With the transition towards a permanent and intensive agriculture for national development, these policies have led during the last ten years to the eradication of opium production, the Land Allocation programme and finally the banning of shifting cultivation — which most of the highlanders rely on for their subsistence. They impose constraints on people's lives to such an extent that they decide to move, seemingly of their own will.[1]

The impact of this displacement is striking. National censuses of the last twenty years immediately draw attention to the significant changes affecting the demographic distribution of population within the territory and the resulting imbalances. The latter will be discussed by focusing on the different territorial scales (province, district), then the data from the 1995 and 2005 censuses will be discussed since they may mask some social realities such as, for instance, the fact that some important migratory flows are sometimes triggered by non-rational economic phenomena.

An Imbalance Affecting Both Demographic Distribution and Occupation of Territory

Considering the 1995 National Census, the main conclusion of Bounthavy and Taillard (2000, p. 46) regarding population dynamics in Lao PDR was that population pressure would increase strongly, with an exacerbation of migratory flow. This trend was confirmed ten years later. The demographic data from the 2005 National Census clearly indicate the unequal dynamics of demographic growth. This is first and foremost exemplified by the demographic imbalance between provinces. The 1995 Census showed a negative migratory balance in the North and North-East provinces, whereas some provinces and districts in the North-West were very attractive, such as Müang Sing and Namtha districts.

Between 1995 and 2005, the annual growth rate of the population was 2.1 per cent per year (see Table 17.1). Oudomxay and Luang Namtha provinces matched this national rate, while Phongsaly province was far behind. These data also indicate a demographic movement towards the North-West provinces — Oudomxay, Luang Namtha, Bokeo — while Phongsaly

Table 17.1
Population Distribution in Northern Provinces in 1995 and 2005

	Phongsaly	Bokeo	Oudomxay	Luang Namtha
1995	152.848	113.612	210.207	114.741
2005	165.947	145.263	265.179	145.310
Annual growth rate	0.8%	2.2%	2.61%	2.1%

was continually losing a large part of its population. This can be understood as an effect of the attractivity of the North-South corridor, crossing Luang Namtha and Bokeo provinces, but also as an effect of other developed transnational roads, such as National Road No. 13, crossing Oudomxay province from the Chinese border to Luang Prabang and Vientiane.

The weak growth rate of some provinces can be understood by looking at other data as the demographic imbalance is also considerable within the provinces. The Map 17.1 designed by Bounthavy and Taillard after the 1995 Census (2000, p. 57) shows important migratory flows from the highlands to the lowlands and the cities (within Oudomxay or Luang Namtha provinces, the flow of migration is towards the provincial capital cities). In Phongsaly province, for instance, this internal demographic imbalance is particularly relevant. Since the 1990s, this province, like most Laotian provinces, has been affected by various policies mentioned above.

As a result, flows of highlanders have come to the lowland districts of Boun Neua and Boun Tay to cultivate rubber — as tea, the main cash crop cultivated in the mountains, is no longer an economically viable opportunity. These two districts are now facing a growth in their population while the mountainous districts of Phongsaly and Samphanh are continually losing their population (it is an ongoing process, as some villages of mountainous Samphan district are still resettled in the Chinese concessions of Boun Tay district). For instance, the population of the Phongsaly district was 24,351 inhabitants in 1995 and shifted to 25,213 in 2005, a growth of 0.2 per cent per year. During the same period, the Boun Tay district had an annual demographic growth of 2.8 per cent. There is also an imbalance within the districts as shown by the example of the district of Phongsaly in the two Maps 17.2 and 17.3.[2]

Most villages located far from the road have disappeared while the number of villages resettled along the road has increased. We can also see

Map 17.1
Balance of Inter-provincial Migration (excluding Vientiane)

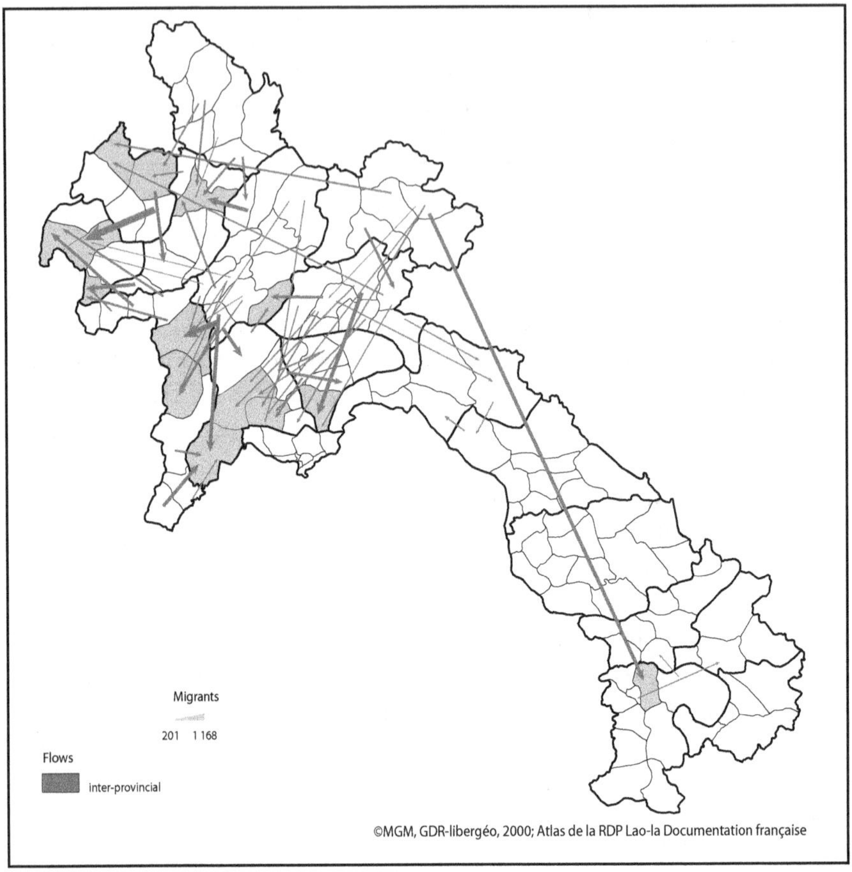

Source: 1995 census, NSC.

that the main city of Phongsaly has grown. Nevertheless, the growth rate is now very weak as land pressure and the lack of economic opportunities make these places less attractive for migrants. Several other districts such as Boun Tay district (Phongsaly province), Namtha district (Luang Namtha province), etc. are facing a similar situation.

In the current context of high rural mobility, which is linked both to measures taken at national level (displacements along major roads and regrouping of several villages in order to form larger administrative

Maps 17.2 and 17.3
Distribution of Villages in Phongsaly District in 1995 and 2005

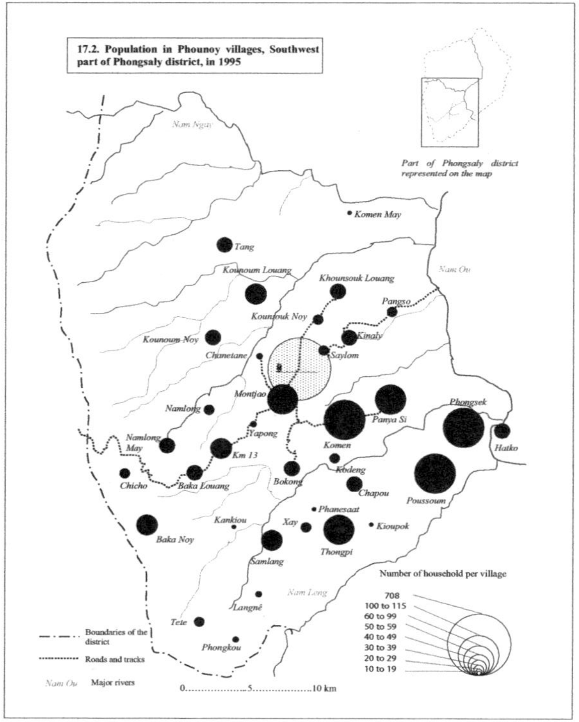

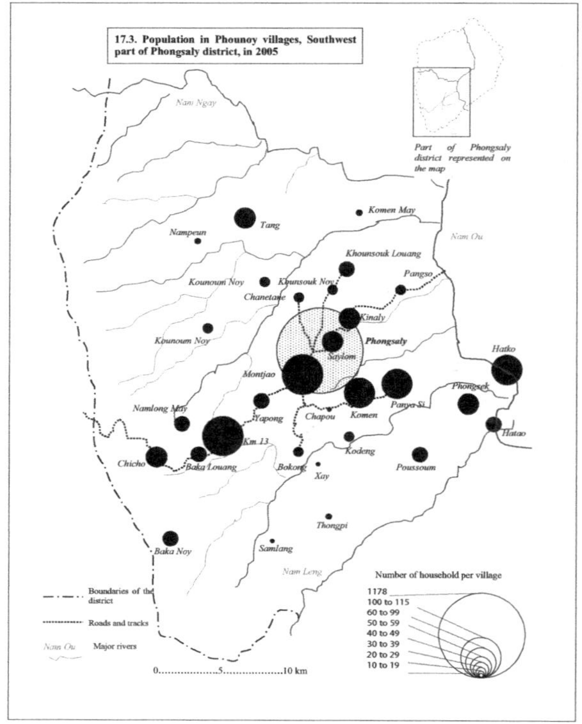

Source: Phongsaly district office of Statistics.

units) and to the economic attractiveness of some provinces (as a result of the opening of borders and an increasing amount of foreign investment) two new interconnected phenomena have now appeared: the development of new urban centres and large scale rural migration towards those urban areas.

Between 2005 and 2010, the urban population growth rate was 5.6 per cent, more than twice the national population growth rate which was 1.8 per cent, and far greater than the population growth rate in rural areas which was 0.1 per cent and may indicate by itself the movement of populations away from the countryside and towards cities.[3] These demographic data show us a substantial increase in the flow of migration towards the lowlands and the emergence and development of new kinds of urbanization.[4]

As opposed to former migrations in the history of Laos, rural migration towards urban centres is a new and largely growing phenomenon. The urban population represented 17 per cent (of the total) in 1995, 27 per cent in 2005 (equivalent to 1,220,374 people) and, according to estimates, will reach 38.2 per cent in 2030.[5] Over the last ten years, growing numbers of farmers have gathered in provincial capitals, in district administrative centres located in the lowlands, or in small, booming towns.

This was a trend already mentioned by Bounthavy and Taillard for the 1995 census, noting that movements to and from the towns (largest and smallest ones) included more than two-thirds of all inter-district and inter-province population movements (in fact, the census does not take into account movements within districts, particularly the resettlement of highlanders in the lowlands). In 2005, this trend was confirmed and growing: over a ten-year period (as shown in the Table 17.2) the size of some provincial capitals nearly doubled.

Table 17.2
Urban Population in Northern Provinces, 1995–2005

	Phongsaly	Houaphan	Oudomxay	Luang Namtha
1995	5.7% (8,658 hab.)	5.9% (14,404 hab.)	15.1% (31,678 hab.)	17.1% (19,621 hab.)
2005	12.6% (20,854 hab.)	11.8% (33,022 hab.)	15.2% (40,409 hab.)	21.8% (31,734 hab.)

Sources: State Planning Committee (1997); Steering Committee for Census of Population and Housing (2006).

Data Masking Some Social Features

The flow of migrants and the related apparent attractivity of the lowlands may not mask social features: for peasants, migrating is much more a choice by default than a real economic rational choice. With the banning of shifting cultivation, transition to commercial agriculture is quite unavoidable. But in reality, cash crops are not economically viable for the peasants in most mountainous areas: the resettlement and the concentration of villages along the roads have limited the amount of land available for agriculture; and villagers have difficulty getting enough income from tea or maize bought at low prices by Chinese traders. Moreover, there is a snowball effect, as I showed for Phongsaly district elsewhere (Bouté 2005)[6]: large-scale forced migrations in Phongsaly district had disastrous effects on those who stayed in their villages. With the disappearance of social links, people staying in the old villages experience a lack of labour forces and solidarity networks, ageing of the population, difficulties for unmarried people to find a partner, etc.

Therefore, in mountainous areas, migration to the lowlands by highlanders having no economic and social alternatives in their villages, or anticipating these problems, is continually increasing. But what kind of economic and social alternatives do they really have in the lowlands? In order to understand the demographic attractivity of the lowlands, it is important to consider the logic of migration and the perceptions of development.

Peasants' strategy regarding migration is, all in all, more influenced by factors such as family reunification than economic considerations. Surveys conducted in Namtha district and Nateuil clearly show that flows of people settling in these areas are not strictly controlled. Officially, families have to get permission from the authorities in both their former district and their new location before leaving. They have to provide proof they will be able to ensure their livelihood in their new settlement. In fact, village headmen have few means of controlling the newcomers, or refuse to do so. Firstly, because the flow of migrants coming and buying land is a real source of income for the inhabitants of the place, and also because district and provincial authorities also put pressure on village headmen to increase the demographic balance of small towns and large villages. For instance, Nateuil's authorities welcomed all migrants without consideration for their economic conditions as they are hoping for a large increase in the size of the village so it can become a secondary district city (*müang noy*). Apart from these elements, some village headmen and some district authorities have also totally integrated the idea that "mountain is misery" and are thus unwilling to refuse to admit new settlers.

From the migrants' point of view, joining family members will allow them to strengthen social networks — weakened in their former villages because of the exodus of villagers to new settlements. Family members are expected to help them to find and get land, to share the same house or land occupancy, to provide mutual help. It is the most important factor of migration, more important than the economic one: surveys show that many families are moving in order to join their family network without having the appropriate economic capacities to do so (no fields, no land, etc.).

The flow of poor migrants to the lowlands, without any economic prospects, can also be understood through the representation of development and modernity. Shifting cultivation has been disparaged for several decades by both the government and international institutions; highlanders are often still considered as backwards, belonging to another time. And the impact of these perceptions among the highlanders should not be underestimated, since it motivates their wish to resettle (see Bouté 2005, Lyttleton 2005). In an article, agronomist O. Ducourtieux (2006) argues that, in Phongsaly district, remote villages had more natural resources and economic potential than villages cultivating tea resettled along the roads. Despite his observation, and despite any economic rationality, inhabitants of remote villages in Phongsaly district had left their villages until they were totally empty and moved to many settlements along the roads (see previous maps).

Accounts by villagers recently and voluntarily resettled in roadside villages are also a good illustration of this paradoxical attitude. People admit they had enough land to live in their former villages while also acknowledging they have no lands in their new homes (they sell their labour) or not enough resources to buy rice. I quote:

> "In the old village, life is always the same. It's too backward and sad. Here, we can watch the cars on the road".
>
> "We have nothing here. But we are on the road, we may have economic opportunities (what kind? They didn't know). So we are waiting".
>
> "We are poorer than before but at least we live by the road. It will be better for our children".

And indeed, surveys conducted over the last ten years in this area reveal substantial impoverishment of families. Thus, on the one hand, migration to roadside localities can be understood as the result of the representations of development (and injunctions to leave remote villages) given by the authorities, radio, and so on. On the other hand, the settlements in these new villages have made few families much wealthier than before. There is a "snowball" effect: some families moved, expecting to experience the same kind of economic development as some wealthy families. But they did not really consider the economic conditions of these wealthy families before moving (important sales of opium, plenty of cattle, efficient political and family networks), conditions that helped them to make investments in the new place.

A similar paradoxical attitude can be observed in villages located along the development corridors or in lowland villages: in Nateuil, more than half the population come from Phongsaly province and most of them are facing economic issues: — more than half of the newcomers, in the last five years, have no lands for agriculture (they rent the land and cultivate rice around the rubber trees of the owners); — the other half have small rubber fields but no income yet. They fear the loss of these fields in the short-medium term as some farmers are forced to sell their lands due to debt.

2. EMERGENCE OF NEW SOCIAL STRATIFICATIONS

Imbalances caused by migration have considerably transformed the demographic landscape of Northern Laos. On the one hand, poverty has increased in the highlands with the weakening of social links and growing resettlement along the roads, with paradoxically, at the same time, large pieces of land left vacant and no longer developed. On the other, land and demographic pressure are high in the lowlands.

Populations in Northern Laos are facing some dramatic changes and, from a formerly self-sufficient society organized on an ethnic and village basis, practising shifting cultivation, they are rapidly turning into a society scattered among small multi-ethnic towns located in lowlands, moving towards market-based production. Considering cities and small towns as crucial elements for understanding the ongoing development of the Northern area of Laos, I will focus here especially on small towns and cities, that is to say the point of arrival of peasant migration. Due to migration to the lowlands, pluri-ethnic villages grow in numbers. Here ethnicity *per se* is not the issue but rather the fact that the inhabitants of several localities are living together without the solidarity mechanisms they used to rely on in their former villages. In other words, it means a weakening of social links with, at the same time, an increase in social stratification, a point we will pay particular attention to here.

Increasing Social Stratification

The appearance and development of social stratification in cities and towns in Northern Laos is quite new. While the main cities of Northern provinces have existed for a long time, the small towns of the districts have only really grown in size since the 1980s or 1990s.

There were three important periods in the development of most Northern towns such as Phongsaly, Oudomxay, Boun Tay and Nateuil: first of all there were the inhabitants of the first villages established on the site, before its urbanization. They were mainly peasants, and it was their land that the State appropriated after 1975, to build new administrative areas (buildings and housing for civil servants). Some sold their land to newcomers wishing to acquire land for building or farming. The first occupants were Tai or Lao in most urban areas (Phongsaly, inhabited by Phounoy and Ho, is an exception; it was the only provincial capital — with Sekong, in the south — to be mainly inhabited by a non Tai-speaking population). This settlement by Tai-speaking population is not surprising since towns in Lao PDR were nearly all originally centres of power (*müang*) controlled by Tai population. The first inhabitants, either because they maintained close contacts with the colonial administration, or because they had some knowledge of the Lao language due to their ethnic origins, were among the first to hold the administrative positions created in these places. Another characteristic of the first inhabitants of provincial capitals is their mobility. Few of them live today in their birthplace: some have gone to live abroad, and others have moved to major cities in Laos.

First, there were two or three major old villages. Their inhabitants were progressively joined by an important settlement of civil servants, from the beginning of the 1990s until the beginning of the 2000s — since these civil servants encouraged their relatives to come and settle in town with them. Today, civil servants represent 60 per cent of the population of Phongsaly, and 30 per cent of that of Oudomxay.[7] Later, some of them decided to quit their occupation as civil servant to engage in trade. The demographic weight of this arrival of civil servants and their families has led to the transformation of large villages or townships into urban centres with a pluri-ethnic population (since the civil servants originally came from different groups). New activities have also been created — development of shops or building firms — linked to the influx of populations who, unlike the peasants, have ready money provided by their salaried jobs.

Indeed, since 1995 and up to 2000, the strong new influx of immigrants to the cities consisted mainly of peasants arriving from neighbouring or remote villages (a flux which stopped in 2004–05 because of strong land pressure). Farmers now form between 20 per cent and 30 per cent of the population of the provincial capitals,[8] and 50 to 70 per cent of the population of the district capitals, or big villages such as Nateuil.

Thus, these towns have a new profile, and social and economic surveys allow us to identify five categories from their population: families of traders (e.g. those with one member involved in a commercial activity), families of civil servants (both husband and wife are civil servants),[9] mixed families (one member of the couple is a civil servant, the other is a peasant), farming families (both husband and wife are peasants).

The annual income of these mixed and farming families range between: 3 to 5 million kips (US$370 to US$650) for the farming families most recently settled in town, 8 to 20 million kips (US$1,000 to US$2,500) for some mixed families[10] and farming families, 15 to 25 million kips (US$1,800 to US$3,100) for the wealthier mixed families, and more than 80 million kips (US$10,000) for some trading families

The existing social and economic stratification has emerged in the last decade and will increase more and more in the future. It is particularly obvious in the cities and towns I targeted but it is also present in villages concerned by similar waves of migration (Lyttleton 2005, p. 55).

Land Pressure, Rising Prices: Changing Livelihood in Northern Towns

These elements were already factors explaining the impoverishment of the migrants. The agricultural policies of the previous decade in Lao PDR had already generated a social differenciation between the farmers, as Ducourtieux et al. (2005) argued; they noted that the Land allocation programme, by reducing drastically the fallow areas, was impoverishing farmers who did not have access to other means of production. On the other hand, the wealthiest farmers could take the risk of investing in the new permanent crops and were thus able to increase their income. So the growing of cash crops accelerated this process of economic differentiation based on land access as previously noted in several rural areas in Northern Laos. There is an economic spiral: wealthy families are getting wealthier by investing in cash crops, poor families without any land ownership are getting poorer by progressively selling their few properties (livestock, land). Among those families, one can distinguish between families with no land, and families with land. Among the latter, some are already in debt because they invested in cash crops for their fields.

Another explanation for existing social differenciation is the cost of resettlement which prevents farmers from saving money to buy fields or invest in cash crops. In small towns such as Boun Tay or Nateuil, the total cost for resettling one family may include the settlement tax to be paid to the village headman and district authorities (around 450,000 kips (US$56) in a village like Tchaho (5 km from the district city) or 500,000 kips (US$62) in Boun Tay city); moving expenses (family possessions, bus tickets) are between 700,000 and 2 million kips (US$87 to US$250); money to get land to build on: 10 million kips (US$1,250) or more, and then house building (from US$375 for a bamboo wooden house). So without buying any fields, the average cost for resettling is around 20 million kips (the price of seven or eight buffaloes), a huge amount when considering the average of approximately one or two buffaloes per family in Phongsaly district. This amount for one family's resettlement matches the annual income of the wealthier families in Phongsaly city.

So land access issues differenciate old settlers from recent ones in these towns. Former inhabitants of the oldest urban villages and those who came fifteen years ago are those who usually have a salaried activity, and at the same time, own a good deal of land (inherited from their parents or vacant land that they could occupy first). They mostly sold their previous lands (and moved to another province) or cultivated them, hiring migrant workers. Their economic situation is quite the opposite of families having the lowest annual income of 3 to 5 million kips. They form, in every village of Phongsaly city, the poorest part of the population (5 to 10 per cent of the total). Their income does not exceed their expenses so they cannot buy any plots of land; or, if they already have some land, they do not have enough money to buy tea trees. The third category is composed of poor mixed families or famers settled in town for a long time, with an income ranging from 8 to 20 million kips. They have plots of tea trees. All their savings were invested in land to build a house, in the house-building or in a tea plantation. But even if they have a higher income than the former category, they cannot save enough money for other investments. At present, most of them have difficulty paying for their children's schooling, or are spending all their income on it.

The existing social and economic stratification between these different categories of populations will probably increase in the future as the price of land (housing, plots) is growing in a context of land and demographic pressure. Price increases are considered to be greater in areas where infrastructures are being developed than in towns where infrastructures already exist. They are also more common in peri-urban areas (Onmanivong et al. 2006). In other words, the increase in land prices especially affects areas where there are the greatest numbers of recent migrants, often burdened by moving costs. Similar disparities among urban populations have often been noted "Buyers and sellers in a city centre seem to have a good understanding of current and future land values, while the poor living mostly in peri-urban areas often don't know the value of their land and more particularly, appear unaware of possible increases in value. They are therefore easily cheated" (Onmanivong et al. 2006, p. 9).

But the social segregation regarding access to land occupancy can be also linked to the implementation of the Land Titling programme published in 2005. This law gives everybody official land titling documents. It has already been implemented in some provinces and in 2009 was about to be completed in Northern provinces. The policy is supervised at the highest State level (Prime Minister's Office) and put into practice according to a standard national procedure, at the local level by the provincial and district land tenure service, under the supervision of the provincial governors.

In Oudomxay and Phongsaly provinces, the Land Titling programme will first concern housing land (for provincial capitals, district main towns and then housing land in villages). It will then be extended to agricultural lands in a second stage. Prices are fixed according to many criteria such as attractivity: location of a plot in a provincial capital, in a district main town, in a village, near a road, and so on. But the difference — and almost the inequality — between the two provinces depends on the following point: in Phongsaly, the provincial authorities have identified two kind of urban lands — lands which are the property of the citizens (inherited from their parents); and government-owned lands, allocated in the 1980s and 1990s to the new provincial offices and where the civil servants built their houses. Regarding the decision of the provincial authorities, half of the urban villages in Phongsaly city are considered to be "government lands". Even if the present settlers have already bought their land from the former occupants, have received sale certificates, and regularly pay taxes for this land, according to the Land Titling, they will have to buy back the land from the provincial services in order to be considered as legal owners. Nevertheless, civil servants benefit from price reductions (they will pay 30 per cent of the total amount if they have already built their house; 25 per cent of the amout for a plot without a house). Other people, like farmers or traders, will have to pay 100 per cent of the total amount.

For example, Khamtao is a civil servant living 3 km from the centre of Phongsaly city. He has built his house on former agricultural land belonging to his wife. Her parents had cultivated this land for three generations. To receive an official land ownership title (*bay tha din*), he has to pay 50,000 kip per square metre. Since he has 229 square metres, the total amount is 11,450,000 kip. Minus the 70 per cent reduction allowed for civil servants, the total amount is 3,435,000 kip. Thitpan is a farmer who came from the village of Thongpi (7 hours walk from the city) and lives in one of the nine urban villages of Phongsaly city. He came in 2000, after the district authorities declared that all inhabitants of this village had to move in order to reduce shifting cultivation. To finance his move, he sold all his goods (mainly cattle) and bought from a former inhabitant of his new village a 400 square-metre plot to build his house, for a cost of 4 million kips. He also borrowed money from relatives to buy a small plot where he grows tea for a living (5 million kips). In order to receive the land ownership document, he will have to pay a total amount of 20 million kips just for the land on which his house stands, a sum he cannot afford.

In contrast, in Oudomsay, provincial authorities decided not to distinguish between government and non-government lands. By doing so, they do not oblige people to buy land from the State if they have already bought it from the former occupants. In order to get the land ownership document, farmers, traders or civil servants will only have to pay a tax from 0.5 to 3 per cent of the purchase price. Then, if Thitpan lived in Oudomsay, he would have only to pay 600,000 kips, which is 33 times cheaper. In the small town of Boun Tay, in 2009, 11 families, among the poorest and most recently arrived, were confronted with the same problem; the land which they had bought a few years previously from a wealthy local trader was declared "government land", and they were faced with a dilemma — either pay for their land a second time, or leave. Vandergest (2003, p. 49), using an example from Thailand, noted that "Land titling has been widely criticised for the way that titles might lead to loss of control of land and growing class inequalities as small farmers are forced to sell their land due to debt." Our example shows clearly the impoverishment that is, or will be the result of applying this law.

A similar phenomenon can be observed for farmland; price increases make it increasingly inaccessible for some categories of town dwellers, the poorest, who are often peasants. Since the year 2000, there has been a very substantial increase in the price of agricultural land, This results mainly from the shift from subsistence farming to cash crops cultivation imposed by agrarian reforms.[11] Since in Northern Laos, these crops are mainly destined for the Chinese market — either because Chinese traders come to buy the crops, or because Chinese investors obtain agricultural concessions employing whole families of peasants or even villages — pressure on land is particularly strong in regions adjacent to the Chinese frontier, and especially along the North-South corridor, which is particularly favourable to trade. The consequences of this new "green gold-rush" are large-scale privatization of land, strong land pressure and, as a corollary, considerable inflation of land prices. The most attractive locations are also subject to demands from outside the province, for example in Oudomxay province, where 80 per cent of investors and purchasers of land come from other provinces (Chantalasy et al. 2005, p. 9).

The rise in price of agricultural land that is supposed to be the peasants' main source of income therefore has direct and dramatic consequences for migrants of rural origin, who most frequently set up as farmers — or hope to do so. In the small capital of Boun Tay District, since the rubber planting boom, prices for plots of land have increased tenfold in ten years. If at the beginning of the 1990s fallow land was not worth much, in 1998 it began to be sold for between 30,000 and 100,000 kips per hectare, and then cost more than one million kips per hectare in 2005. Land values are rising significantly in other localities such as Oudomxay, Nateuil, Phongsaly and other cities in Laos.[12]

The increase in land prices leads to a speculative bubble harshly affecting the farmers. For instance, village zoning and land allocation divide land into agricultural and non-agricultural land (such as forests), but in areas where pressure on ressources is high due to population density and external demands, land reform is not respected and the rights of the farmers are not considered. It occurs especially in demographically growing places such as the small towns of the district. For instance, all around Boun Tay city, newly-protected forest (where farmers used to practise swidden agriculture but have not been allowed to do so for the last few years) has been allocated to rich traders' families or civil servants from Phongsaly city in order to develop rubber plantations. In Nateuil as well, though plots of land were allocated to newcomers to grow sugar cane in the 1990s, in 2006 they were taken away from them by a civil servant and given to a rich Lu-Chinese trader instead. This trader is now making the largest rubber plantation in the area.

To sum up, agriculture is turning into a business from which farmers are excluded. The paradox is the following: traders and civil servants have at present more land than farmers. For instance, some of the traders or civil servants in Phongsaly city have more than 150,000 tea trees compared with only 30,000 trees for a peasant family. Some others also invested in the growing of other cash crops, such as teak and eucalyptus trees, or rice fields and rubber plantations in the lowlands. The situation is quite similar in Boun Tay and Nateuil. Families having more than two hectares of cash crops are systematically working as traders or civil servants. This fact has already been established for other districts of Luang Namtha province, as the *Thai Prachatai* newspaper published a sarcastic remark that the construction of Road 3 — the part of the North-South Corridor crossing Luang Namtha province — had created an economic corridor only for China, and that rubber plantations had enriched the elite class of Lao PDR while producing many poor people who have to work in the plantations and factories.[13] So there is the emergence of a new social category, the "landless farmers", who, I imagine, will make up a growing part of the urban population as is already the case for some populations in Northern districts.[14]

CONCLUSION

In Northern Laos, population distribution within the territory has been changing significantly over the past few years, with a concentration of populations along roads in general, and also a population increase in the lowlands and the north-west provinces, where there is a growing development of exchanges.

A certain number of studies have emerged in recent years concerning the opening of roads, with strongly contrasting points of view: some maintain that a road provides development for people resettled along it (Leebouapao et al. 2005), whereas other studies are more critical. They consider that the road could produce negative effects, while, at the beginning of the twenty-first century it is seen as a necessary tool to give impetus to livelihoods (Rigg 2002). It is not our role here to give our opinion on this debate; this article is mainly concerned with observing the social implications that are certainly favoured by the opening of roads (spontaneous migration), but also by national policies: population movements, the change to the growing of cash crops and the opening of trans-border exchanges. To an increasingly unequal demographic and spatial distribution over the provinces of Northern Laos as a whole, we may add growing land pressure, increasing competitiveness between actors, and especially — this is the main subject of our study — new forms of social differentiation, with the appearance of landless families, economically dependent on others, while other actors — civil servants, traders or certain "rich" peasant families — benefit from the effects of corridors and foreign investments.

Questions could be raised about the impact of corridors, development and roads on this differentiation. On the one hand, the attractiveness of certain locations — situated in the lowlands — close to major roads built explicitly to encourage trade, is obviously an important factor leading to strong demographic pressure and consequently land pressure which does not leave the most fragile populations unscathed. On the other hand, the example of the Land Titling programme for building land in two provincial capitals — one situated at a major road junction, the other enclosed to such an extent that the transfer of the capital to the lowlands is now planned for the near future — may in future relativize the observation that pressure is probably stronger on populations in the most economically attractive locations. Inequalities with regard to access to land have grown in the town of Phongsaly, which is less well situated and less attractive than Oudomxay, because of local policies, which may aim to make up for the small amounts of local taxes collected, mainly on account of the mass migration emptying the province. But, whatever the case may be, these two situations — agricultural lands under pressure along the corridors, and building land becoming more expensive for the poorest people in less attractive locations — show that the economic situation of part of the population of Northern Laos is becoming alarmingly fragile.

In this type of economic and social context, it is possibly time to think about other forms of development, taking greater account of the social dimension, and aiming to protect populations made fragile by development policies. A remedial solution could include strengthening State commitment, since, for the time being, the State seems to be out of phase with the unequal social aspects of these changes.

Notes

1. For these reasons, a certain number of population movement analyses point out that it is tricky to distinguish between voluntary and forced migration in the context of a relatively undemocratic state, since both result from the same causes, i.e., development policies ill-adapted to the lifestyle of the populations concerned (see, e.g., Goudineau 1997 and Evrard and Goudineau, 2004).
2. See also maps of Luang Namtha district for similar comparison in Thongmanivong et al. (2009).

3 According to UN figures (<http://data.un.org>).

4 This refers to provincial capitals but above all to the main towns of the districts.

5 Source: UN common database; UN Population Division estimates.

6 See also Lyttleton and al. for similar assessment in Muang Sing district (2004, p. 58).

7 These figures indicate civil servants still working (according to statistics for Oudomxay and Phongsaly districts, gathered during surveys carried out among headmen in the urban villages of the Oudomxay and Phongsaly areas). A great many former civil servants in the towns of Oudomxay and Phongsaly have taken to trade and so do not appear in these statistics. Also, these figures indicate an average: in certain urban villages in Phongsaly (e.g. Phonesahat) or Oudomxay (Longthana), 90 per cent of the population consists of civil servants' families.

8 From 30 per cent for the town of Phongsaly to 20 per cent for the town of Oudomxay which is more recent, since it was rebuilt after 1975. Only families whose main income is from farming are counted as farming families, based on statistics from Oudomxay and Phongsaly districts. Figures for district main towns are those for the districts of Boun Tay, Boun Neua (Phongsaly province), and Namo (Oudomxay province).

9 The proportion of civil servants is lower in small towns such as Boun Tay or Nateuil where only categories 1, 3 and 4 are represented.

10 Families where one of the couple is a civil servant and the other a peasant are counted as "mixed".

11 Concerning this agrarian transition, see e.g., Evrard (2003), Tanaka et al. (2007), Cohen (2005) and Ducourtieux (2006).

12 Prices vary according to the distance of the field from the road, its exposure, the slope, etc. For other locations, see the study report of Rabé et al. (2007, pp. 28–30).

13 Thai Prachatai (article on 21 September 2007 of the daily web newspaper <www.prachatai.com>), op. cit. in Tsuneishi (2009).

14 See Lyttleton et al. arguing that large numbers of highlanders recently resettled in the lowlands of Sing and Long districts (Luang Namtha province) will be highly dependent on wage labour in the foreseeable future (2004, pp. 25 and 59–61).

References

Bounthavy, Sisouphantong and Christian Taillard. *Atlas of Laos: The Spatial Structures of Economic and Social Development of the Lao People's Democratic Republic.* Chiang Mai: Silkworm Books, 2000.

Bouté, Vanina. "Des gardiens des confins aux bâtisseurs des plaines: parcours d'une populations tibéto-birmane du Nord-Laos" [From Mountain Border Watchdogs to Plains Wet-rice Cultivators: The Long Route of a Tibeto-Birman People of Northern Laos]. *Moussons* 8 (2005): 35–60.

Chantalasy, Anothai, Lintzmeyer Florian, Rock Florian, Soulivanh Bouakham, and Souphida Phounsavath. *Study on Land Markets in Urban and Rural Areas of Lao PDR*. Land Policy Study n°2. Vientiane: GTZ, 2005.

Cohen, Paul T. "The Post-opium Scenario And Rubber in Northern Laos: Alternative Western and Chinese Models of Development". *International Journal of Drug Policy* 20, no. 5 (2005): 424–30.

Ducourtieux, Olivier. "Is the Diversity of Shifting Cultivation Held in Enough Esteem in Lao PDR?". *Moussons* 9–10 (2006): 61–86.

Ducourtieux, Olivier, Jean-Richard Laffort and Silinthone Sacklokham. "Land Policy and Farming Practices in Laos". *Development and Change* 36, no. 3 (2005): 499–526.

Évrard, Olivier. *La mise en œuvre de la réforme foncière au Laos: impacts sociaux et effets sur les conditions de vie en milieu rural* [Rural Land Planning in Laos: Social Impacts and Effects on Rural Livelihood]. Livelihood Support Programme, Roma, Organisation des Nations-Unies pour l'Agriculture et l'Alimentation (FAO), 2003.

Évrard, Olivier and Yves Goudineau. "Planned Resettlements, Unexpected Migrations and Cultural Trauma". *Development and Change* 35, no. 5 (2004): 937–62.

Goudineau, Yves, ed. *Resettlement and Social Characteristics of New Villages: Basic Needs for Resettled Communities in the Lao PDR*, 2 vol. Vientiane: UNESCO/ United Nations Development Programme/Orstom, 1997.

Leebouapao, Leeber, Souksavath Phonesaly, Sone Phetsamone, Darachanthara Souphith, and Norintha Vanthana. "Lao PDR's Cross Border Economy: Case Study in Dansavanh, Savannakhet, and Houaysai/Tonpheung, Bokeo, Lao PDR". In *The Cross Border Economies of Cambodia, Laos, Thailand, and Vietnam*, edited by Development Analysis Network. Phnom Penh: Cambodia Development Resource Institute, 2005.

Lyttleton, Chris. "Marked Boud: Relocation and Disjunction in Northwest Laos". In *Migration and Health in Asia*, edited by B. Yeoh, S. Jatrana and M. Toyota. London and New York: Routledge, 2005.

Lyttleton, Chris et al. *Watermelons, Bars and Trucks: Dangerous Intersections in Northwest Laos*. Vientiane: Institute for Cultural Research of Laos/Macquarie University, 2004.

Onmanivong, Khamdeng, Soulivanh Bouakham and Babette Wehrmann. *Study on Urban Land Markets in Lao PDR*. Land Policy Study No. 5. Vientiane: GTZ, 2006.

Rabé, Paul, Thongbonh Thenekham and Vongsiharath Vangdeuane. *Study on Urban Management and Planning in Lao PDR*. Land Policy Study No. 10. Vientiane: GTZ, 2007.

Rigg, Jonathan. "Roads, Marketization and Social Exclusion in Southeast Asia. What Do Roads Do to People?". *Bijdragen tot de Taal-, Land- en Volkenkunde* 4 (2002): 619–36.

State Planning Committee, National Statistical Centre. *Results from the Population Census 1995*. Vientiane, 1997.

Steering Committee for Census of Population and Housing. *Results from the Population and Housing Census 2005*. Vientiane, 2006.

Tanaka, K., S. Yokoyama and K. Phalakhone. "Land Allocation Program and Stabilization of Swidden Agriculture in the Northern Mountain Region of Laos". In *Shifting Agriculture in Asia: Implications for Environmental Conservation and Sustainable Livelihood*, edited by K.G. Saxena, L. Liang, and K. Rerkasem, Dehra Dun (India): Bishen Singh Mahendra Pal Singh, 2007.

Thongmanivong, Sitphong, Fujita Yayoi, Phanvilay Khamla and Vongvisouk Thoumthone. "Agrarian Land Use Transformation in Northern Laos: From Swidden to Rubber". *Southeast Asian Studies* 47, no. 3 (2009): 330–47.

Tsuneishi, Takao. *Border Trade and Economic Zones on the North-South Economic Corridor: Focusing on the Connecting Points between the Four Countries*. IDE Discussion Paper No. 205. Institute of Developping Economies, 2009.

Vandergeest, Peter. "Land to Some Tillers: Developement Induced Displacement in Laos". *International Social Science Journal* 55, no. 1 (2003): 47–56.

18

CHINESE NETWORKS, ECONOMIC AND TERRITORIAL REDEFINITIONS IN NORTHERN LAO PDR

Danielle Tan

Luang Namtha, Oudomxay and Bokeo provinces, located in the North-Western part of Lao PDR, have benefited long ago from a strategic position at the crossroads between Thailand, China, Myanmar and Vietnam. They were part of the caravan trade network that thrived between the thirteenth and nineteenth century, and then, of the Golden Triangle opium trade. Due to its geographical assets and its cross-border trade linkages, Northern Lao PDR is once again at the forefront of the regional stage through the implementation of the GMS North-South Economic Corridor (NSEC), which links Kunming to Bangkok by crossing the Northern provinces of Lao PDR. However, this regional integration process also means dramatic socio-economic changes in the remote and impoverished Upland Lao PDR. Will Northern Lao PDR be able to reap the benefits from the NSEC eventually? How is the NSEC impacting on local livelihoods, and who are the main drivers of change?

This research aims to examine the Asian Development Bank's argument that Lao PDR could become the crossroads of the trade routes in the GMS if this landlocked country manages to turn to its advantage the migration and transnational networks passing through its territory. I will explore the role of Chinese networks in regional trade, since the increasing flow of Chinese newcomers into Lao PDR is obviously metamorphosing the socio-economic landscape of the region. The present research is based on empirical data —

Map 18.1
The Economic Quadrangle

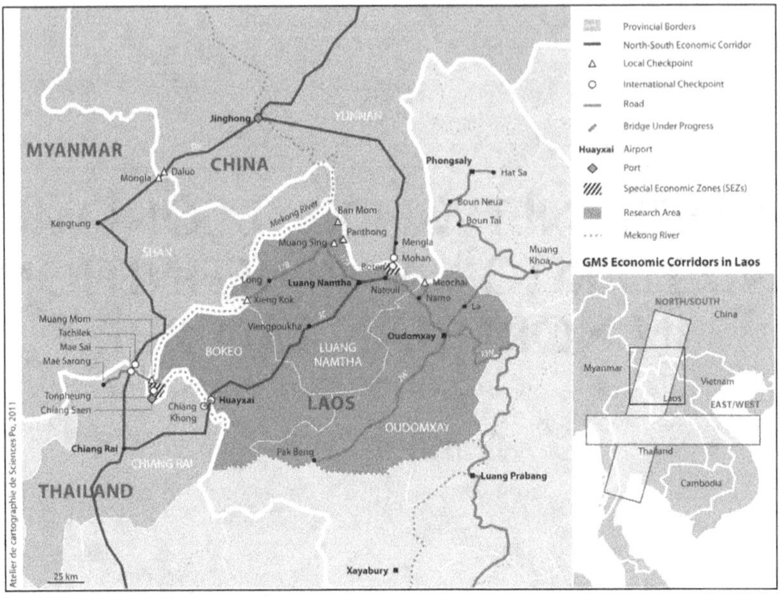

Source: Data compiled by Danielle Tan (2011).

interviews with Chinese migrants and Lao officials as well as analysis of socio-economic statistics and reports — collected between 2007 and 2009 during several fieldwork studies carried out in the main cities of three Northern provinces of Lao PDR (Oudomxay, Luang Namtha, Bokeo) affected by the NSEC (see Map 18.1).

First, I will describe what is at stake since Lao PDR has been involved in the GMS programme, and specifically in the case of implementing the NSEC. Then, I will describe how Chinese newcomers are gradually weaving their transnational networks into Lao economy and how they are accelerating the pace of change in the region.

1. THE NSEC'S CHALLENGES FOR LAO PDR

At the Origin of the NSEC: The "Sino-Thai Strategy"[1]

The concept of Economic Quadrangle encompassing the northern borderlands of Southeast Asia first emerged in the early 1990s under the initiative of Thailand in order to strengthen economic cooperation with China's Yunnan province, Lao PDR and Myanmar. Sino-Thai cooperation began in 1993 with the objective of improving Upper Mekong navigation and railways, thereby reviving the French colonial dream of a "river road to China" (Osborne 1975) as well as the ambitious British plans for a rail link between Thailand and Yunnan.

However, progress of this Economic Quadrangle strategy relayed by the GMS initiative went very slowly, delayed by the 1997 Asian financial crisis. As a result, the ADB has promoted a new integrated development approach, based on the economic corridor concept as a crucial junction for trade, tourism and investment. While the ADB has attempted to make the East-West Economic Corridor the cornerstone of the GMS strategy, the North-South Economic Corridor project has gained strong support from Thailand and Yunnan, becoming the priority of the Sino-Thai economic development strategy.

China has perceived the benefits of promoting the GMS corridor for its own national and provincial strategy to overcome the economic disparity between the East Coast and the hinterland provinces of western and south-western parts of the country. Hence, Yunnan was promoted as the gateway to the GMS. The furthering of the NSEC also matches the Chinese national policy by opening an avenue through Southeast Asia for importing natural resources and exporting manufactured goods, the ultimate goal being to link Kunming to Singapore by railway. At the same time, the NSEC makes it possible to gain new strategic footholds via access to the Andaman Sea, the Gulf of Thailand, and various ports in Cambodia. In this context, Lao PDR is not only perceived as a key corridor but also as an abundant land reserve.

As far as Thailand is concerned, the concept of economic corridor was an opportunity to become the transportation hub and investment centre of the GMS. The NSEC, oriented towards the Chinese market, was intended to foster development of the remote areas of Northern and Northeastern Thailand, and raise the livelihoods of rural communities by creating economic linkages. Thai participation in the GMS was further encouraged in 1988 by Prime Minister Chatichai Choonhavan's policy of turning Indochina "from battlefield to a market place". He envisioned the revival of the *Suvannaphum* myth[2] — the golden land — where Thailand would be at the centre of the Mekong, expecting that the vast natural resources of Indochina and Myanmar could bolster Thai economic power.

The Competition between Myanmar, Lao PDR and the Mekong River Transportation

The NSEC was supposed to make Lao PDR move forward with the sub-regional integration process. Actually, Lao PDR still has to take up many challenges to build on the economic opportunities offered by the corridor, and the promises of the new road are not clear for the moment. At the present time, there are three different means of transportation to expand trade between Thailand and Yunnan, competing with each other: the Lao and the Myanmar overland branches of the NSEC, as well as the navigation along the Mekong River.

One important point to be remembered is that in the NSEC plans, the route through Myanmar is the shortest and the potential for trade and tourism is higher than that of Lao PDR. That is why China first helped to improve the Daluo-Chiang Tung section in 1993, while Thailand provided a loan of 300 million baht for the Chiang Rai-Tachilek-Chiang Tung section (164 km). Because of security problems with ethnic minorities' rebellions, the road was often closed, heavily affecting border trade and tourism in Mae Sai. Therefore, Thailand and China started to think about the Lao PDR option. China spent US$4 billion to build the highway to the Lao border (831 km). Then, the ADB, Thailand and China put equally US$30 million each to build the Lao section of the NSEC (R3E — 228 km). However, the financial crisis of 1997 as well as problems between Laos and Thai private companies has delayed this project for many years; the final agreement was only sealed in August 2002. China ranks this road as a priority project to expand trade with Southeast Asia. That is why, in 1993, China started to build on the Lao PDR-China border the Mohan-Boten international port — a large international trade centre for overseeing future international trade, investment and tourism. The journey from Boten to Huayxai on the old, mostly dirt road took as long as two days. Today, the trip takes only five to six hours with the new road. The NSEC was inaugurated during the 3rd GMS Summit held in Vientiane on 31 March 2008. To complete the NSEC plans, Thailand and China also agreed to share the cost of the US$43 million-bridge linking Huayxai (Lao PDR) to Chiang Khong (Thailand). Construction was expected to start in 2004 and be completed in 2007, but the recent political turmoil in Thailand has delayed the project. It was not until 2009, during the 15th ASEAN Summit, that China, Thailand, and Lao PDR signed an agreement. The work began in March 2010 and should be completed in thirty months.

In the meantime, Yunnan province actively pushed for shipping on the Mekong River. The Chinese province invested 40 million Yuan in improving Jinghong port, located 380 km from Chiang Saen. Since the official opening of the port in June 2001, it takes two days for Chinese cargo vessels to transport goods from Chiang Saen to Jinghong, made possible by a first phase of blasting on the Upper Mekong rapids. Currently, 95 per cent of China-Thailand trade is conducted through shipping from Bangkok to China; it takes ten to fifteen days to reach Guangdong. However, since the improvement of water navigation on the Mekong River, trade between Yunnan province and Thailand has been increasing (see Table 18.1).

Though the Burmese section of the NSEC is still unstable, China is Myanmar's second largest trading partner, after Thailand. Bilateral trade between China and Myanmar amounted to US$2 billion in 2007 — trade with Yunnan accounting for more than half — while the volume of trade between Thailand and Myanmar amounted to US$3.25 billion in 2007. China-Myanmar trade outweighs the bilateral trade between Lao PDR and China (US$263.86 million in 2007) as well as trade between Lao PDR and Thailand (US$1.78—billion in 2007) combined with the transit trade through Lao PDR (only US$1.7 million in Bokeo and US$7.7 million in Luang Namtha in 2007) (see Figure 18.1).

The economic benefits of the road to Lao PDR, meanwhile, are less promising than expected because of the missing bridge, but also due to complex administrative procedures that create trade bottlenecks at the borders. Chinese entrepreneurs thought that international trade would thrive thanks to the new road, but they complain that it takes at least one month to reach Bangkok, instead of the two or three days expected (*Xinhua* 2009).

Lao PDR is blamed for being insufficiently involved in the implementation of the Cross-Border Trade Agreement (CBTA) signed in 2007 to simplify the procedures. However, there are reasons for this so-called passivity: Lao PDR is reluctant to give up its tariff revenue[3] because it fears it will not get the benefit of the transit trade. Indeed, the transition to the new international trade system raises the crucial issue of providing traffic services and facilities along the R3E (import-export services, warehouses, transportation services). It also questions the Lao government's capacity to seize tourism opportunities by creating more favourable conditions for local businesses and by attracting foreign investment, which would utilize local resources. These issues must be

Table 18.1
Trade by the Mekong River Route at Chiang Saen Port, Thailand, 2001–08

	With Myanmar				With Laos				With Yunnan				Total		
	Exports	Imports	Total (million baht)	% Compared to the Total Volume of Trade	Exports	Imports	Total (million baht)	% Compared to the Total Volume of Trade	Exports	Imports	Total (million baht)	% Compared to the Total Volume of Trade	Exports	Imports	Total (million baht)
2001	63.60	1.60	65.20	2.04	125.40	225.20	350.60	11.00	2,225.60	547.00	2,772.60	86.95	2,414.70	773.90	3,188.60
2002	31.40	0.00	31.40	0.89	82.10	212.10	294.20	8.35	2,735.70	460.60	3,196.30	90.75	2,849.20	672.80	3,522.00
2003	19.70	1.60	21.30	0.49	29.50	81.60	111.10	2.55	3,129.10	1,087.40	4,216.50	96.96	3,178.20	1,170.60	4,348.80
2004	197.20	1.20	198.40	5.11	269.70	65.20	334.90	8.62	2,110.30	1,242.40	3,352.70	86.28	2,577.10	1,308.80	3,885.90
2005	1,325.70	44.10	1,369.80	19.32	453.10	46.50	499.60	7.05	4,171.80	1,047.30	5,219.10	73.63	5,950.60	1,137.90	7,088.50
2006	1,636.90	32.30	1,669.20	21.52	398.80	20.10	418.90	5.40	4,587.30	1,082.40	5,669.70	73.08	6,622.90	1,134.80	7,757.70
2007	1,418.30	3.60	1,421.90	19.74	288.30	13.00	301.30	4.18	4,618.90	862.40	5,481.30	76.08	6,325.40	878.90	7,204.30
2008 (January–July)	934.70	0.60	935.30	21.89	99.20	9.20	108.40	2.54	2,719.10	509.60	3,228.70	75.57	3,753.00	519.20	4,272.20

Source: Customs Department of Thailand (Tsuneishi 2009), p. 45.

Figure 18.1
Transit Trade in Luang Namtha and Bokeo Laos (2004–08)
(US$)

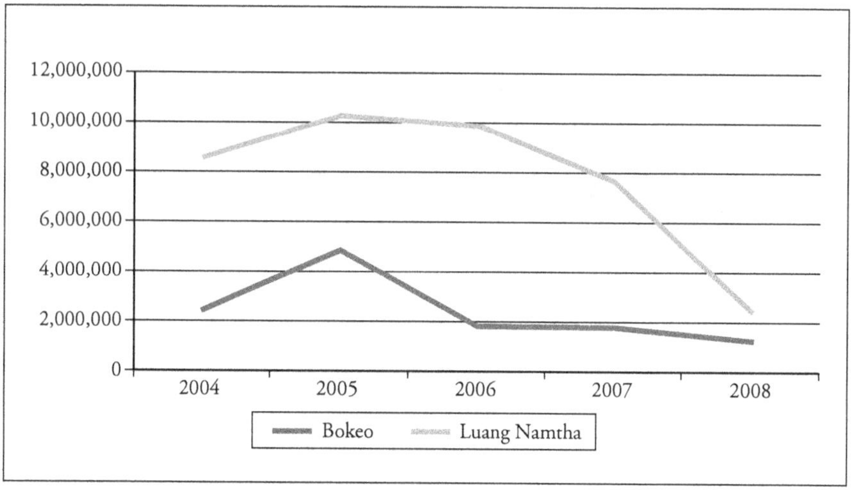

Source: Department of Industry and Commerce of Luang Namtha and Bokeo Provinces (2009).

addressed if the country does not want to limit itself to "watching the trucks go by" and paying for the maintenance of the road. Otherwise, the new corridor is more likely to fortify the stronger economies, such as Thailand and China, while Lao PDR, as a transit country, may not only gain less, but it will be disproportionately burdened by the social and environmental impacts related to the road.

2. THE ROLE OF CHINESE NETWORKS IN SPEEDING UP THE PACE OF CHANGE IN NORTHERN LAO PDR

Things are already changing rapidly in Lao PDR. In just a decade, Chinese networks have radically transformed the political economy of Northern Lao PDR. The North-South Economic Corridor has contributed to accelerate Chinese migration and investment. In this second part, I will emphasize the role of Chinese networks in speeding up the pace of change in Northern Lao PDR. While they are the main drivers of change, they are also the main "winners" on the verge of dominating the local economy by weaving their transnational networks into trade, agri-business, transport and tourism sectors.

Overview of Chinese Investment in Northern Lao PDR

Since early 2000s, Northern Lao PDR has been undergoing a dramatic transition from subsistence farming based on shifting cultivation to market-oriented agriculture. Road improvement and regional market forces, fuelled by skyrocketing Chinese investment, have driven this shift.

The majority of Foreign Direct Investment (FDI) to Northern Lao PDR comes from Mainland China and Hong Kong (96 per cent, 87 per cent, and 76 per cent, respectively, in Luang Namtha, Oudomxay and Bokeo provinces) (see Table 18.2).

The main Chinese investors are state-owned enterprises and large private companies. While the former make heavy investments to develop agriculture, industry, hydropower plants or mining (Rutherford et al. 2008), the latter have chosen the tourism sector, and particularly the lucrative sector of casinos. Chinese investment is mainly concentrated on agriculture (48 per cent in Luang Namtha and 53 per cent in Oudomxay), except in Bokeo (only 15 per cent), where tourism megaprojects, such as *Dokgniewkham* Casino in the Golden Triangle Special Economic Zone, absorb the majority of Chinese investment in the province. The industry and commerce sectors (rubber and tobacco processing factories, warehouses, motorbike assembly factories, etc.), and to a lesser extent, mining are also targeted by Chinese investors.[4] Although agriculture remains the principal economic sector in Northern Lao PDR, the importance of hydropower and mining revenues is increasing, following the same pathway occurring at the national level. Hydropower and mining development have been identified as the highest priority sectors by the government, given their potential to stimulate economic growth and lead to greater revenues. As a result, Northern Lao PDR is becoming a growing mineral and hydropower centre: 54 hydropower plants are planned in the Northern provinces, including 9 in Oudomxay, Luang Namtha and Bokeo provinces (World Bank 2008).

At the national level, China's engagement ranked as the top foreign investor for the first time in 2007, surpassing Thailand and Vietnam (see Table 18.3). Over the period 2000–10, Chinese firms have invested in 397 projects worth about US$2.71 billion, making China the second largest foreign investor after Vietnam (US$2.77 billion), while Thailand has downgraded to the third place (US$2.6 billion). Chinese investment covers hydropower plant development, mining, shopping centres, banking, restaurants, hotels, rubber plantations and telecommunications. Bilateral trade and investment between Lao PDR and China will most likely see a big surge if the US$7 billion Laos-China railway project is eventually

Table 18.2
Chinese Investment in Luang Namtha, Oudomxay, and Bokeo Provinces
(US$)

	Luang Namtha (1993–2008)	Oudomxay (1992–2009)	Bokeo (1995–2010)	Total
Total Registered Investments	36,125,772	108,826,766	81,787,584	226,740,122
Domestic Investments	5,307,592	23,133,040	13,925,284	42,365,916
Registered FDI	30,818,180	85,693,726	67,862,300	184,374,206
Other FDI	1,236,250	11,062,922	16,305,000	28,604,172
Chinese FDI	29,581,930	74,630,804	51,557,300	155,770,034
% Chinese FDI/Total FDI	96	87	76	85

Source: Department of Planning and Investment of Luang Namtha, Oudomxay and Bokeo Provinces, 2009–10.

Table 18.3
Chinese Investment by Sector in Luang Namtha, Oudomxay et Bokeo Provinces
(US$)

	Agriculture	Industry and Commerce	Mining	Services	Total
Luang Namtha (1993–2008)	14,165,060	10,444,787	3,604,000	1,368,083	29,581,930
Oudomxay (1992–2009)	39,620,500	20,844,892	12,554,912	1,610,500	74,630,804
Bokeo (1995–2010)	7,950,000	12,425,100	702,200	30,480,000	51,557,300
Total	61,735,560	43,714,779	16,861,112	33,458,583	155,770,034

Source: Department of Planning and Investment of Luang Namtha, Oudomxay and Bokeo Provinces, 2009–2010.

implemented. Indeed, Lao PDR and China agreed that construction of the high-speed train linking Boten to Vientiane would begin in April 2011 to mark the 50th anniversary of Lao PDR-China diplomatic relations and be completed within four years. However, at the last minute, the inauguration ceremony was postponed indefinitely because of the corruption scandal that led to the removal of China's Minister of Railways Liu Zhijun in February 2011. To put an end to rumours about the cancellation of the railway, the Lao government has confirmed that the project was still on the agenda (*Vientiane Times* 2011).

The rapid influx of Chinese investment in Lao PDR would not have been possible without the working of transnational business networks built by the newcomers as well as the go-between role of some ethnic minority groups, belonging to the Chinese cultural sphere, such as the Chin Haw, the Hmong or the Akha.

The Rebirth of Chinese Communities in Lao PDR

Over the last decade, the socio-economic landscape of Northern Lao PDR has been deeply reshaped by the arrival of Chinese migrants in search of business opportunities. In fact, this new migration resumes a rebirth of Chinese communities in Lao PDR, which used to be small compared with other Southeast Asian countries due to a terrain unfavourable to the development of commerce. Laos has the particularity — along with Thailand and Myanmar — of harbouring two types of Chinese communities (Halpern 1961; Rossetti 1997). In the North, the early overland migration was characterized by caravan drivers, dealers, traffickers and a few shopkeepers from neighbouring Yunnan province called Haw or Chin Haw[5] — Yunnan's Chinese. They were followed in 1949 by Kuomintang soldiers fleeing the People's Liberation Army (PLA). In contrast, larger waves of overseas Chinese (*Huaqiao*) from Guangdong, Fujian, and Hainan settled in the main cities of Central and Southern Lao PDR during the colonial period. Originally arriving as coolies to develop the European empires, the overseas Chinese in Lao PDR and throughout Southeast Asia rapidly dominated the main economic sectors due to their networking ability (Purcell 1965; Pan 1999). In fact, those who ended in Lao PDR had failed to prosper in Thailand, Vietnam, and Cambodia.

The 1960s saw the massive arrival of PLA troops who came to set up a vast road construction project in Northern Lao PDR. But the cooling Sino-Lao diplomatic relations at the end of the 1970s resulted in the withdrawal of Chinese military workers and the few remnants of community life left in Luang Prabang, Oudomxay and Xiang Khoang, while in Central and Southern Lao PDR, the political changes of 1975 led to a massive exodus of most of the Chinese community, which dropped from 100,000 people to about 10,000 in the early 1990s.[6]

Present-day Northern Laos harbours a sizable and complex assemblage of Chinese communities with varying tenures of residence. According to Chinese official data, there are 30,000 Chinese residing in Lao PDR but their number could be ten times higher (*The Nation* 2007). The estimation of the Chinese population in Northern Lao PDR is difficult since Lao official data at the national and provincial levels are divergent.[7] Lao provincial officials recognize that the current administrative system cannot precisely assess their presence because they are very mobile. Those who live at the Chinese border can enter Lao PDR with a ten-day border pass.[8] The newcomers often run their businesses under a Lao name or settle illegally, and a very small number legalize their stay by marrying Lao women.[9] Strict control of cross-border migration in Northern Lao PDR would be in no way possible due to porous borders naturally formed by the mountains and the Mekong River, and because of the difficult access to some areas during the rainy season.

While Chinese migrants in the 1990s were mostly males (Rossetti 1997), I observed during my fieldwork that the pattern of migration has increasingly become feminized. Most of the newcomers I met run their business with their families, so they can go back to China regularly without having to stop their activities. The families often seize the opportunity to raise more than one child in Lao PDR. When they have children of school age, they prefer to leave them with the grandparents to receive a Chinese education. I encountered many single or married women who ran their own businesses, leaving their family behind. In some cases, migration can be an avenue of empowerment for women. However, since they stay with their peers, this group has achieved limited integration in Lao society. They often employ Haw or Akha staff as translators, middlemen or business partners to deal with Lao customers and officials. Regardless of how long they have been or plan to be in Lao PDR, they see their tenure as temporary and strictly for the purpose of economic advancement.

Given that this circular and chain migration is still recent, the newcomers have not yet organized themselves into associations, unlike the overseas Chinese during the colonial period.[10] Once the new pioneers have established a foothold, they make contact with the home community to recruit labour and business partners. Therefore, transnational networks are crucial to understanding migrant settlement and community building in Lao PDR.

Table 18.4
Changes in Production of Rice, Maize, Sugar Cane, and Rubber

	1,000 tons				Hectares		
	Rice				Rubber		
	Upland	Lowland	Maize	Sugar Cane	Area Planted in 2007	Area Planted in 2008	Plan for 2010
Luang Namtha							
1995	22.7	20.3	1.3	10.5	8,770	21,700	20,000
2005	12.8	47.6	7.6	54.2			
2009	11.1	44.1	10.0	137.7			
variation 1995/2009 (%)	–51.1	117.2	693.7	1,211.4			
Oudomxay							
1995	42.5	27.8	6.6	3.6	4,500	17,100	21,000
2005	35.1	47.0	70.9	6.5			
2009	19.2	60.2	107.3	3.7			
variation 1995/2009 (%)	–54.8	116.5	1,530.7	3.4			

Table 18.4 (*Cont'd*)

Bokeo							
1995	12.9	24.8	1.3	1.6	700	9,800	25,000
2005	8.6	52.7	30.1	0.0			
2009	19.2	63.3	69.5	0.0			
variation 1995/2009 (%)	48.8	155.2	5,372.4	–100.0			
Lao PDR							
1995	296.1	1,071.3	50.4	62.3	28,205	140,550	249,360
2005	214.8	2,082.1	372.6	196.1			
2009	224.0	2,193.4	690.8	856.9			
variation 1995/2009 (%)	–24.3	–31.8	2,151.7	1,274.8			

Source: Ministry of Agriculture and Forestry, 2009.

The Transnational Networks of the Agri-business Boom

The agri-business boom represents the most important change in Northern Lao PDR's economy. Large-scale rubber plantations and other commercial cash crops, such as lowland rice, maize, and sugarcane are currently replacing upland rice farming (see Table 18.4).

When the borders reopened in the 1990s, the foundations for a new economic cooperation with China were already laid thanks to infrastructures set up by the PLA twenty years ago. First, small waves of Lao Akha and Hmong refugees from the American War were repatriated from the border villages of Xishuangbanna to Mom cluster in Sing district and Ban Hat Ngyao in Namtha district (Luang Namtha province). Under the influence of their Chinese *pi-nong* (relatives) who have experienced better standards of living by planting rubber and because they used to work in the state farms in Xishuangbanna, they were the first to plant rubber in the mid-1990s. Though this community is limited in size, it served as a crucial transnational link in the smallholder rubber value chain, as well as providing examples and an essential source of knowledge for other villages.

In 2003–04, Luang Namtha began receiving an influx of Chinese investment from Yunnan. Chinese rubber companies seized the opportunity of acquiring cheap land with a low-risk investment. They started to register formally because they were favoured by the opium replacement policy that eases labour restrictions, provides financial incentives as well as a special fund of 250 million Yuan to assist businessmen through grants and loans. This strategy also reflects the broader Chinese policy direction that aims to encourage Chinese investment abroad named "*zou chu qu*" (literally translated as "go out" or "going global" policy). Though companies are typically headquartered in Mengla, Jinghong, or Kunming, investment can come from as far away as coastal China. These companies have strong governmental ties and usually subcontract with the existing local Chinese communities (mostly Sichuan and Hunan migrants engaged in miscellaneous trade and who have been converting to rubber in recent years) and employ Lao Akha or Lue personnel, as translators and supervisors.

Along with the large companies, a myriad of smallholders — Chinese entrepreneurs from neighbouring towns in Xishuangbanna as well as Lao villagers benefiting from the experience of their transnational networks (*pi-nong*) across the border — are now engaged in the "rubber boom". According to Weiyi Shi (2008), in 2006, the villagers of Luang Namtha province alone planted 11,000 ha out of the 12,600 ha, the small remaining area being planted by the companies through contract farming or concessions.

These figures are likely to be underestimated since the villagers may under-report their plantation areas in fear of taxation.[11]

However, over the last couple of years, social and environmental concerns have grown on large-scale rubber concessions (farmers losing access to their land, monoculture, forest loss, etc.). The "2+3" model of contract farming has been heavily fostered in Lao PDR, as it is seen as the best way to promote rural development. Under this system, the company provides capital, technology and a market for the final product, and farmers provide land and labour. Profits are split, generally with 70 per cent to go to the farmer and 30 per cent to the investor. However, it often slips into a "1+4" system. Since the farmers cannot afford the prolonged uncompensated labour input (at least seven years), they contribute only land, in exchange for 30 per cent of the future plantation and current wages if they work for the company. Ultimately, the "1+4" system is closer to the concession model. Although the northern provinces favour contract farming,[12] concession-style plantations still exist, such as the most controversial 300,000-hectare plantation operated by *China-Lao Ruifeng Rubber* in joint venture with the Lao army.[13] As a result of the shortage of Lao labour, it is likely that the labour force will be imported from China. The labour force needed for a rubber plantation is estimated at one to one and half workers per hectare. A plantation of 30,000 hectares will require at least 30,000 to 45,000 workers, while one only finds in the area an average of five villages inhabited by 200 to 300 people.[14] During the preparation of the 7th National Socio-Economic Development Plan (2011–15), the government acknowledged a shortage of 300,000 people, mostly in agriculture and skilled trades (MPI 2010).

The other main trend of the agri-business boom is the commercialization of cash crops such as maize, sugar cane or cassava. I will focus on sugar cane in Muang Sing, which was one of the main precursor districts. The cultivation of sugar cane in Muang Sing began in the mid-1990s for local consumption and the surplus was exported to the Chinese border district of Mang. The first villagers who planted sugar cane were Akha. At that time, they had to find themselves markets in China. Some growers frequently could not earn income or even lost money because they could not find any markets and they had to deal with bad transport conditions (small trucks and dirt roads becoming impassable during the rainy season).

Then Chinese traders realized the potential of sugar cane in Lao PDR. Representatives of *Phong Sugar Factory*, a Chinese company based near the border, approached Lao farmers to discuss the possibility of growing sugar cane commercially. Credits with interest were given to the farmers, and in 1997 Lao authorities intervened to encourage the farmers to grow sugar cane, through an agreement with Chinese traders. In spite of inappropriate planting techniques and no pick-up system because of the bad state of the roads, Lao farmers kept planting sugar cane with a 50/50 split with Chinese companies. At the time of NERI's survey (Khontaphane 2006), Chinese middlemen working for the sugar factories in Yunnan as well as Lao registered companies controlled the sugar cane exports. They provided all the services, from supplying the seedlings and the fertilizers to Lao farmers, to the transportation, as well as customs duties services. Lao-registered companies were mainly used for help with the documentation for exporting through Panthong checkpoint, not far from Muang Sing. The trucks were weighed in China. Lao farmers were often suspicious about their final payment because they felt cheated by the middlemen, seldom receiving money in time or not at all. In 2001, to solve the problem, a weighing station was built by an agent of the *Phong Sugar Factory* and by local officials. Then in 2003, an official contract was signed between the sugar companies and Lao authorities to give greater transparency.

In June 2009, during the last field trip to Muang Sing, it appeared from the interview with the DAFO (District Agriculture and Forestry Office) that there were no Lao middlemen anymore. Four representatives of the sugar companies are permanently based in Muang Sing to deal with the farmers. Eventually, development of sugar cane has enabled some Akha villages to improve their livelihoods. According to Lyttleton et al. (2004), and confirmed by NERI's survey, successful growers can make upwards of 20 million kip/year (US$2,000). When I visited the Akha village surveyed by NERI in 2006, the selling of sugar cane had enabled children to be sent to study in Vietnam and China, and to plant rubber without contracting with Chinese companies. The case of sugar cane in Muang Sing demonstrates how Chinese networks are entering a new phase of consolidation after many years of pioneering.

Long-Distance Trade: From Lao Monopoly to the Hegemony of Chinese Transport Networks

Besides Chinese from neighbouring Yunnan coming to Northern Lao PDR to work in construction, industry and agriculture, what is striking is that the movement of the last ten years characterizes mostly Han migrants from Sichuan, and especially Hunan, in search of better livelihoods. They typically engage in miscellaneous trades in household supplies, hardware, motorcycle repair, cellphone shops, restaurants, hotels, and they occupy more and more room in the local markets. They have settled throughout Northern Lao PDR, in all provincial and district capitals, to supply cheap Chinese goods to the urban and rural populations of the region. They are currently on the verge of supplanting Lao traders and transporters, and especially Lao women.

In his outstanding description of North-Western Lao PDR during the emergence of the Economic Quadrangle era in the 1990s, Andrew Walker (1999) demonstrates how Lao women gained a foothold in the long-distance trade. The situation was very unusual in Southeast Asia since long-distance trade is typically the domain of men, and especially Chinese men, because of their extensive network of contacts in provincial towns and capital cities (Halpern 1961, pp. 5–7). When the borders reopened, the *mae kha* (Lao women traders) were naturally in the foreground to seize the new trade opportunities. They bought goods in Chiang Khong, the most popular supply centre for long-distance trade throughout Northern Lao PDR, and headed down river from the Huayxai cargo port to Pakbeng, where trucks were waiting to load the goods to continue their journey to Oudomxay, one of the main selling places. Some of these women continued their trip to Mengla, the first town across the Chinese border. However, the trade was on a smaller scale than in Chiang Khong because the assistance in negotiating with the Chinese customs was much more valued by Lue intermediaries, who share linguistic and cultural similarities with the Lao.

During my fieldwork between 2007 and 2009, I found a completely different situation. I saw mostly Chinese and Thai trucks transporting goods along the main roads of Northern Lao PDR. What Andrew Walker predicted in the conclusion of his book came true: with pressures to lift the restrictions on the operation of foreign trucks and with improved road connections, Thai and Chinese transport firms would be in a position to provide a more rapid and streamlined service than Lao operators could offer with their outdated vehicles. Moreover, R3E is also a threat to boat operators since it captures most of the river traffic that travels to and from the northern towns via the ports at Pakbeng and Xiengkok. Improvements in transport infrastructures and trading conditions have encouraged the emergence of powerful trading-transport alliances, which exert great influence over market prices and commodity volumes. By transporting large volumes of goods to the major townships of Northern Lao PDR, Thai and Chinese companies have achieved economies of scale that threaten the viability of independent Lao traders and transporters. In brief, local populations "*play no entrepreneurial roles in an economy where the exchange system is controlled by external metropolises*" (Walker 1999, p. 190).

With strong connections to China and Yunnan in particular, and easy access to capital and credit, Chinese traders are better positioned to engage in importing cheap Chinese goods, for which there is a strong local demand (tractors, electrical goods, food, clothes, etc.). The recent trade liberalization policies allowing virtually any individual to conduct import and export activities have resulted in more small and medium sized traders and a consequent decline in the business of the registered companies. Thus, at every checkpoint, whether local, traditional or international, we can find this "army of ants"[15] vigorously engaging in trade. These groups are neither registered trading companies nor *bona fide* passengers.

In practice, the passenger-traders usually bring in goods at a negotiated reduced duty rather than with complete avoidance. They sometimes purchase NTFPs (Non-Timber Forest Products) like wild mushrooms and bamboo, and some seasonal agricultural produce. These Chinese passenger-traders hire Lao or Chinese small trucks or are transported either by daily Chinese public buses, which come from Jinghong or Meng La and stop in the main cities along the R3E (Huayxai and Luang Namtha), or by private buses coming from Kunming or Meng La and heading to Oudomxay, Luang Prabang and Vientiane along the R13.

To import larger volumes of goods, Chinese traders mainly use Lao registered companies to clear the documents. Frequently, an ethnic Chinese (usually Haw because of common language) serves as a middleman between the Chinese traders in China and Lao PDR, the Lao registered company and the customs. The Chinese traders in Lao PDR purchase products from Chinese traders in China (most of the time, in Kunming or Chongqing) who acquire them in the factories at a very low price. The goods are transported to *Hong Long*'s warehouse[16] established at the China-Lao PDR border. The Haw middleman goes to Mohan to get the list of orders written in Chinese, negotiates with the Chinese clients on the price of the package service (paperwork, customs duties) as well as with the customs — they agree on less taxes in exchange for "tea money". Then the Haw middleman translates the papers for the Lao import/export registered company, which provides the necessary documents.[17]

Regarding long-distance trade, it appears that there are very few Lao registered companies. Most of them are both import/export and carrier operators. In Luang Namtha, three registered companies cover 50 per cent of the market, companies settled in other provinces share 20–30 per cent, and small-scale operators take the remaining business. *Lao-Singapore Import-Export Company* is one of the major registered companies in Luang Namtha. In 1997, the joint venture company first settled in Vientiane. It opened an office branch in Luang Namtha in 1999, and another one in Nateuil at the border in 2000. Yet, in Luang Namtha province, the company only owns eight trucks. Thus, it is regularly obliged to rent trucks from other carriers when demand is higher, which has the effect of raising prices because it can take up to two weeks to wait to fill the truck to come back. Therefore, the cost to Vientiane for a 15 ton-truck climbs from 7 million kip (about US$800) to 9 million kip (US$1,050). The joint venture mainly imports consumer goods, agricultural and electrical equipment, and exports wood, as well as agricultural products (corn from Oudomxay) and NTFP (from Luang Prabang and Xayabury) that it buys from Lao farmers. However, in 80 per cent of cases, the company simply provides the necessary documentation for export to Chinese middlemen.

The other major import/export and transport operator is *Bolisat Khongaloun*, a Lao registered company. The owner is an ethnic Chinese who can speak Chinese. Although the company was registered in October 2008, it has been operating for six–seven years by using the quota licence of other companies. Originally, the implementation of quota policy aims to control imports. In theory, traders need a quota licence issued by the province to import goods. Nevertheless, given strict administrative constraints, Lao registered companies are compelled to buy unused quotas from other companies, while small traders operating informally do the same with companies officially registered. Although the transfer of quota is not allowed, this practice is common and tolerated by the Lao administration. In 2008, *Bolisat Khongaloun* registered 23 trucks. In May 2009, the Lao company did 122 truck-trips, mainly from the Chinese border to the capital Vientiane, and it started to engage in transit trade operations along the R3 (67 truck-trips from Huayxai to Boten/Mohan and 55 truck-trips from Mohan to Huayxai). Ninety per cent of its customers are Chinese. The company is currently in the process of setting up a joint venture with Chinese investors.

The third major import/export operator in Luang Namtha is a state-owned company that is not engaged in transport, which considerably limits its activity. The director acknowledges that, even if they had trucks, they could neither compete with large Chinese and Thai companies, nor with small Chinese traders who operate informally. The state-owned company mainly imports building materials, electrical equipment and machines, and exports rice and timber. Although the competitiveness of the company is relatively weak, it is in charge of an important regulatory function. When the state fails to meet its objectives in terms of revenue, state-owned enterprises as well as military-run companies are allowed to export raw wood, a right exclusively reserved by the central government since 2007.

The absence of large Lao transport companies can be explained by the fact that this sector needs a lot of capital for a very low profit so far, because the cost of transportation is too high. First, the traffic flows are too small and unbalanced; too many trucks come back empty from Vientiane to the North. Second, the roads are in poor condition, and there is no insurance system in case of accident. Third, the trucks are not equipped with international standard facilities. Lao companies still need to invest in cooling systems for example to transport fruit and vegetables. According to Lao registered companies interviewed in Oudomxay, they cannot compete with Chinese traders who operate informally. They buy trucks under a Lao name in Vientiane and transport most of the imported goods intended for the new Chinese market in Vientiane, without being registered as a company. If the trucks are overweight, the Chinese traders pay the police to let them circulate. Regarding the transit trade, profits are too low, which is why Lao registered companies only clear the documents for Thai and Chinese companies, which transport goods with their own trucks. As a result, the rapid improvement of Chinese transport networks contributes to increasing the inequality of resources between Chinese and Lao traders, whether small entrepreneurs or large companies.

Markets, Guesthouses and Restaurants: The Trader's Key Stops

Settled both in rural and urban areas throughout Northern Lao PDR, Chinese traders are currently on the verge of dominating the local economy by weaving networks between important trade nodes, such as markets, restaurants and guesthouses. Commercial connections appear to have been growing with China and at times replacing previous Thai connections. To control the massive arrival of Chinese traders, the Lao provincial authorities have decided to put them in the same place.[18]

In Huayxai, the first Chinese traders arrived twenty years ago and settled informally on the main road along the Mekong River. Shortly after, they were resettled on the outskirts of the city, near the former bus station, but the place was still not large enough to gather all the Chinese traders. In 2005, a wealthy Chinese entrepreneur from Hunan, who made his fortune in the wood business,[19] invested US$500,000 in the construction of the new market, located 4 km away from the centre, in the direction of the future bridge. In 2008–09, 40 shops and 61 traders were registered at the Department of Labour. The Head of the Department of Industry and Commerce in Bokeo acknowledged that this figure is probably less than the reality:

> We regularly visit the Chinese market to check the registration but it is hard to talk with them. They do what they want, even if we come with the police. Just over the last 6 months, 27 shops have opened. They ask their relatives to come and set up a new shop. They should gather under an association, it would be easier for us.[20]

Huayxai is the only city in Northern Lao PDR where the Chinese market is not in the same location as the Lao market because the province is still under Thai influence. No Chinese products are found at the Lao market, and the wholesalers are mostly Thai and Lao. Because of the proximity with Chiang Khong, Lao traders can buy Thai goods cheaper as well as Chinese goods of better quality coming from Tachilek at the Thai-Burmese border. However, some Thai customers cross the Mekong River to buy electronic items at the Chinese market in Huayxai.

In 2005–06, Haw investors built new markets in Oudomxay, Muang Sing and Luang Namtha.[21] They gather Lao and Chinese traders, each occupying a specific area in the market building. The goods sold are quite similar — clothes, cosmetics, daily consumer goods, electrical appliances, tools and some foodstuffs, but the Chinese are more specialized in electronic items and other high-price items (cell phones, TVs, radios, etc.). Each group of traders often has its own wholesale network. Lao traders deal mostly with Lao and Thai wholesalers, selling mainly Thai products, and some Vietnamese products (clothes, candies, beauty products) coming from Vientiane or Huayxai. Although Huayxai is closer, it is cheaper and faster to order from Vientiane.[22]

The Haw are dominant among the Lao merchants in these three cities and play a crucial role of go-between with the Chinese. Haw traders also sell Chinese products that they either get in Mohan, because they are cheaper and there is more choice, or they buy from Chinese wholesalers, who often have their shops in the market and purchase large volumes of goods for re-sale to local merchants. Usually, Chinese traders have strong connections with wholesalers in La or in other districts of Yunnan. Although a large majority of the Chinese traders in Northern Lao PDR are from Hunan, they often used to have the same activity in Yunnan or they still have relatives who operate over there.[23]

In Oudomxay, Muang Sing and Luang Namtha, most of the sellers at the fresh food market are Lao, mainly Haw, with a few Chinese sellers from border districts. Nevertheless, Chinese traders rule the food market and have strong connections with the Chinese restaurants in town because most of the vegetables are locally grown and imported by them. Lao farmers do not produce vegetables and fruit anymore because Chinese products are cheaper. The incentives for using local checkpoints (Panthong, Meochai) are high since this avoids the cumbersome official procedures at international checkpoints (Nateuil/Boten).

Numerous Chinese-owned shops (selling cell phones, large electrical items, car and tractor engines, televisions and home furniture, or barbershops, beauty salons, etc.) are usually settled around the market or along the main streets, creating a kind of new Chinese neighbourhood. After 6 pm, when Lao shops close, Chinese shops remain opened, and the Chinese language is commonly heard in the streets. These shops are easily recognizable thanks to their red and yellow Chinese characters and lanterns posted on shop fronts. Most of these small and medium-sized traders are registered under Lao names. Only the large companies are registered under Chinese names. In fact, formal and informal sectors are intertwined. For example, in Luang Namtha, the beauty shop is officially registered but the owner is also engaged in informal activities such as importing agricultural products, while the main business of the motorbike dealer is a banana plantation.

Besides these shops, a growing number of hotels/guesthouses and restaurants have been set up, becoming key stops for Chinese traders, workers, and tourists. Actually, they represent crucial nodes of the regional business networks because they play multiple roles, acting as information centres, gambling houses (to play Mah-jong parties), bus ticket offices, small warehouses, intermediaries between traders and private buses, meeting rooms for Chinese companies, etc. In fact, restaurants and guesthouses have been set up wherever Chinese buses stop over. They are often engaged in many activities such as running a shop in another city, even as far away as Vientiane, a transportation business, clinics or rubber production. Due to capital constraints, they tend to have only smallholdings, but many now also serve as subcontractors and supervisors for large Chinese investors.

Even though these restaurants/guesthouses mainly receive Chinese customers, more and more Lao people and foreign tourists have started stopping there. Since most of the Chinese traders cannot speak Lao, they hire Haw personnel to overcome the language barrier. As a result, Chinese markets, shops, restaurants and guesthouses have become the most visible markers of urbanization in producing a new form of urban centrality that Lao cities were previously lacking.

The "Casino Strategy" as the New Pillar of Development

In the perspective of welcoming more tourists in Northern Lao PDR, the Lao government has encouraged Chinese investors to build luxury hotels and casinos in the Special Economic Zones (SEZ) implemented at the Chinese and Thai borders. The idea is to attract foreign tourists, specifically Chinese and Thai, since gambling is strictly forbidden in their countries, and stimulate a development dynamic by providing tax advantages.[24]

Long-term land concessions have been granted to Chinese investors to build "Golden Boten City" in Luang Namtha province and "the SEZ of the Golden Triangle" in Bokeo, located at both ends of the R3E. They should have become the most internationally modernized cities in Lao PDR. In fact, they have turned into Chinese enclaves in Lao territory, which benefit from a regime of extra-territoriality, ironically recalling the concessions imposed on China in the mid-nineteenth century by the colonial powers (Nyíri 2009).

The objective of the first phase of Golden Boten City, worth US$103 million, was to build a large tourist complex, a shopping centre and a regional hub for transit goods. In 2002, high-ranking Lao officials granted *Fu-Khing Industries Hong Kong Corporation* a thirty-year concession, renewable twice. The concession of 1,650 hectares was officially inaugurated in November 2004. In fact, in late 2006, instead of luxury resorts, it was a public housing-like casino that had sprung up in the middle of the rainforest, surrounded by shops and warehouses, mostly unoccupied. A few thousand Chinese traders, dealers and prostitutes have settled in this "casino-city", against a handful of Lao. In 2009–10, construction accelerated. Brand new buildings and hotels appeared, while other projects were under construction. The currency used is Chinese Yuan, all the products consumed locally are imported from China, everything is written in Chinese, and even the time zone is set according to China. Pornographic items, still banned in Lao PDR, can be found in Boten, as well as cabaret shows performed by "*Kathoey*" (male-to-female transgender persons) from Thailand. Most of the gamblers are Chinese and also professionals. They are equipped with headsets for receiving instructions from their bosses based in China, who can follow the game on the Internet.

In fact, the main investor in Golden Boten City is Lin Mingxian, one of the most famous drug lords of the Golden Triangle. Better known as U Sai Lin, he still commands a 2,000–3,000-strong militia known as the National Democratic Alliance Army (NDAA) in the Shan State Special Region No. 4. Born in Yunnan, Lin Mingxian used to be a member of the Red

Guards and was sent during the Cultural Revolution to fight alongside the Burmese Communist Party. In the 1990s, he began investing in Mongla, his fief at the Sino-Burmese border, in the construction of gambling houses with drug money. In January 2005, after the closure of his casinos by Chinese authorities who wanted to stop the haemorrhage of capital diverted by corrupted officials, he came to invest in Boten. However, in January 2010, Chinese journalists revealed the Mafioso atmosphere in Boten. They reported how the gamblers were trapped, tortured, and sometimes murdered if their families did not pay their debts. Under the pressure of Peking, the Lao government decided to close the casino in April 2011. Since then, traders and prostitutes have been deserting Golden Boten City to join the new casino located in Tonpheung district.

While the project began in 2008, the "Special economic zone of the Golden Triangle" was inaugurated just a year later, on 9 September 2009. The Lao government, which owns 20 per cent of the project, hopes to attract one million tourists as well as 300 companies. Registered in Hong Kong, *Dokngiewkham Company* was granted a ninety-nine-year concession covering an area of 3,000 ha. The first phase of the project cost US$86 million. The company plans to invest 2,240 million dollars between 2010 and 2017 to develop infrastructures and create a truly modern city. The city renamed Kapok should accommodate more than 200,000 Chinese workers and their families (*RFA* 2011). It seems that the Lao government has learned the lessons of Golden Boten City, whose development was chaotic and unsatisfactory. By owning shares in the new project, the Lao government maintains better control of operations and directly recovers profits if the operation is successful. According to Thai journalists, Lin Mingxian is also among the investors in Tonpheung (Lintner and Black 2009).

These two megaprojects illustrate the overlap between legal activities (development of tourism, Special Economic Zones) and illegal practices (gambling, prostitution, drug money laundering, etc.), on which the Lao state based its poverty alleviation policy. While newspapers in the region begin to describe the SEZ of the Golden Triangle as a gigantic machine for laundering drug money, the director of *Dokngiewkham Company* Zhao Wei rejects these accusations. As far as Lao authorities are concerned, they ensure that they control the situation and claim that they have no information about Zhao Wei.

Nevertheless, the manhunt against the Shan rebel Naw Kham reminds us that the borderlands of the Golden Triangle still escape from the control of Lao and Chinese centres. Naw Kham is considered by local people as a kind of "Robin Hood of modern times" in attacking rich Chinese positions in the Golden Triangle. The casino is suspected of being a transit point for drug smuggling into Thailand. Naw Kham's militia has been "taxing" both Chinese cargo ships along the Mekong River and the trafficking in heroin and methamphetamine run by the UWSA (United Wa State Army) crossing its territory. Beijing has demanded his head after Naw Kham seriously wounded Chinese policemen in a shooting involving the Chinese and Burmese armies. In April 2011, he kidnapped thirteen Chinese in the Golden Triangle SEZ (*Democracy for Burma* 2011). They were released after the director paid a ransom of 25 million baht (US$8.3 million). In October 2011, tensions reached a climax after thirteen sailors were killed on two cargo ships attacked in the Golden Triangle. Since then, China has decided to suspend shipping on the Mekong River (*Channel News Asia* 2011).

The Lao government still believes that setting up new economic zones in remote parts of the country will help to boost investment in these areas. However, because of the current tensions in Boten and in the Golden Triangle, the Lao government has decided not to authorize further casinos in the upcoming SEZ (KPL 2010).

CONCLUSION

A couple of years after the opening of the R3E, it seems that the traffic along the Lao branch of the NSEC is still relatively low because of the missing bridge in Huayxai and customs procedures that create trade bottlenecks. Moreover, the recent troubles in Boten and the Golden Triangle, as well as the uncertainty of Chinese investment in the high-speed railways seriously undermine the connectivity and the functioning of the NSEC.

Nevertheless, Chinese investment and traders have been profoundly transforming the socio-economic landscape of Northern Lao PDR, both in urban and rural areas. So far, Chinese entrepreneurs are the main winners of the regional integration process since they manage to create business opportunities along the corridor, thanks to the complementary of their transnational networks.

One of the most striking effects of Chinese networks is the creation of a new urban hierarchy organization among the provincial cities of Northern Lao PDR. Although Huayxai is one of the main nodes of the NSEC, its urban development will strongly depend on the economic dynamic engaged after the construction of the new bridge. Otherwise, the provincial capital city of Bokeo may be relegated as a secondary city. Luang Namtha, at the opposite end of the R3E, will be likely to take advantage of the increasing cross-border trade with China. Interestingly, Oudomxay, which is not located along the R3E has become one of the most dynamic urban centres in Northern Lao PDR, mainly due to the business interactions between ethnic Chinese (Chin Haw) and mainland Chinese newcomers. The cases of Muang Sing and Viengpoukha, the former located at the Sino-Lao border and the latter along the R3E, show that the implementation of economic corridors requires more than just improved infrastructures to be beneficial to the local economy. Regarding the problems occurring in Boten and Tonpheung, one may question the sustainability of the "Casino strategy" in the Special Economic Zones.

The preliminary assessment shows that the NSEC will benefit the stronger economies such as China and Thailand, while Lao PDR could pay an expensive social and environmental cost. The dramatic transition driven by national policies and market forces accentuates the challenge that the Lao central and provincial authorities have to take up to mitigate the negative impacts of economic liberalization. This chapter has aimed to highlight the complex relationships between trade and human development, showing that though international trade is an important factor to increase economic growth and remains necessary for overall human development, the linkages between the two are neither automatic, nor guaranteed.

Notes

1. For further details, see Masviriyakul (2004).
2. *Suwannaphum* is the name of a country mentioned in ancient texts of the Theravada Buddhist tradition. Its location is still controversial among historians. Several countries in Southeast Asia claim this legacy, including Thailand, which recently named its new international airport by that name.
3. Lao PDR has committed to implementing tariff reductions under AFTA (ASEAN Free Trade Area) by 2008. The country is also at the final stage of the World Trade Organization (WTO) accession process. The ASEAN-China Free Trade Agreement (ACFTA) will come into effect for Lao PDR in 2015.
4. These figures should be treated with caution because several megaprojects such as the "Golden Boten City" or the dams are not registered at the provincial level.
5. We need to go back to the Thai chronicles of the thirteenth century to trace the origins of the name Chin Haw, which designated in Tai languages

the overland Chinese coming to Thailand (Panthay in Burmese). This name was widespread in the sixteenth century to describe the Chinese Muslim caravans plying the mountains of the Upper Mekong. For the Chinese, the Haw are known by the term *Hui* (Chinese Muslims). They were long thought of as hill tribes or Muslim, but since the twentieth century, this term became ambiguous because of the immigration of Yunnan's Han Chinese into the Golden Triangle, after the defeat of the KMT. Anyway, these overland Chinese reject this name (Hill 1998; Forbes and Henley 1997). The Lao also give this name to the Han Yunnanese who had settled for five or six generations in the North, especially in Phongsaly. They consider themselves as ethnic Chinese and Lao.

6 The Association of overseas Chinese in Vientiane proposed the figure of 100,000, while most official sources had recorded 40,000 people (Rossetti 1997, p. 1).

7 According to a GMS Migration Policy Briefing, there are 80,000 Chinese estimated to be working for Chinese led projects (ADB 2010).

8 A border pass, provided at the international and local checkpoints, allows all Chinese citizens to travel in four northern provinces (Oudomxay, Luang Namtha, Bokeo and Phongsaly) for a period of ten days, extendable for another ten days. It is not necessary to be resident in Yunnan to qualify for this border pass. Each crossing only costs 10,000 kip (just over a dollar). Chinese citizens can also extend their stay with a three-month business visa for RMB 160 (US$25), while the resident card costs RMB 2,000 (US$300) per year. Moreover, tourist visa can be delivered at the international checkpoint of Boten, where the procedure is rapid. The visa costs US$20, and it is valid for thirty days.

9 Interviews with the Heads of Department of Industry and Commerce in Luang Namtha, Bokeo, and Oudomxay provinces, 2008–09. In Oudomxay province, 24 marriages between Chinese men and Lao women were registered in 2005, 15 in 2006, 28 in 2007, and 19 in 2008 (Immigration Department of Oudomxay province 2008).

10 The Chinese in French Indochina used to organize themselves into associations called *bang* in Chinese and *congrégations* in French, in accordance with their dialect and/or province of origin. They aimed to self-govern the Chinese communities. They provided mutual aid assistance, mediated internal disputes, and assisted the government in implementing laws and collecting taxes.

11 Interview with GTZ, September 2008.

12 In October 2005, Luang Namtha, Bokeo and Oudomxay provinces formed an official agreement not to issue land concessions for rubber development but rather to promote "2+3" contract farming (Shi 2008).

13 The concession is not registered in these terms on the official list of FDI in the Department of Planning and Investment of Luang Namtha province.

14 Interview with GTZ, September 2008.

15 Interview with the Head of Department of Planning at the Ministry of Agriculture and Forestry, Vientiane, July 2008.

16 Hong Long is a Chinese company registered in 2002 with US$900,000 of capital.

17 Interview with the head of a state-owned import/export company in Luang Namtha, June 2009.

18 Interviews with the ethnic Chinese owner of the market in Oudomxay, with the Head of Department of Industry and Commerce in Bokeo, and with Chinese traders in Luang Namtha, September 2008 and June 2009.

19 Interview with his relative, Huayxai, July 2008.

20 Interview, Huayxai, June 2009.

21 Haw entrepreneurs invested US$746,873 in the market in Muang Sing, US$816,575 in Luang Namtha, and US$1,166,789 in Oudomxay, data from the Department of Planning and Investment of the three provinces, 2008–09.

22 It takes three days from Vientiane and the transport costs 20 baht/package, but the traders have to pay cash, while from Bokeo, they can pay later but it is more expensive, around 25 baht/package, and it takes longer to arrive, interviews with Lao traders, 2008–09.

23 To have an exhaustive description of the different waves of Hunanese migration to Yunnan since the 1950s, see Hansen (1999).

24 This development strategy is not specific to the North but is also found at the national level. The first casino, *Dansavanh Casino Resort* opened in 1999 near Vientiane. *Syuen Group*, a Chinese Malaysian company, invested US$200 million and manages the casino with the Lao army, which owns 25 per cent of the joint venture. The company completed in 2008 the construction of *Dansavanh Hotel*, a hotel/casino in Oudomxay. In April 2009, *Sanum Investment Ltd.*, a Chinese company based in Macao, opened Savan Vegas, a US$50 million-casino in the Savan-Seno SEZ in Savannakhet, located on the East-West Economic Corridor.

References

Asian Development Bank. "Build It and They Will Come. Lessons from the Northern Economic Corridor: Mitigating HIV and Others Diseases". Manila: ADB, 2009.

———. "Toward Sustainable and Balanced Development: Strategy and Action Plan for the Greater Mekong Subregion North-South Economic Corridor". Manila: ADB, 2010.

———. "GMS Migration Policy Briefing". Extracts from an ADB Project entitled "Forecasting Migration Flows: The Relationships among Economic Development, Demographic Change and Migration in the Greater Mekong Subregion". Manila: ADB, April 2010.

"Shaking Up the Neighborhood". *Bangkok Post*, 4 May 2008.

Banomyong, Ruth. "Logistics Development Study of the North-South Economic Corridor". Manila: ADB, 2007.

———. "Benchmarking Economic Corridors Logistics Performance: A GMS Border Crossing Observation". *World Customs Journal* 4, no. 1 (March 2010): 29–38.

"China Halts Mekong Shipping after Deadly Attack". *Channel News Asia*, 11 October 2011.

Cohen, Paul T. "The Post-Opium Scenario and Rubber in Northern Laos: Alternative Western and Chinese Models of Development". *International Journal of Drug Policy* 20, no. 5 (September 2009): 424–30.

Dai, Xiaoxiang and Karl Goldnick. "Survey of Economic Activities of Chinese Migrants in the Province of Luang Namtha". Luang Namtha: GTZ-CIM, January 2007.

"Golden Triangle Godfather Naw Kham Releases Chinese Abductees". *Democracy for Burma*, 11 April 2011.

Diana, Antonella. "Roses & Rifles: Experiments of Governing on the China-Laos Frontier". Ph.D. dissertation, Department of Anthropology, Australian National University, 2009.

Diokno, Maria S. and Van Chin Nguyen, eds. *The Mekong Arranged and Rearranged.* Chiang Mai: Mekong Press, 2006.

Dywer, Mike. "The Internal Frontier: Chinese Extraterritoriality in Northern Laos?" *New Mandala*, 11 February 2011. Available at <http://asiapacific.anu.edu.au/newmandala/2011/02/11/the-internal-frontier-chinese-extraterritoriality-in-northern-laos/> (accessed February 2011).

Evans, Grant, Christopher Hutton, and Kuah Khun Eng, eds. *Where China Meets Southeast Asia: Social and Cultural Changes in the Border Regions.* Bangkok: White Lotus, 2000.

Forbes, Andrew and David Henley. *The Haw: Traders of the Golden Triangle.* New Zealand: New Zealandfilm House, 1997.

Fujita, Yayoi and Sithong Thongmanivong. "Recent Land Use and Livelihood Transitions in Northern Laos". *Mountain Research and Development* 26, no. 3 (2006): 237–44.

Gluckman, Ron. "Golden Boten City. Bungle in the Jungle". *Forbes Asia*, August 2011.

GMS Studies Center. "A Study on China-Laos Economic Relations". Final Report. Kunming: Yunnan University, July 2009.

Gray, Dennis D. "China Exerting More Influence over Laos". *Associated Press*, 3 April 2007.

Halpern, Joel M. "The Role of the Chinese in Lao Society". Santa Monica: The RAND Corporation, March 1961.

Hansen, Mette Halskov. "The Call for Mao or Money? Han Chinese Settlers on China's South-Western Border". *The China Quarterly*, no. 158 (1999): 394–413.

Hill, Ann Maxwell. *Merchants and Migrants: Ethnicity and Trade among the Yunnanese Chinese in Southeast Asia*. New Haven, Connecticut: Yale Southeast Asia Studies, 1998.

Khontaphane, Sirivanh, Sathabandith Insisiangmay and Vanthana Nolinth. "Impact of Border Trade at the Lao-Chinese Border in Local Livelihood: Lao-Chinese Border Trade in Luang Namtha and Oudomxay Provinces". Technical Background Paper for the Third National Human Development Report. Lao PDR, Vientiane: UNDP, 2006.

"No Casino in New Special Economic Zones". *KPL Lao News Agency*, 18 October 2010.

Lazarus, Kate et al. *An Uncertain Future: Biodiversity and Livelihoods Along the Mekong River in Northern Lao PDR*. Bangkok, Thailand and Gland, Switzerland: IUCN, 2006.

Lintner, Bertil and Michael Black. *Merchants of Madness: The Methamphetamine Explosion in the Golden Triangle*. Bangkok: Silkworm Books, 2009.

Lyttleton, Chris et al. *Watermelons, Bars and Trucks: Dangerous Intersections in Northwest Laos*. Vientiane: Rockfeller Foundation, Lao Institute for Cultural Research, 2004.

Masviriyakul, Siriluk. "Sino-Thai Strategic Economic Development in the GMS (1992–2003)". *Contemporary Southeast Asia* 26, no. 4 (2004): 302–19.

Ministry of Planning and Investment (Lao PDR). "7th National Socio-Economic Development Plan (2011–2015)". Draft. Vientiane: MPI, May 2010.

Nivelle, Pascale. "Laos Vegas". *Libération*, 17 April 2007.

Northern Laos Industrial Economic Development and Cooperation Planning Preparation Group. "Planning for Industrial Economic Development and Cooperation in Northern Part of Lao PDR". Mid-term Draft. Vientiane: NERI, May 2008.

Nyíri, Pál. "Extraterritoriality. Foreign Concessions: The Past and Future of Form of Shared Sovereignty". *EspaceTemps.net*. Mensuelles, 23 November 2009. Available at <http://espacestemps.net/document7952.html> (accessed January 2010).

Osborne, Milton. *River Road to China: The Mekong River Expedition, 1866–1873*. New York: Livewright, 1975.

Pan, Lynn, ed. *The Encyclopedia of the Chinese Overseas*. Cambridge, Massachusetts: Harvard University Press, 1999.

Purcell, Victor. *The Chinese in Southeast Asia*. London: Oxford University Press, 1965.

"Casino Operations are Bolstered in an Area Reserved for Chinese investments". *RFA*, 12 January 2011.

Rossetti, Florence. "The Chinese in Laos: Rebirth of the Laotian Chinese Community as Peace Returns to Indochina". *Chinese Perspectives*, no. 13 (1997): 26–39.

Rutherford, Jeff et al. "Rethinking Investments in Natural Resources: China's Emerging Role in the Mekong". Cambodia, Denmark: Heinrich Böll Stiftung, WWF, International Institute for Sustainable Development, 2008.

Shi, Weiyi. "Rubber Boom in Luang Namtha. A Transnational Perspective". Vientiane: GTZ, 2008.

Stuart-Fox, Martin. "The Chinese Connection". *Southeast Asian Affairs* (2009): 141–69.

Summers, Tim. "China and the Mekong Region". *China Perspectives* 3, no. 75 (2008): 68–77.

Taillard, Christian. "Corridor Linkages in the Greater Mekong Subregion: New Implications for Peninsular and Regional Powers". In *New Dynamics between China and Japan in Asia: How to Build the Future from the Past?*, edited by G. Faure. Singapore: World Scientific Publishing Co. Pte. Ltd., 2010.

Thiesmeyer, Lynn. "Informal and Illegal Mouvement in the Upper Greater Mekong Subregion. Costs and Benefits of Informal Networks for Goods and People". Occasional Paper no. 3. Observatory Series. Bangkok: IRASEC, October 2010.

Transnational Institute (TNI). "Alternative Development or Business as Usual? China's Opium Substitution Policy in Burma and Laos". *Drug Policy Briefing*, no. 33 (November 2010).

Tsuneishi, Takao. "Development of Border Economic Zones in Thailand: Expansion of Border Trade and Formation of Border Economic Zones". IDE Discussion Paper. Japan: IDE-JETRO, May 2008.

———. "Border Trade and Economic Zones on the North-South Economic Corridor: Focusing on the Connecting Points between the Four Countries". IDE Discussion Paper. Japan: IDE-JETRO, July 2009.

"Laos-China Railway on Track, Despite Rumours". *Vientiane Times*, 20 May 2011.

Walker, Andrew. *The Legend of the Golden Boat. Regulation, Trade and Traders in the Borderlands of Laos, Thailand, China and Burma*. Richmond: Curzon Press, 1999.

———. "Conclusion: Are the Mekong Frontiers Sites of Exceptions ?" In *On the Borders of State Power. Frontiers in the GMS*, edited by M. Gainsborough. New York: Routledge Studies in the Modern History of Asia, 2009.

Walsh, John. "The Rising Importance of Chinese Labour in the Greater Mekong Sub-Region". *The Asia-Pacific Journal* 12, no. 2 (March 2009).

World Bank. "Lao People's Democratic Republic: Policy, Market, and Agriculture Transition in the Northern Uplands". Washington, D.C.: The World Bank, 2008.

"Trade Bottlenecks Hinder Development in Mekong Region". *Xinhua*, 8 June 2009.

CONCLUSION:
Comparing the Transnational Spatial Dynamics and Stakeholders

19

CORRIDORS AND CITIES: Connectivity or Integration Process in Southeast Asia?

Christian Taillard

INTRODUCTION: COMPARISON OF TRANSNATIONAL STRUCTURES EMERGING IN THE GREATER MEKONG REGION AND THE MALACCA STRAITS

In conclusion to this work, before undertaking the comparison of transnational integrations in the Greater Mekong Subregion (GMS) and the Malacca Straits (triangular cooperation between Indonesia, Malaysia and Thailand and Singapore, Johor and Riau, respectively referred to as IMT-GT and SIJORI), it is worthwhile remembering the basic elements justifying the comparison of the regions chosen by the Transiter programme (Transnational Dynamics and Territorial Changes.

The first basic element in the comparison of these two emerging integrations in the context of regionalization as part of the process of globalization, concerns their location at the crossroads between major North-South and East-West international and subregional trade routes. Southeast Asia is at the junction not only between Northeast Asia and South Asia on the world container route, but also between its two components: continental for the Greater Mekong Subregion and maritime for the Malacca Straits.

These regions, located at the crossroads of world and subregional traffic, also share spatial profiles with a north-south structure resulting from the respective positions of sea coasts, rivers and highlands. The Himalayan river system, in the corner of the Asian continent, is framed by the North-South shorelines of the Indochinese peninsula. The Malacca Straits, situated between two oceans, are bordered by island and peninsular spaces structured by highlands with a north-south orientation. These two North-South profiles,[1] combined with the abovementioned connections between world and subregional traffic, lead to a dynamic of spatial recompositions where emerging transnational North-South corridors play an essential role.

To these spatial considerations forming the basis for comparison we may also add the parallelism in the time-scale of the transnational dynamics implemented during the last two decades with the help of the Asian Development Bank (ADB). The approach by corridors in the Greater Mekong Region and growth triangles in the straits, adopted in the first decade (1992–2001) has been superseded by a regional network approach structured by a mesh of corridors during the second decade (2002–12).

To conclude this comparison of emerging transnational structures, two approaches are proposed: one of them, in this concluding text, gives prominence to the main spatial forms produced: at corridor level as well as at the level of the regional network resulting from their combination, and that of the new urban structures to which they give rise. A second conclusion will take as its main theme the processes and the multiplication of actors involved in ongoing recompositions.

1. COMPARISON OF REGIONAL STRUCTURES PRODUCED BY CORRIDORS

There is nothing new in the use of development axes in territorial planning, this has long been one of the main tools of planners at both national and infra-national levels. The innovation of international bodies — United Nations and regional development bank programmes — is to have made it the main instrument for developing commercial exchanges in the post-Cold War period, especially at subregional level where they are part of the regionalization process of globalization. In this context, the definition of transnational corridors given by United Nations planning experts has to be reminded: "A transnational corridor is a new geographical space where strategies of urban development and competition are displayed … opening a new type of polycentric and geographical space and new regions to development and investment."

It is in this perspective that we will compare transnational connectivity strategies deployed in Southeast Asia, which rely on the setting up of structural corridors according to two approaches. The first compares two different types of transnational corridors: that presented by Ruth Banomyong in his chapter and that formulated from other discussions related to this work. The second approach compares, at different levels, the role of these corridors in spatial redefinition; first at the level of the corridors themselves, then at the level of their combination into a regional network, and finally at that of the reticular polycentric organizations resulting from them.

Dynamic Typology of Corridors Presented by Ruth Banomyong

In the classification proposed by Ruth Banomyong, economic corridors — the name accepted by the ADB — only come into existence after developments enable the transport corridor, via intermediate stages (multimodal and logistic corridors) to finally become an economic corridor. This dynamic approach has the advantage of emphasizing that the terminology used by the ADB is formulated in terms of medium-term objectives, without taking into account the stage reached by the corridors that it finances, in terms of both economic potential and territorial structure. This prejudice has been the source of many misunderstandings, confusing reality and objectives, and has led to the author being commissioned to provide an assessment of the level of development reached by these corridors.

Transport corridors, the first type of classification, link the most productive areas and favour exchanges between them by lowering transport costs. At the same time, these corridors tend to increase territorial inequality by increasingly marginalizing neglected areas. Multimodal corridors, which are a variant of the first type, dispose of infrastructures providing, at certain nodes, interconnectivity between at least two different means of transport, managed by an operator providing an efficient (just-in-time and door-to-door to the end-user), faster and less costly service. Logistic corridors, the third type, presuppose the conjunction of private and public actors to optimize the flow of goods through the corridor and arbitrate between the interests of the different users. Economic corridors, the fourth type, aim to attract investments to diffuse new activities throughout the corridors by strengthening the hierarchical nodes that provide their structure and optimizing both material infrastructures, hardware, and their regulation, software, especially at border crossing points.

This evolutionary typology has the advantage of being open-ended, since stages can be omitted and extra stages added. In this way, some Japanese authors place industrial corridors between logistic and economic corridors, especially by the development of industrial clusters at structural nodes. For its part, the ADB mentions urban corridors structured by city clusters in its Urban-led Development Strategy. However, this typology does not prove fully operational in the present context. The performance of the corridors as a whole is aligned with that of their weakest link, taking the example of the supply chain. The result obtained means that almost all the corridors evaluated fall into the first type, that of transport corridors.

This result is explained by the delay in the implementation of free circulation which should, however, influence the release of funding for transport infrastructures, corridor by corridor. This initial agreement in principle should have enabled the AFTA to be anticipated, like the many agreements signed later dealing with the liberalization of border crossings, single customs inspection and computerized management of flows, although, with a few exceptions, these are not applied, even at the end of the second decade of these regional cooperation programmes! The initially planned simultaneous investments in hardware and software therefore did not take place, even though it was obviously a condition for success, especially in the Greater Mekong Subregion, where it was necessary to resume the flows interrupted by decades of conflict during the Cold War. The obstacles encountered will, however, soon be removed since the application of the AFTA in 2015 between all ASEAN members will make any dilatory manoeuvres impossible. It is at this point that the classification proposed by Ruth Banomyong will become truly operational.

Typology of Corridors Based on the Position of Command Centres and the Scale of Integration Produced

In Vientiane discussions, Manuelle Franck has outlined another typology distinguishing between North-South corridors, which give the most structure at national and regional levels, and East-West corridors, providing far less at both levels, on account of the meridian tropism which characterizes regional construction. The addition to the orientation of flows, whether

North-South or transversal, of criteria taking into account the position of governing centres along the corridor, and the scale of organization produced (infra-national, national or subregional) can give increased complexity to this outline by duplicating each of the two types identified.

First Type: North-South Corridors with an Almost Exclusive or Mainly Dominant National Function, Governed by Metropolizes Situated at Either End

There are two corridors of this type: Medan-Palembang for the Malacca Straits region, and Hanoi-Ho Chi Minh City for the GMS, although an intermediate regional capital is emerging in each case: Pekanbaru and Danang, respectively.

The ultimate purpose of the eastern trans-Sumatra, governed in the north by Medan and in the south by Palembang, from the hyper-centralized regime of the New Order to today's decentralization policy, has remained Sumatra's integration with Java. This centripetal strategy aiming to integrate the two most densely populated areas is perceived as a necessity for national development within an archipelago State. Thus, the projected bridge crossing the straits separating the two islands of Java and Sumatra is a national priority. This requirement prevents the corridor from being included in a subregional plan in the eyes of Indonesian planners, whereas for the ADB it is a component of a trans-Asian route which should link Jakarta to Bangkok after crossing the straits between Medan and Penang by boat (ie. the chapter by Muriel Charras). This strong national tropism and the duplication of this North-South corridor between the former interior trans-Sumatra following the highlands and the new one located behind the marshy east coast has not favoured the construction of a high-quality infrastructure. The limited funding available on account of the 1997 economic crisis in Asia and today's worldwide economic situation, has had to be divided between the two parallel corridors! Today, north-south traffic has to take alternate sections of both roads according to their condition! On this account the eastern trans-Sumatra's power of integration remains limited.

The coastal road and rail corridor between Hanoi, Danang and Ho Chi Minh City also has a strong national tropism since it links the two Vietnamese metropolises governing the deltas of the Red River and the Mekong, via the string of narrow plains along the central coastline. It is paradoxical that this backbone of the country does not feature in the GMS list of corridors, since it has a much more strongly defined transnational role than the eastern trans-Sumatra. This corridor is continued to the north by two corridors linking the Vietnamese capital to Kunming and Nanning, respectively, and to the south by the Southern corridor linking Ho Chi Minh City to Bangkok, especially since this corridor, only a road link today, will soon also be a rail link since China has released funding for the missing section between Ho Chi Minh City and Phnom Penh.

Second Type: North-South Corridors Combining National and Transnational Functions, Governed by Three Metropolizes of Comparable Size at Each End and in a Central Position

This category comprises four corridors, one from the Malacca Straits: Singapore-Kuala Lumpur-Bangkok and three from the GMS: Kunming-Mandalay-Rangoon, Nanning-Hanoi-Bangkok and Kunming-Jing Hong-Chiang Rai-Bangkok.

The Singapore-Kuala Lumpur-Bangkok multimodal corridor belongs to this group, even though the section going along the west coast is also the backbone of the Malaysian peninsula. These transnational connections are more strongly marked today than in Vietnam, as can be seen by the existence of twelve trains a day with a capacity of 8,000 EPV containers running between Songkhla and Medan, as mentioned by Nazery Khalid.

The Kunming-Mandalay-Rangoon corridor is also a strong economic reality, in view of its senior position. It was set up by the Chinese long before the planning of the Greater Mekong Subregion corridors (into which it was only integrated during the second decade), to have an outlet to the Indian Ocean. This corridor is secured by the dominant economic role played by China, not only in the central regional capital, Mandalay, but also in Rangoon. It is also a geostrategic reality for Myanmar: firstly because its southern part links Burmese territories centred on the Irrawaddy valley, and secondly because it makes the central region a link between China and India (ie. Ella Vignat's contribution).

The North-South corridor between Kunming and Bangkok links the two main regional economies of Thailand and China via the province of Yunnan by a modern road, mostly four-lane, apart from its split into two branch roads, a two-lane road via Myanmar and a generally three-lane road via Laos. Although the road infrastructure is finished, it will not be operational as a transnational link until 2013. The completion of the bridge over the Mekong between Chiang Khong and Huay Xai will give the Laotian branch a definite advantage in economic and safety terms compared to the Burmese branch. Jing Hong and Chiang Rai, which govern the two branches of the duplicated corridor, provide a relay for the command posts of Bangkok and Kunming, at each end of the road. Finally, at Phitsanulok, the corridor joins the heavily frequented main axis of Thai national development which links the Northern Region to the Bangkok area via the Chao Phraya valley.

The Nanning-Hanoi-Bangkok corridor, the last to be set up during the second decade of the GMS, is only operational in the multimodal Nanning-Hanoi section. The intermediate section between Hanoi and Vientiane, which is included in Laos's 8th Five-Year Plan (2011–15), is not yet implemented. This corridor will put an end to Vientiane relative isolation, which is the only GMS capital not yet situated on a corridor linking the dominant regional economies (ie. Christian Taillard's contribution). The thai section of this corridor also links the less developed Northeastern Region to the hyper centralized Bangkok Metropolitan Region.

Third Type: Transversal Corridors with Powers of Organization at a Regional Level, Governed by a Single Central Node

The third type — and the fourth — regroups two transversal corridors. The Ho Chi Minh City-Phnom Penh-Bangkok-Rangoon corridor gives structure to the southern part of the GMS, and the Medan-Penang-Songkhla corridor has a similar role in the northern part of the Malacca Straits. Both these corridors are governed by a node in central position, a national capital for the first and an infra-national capital for the second.

The Ho Chi Minh City-Phnom Penh-Bangkok corridor is being continued to Rangoon. It will soon link all the metropolises in the southern part of the Indochinese peninsula, and for this reason has great potential for developing exchanges, on account of the congestion in the region's seaports. However, as we have already seen, it is necessary to wait for the implementation of the AFTA in 2015 to solve the problem of border crossing, the most serious cause of bottlenecks. Bangkok occupies a pivotal position between Rangoon and Ho Chi Minh City.

The Medan-Penang-Songkhla trans-straits and trans-peninsular corridor, thanks to port infrastructures on the one hand and two rail and road dry ports on the frontier between Malaysia and Thailand on the other, has the merit of linking the most dynamic coastal areas of the Malacca Straits and peninsular Thailand. Here also, Penang plays a central governing role since it gives the nodes governing both its entrances the possibility of a link to the trans-world container shipping route.

Fourth Type: Transversal Corridors with Little Structuring Power, Governed by a Single Metropolis, Where Trans-border Exchanges Are More Important than Transnational Flows

Two corridors belong to the fourth type: the East-West Danang-Khon Khaen-Phitsanulok-Moulmein corridor in the GMS region, and the Dumai-Melaka trans-straits corridor.

The East-West corridor, linking both sides of the Indochinese peninsula, was chosen as a pilot corridor by the GMS since it does not go through any national capitals, and since consequently, the stakes are less important and the economic flows much more limited (ie. Christian Taillard's chapter). Thus, trans-border exchanges between the four countries crossed are the most important. The limited transnational flows using part of this route are operated by Japanese companies, and are asymmetrical: they leave from Bangkok and use part of the corridor to get to Hanoi and Ho Chi Minh City in Vietnam (ie. Ruth Banomyong's contribution). These are therefore temporary flows, and will disappear when the Bangkok-Nanning corridor serving Hanoi is completed.

The Dumai-Melaka corridor operates at straits level, with the exception of the shipment of petroleum products and palm oil from the port of Dumai. Other flows, whether of passengers or goods, especially barter trade goods, are entirely comparable to trans-border traffic in the GMS, operated by individuals who regularly cross the border to exchange goods of a value limited by the customs authorities for each trip. It is significant that this corridor, for ADB planners, is limited to both shores of the straits, and is not taken into account by Indonesian planning, even though this trade is beginning to be too much for it. Malaysian entrepreneurs are investing, especially in Padang province on Sumatra's west coast, and in Pekanbaru on the east coast (ie. Manuelle Franck's contribution).

New Types of Regional Organization Linked to Corridor Planning

The structuring of corridors by nodes, as we have seen in the typology presented above, works equally well for the GMS and the Malacca Straits.

Three combined processes enable the new types of regional architecture produced to be analysed.

A New Hierarchy of Nodes and a Governing Role of Corridor Heads

In both cases there appears a new positioning and a new hierarchy of the nodes structuring the corridors. Provincial capitals are usually located in the centre of their provinces, both in Sumatra and in the Indochinese peninsula, and the same is true for State capitals in the Malaysian Federation. Between these inherited nodes, a new type of node is introduced, in a peripheral position with respect to political boundaries. Small border towns, formerly forgotten, become strategic places now competing with middle-sized cities located close to the interfaces between provinces or states on national sections of the corridors. Despite their smaller size, border towns often form nodes with greater structural power than the secondary cities on these corridors, due to their role in the movement of transnational flows. Their weight is increased in the case of twin cities on each side of the border, a profile whose dynamics are linked to the planning of the corridor, and which will be studied in the second part of this conclusion.

Another common feature seen in the second type of corridor presented is the governing role of corridor heads, which are generally located in the ports providing a link between continental traffic and the maritime routes that dominate international exchanges. The only exceptions are Kunming and Nanning in the GMS, which can be explained by China's dimension as a continent-state. They are trying to assert themselves as entry-points to the corridors that they govern, thanks to a generous provision of infrastructures by the Chinese government to make up for their distance from international traffic.

However, there is an essential difference between the land and sea corridors studied, resulting from their function of concentrating traffic. The tunnel effect brought about by bridges (over rivers and narrow straits on the borders) and land border posts is much stronger than in the ports governing the maritime corridors of the Malacca Straits. These combine the functions of concentrating and dispersing flows. On one side, the roads leading to the port converge, and on the other, coastal traffic multiplies towards neighbouring ports on the same shore or cross-straits towards ports on the opposite side (ie. Map 3.4 of Nathalie Fau's chapter on Malacca Straits).

Regional Network Organized in Strips or Bundles

The planning of the corridor network, introduced during the second decade of these regional cooperation programmes, has modified the regional architecture, with noticeable differences between the GMS and the Malacca Straits, according to the position of the strategic North-South corridors at the heart of the arrangement, in relation to the demographic and economic concentrations structuring national development.

In the Malacca Straits, the backbone of the regional network consists of the international maritime corridor using the straits and the North-South corridors bordering it on both shores. They polarize the population and economic activity of both the island of Sumatra and the Malaysian peninsula, even though there are great economic differences between them today. These differences are the result of the peripheral position of these corridors in relation to national development in the first case, and their central location in the second. For this reason, the regional network is organized around a North-South strip centred on the straits and bordered by economically dynamic areas on both shores. This central strip is framed by two symmetrical strips of highlands which are associated with this dynamic, albeit to a lesser extent, especially for perishable goods and agro-industrial products. The exterior coastal strips, in the west of Sumatra and the east of the Malaysian peninsula, occupy a peripheral position, even though they are an integral part of both the regional cooperation programmes developed at the northern and southern entries to the straits. The economic dynamism of the west coast of the Malaysian peninsula and the growth of Malaysian investment to valorize natural and tourist resources on the east coast of Sumatra have led to an integration process which should eventually justify the extension of the Melaka-Dumai corridor as far as Padang, and that of the Songkhla-Penang-Medan corridor as far as Sibolga. However, the question of integrating the east coast of the Malaysian peninsula is not posed at straits level, but rather at that of the development of the Malaysian Federal State. This coast, aside from its oil resources, is in a pivotal position between the dominant economic axis of the peninsula and the States of Sabah and Sarawak on the opposite shore of the South China Sea.

The GMS network, in a continental position, will be organized around the two strategic corridors of the GMS grid, Bangkok-Kunming and Bangkok-Nanning, linking the two main economic powers of the subregion, Thailand and the two Chinese provinces. The network takes the form of two north-south bundles, each consisting of several corridors, converging towards Bangkok, thus strengthening its position as the main entry point to the GMS. Rather than following them, these two bundles cut across the Mekong and Chao Phraya valleys which occupy a central position at the level of the peninsula and concentrate the economic activity of the neighbouring countries. These bundles relegate the two sea coasts of the peninsula to a peripheral position, especially the coast of Vietnam, even though it includes two centres of activity that count at GMS level. It is true that this country is much more strongly oriented towards the traffic of the trans-Asian maritime route following its coast. The new projected Chinese Kunming-Vientiane-Bangkok high-speed line is a good illustration of the regional tropism linked to these two bundles at the centre of the regional network, since it successively follows sections of both. The route first runs alongside the Kunming-Boten section of the western North-South corridor, then the Boten-Vientiane section of the central North-South corridor parallel to the Mekong, and finally the Vientiane-Bangkok section of the North-East corridor (ie. Map 2.4 of Christian Taillard's chapter on GMS).

Transnational Polycentric Networks Governed from the Central Zone or the Periphery

The main backbone of the network, grouping or dividing the main economic centres of the GMS and the Malacca Straits, leads to transnational development, reticular and polycentric, differentiated by their position and the hierarchy of their main command centres, in a central or peripheral position, and by their means of linkage, direct or indirect, to world shipping routes.

The polycentric network of the Malacca Straits is introverted: the centres of government are all located along or close to its two shores, at both entries to the straits and in a central position between then. On the east coast, there is a clear hierarchy. Singapore, the city state which for many years monopolized the role of internationalized multimodal centre at the southern entry to the straits has had to share this role, first with Port Klang, the port of Malaysia's capital Kuala Lumpur, to export the industrial goods produced in the Klang Valley and the central area of the Malaysian peninsula, then with Penang's port, Butterworth, for the export of electronic goods from the free zones and industrial clusters in the northern regions of the peninsula. Finally, since the beginning of the first decade of the twenty-frst century, it has had to share its role of managing containers in transit with Tanjung Pelepas, one of Johor's two ports, located on the other side of the eponymous narrow straits separating it from Singapore. The Malaysian shore of the straits is therefore governed by two national capitals (Singapore and Kuala Lumpur), relayed by two regional capitals in the north and south of the Malaysian peninsula (Penang and Johor Bahru). The other, less important nodes are either dry ports which are scattered along the three sections of the economic corridor of the west coast of the Malaysian peninsula, specialized feeder ports supplying the three main Malaysian ports, or ports engaging in barter trade with ports on the Sumatran side.

In Sumatra, polycentrism is organized around the three main nodes governing the three regions linked by the eastern trans-Sumatra. Medan and Palembang, situated at each end, owe their importance to the roles they have played in Indonesian national development. Belawan, Medan's port, was for a long time the only Sumatran port open to international traffic, and maintained shipping links with Penang, on the opposite side on the straits, long before it became the end of the Songkhla-Penang-Medan corridor. Palembang, the main city-port of the Sumatra Southern part, is still dependent on the Javanese central power and an essential relay on the Sumatra Eastern corridor. Much later than Medan, Pekanbaru is becoming the capital of the central Sumatran region in the first decade of the twenty-first century. Its position as a regional node on the eastern trans-Sumatra was strengthened by its link with the deep-water port of Dumai, placing it at the outlet of the second corridor crossing the straits from Dumai to Melaka. The secondary nodes are either provincial capitals located along the trans-Sumatra, or coastal ports cut off from it, existing by barter trade with ports on the other side of the straits.

The polycentric network of the GMS is governed from peripheral areas, either sea ports at the head of corridors on the three sea coasts of the Indochinese peninsula (South China Sea, Gulf of Thailand, Indian Ocean), or the two Chinese provincial capitals which provide the region's attachment to the Asian continent. As we have seen, three metropolizes (Bangkok, Kunming and Nanning) govern the two structural bundles (each with three or four corridors) and play a predominant role, even though the other national or provincial port capitals at the head of corridors multiply the number of possible entry points to the GMS. As well as this hierarchy of governing centres at the entrance to corridors, there is another hierarchy differentiating between cities in the interior located at intersections between north-south and transversal corridors. Their importance depends on their rank in national urban networks: national (Hanoi, Vientiane or Phnom Penh), regional (Mandalay, Khon Kaen) or provincial capitals, on the number of corridors served by the node (two or three), or on the intensity of traffic on these corridors that today are often national rather than transnational (Phitsanulok, Nakhon Ratchasima). Other nodes are found at points where two branches of a corridor converge or separate, such as Jinghong and Chiang Rai on the Kunming-Bangkok corridor.

The polycentric organization governed from both sides of the Malacca Straits appears more linear, on account of its relative symmetry on each side of the straits, than that governed from the continental and maritime periphery of the GMS. The larger size and continental nature of the latter gives the reticular organization a greater capacity to regenerate peripheral border regions crossed by corridors. In doing this, it also increases inequalities with neglected peripheral areas falling between the meshes of the network, and which can only be integrated at the level of national planning.

The two polycentric organizations also differ greatly with regard to their links to world shipping routes. Despite the existence of ports as command centres on all three sea coasts of the Indochinese peninsula, the GMS does not benefit from a direct link to the trans-world container route similar to that of the three multimodal port complexes on the eastern shore of the Malacca Straits. The Thai ports on the eastern seaboard and the Burmese ports on the Indian Ocean are off-centre from this worldwide container shipping route. Vietnamese ports, on the contrary, are close to this route, since it passes along the Vietnamese coast, but there are no international multimodal port complexes between the Taiwan and Malacca Straits. On this account, the GMS can only have an indirect link to world shipping routes, and this necessitates the use of feeder ships to provide a link with port complexes dealing with containers in transit. On the positive side, however, the region enjoys a greater potential for assertiveness in the regionalization process of globalization.

2. COMPARISON OF NEW TYPES OF URBAN STRUCTURE LINKED TO CORRIDORS: TWIN AND PAIR CITIES

The innovation in urban structure introduced by the transnational integration strategy via corridors, proposed by the ADB to the countries of the Greater Mekong Subregion and the Malacca Straits, is essentially found in the emergence of two new types of dual urban structures, twin cities and pair cities (identified by Manuelle Franck in her contribution), according to the distance between the two components. These dual urban sructures are included in national urban networks at places where corridors cross land or sea borders.

Criteria Chosen for Comparison of the New Urban Structures Identified

These new urban structures, whether twin or pair cities, are found on all urban scales, ranging from small border towns relatively unchanged by the corridor passing through them to the urban metropolises on either side of the Malacca Straits, via the most numerous group, medium-sized but extremely dynamic urban pairs.[2]

Seven criteria were chosen for this comparative and typological approach:

1. As these are border towns, the first criterion takes into account the type of border concerned (land, river or sea), and the position of this border area in national development, according to whether it is in a peripheral or a central position.
2. The difference between the two cities on each side of the border, measured in terms of demographic, economic and political weight, is a second criterion with a strong influence on the relationship between the two components: complementarity or rivalry.
3. The third criterion taken into account is the age of the cities and their hierarchical position in their respective national urban networks, especially their dependence on one or several command centres.
4. A fourth criterion, central to this study, is the weight of the new transport infrastructures linked to the corridor in the cities' respective urban dynamics, especially in relation to the change in rhythm and scale of urbanization.
5. The fifth element taken into consideration is the degree of commitment of the central power, in terms of both planning and investment, conditioning the size of installations and the attractiveness to private investment, either domestic or foreign.
6. The sixth criterion, linked to the one before, takes into account the level of institutionalization of cooperation between the two cities, which is the only element that will ensure the permanence of the trans-border twin's dynamism.
7. The final criterion refers to new forms of urban renewal introduced for the emerging conurbations and urban regions, and takes into account the degree of internationalization of these forms due to the access to new transnational functions.

The hierarchy of these criteria in each dual urban structures studied enables us to establish two typologies common to the two transnational dynamics studied, GMS and Malacca Straits, the first concerning twin cities, the second pair cities.

Attempted Classification of Twin Cities

Twin cities are much more prevalent in continental Southeast Asia, where land or river borders are more frequent, whereas in the Malacca Straits, maritime borders are more common. However, twin cities can be found at each end of the Malaysian peninsula, especially on each side of the Johor Straits, which are more like a river border on account of their width. Three types of twin cities can be identified according to the dominant dynamic linking the two cities.

First Type: Dominant Urban Dynamics Linked To Corridors

The first type concerns cities located on either side of land or river borders, in a peripheral position in their respective national territories, whose development is generated far more by their desire to benefit from the trans-border dynamics provided by the corridors than by any imperatives linked to national development. Beyond the specific characteristics of each, this type includes the Danok-Bukit Kayu Hitam pair (studied by Abdul Rahim Anuar, Muszafarshah Mohd Mustafa and Amel Farhat), cities located at the point where the Bangkok-Kuala Lumpur-Singapore road corridor crosses the land border with peninsular Malaysia, and the Savannakhet-Mukdahan pair on the GMS East-West corridor (studied by Elsa Lainé), which associates two cities located on either side of the Mekong River border separating Laos and Thailand. These are modest sized cities on the national scale, generally with less than 50,000 inhabitants, with the exception of Savannakhet (renamed Kaysone Phonevihane), which, with 84,000 inhabitants, is the second largest city in Laos, a relatively little urbanized country.

The lack of symmetry between the components of the urban pair on the Mekong may be explained by an inverse development ratio at a local level: the city of Savannakhet (Kaysone Phonvihane district) is a dynamic centre in Laos compared to Mukdahan, which suffers from its peripheral position in the Thai urban network and *vis-à-vis* of the Bangkok economic domination on a national scale. On the contrary, the economic differences existing on a cross-border level are in favour of the Malaysian element in the second example of twin border cities, which possesses an industrial economy oriented towards exports, far more advanced than the specialization in rubber-growing and tourism of its Thai counterpart. In this case, the difference can be explained by the economic macrocephaly of Bangkok and its metropolitan area, which concentrates all the country's wealth and weakens the development of peripheral regions, especially the southern peninsular region.

Whether these cities existed before the construction of the corridor, like the Mekong border cities, or were created by the emergence of new border towns in liaison with old administrative centres situated on the corridor but at some distance from the border (e.g., Sadao on the Thai side and Jitra on the Malaysian side), the gathering speed of urbanization and the extension of city limits date in both cases from the building of transport infrastructures linked to the corridors: the international bridge over the Mekong or the border logistic centre and dry port at Bukit Kayu Hitam. Even so, the two urban components have not gained equal benefit from these installations, on account of the difference in commitment of their central governments.

On the Malaysian side, the central power has undertaken planned development, including the logistic node and the industrial cluster of Bukit Kayu Hitam in a polycentric north-south urban area, consisting of the administrative centre in Jitra to the south, the commercial and leisure areas in Changlun and the university campus in Sintok, situated in an intermediate position. At the same time, on the Thai side, the bipolar development around Sadao — an administrative city with agricultural activities, and Danok — a border city devoted to trade and tourism, separated by about fifteen kilometres, occurred spontaneously in response to private investment. On the Mekong, the Laotian central government, with the aid of foreign investors, has financed the construction of the special economic zone at the Savannakhet bridgehead, which has enabled this city to develop as a new centre in rivalry with the colonial city. These two entities are today linked by an intermediate zone organized around the new market and the new administrative centre, transferred from the historic centre shortly before the bridge was built, in the context of the national programme for the development of secondary cities financed by a World Bank loan. However, the free economic zone planned for the Thai side has not been built, since the building of infrastructures was left to the private sector which then failed to provide funding.

This difference in the central power's commitment leads to other features common to the cities dominating these twin cities. First of all, there is a much greater internationalization of urban development linked to the presence of foreign investors in the Savannakhet free zone, for example, the Savan Vega Casino financed by a Chinese investor from Macao, or the presence of a factory making Boeing aircraft parts in the Bukit Kayu Hitam industrial cluster, whereas their respective urban counterparts remain dominated by Thai stakeholders. The greater advances in internationalization in the Malaysian polycentric urban area can also be seen in the shape of the city, with innovative elements such as the industrial cluster, the high-tech complex linked to the university or the new residential area organized as a gated community. The urban landscape is more traditional in city of Savannakhet, although the shopping centre, inspired by the mall and the renovation of the colonial heritage in the city centre to encourage tourism are part of this internationalization policy. On the other hand, the urban landscape of the Thai elements of these two border pair is not much different from that of neighbouring cities of comparable size.

A final characteristic common to these two twin border cities is the low degree of institutionalization of cooperation between the two urban components. The regulation of border crossings remains in the hands of the central powers and the local actors are left little room for manoeuvre, even though this might help remove administrative hurdles. In particular, there is an absence of planning in the features shared by both elements, since the regulation of flow is of greater importance than the orientation of urban development. From this point of view, however, the Savannakhet-Mukdahan border cities appears more advanced since there are cross-border shuttles between the two sides serving the casino, and a project for a shared airport has been agreed on, and although the practicalities of its implementation have not yet been worked out, the project could soon become a reality as part of the funding scheme of the Irrawaddy-Chao Phraya-Mekong Economic Strategy (ACMES), steered by Thailand. This scheme could also provide, at transnational level, the investments that were denied to Mukdahan at national, or even at northeastern regional levels. For the border twin cities between Malaysia and Thailand, no projects seem to be emerging on a transnational scale, probably because of Penang's economic domination of the North Corridor Economic Region of Malaysia and the southern peninsular region of Thailand.

However, two differences are noteworthy between these two examples of twin cities. The first concerns the size and direction of the flow of people between the two border cities. At the Thai-Malaysian border, people flow into the cities in both directions, alternating between Thai workers going to work in the Malaysian industrial cluster, and Malaysian tourists, especially at weekends, using the Thai tourist facilities in Danok. There is no migratory movement on this scale between the two Mekong border cities, apart from the flow of Thai gamblers going to the Savan Vega casino. Also, at an urban level, there is a difference in the orientation of town planning, due to the border. In the first case, the orientation is clearly linear, following the road corridor. In the second case, two structural axes are combined, one following the river with a strong upstream tropism towards the international bridge, the other transversal, following the East-West corridor, to the west towards Khon Kaen, and to the east towards Seno and the intersection with the Laotian central North-South corridor. Finally, the example of the Thai and Malaysian twin cities has the advantage of emphasizing that the border pair does not exist only at the level of two separate components, but also at the multipolar level of urban development on both sides of the border.

Second Type: Cities Resulting from a Combination of National Geostrategic Stakes and Transnational Dynamics

Like Savannkhet and Mukhdahan, Hekou and Lào Cai (discussed, respectively, by Sebastien Colin and Marie Mellac) are separated by a river border, the Red River and Nanxi, and where the North-South corridor, linking Kunming in China and Hanoi and Haiphong in Vietnam, crosses the China-Vietnam border. Like the twin cities studied on the Mekong, these two riparian cities are of unequal weight (numbering, respectively, 70,000 and 100,000 inhabitants), displaying a demographic ratio in inverse proportion to the economic weight of the two countries. Beyond these similarities, however, the transnational dynamics of Hekou-Lào Cai are very different to those of Savannakhet-Mukdahan, with greater importance given to geostrategic concerns, linked to old rivalries inherited from a thousand years of Chinese occupation in Vietnam, followed by the brief period of armed conflict between the two countries in 1979. These strategies have thus developed in a common context of absence of regional relay points between these border cities and the national or provincial capitals on which they depend. This double historical and geographical constraint has increased the dominant role of states that maintain tight control over the trans-border relations that have resumed since the regularization of relations between the two countries in 1991.

Vietnamese strategy consists of containing and keeping at arm's length powerful Chinese neighbours, while benefiting from the economic flows and relocations that they generate. It is a question of anticipating the potential profits of the Chinese Go West policy which is supposed to boost the economy of Yunnan, even though exchanges along the corridor are still modest in size. The start begins with a change of status of Lào Cai from a simple provincial capital on the Chinese border to a structural node of importance at corridor level, which could enable it to become more important than Hekou, which is only a district capital dependent on Honghe prefecture, located a long way behind the border. The Vietnamese government's strategy is not without ambiguity. On the one hand, the central power initiates and finances the installations structuring trans-border logistics and changes in scale of urban infrastructures. On the other, it applies a brake to policies of trans-border cooperation undertaken by the city authorities in order to retain control over the border, considered as a strategic national stake. The central power also instrumentalizes transnational GMS programmes so that these can finance part of the transport infrastructures: the 244 km of the Hanoi-Lào Cai section of the corridor are financed by the ADB.

China has also defined a new policy, directly enforced from Kunming, the provincial capital, in order to take advantage of its difference in economic weight with its southern neighbour. It has thus invested in the rebuilding of the Chinese section of the Kunming-Hanoi-Haiphong road corridor and the construction of a new border bridge to facilitate the crossing of the Red River, and also in the renovation of the Chinese section of the railway and its re-opening to cross-border traffic. By doing so, it has given Yunnan a third maritime access to the Gulf of Beibu (Gulf of Tonkin), in addition to existing accesses to the Gulf of Thailand and the Gulf of Martaban in Myanmar. Like Vietnam, it has made parallel investments in urban infrastructures, even though urban development is on a smaller scale than in Lào Cai.

In Lào Cai, the urban development is oriented by urban plans based on functional zoning and the introduction of new forms of internationalized urban infrastructures. The city of Lào Cai is organized around the two old bridges — rail and road — over the Nanxi River, linking the colonial city centre and Hekou. Thanks to another new bridge, the administrative and social facilities of the old city centre are divided between the two sides of the Red River. The centre also has a block of buildings containing a supermarket, a hotel and offices, situated close to the first border zone equipped with logistic and warehousing services. Today, goods still cross the international bridge in handcarts, to avoid the customs inspections imposed on motor vehicles crossing the new bridge. This central zone is currently being redesigned, using a double functional delocalization process: trans-border logistics are being moved to the new district at the head of the motorway bridge over the Red River and administrative offices are being concentrated in a new district to the south. The space thus vacated is devoted to high-level economic, financial and tourist services, and the river frontage has been developed as a promenade lined with six or seven-story buildings, creating a skyline revalorizing the city's image seen from Hekou. Upstream from the old city centre, the new trans-border logistic installations for dealing with road traffic have been regrouped next to a tax-free shopping area and a light industrial zone, and downstream, apart from the new administrative district, residential areas have been built, with houses for the upper classes and blocks of flats for the middle and working classes. In the eyes of the Vietnamese authorities, this functional zoning has the advantage of removing residential areas from the border and placing them at a distance, making them easier to control. Its result from these reorganizations a real change of scale of polycentric urbanization process and the emergence of a conurbation on the Vietnamese bank of the Red River.

On the other side of the rivers in Hekou, urban development is less extensive; and is different from that of small towns of comparable size in Yunnan since the town has grown up along the Red River towards the new bridge, where a new district had been developed, including a special economic zone and an administrative centre with a hospital and schools. This duplication of the city and its facilities between the old town and the new district, similar to urban development observed in Savannakhet, does not have the extent of the urban development of Lào Cai, and has not benefited from the financial means allocated by the Vietnamese State for this project, designated as a strategic objective of national construction in the context of GMS-initiated transnational dynamics.

A final important difference from the Mekong twin cities studied is the attempted institutionalization of cooperation between Hekou and Lào Cai, first by means of an annual informal trade fair, held since 2001 in alternate years in each city, then with the creation of the Beishan-Kim Than trans-border cooperation zone in 2005, and the signing of a Memorendum of Understanding (MOU) in 2011, setting up a Sino-Vietnamese steering committee. These renewed attempts show the desire of the local authorities on each side of the river to pursue regional integration. Such initiatives, however, are often met with a flat refusal by the central government, which wants to remain in charge of the process and preserve higher national interests in bilateral exchanges!

The twin cities of Ruili-Muse also belong to this type. They are situated on the Sino-Burmese border, which has not been studied in this work, even though the first of these is mentioned in Sébastien Colin's chapter. Crossed by the Kunming-Rangoon corridor, which was set up long before those of the GMS, these twin cities are the result of China's wish to gain access to the Indian Ocean, while bypassing the Malacca Straits, and that of Burma's long-isolated military junta to favour relations with China, its only protector. In these conditions, the more important of the two is, of course, Ruili, which, apart from the building of the road — and soon the rail — corridor, has benefited from an urban development out of proportion with that described for Hekou, closer to that experienced today by Lào Cai. A trade fair has been organized in alternate years in both cities for many years, and since 1999 the city of Ruili has incorporated the neighbouring town of Wanding, thus creating the largest urban border zone in Yunnan, with 120,000 inhabitants. Finally, this border is also closely monitored by China on account of the very important Sino-Burmese exchanges that cross it, such as the gas pipeline which will serve Yunnan and the southwestern provinces of China from the Burmese port of Sittwe. In addition, this border is a major crossing point for the drug trafficking that the Chinese government is now attempting to eradicate.

Third Type: The Many Forms of Border Enclaves

Although it has not been developed as such in this work, a third type of urban development on the borders has been cited several times, and deserves a mention in this typology: urban development as a border enclave linked to cities located further away.

The border enclave at Boten, located on the Laotian branch of the North-South corridor, has a clearly institutionalized status. A thirty-year concession, renewable twice, was given by the Lao People's Democratic Republic (ie. Sebastien Colin's contribution) to a Hong Kong company to build a casino, next to a shopping and leisure centre, both situated in the depths of the forest! These were built after the closure, at the request of the Chinese authorities, of a casino built at Mongla, a Burmese border town on the western branch of the North-South corridor. The Boten casino, opened in 2008, was then closed in 2011 for the same reasons: violence perpetrated by the manager against Chinese customers in order to recover their gambling debts. The future of this urban enclave, linked to the town of Mohan on the other side of the border, is thus more than uncertain.

China and Thailand, which do not have a common border, have undertaken an original type of cooperation to set up a delocalized Sino-Thai free zone on the Thai side of the Mekong in the golden quadrilateral, in order to promote the development of bilateral exchanges (ie. Christian Taillard's contribution). This free zone, set up in partnership with the

National High-Tech Industrial Development Zone of Kunming, includes an industrial zone, a logistics park, business park and a cultural centre, modelled on those that Yunnan has constructed in Ruili on the Burmese border. The Thai planning authority, the National Economic and Social Development Board, has prepared a study identifying 112 projects for an investment potential of 476 million dollars at Chiang Saen, and also Mae Sai and Chiang Khong on each branch of this quadrilateral. Lately, this authority decided to relocate the Sino-Thai economic zone project in Chiang Khong close to the bridge under construction on the Mekong River completing the Lao branch of the North-South main corridor. With this Sino-Thai free development zone Thailand is trying to consolidate its dominant position over Huay Xay, the twin city of Chiang Kong, at the other head of the bridge.

The third type of town belonging to this category is completely informal: it is unauthorized building in the buffer zone separating two border posts, Thai and Cambodian, in the Southern corridor linking Bangkok to Ho Chi Minh City via Phnom Penh, studied by Thierry Lejard in a programme, presently in the course of publication, undertaken by the Research Institute on Contemporary Southeast Asia (IRASEC) of Bangkok. In the no-man's land between the little Cambodian border town of Poipet and the Thai town of Aranyaprathet further away from the border six market-workshops have grown up and nine casino-hotels, housing all kinds of traffic. These are tolerated by the local authorities who derive extra income from them since they attract flows of workers who have to pay to use the shops in the market-workshops employing them. This no-man's land is qualified by the author as "non-existent" in reference to its unofficial character, but it nevertheless represents considerable economic stakes on one of the most active corridors in the GMS since it links the southern capitals of the Indochinese peninsula.

Attempted Typological Classification of Pair Cities

Pair cities are a frequent occurrence in Southeast Asia, strongly differentiated from twin cities by the distance separating them and by the complementarity of their functions, made possible by a strategic transport route linking their development. In the GMS, the most common form is the pair represented by a river capital and the seaport linking it by road and rail to international trade routes. Hanoi and Haiphong on the Red River delta, Bangkok on the lower Menam Chao Phraya and the Thai Eastern Seaboard ports in Thailand, Phnom Penh at the head of the Mekong delta and the port of Sihanouk City on the Gulf of Thailand are all good examples. In the Indonesian archipelago there are the pairs formed by regional capitals inherited from agrarian kingdoms and their port outlets on the northern coast, such as Jogjakarta and Semerang in Java.[3] Often, the transport infrastructure linking the two cities of the pair is at the origin of an urban corridor which is not unlike modern corridors apart from the fact that the distances between cities are much smaller.

In the context of this attempted classification, we will confine ourselves to a typology of urban dynamics of pair cities linked to the recent introduction of economic corridors financed by the ADB. Three types can be seen, according to the pair's dominant dynamic.

First Type: Pair Cities Associating an Inland City and a Port Each Side of the Malacca Straits with Strong Support from the Central Government

This first type includes urban pairs whose main feature is that they are situated within national contexts on each side of the straits, and can mobilize transnational dynamics thanks to the corridors linking them to sea routes using or crossing the straits. The lack of symmetry between economic dynamics on each side of the straits and the degree of completion of the corridors concerned differ according to whether the dominant city is well inland, such as Palembang, Jambi and Pekanbaru, located at the point where the eastern trans-Sumatra crosses the rivers, all hoping to obtain a seaport, or in a near-coastal position via Port Klang for Kuala Lumpur which has forged currently exclusive links with the dry port of Ipoh on the west coast of Malaysia.

The Pekanbaru-Dumai pair in Central Sumatra (studied by Manuelle Franck) links the capital of Riau province with a new deepwater port located 150 km away on the west coast of the Malacca Straits. The emergence of the pair is relatively recent; it started with the exploitation of a hydrocarbon deposit in the 1950s; the head office of the licence-holder Chevron is in the provincial capital and a pipeline brings the oil to the seaport of Dumai which exports it. This port also exports palm oil from the plantations that have rapidly developed in the province, especially since the opening of the eastern trans-Sumatra. However, the poor condition of this section between the two cities has limited the developing synergy only to these two economic activities. The emerging dynamic of these pair cities dates essentially from the first decade of the twenty-first century, which marked the end of Jakarta's New Order hyper-centralization policy, the reopening of the province towards the Malacca Straits and the implementation of decentralization which gave the province the means to pursue its own economic strategy thanks to the royalties derived from its two main resources. To this new context should also be added the redesigning, in the context of national planning, of the eastern road corridor which will be doubled in the long term by a rail corridor in order to complete Sumatra's integration with Java. The upgrading of the eastern trans-Sumatra will at last make it possible to create the missing high-performance link between the two pair cities. We should also mention, in the medium term, the westward extension of the Dumai-Melaka trans-straits corridor, on account of increased Malaysian investments, not only in Riau province of Central Sumatra, but also in West Sumatra, which were not foreseen by the ADB.

The historic association of the Ipoh-Kuala Lumpur pair, associating a dry port and the federal capital (studied by Amel Farhat) also originates in the mining economy. It has a longer history, however, since it begins with tin mining in Ipoh during the colonial period, justifying the removal of Perak state capital from the coast to the Kinta Valley deposits and the building of road and rail transport infrastructures to export tin via the ports of Penang in the north and Kuala Lumpur in the south; both 150–200 km from Ipoh. However, its history was marked by two breaks, which influenced the destiny of the urban pair. The first one was economic with the collapse of tin prices at the beginning of the twentieth century, faced with competition from plastics, leading to a major economic crisis since until then the city had specialized exclusively in mining industries. Ipoh had to wait until the 1990s, with the federal policy aimed at industrializing the nodes situated along the Malaysian west coast economic corridor and relocating logistic and financial port installations to a series of dry ports strung along the corridor (studied by Nazery Khalid's contribution), to resume its growth. The city benefited from an efficient multimodal transport corridor, associating road and rail transport, and an industrial workforce which could be redeployed in export activities with greater added value, the spearhead of Malaysia's global integration strategy. The second break was political, and occurred in 2008 when the opposition won the Penang State election, leading the majority in power at the head of the federal State to redirect the entire trade flow managed by Ipoh dry port to Kuala Lumpur and integrate Perak State into the federal metropolitan region, whereas at the same time the ninth plan placed the Kinta Valley in the Northern corridor economic region! In this way, Ipoh lost its function of intermediary node on the North-South corridor linking

Penang and Kuala Lumpur, and formed an urban pair with the federal capital.

Thanks to this changing context, mostly at the initiative of central governments, the complementarity within each of the pair cities studied has changed rhythm and scale from the economic and urban point of view. Pekanbaru has become the emerging metropolis of the central region of Sumatra, catching up on its late start compared to the other two regional capitals at each end of the eastern trans-Sumatra (Medan in the North and Palembang in the South). It is the main road transport hub at the intersection between the North-South and East-West Sumatranese corridors, and is also an air transport hub serving Sumatran provincial capitals, but also Jakarta and the main cities in the Malaysian peninsula, thanks to flights operated by low-cost companies.[4] With its 900,000 inhabitants, it is equipped with high-level services and tertiary installations characteristic of second-generation metropolizes: business centres, head offices of provincial companies and regional offices of the main banks, malls and residential estates, internationalized universities, hospitals and tourist facilities, Malaysian and Singaporean consulates. Urban projects such as the restructuring of the city around a new North-South main road, an increased number of bridges, the development of the waterfront and the creation of three high-capacity bus services are good illustrations of the city's change in status. This emerging metropolis between Medan and Palembang forms part of the urban network along the eastern trans-Sumatra.

Ipoh, on the contrary, has led to an urban development on the scale of the integrated urban region of the Kinta Valley, which today numbers 700,000 inhabitants, and is developing in unusual ways. The urban centre revolves round three poles: first the colonial heritage centre around the central station and the dry port which were at the origin of the urban renewal, then Green Town, adjoining it to the East, which has acquired the function of business district and high-ranking administrative centre, but in the form of a garden city inspired by the new towns developed in the federal capital, and finally the new complex containing high-tech industries and the residences of expatriates and the upper classes of Meru. This is situated at the city's Northern entrance where the North-South Malaysian corridor splits into two branches bypassing the conurbation to the east and west. On both branches of the corridor and further North in the Kinta valley, this central tri-polar zone governs relay towns with specialized industrial clusters (mining and chemical industries to the East, electronic and pharmaceutical industries to the West, automobile industry to the

North) and residential estates to house their employees. This is a polycentric organization on the scale of the urban region.

In both pairs of cities, complementarity sometimes gets into difficulties, which have to be removed for their dynamics to become lasting. For the Pekanbaru-Dumai pair, the future is linked to the efficiency of the long-awaited motorway corridor linking them. Dumai is already asserting itself as an increasingly diversified industrial city, adding to the oil and agro-food industries new specialities such as fertilizer manufacture, shipbuilding (still on the drawing board) and various export industries concentrated on five industrial estates. Its future prospects will also depend on the development of its port infrastructures, especially the shuttle ferry service carrying trucks from Dumai to Melaka, and in the longer term, a projected bridge across the straits linking the two cities via Rupat Island. The dynamics of the Pekanbaru-Dumai pair seem to have got off to a good start, but their future nevertheless depends on continuing investment in both land and sea transport to strengthen the emergence of a central region that is important on the Sumatranese scale thanks to the intensification of cross-straits traffic.

For Ipoh-Kuala Lumpur, the dry port is a strategic element in the dynamics of the pair cities. Its location in the historic centre of Ipoh, while providing a multimodal link with road and rail corridors, initially favoured the city's economic rebirth. Today, it acts more as a brake to it, due to lack of space to develop. So, the port has embarked on a new phase of relocation of its logistic and financial functions by creating annexes in the main industrial clusters structuring the urban region. Ipoh's long-term dynamics are nevertheless conditional on its recovery of its position as an intermediary between the two export ports of Kuala Lumpur and Penang, once the political constraints imposed on it are removed. The diversification of governing centres is the only way to enable Ipoh to become an important node connecting the two regions — northern and central — structured by the North-South economic corridor on the West coast of Malaysia.

Second Type: Pair Cities Formed by a Provincial or State Capital and A Border City on the Same Corridor

The second type of pair cities has not been specifically studied in this work, but several contributions refer to it since these cities, as we have seen in the first part, contribute significantly to the spatial organization of the corridors in both the Malacca Straits and the GMS.

Sebastien Colin's contribution showed the small size of cities at the Yunnan border on the main GMS North-South corridor. The growth of Daluo, an agricultural Chinese town of 20,000 inhabitants situates on the Burmese branch, which used to exist thanks to tourists visiting Myanmar, was halted first after the closure of the Mongla casino in 2006 on the other side of the border at the request of the Chinese Government, and then due to the recurrent closure of the border for political reasons, especially since 1989. Mohan, on the Laotian branch of the same corridor, a small town of less than 50,000 inhabitants, benefited from its position opposite the Chinese enclave of Boten on the Laotian border. This branch of the North-South corridor, on the contrary, stayed open, so the town benefited from the logistics of the Chinese border post, and, until the closure of the Boten casino, from tourism linked to gambling. The small size of Daluo and Mohan border towns, compared to that of the above-mentioned cities of Ruili-Wanding (on the Kunming-Rangoon corridor, with nearly 120,000 inhabitants), and Hekou (on the Kunming-Hanoi-Haiphong corridor, with nearly 70,000 inhabitants) could be linked to the splitting of the corridor into two branches and to the weight of Jinghong, the capital of Dai autonomous prefecture (Chinese term designating the Tai), with nearly 350,000 inhabitants and an important tourist centre in Yunnan, relatively close to the Burmese and Laotian borders (about 100 km).

Research of this type would be even more interesting because Elsa Lainé has just completed a thesis on the twin cities situated on either side of the Mekong in the border sections with Myanmar and Laos. Her research seems to confirm the theory regarding the weight of governing centres located behind the border. This theory appears valid for Chiang Rai (70,000 inhabitants in the city and 105,900 in the conurbation), a provincial capital that has become the second largest urban area in the Northern region of Thailand (after Chiang Mai which has nearly 200,000 inhabitants) thanks to its position on the main North-South corridor. Its position is symmetrical with that of Jinghong, at the point where the two branches of the corridor converge, easily dominating the Thai border towns, Mae Sai on the western branch (35,600 inhabitants) and Chiang Khong on the eastern branch of the corridor (25,600 inhabitants). The theory also takes into account the dependence of Mukdahan (44,000 inhabitants) from Khon Kaen (150,000 inhabitants), the Northern Thai capital which possesses the only dry port in the region. Finally, the size of Ubon Ratchathani (128,000 inhabitants), also in the North-East, seems to explain why there is no urbanization, but merely a border post, on the Thai border of the second East-West corridor linking Ubon Ratchathani to Pakse in Laos.

In their chapter, Abdul Rahim Annar, Muszafarshah Mohd Mustafa and Amel Farhat have analysed the governing role of Penang, a second-generation metropolis, straddling the eponymous straits, with a population of 1.3 million. Thanks to its multimodal logistics centre at Butterworth, providing a direct link to the world container shipping route, it dominates the north corridor economic region of the Malaysian peninsula, especially the industrial and technological zones of Alor Setar, the capital of Kedah State (a conurbationof 300,000 inhabitants), and the dry port of the border town of Bukit Kayu Itam which processes containers coming from its own industrial zone as well as those from the southern region of Thailand. Although the latter has a capital, Hat Yai (800,000 inhabitants) and a seaport, Songkhla, the major part of exports from this region transit through Butterworth rather than through the Thai Eastern Seaboard ports, which do not benefit from a direct link to the world shipping route.

Third Type: Pair Cities Located on Either Side of the Malacca Straits

The seaports situated opposite each other on the two shores of the straits are the final type of pair cities. A major difference distinguishes them from the preceding types: their relations are far from exclusive since the straits, as we have seen, enable each to maintain connections with other ports on the opposite shore. This feature characterizes the main freight and passenger ports of Medan, Tanjung Balai and Dumai on the Sumatranese side and their opposite numbers, Penang, Port Klang and Melaka on the Malaysian peninsula, but also the minor ports dealing in barter trade and smuggling, especially Tanjung Balai, Bengkalis and Kuala Tunggal which are the hub of Sumatra, like Muar and Batu Pahat situated opposite the straits (ie. Map 3.4 in Nathalie Fau's chapter).

The transnational dynamics of the 1990s linked to the two growth triangles dominating the entries to the straits and the planning of the three transversal economic corridors in the first decade of the twenty-first century did not favour in equal proportions the changing rhythm of development in the pair cities and ports identified. The pair that works best associates in a strongly hierarchical way the two biggest cities in the north of the straits, Medan and Penang. The federal Malaysian government has provided Penang with the most varied port terminals: a regional container hub linked to the world shipping route, bulk and barter trade terminals serving the main ports of North Sumatra, a port of call for cruise ships on the route between Singapore and Phuket or Krabi. Also, thanks to the presence of Chinese entrepreneurs from Hokkien, trade links between Penang and Medan are long-established and could become stronger as the IMT-GT regional cooperation programme gathers momentum. In a central position on the Medan-Penang-Songkhla corridor, Penang has, as we have seen, benefited from the attraction of container traffic from North Sumatra and from the Northern corridor region of the Malaysian peninsula and Southern Thailand, to become the command centre of the entire corridor.

The Dumai-Melaka pair on the second cross-straits corridor is much less active at present, even though Dumai has the most varied port facilities with the ports situated on the Malaysian shore in the section between Port Klang and Johor, on account of the narrowness of the straits at this point. Passenger traffic predominates thanks to the ferries to Melaka; only exports of hydrocarbons and palm oil are more important than barter trade, which is modest in weight. The pair's development potential nevertheless depends on an increase in Malaysian investments, in both Dumai and Pekanbaru. It is also conditioned by the projected bridge linking the two port cities via the island of Rupat, which would lead to a tunnel effect in the straits comparable to that of the bridges where the corridors cross the Mekong, linking twin border cities on each side.

CONCLUSION: FROM NETWORK CONNECTIVITY TO TRANSNATIONAL INTEGRATION

At the end of this work, it appears clearly that, in spite of the difference between corridors: mainly continental in the GMS and mainly maritime in the Malacca Straits, the spatial patterns produced by the strategy of network creation by development corridors are easily comparable, at all levels studied: that of corridors structured by the hierarchy and position of nodes, and finally, that of the urban development accompanying them.

It is significant that planners give preference to the concept of connectivity to describe the ongoing dynamics of regionalization. This term has the advantage on one hand of translating the priority of the economic factor in the open-ended regionalization process aiming to create optimal conditions for the spatial deployment of firms, and, at another hand, to create a network of dynamic economic areas that are important on a world scale. It should be noted, however, that the transport infrastructures created in the corridors, and the urban development to which they have given rise, involve regional dynamics that go well beyond the original intentions, even though they are not strongly institutionalized, as seen for example for the integration of the Asian Pacific seaboard in the work entitled *Intégration Regionales en Asie Orientale.*[5]

The term of "connectivity" also has the advantage, for political actors, of erasing an important question, that of taking control of the integration process once interconnection between national networks has been achieved. The analysis of the position of command centres, on all spatial scales involved, shows, however, the central importance of this question today. Corridors, and the regionalization networks that they create, are, like territories, stakes of power, both for the countries concerned, as can be seen in the rivalry between Thailand, the two Chinese provinces (Yunnan and Guangxi) and Vietnam, and for regional powers like China, Japan and India, to take the example of the GMS (i.e. Christian Taillard's chapter). This is why, behind the connectivity strategy, there is the involvement of regional integration dynamics which are not only economic, but also political, since States initially played a determining role also in the growth triangle strategy of the Malacca Straits (i.e. Nathalie Fau's chapter related to the Malacca Straits).

The connection between two levels of territorial planning, national or subregional, is strategic for the future of ongoing regionalization. In this respect, the situation is strongly contrasted in Southeast Asia, where Indonesia and Myanmar represent completely opposite situations. Indonesian planning prioritizes national integration, and ignores the regional dimension of the Malacca Straits, with the exception of the enclave of Batam, whereas Myanmar favours integration into the GMS to the detriment of national development, which has been a problem since independence on account of the dissidence of peripheral states controlled by ethnic minorities. Laos is also having difficulty orchestrating its integration into the GMS initiated by the East-West corridors and its national integration supported by the North-South corridor parallel to the Mekong. The Master Plan for the northern region, prepared with Chinese assistance, does not yet combine satisfactorily these two imperatives. Thailand and Vietnam, due to their characteristic centralism, do not have a well-developed regional planning policy, and priority is given to the economic infrastructures of trans-border bridgeheads on transnational corridors so as to obtain comparative advantages in their favour. The two countries that make the best connection between national and regional planning are China and Malaysia. The former has developed an economic projection strategy towards the four countries on its southern border by enhancing the different positions of Yunnan and Guangxi. However, this policy comes up against the fears of hegemony experienced by these partners, as can be seen in the difficulties experienced in setting up the high-speed Kunming-Bangkok railway line via Vientiane. Malaysia has completely assimilated national development and integration into the Malacca Straits, increasing the advantage of the Malaysian peninsula's position at the heart of the straits region. Thanks to its policies and investments, it has been able to challenge Singapore's role as chief regional actor.

Finally, it should be emphasized that the assessment presented here of ongoing regionalization and globalization processes in Southeast Asia should be interpreted with caution within the timescale of the study, and the results observed should be included in a long-term view. During the first two decades of the GMS and Malacca Straits transnational cooperation programmes, the main obstacle to increased flow in the corridors was border crossing. It remains to be seen if this brake will be released, once obstacles are lifted in 2015 with the implementation of the ASEAN free trade agreement (AFTA) and that linking ASEAN and China. It also remains to be seen whether the ADB once more becomes the key agent favouring the interconnection of the high-speed railway lines constructed within national contexts, by resuming, in the third decade of the GMS, the role of facilitator and investor that it played during the first two, which, by setting up regional cooperation for each project, enabled national policies and the imperatives of transnational integration to be harmoniously combined. Therefore further research should be carried out at the beginning of the 2020s to measure the transnational integration process achieved, and provide a better assessment of the impact on the territories and populations concerned.

Notes

1 Although the straits and the isthmus are comparable, they do not have the continental depth separating the two shores which, from this point of view, strongly differentiates them from the Indochinese peninsula.

2 The different scales of urban populations mentioned in this book are approximate since it is difficult to obtain coherent data for all the countries concerned. For small or medium-sized towns, their population, and that of the administrative districts to which they belong, are often confused; for larger cities, illegal migrants are not taken into account, especially for Chinese cities where they are sometimes very numerous.

3 Muriel Charras, "L'Indonésie: le plus grand archipel du monde" [Indonesia: the largest archipelago in the world], *Géographie Universelle*, vol. *Asie du Sud-Est et Océanie*, edited by A. Antheaume et al. (Paris: Bélin-Reclus, 1995), pp. 40–76.

4 High-speed sea passenger shuttles and low-cost flights crossing the straits vary according to demand from one year to the next, or according to the time of year, numbers are especially increased during holiday periods in both countries, when demand for travel is greater.

5 *Intégrations régionales en Asie orientale* [Regional integration in Eastern Asia], edited by Ch. Taillard (Paris: Les Indes Savantes, 2004).

References

Charras, Muriel. "L'Indonésie: le plus grand archipel du monde" [Indonesia: The largest archipelago in the world]. *Géographie Universelle*, vol. *Asie du Sud-Est et Océanie*, edited by A. Antheaume et al. Paris: Bélin-Reclus, 1995.

Taillard, Ch., ed. *Intégrations régionales en Asie orientale* [Regional integration in Eastern Asia]. Paris: Les Indes Savantes, 2004.

20

SUPRANATIONAL, NATIONAL AND LOCAL STAKEHOLDERS IN THE TRANSNATIONAL INTEGRATION PROCESS IN SOUTHEAST ASIA

Nathalie Fau

The second part of this concluding section, after a first part devoted to the main spatial forms of emerging transnational development, deals with the role of the actors in the integration process. The subregions of Southeast Asia (Greater Mekong Subregion, GMS) and the Malacca Straits, intermediate areas between the world system and Nation States, involve various actors intervening at different levels: translational, national or local. "Top-down regionalization", which leaders aim for, mobilizing international or world organizations, comes face to face with "spatial decomposition from the bottom", i.e., on an infra-national scale. The myriad actors involved raise many questions which deserve to be examined in turn. The first question concerns the State's new position: is the State overwhelmed by these flows weaving a network of new territories going beyond the national context, or can it still take the initiative? The second concerns governance: have the new forms of transnational management accompanied the creation of these subregions? The third, which is dependent on the second, questions the connection between the strategies and skills of the different actors: complementarities, indifference or rivalry? The final question, which is central to this book, is whether there are different degrees of involvement between actors in the two subregions — mainland and maritime — in this study.

To deal with these four questions, the second part of the conclusion examines successively the role of supranational, national, and finally local actors in the process of transnational integration, and poses the question of the place occupied by new forms of mobility in the processes studied.

1. TRANSNATIONAL DYNAMICS AND SUPRANATIONAL ACTORS

The Asian Development Bank, a Central Actor in the Continental Section, but a Secondary Figure in the Insular Section

The first observation is the unequal role played by the Asian Development Bank (ADB) in the implementation of the integration process in the GMS and the Malacca Straits region: whereas it is central in the first case, it is highly marginal in the second. Since 1992, the ADB has promoted and accompanied the creation of the GMS in order to favour increased commercial exchanges in the peninsula. Its involvement and commitment have played a central role in making this initiative credible both to the countries involved and to financial backers. On the contrary, the ADB was a late arrival to the IMT-GT project, and it only became involved in 2007, whereas this cooperation zone was created in 1993 following a trilateral agreement between Indonesia, Malaysia and Thailand. It did not play a role either in the creation of the SIJORI growth triangle, which, unlike the preceding project, developed without any formal agreement, spurred on only by the common will of the Singaporean and Indonesian governments. This difference in the ADB's involvement is mainly the result of the different historical paths followed by the mainland and maritime sections. Even though maritime Southeast Asia has lost its former regional consistency, that Robequain (1946) resumed in the term "Malay world" on account of the spatial divisions introduced by the British, Dutch, Spanish and then American colonial powers, exchanges between the two sides of the Malacca Straits have always continued to exist. The period of conflict between Indonesia and Malaysia, which only lasted four years, from 1962 to 1966, did not really interrupt these flows, which sometimes simply changed character, from legality to illegality. On the contrary, the mainland area was fragmented and exchanges interrupted by colonization and decades of war. The ADB's aim in launching the GMS was to re-forge links between the countries of mainland Southeast Asia and benefit from post-Cold War changes in national territorial strategies to promote a regional integration scheme at continental Southeast Asian level.

The second observation is the ADB's mode of operation. In the context of the GMS, it has never behaved as a supranational actor imposing directives on partner States, but rather as a mediator favouring intergovernmental cooperation and pragmatism in order to adapt to economic developments. The ADB does not provide a rigid framework since its role is that of an outside intervening party, advising and playing the role of intermediary between partners and facilitating exchanges. It is present in all phases of the implementation of these cooperation zones. First of all, its aim is to define the fields and potential frameworks of cooperation and the actions to be brought about jointly. It also facilitates the consultative process by organizing consultation meetings at national or supranational level to define the nature of cooperation and define obstacles and sectors that could give rise to joint initiatives. This role of mediator enables disagreements to be resolved and all opinions to be taken into account. Secondly, concrete projects are defined. Finally, the ADB undertakes to find funding and encourage the participation of private international donors and investors. It attracts investments which the participants would not have been able to obtain on their own account. The recommendations made by the ADB in no way prevent each partner from having their own strategic vision of the GMS. Although the ADB has always given prominence to the East-West economic corridor in its strategy, the governments of Thailand, China and Yunnan province give greater priority to the North-South economic corridor which is part of a Sino-Thai economic approach (cf. Danielle Tan). The ADB's choice respects national susceptibilities: although the East-West corridor has been designated as a pilot corridor, it is precisely because by not crossing any large cities, it is not part of national development policies and therefore represents lower stakes and less risk of conflict.

Transnational Management: Nascent in the GMS and Well Defined in the Malacca Straits

This original approach to cooperation promoted by the ADB, which requires no founding agreement between member countries and therefore does not exclude any potential partners, as emphasized by Christian Taillard, encourages the bypassing of political tensions in favour of economic development, but does not allow for any real transnational management of the GMS, which is managed, at best, corridor by corridor in terms of both participating parties and means implemented. There is thus no cooperation agreement on the exploitation of river waters for the river basin as a whole, and countries downstream continue to depend on the goodwill of China, which controls the regions upstream. Its involvement in the GMS encourages the Chinese government to take into account the reactions of its Southern neighbours as can be seen by its decision to stop dredging the upper reaches of the Mekong after the hostile reactions of Thai ecologists and military personnel (cf. Christian Taillard's GMS chapter), but China has nevertheless not stopped building dams on the upper Mekong. In the context of the Mekong Commission, there are, however, institutional agreements preparatory to the transnational management of the middle and lower reaches of the river. The water use charter and the procedures concerning dam building on the main river is a preliminary to a necessary collective agreement with the other countries. In December 2011, the Cambodian and Vietnamese governments requested Lao PDR to postpone the construction of the planned first dam on the middle reaches at Xayaburi so that studies could be carried out on its impact on the main river.[1] The weakness of the Mekong Commission is due to the observer status of China, which refuses to respect the obligations linked to the status of member country (Marie Mellac's chapter).

The transnational management of the Malacca Straits is, in contrast, further advanced. Hampered by article 43 of the 1982 Montego Bay convention, but also by the difficulties of navigating in the Malacca straits, the littoral States (Malaysia, Singapore, Indonesia) have, since 1971, continually improved the transnational management of this maritime area with the creation of infrastructures improving the safety of shipping and the ratification of cooperation agreements (MALSINDO, Eyes in the Sky) aiming to fight together against insecurity at sea (piracy, potential terrorism). A new stage was reached in 2007 with the creation of a cooperative mechanism whose objective is to incite users of the straits (States, but also private companies) to participate financially in the maintenance of safety and security in this area. In this new form of world governance of the management of the Malacca Straits, the neighbouring States continue to take the initiative: each chooses concrete investment projects (Ho 2007, p. 19) (removal of wrecks, replacement of damaged navigational aids, creation of a system of detecting and identifying small craft, etc.) and proposes them to the international community. The Chinese government has thus undertaken to take part in financing the Indonesian project to replace navigational aids destroyed by the 2004 tsunami, the Singaporean measurement centre project and Malaysian projects concerning the creation of reaction capacities for incidents involving harmful or potentially dangerous substances.

Transnational Organizations Strengthening or Weakening ASEAN?

Whether in the GMS or in the growth triangles, the different transnational cooperation zones created are always presented as a means of strengthening ASEAN, and the first stage in the creation of a wide-ranging free trade zone whose limits are gradually going beyond those of ASEAN. In the context of the GMS, the obligation imposed by the ADB to ratify free trade agreements between the partners of each corridor before rebuilding transport infrastructures is a way of anticipating the AFTA. Similarly, the interconnection of electric grids at border crossings in the GMS is taken in charge in the context of the ASEAN 2020 initiative. The recent evolution of the Greater Mekong Region, especially the inclusion of Guangxi in 2004, can also not be dissociated from the official founding in July 2010 of the China-ASEAN free trade zone. In both cases, China has committed itself to multilateral negotiations, whereas until then it had imposed bilateral relations on its neighbours. Sébastien Colin thus shows that the development of cooperation between China and ASEAN in the context of the GMS has always been envisaged by the Chinese government as the first stage of closer ties with Southeast Asian countries. Far from being merely institutional, this link is above all economic: China has moved from sixth largest to first ASEAN partner since the application of the agreements on commercial transactions between China and ASEAN in 2005, while ASEAN has become China's third most important trading partner, ahead of Japan. Also, Marie Mellac rightly emphasizes that the involvement of both China and Vietnam in the GMS has made it possible to "manag[e] bilateral relations in a multilateral context". In spite of political tensions between China and Vietnam, the GMS had given both countries a unique cooperation framework for developing their common cross-border zone. Similarly, Nazery Khalid shows that the construction of inland ports on the West coast of the Malaysian peninsula aims not only to strengthen the cohesion of the Western corridor but also to improve relations with neighbouring countries and ASEAN countries in general, or even with China via the South China-Singapore railway project. The Malaysian sea transport improvement programme also corresponds to ASEAN ambitions to create an "ASEAN Single Shipping Market" in 2015.

The malfunctions of these zones nevertheless show the difficulties of implementing the Charter ratified in December 2008 in which ASEAN partners undertook to set up a single market in 2015, characterized by the free circulation of goods, services, investments and capital between the ten member countries. On numerous corridors in the GMS, the creation of a single customs inspection by a mixed team on each border is having trouble getting started, despite the agreements. Ruth Banomyong thus shows that the weakest link in the corridors is at border crossing points since the ratification of the Cross-Border Transport Agreement (CBTA) for each country does not mean its immediate enforcement. Abdul Rahim also emphasizes that, in spite of the creation of a Joint Development Strategy for Border Areas (JDS) in 2004 and free trade agreements, traffic between Thailand and Malaysia is still hampered by numerous non-tariff barriers: trucks must have both Thai and Malaysian insurance, and also special permission from Kuala Lumpur to enter Malaysia, limits are imposed on imports of rice and fruit from Thailand. Similarly, in the context of the IMT-GT, few legislative tools have been created to facilitate cross-straits flows, and the free circulation of traffic between the two shores is often impeded (cf. Muriel Charras). The advantage of these two cooperation zones is to enable governments to adapt progressively their national legislation, thus anticipating 2015, when the AFTA will come into force, after which date they will no longer be able to resort to dilatory actions.

Finally, we may wonder whether the multiplication of subregional units may not in the long term weaken ASEAN unity. In the mainland area, there are at least five different projects: the GMS, the ASEAN Mekong Basin Development Cooperation, the Golden Square Subregional Cooperation, the Cambodia-Lao PDR-Vietnam Development Triangle, and the Irrawaddy, Chao Phraya and Mekong Economic Strategy (ACMES). Some projects are in direct rivalry with each other. The ACMES and Cambodia- Lao PDR -Vietnam Development Triangle show the rivalry between Thailand and Vietnam to impose their political and economic leadership on Lao PDR and Cambodia (cf. Christian Taillard). Similarly, the integration of Thailand and Myanmar into the Mekong-Ganges programme enables their respective governments to counterbalance the power of China. All these variable-geometry divisions reflect rivalries and power stakes at the level of the GMS.

In the Malacca Straits, the association of the two growth triangles would have favoured the creation of a growth zone including the entirety of both shores of the straits in the same regional space. This transnational redistribution of space would be novel, since the Malacca Straits has never formed a unit of this type. If the two triangles are maintained, it is because of the enduring rivalry between Singapore and Malaysia. The very idea of the Northern growth triangle was Mahathir's desire to compete with the Southern triangle and rebalance Malaysia's growth to the North by strengthening the regional roles of Kuala Lumpur and Penang. Finally, although the ADB has been responsible for both the GMS and the growth triangles in the Malacca Straits, it is surprising to note that the connection between these two transnational projects has never really been considered; at most, a Kunming-Singapore route is envisaged at China's initiative. Nevertheless, the two projects could be in competition with each other; the projected oil and gas pipelines linking the port of Kyaukphyu in Myanmar with Kunming in China are directly intended to short-circuit the hitherto unavoidable passage of the Malacca Straits. Is this opposition between mainland and maritime Southeast Asia still valid today? In her work, Asia's New Regionalism (2008, p. 31), Ellen L. Frost suggests revising this idea:

> In policy circles the word maritime usually refers to the peninsular and insular nations of Southeast Asia, as opposed to those on the mainland. A contemporary atlas, for instance, puts Thailand in mainland Southeast Asia and Malaysia in maritime Southeast Asia even though the two share a long peninsula. In this book, however, maritime Asia is not defined by nation-state boundaries at all. It is the timeless sweep of coastal communities, port cities and towns, and inland trading nodes clustered along ocean-destined rivers not far from the sea. Maritime Asia is where at least 60 to 70 per cent of Asians live.

Transnational Development and Opening of Southeast Asia towards Eastern and Southern Asia

The ever-increasing involvement of powers outside these transnational cooperation zones raises the question of the future of ASEAN in the present context of Asian reconfiguration. The position of Japan is ambiguous. It is not included spatially either in the GMS or the Malacca Straits region, even though it is deeply involved in both projects. It is the greatest investor in the infrastructures of the East-West corridors linking the two seaboards of the Peninsula (cf. Christian Taillard) and the only state outside the neighbouring countries to have contributed financially to the securing of the Malacca Straits since the beginning of the 1960s (cf. Nathalie Fau). The aim of these investments is, however, more to secure Japan's economic future than to favour the creation of a free trade zone that would be extended to ASEAN+3.[2] As Christian Taillard emphasizes, even though Japan's economic interests are numerous within Southeast Asia, its government is lagging behind that of China in developing institutionalized relations with ASEAN. The involvement of Japan in both the GMS and the Malacca Straits enables it to preserve its regional leadership in relation to China until institutionalized agreements are ratified.

Closer ties between ASEAN and India are also tangible in these two transnational cooperation programmes. In 2002, Thailand agreed to extend the GMS East-West corridor towards India in the context of the Mekong-Ganges programme in order to offset its strengthened partnership with China (cf. Christian Taillard). For the same reasons, the ruling junta in Myanmar has increased its bilateral relations with India. But although, as Ella Vignat so rightly says, India "has lost the first rounds and suffered a setback in the energy diplomacy race with China", economic and trade relations between India and Myanmar are developing, especially with the Kaladan Multimodal Transit Transport Project linking the port of Sittwe to the Indian port that should facilitate the transport of goods between Northeast India and Southeast Asia.

Although directly after Independence India was more concerned by its mainland borders, especially with China and Pakistan, the deal changed in the 1990s with the development of India's maritime economy: the country became dependent on hydrocarbon supplies from the Middle East and its economic opening increased its inclusion in the world maritime trade system. Also, Indian maritime strategy changed and the Indian Maritime Doctrine of 2004 affirms the necessity for India to assert itself militarily on the international scene by controlling the shipping routes of the Indian Ocean and its strategic straits but also by deploying its naval force from the Persian Gulf to the South China Sea. India, which thus shows its ambition of becoming a regional maritime power, also considers the Malacca Straits as an extension of the Andaman Sea; it has established a naval presence and modernized military infrastructures in the Andaman and Nicobar Islands situated at the Northern entry to the Malacca Straits (Lawrence and Prabhakar 2009, p. 228). The Indian Maritime Doctrine also recommends the development of naval cooperation between the Indian Navy and navies in the region. In this context, but also in that of the Look East Policy, initiated at the beginning of the 1990s, providing the beginning of closer ties between India and Singapore, and then with ASEAN, the Indian government developed bilateral military cooperation with Southeast Asian countries to patrol the straits. A strategic partnership was signed with Indonesia in 2006 (Ranjit 2010).

These two transnational programmes therefore contribute to redefining the external policies of ASEAN countries by encouraging them to take into account the strategies of the neighbouring Asian superpowers.

2. TRANSNATIONAL DYNAMICS AND STATE ACTORS

Connection between Transnational Projects Promoted by the ADB and National Planning

In view of the ADB's lack of involvement in transnational projects in the Malacca Straits, it is not surprising that the chapters by Amel Farhat, Manuelle Franck and Muriel Charras show a visible deviation between the ADB's plans and Malaysian and Indonesian national planning. These two countries introduce economic corridor projects without really taking into account those proposed by the ADB. In 2010, the Indonesian government launched the MP3EI planning programme (the 2011–15 Master Plan) whose main strategy is to favour the country's development with the help of national economic corridors. Although the definition of the corridors proposed is fairly similar to that set out by the ADB: axes of infrastructures providing links between growth centres (industrial clusters or special economic zones) and attracting investments, their limits, on the contrary, are very different. In the Indonesian planning text there is no mention of the two transversal routes proposed by the ADB to link the two sides of the straits (the Songkhla-Penang-Medan and Dumai-Melaka corridors). The only Straits included are the Sunda Straits. Conversely, the ADB does not take much account of Sumatra's spatial dynamics. There is no suggestion of extending the Melaka-Dumai trans-straits corridor to Padang, which is nevertheless part of the IMT-GT, and maintains increasingly close economic links with the east coast of Sumatra (cf. Manuelle Franck). The only corridor common to both plans is the Sumatran east coast corridor, but, quite apart from the fact that the Indonesian government created it long before the intervention of the ADB, its role and usefulness are perceived in entirely different ways: for the ADB, it is a section of the Asian highway network whose construction is supported by the United nations, and for the Indonesian government it is a means of attaching Sumatra more securely to Jakarta; while one favours transnational dynamics, the other prefers national integration. It is therefore for very different reasons that the ADB and the Indonesian government are both concentrating on the eastern trans-Sumatra, and this consensus should, at least, have the advantage of favouring the development of traffic on this route.

A connection between national plans and those of the ADB is more easily achieved in the mainland areas, especially with the border-zone development policies. Chinese territorial strategies *vis-a-vis* the border regions strongly support those promoted by the ADB for the GMS. They aim at generating economic development in order to raise the peripheral regions, mainly rural and inhabited by ethnic minorities, above the poverty level (cf. Sébastien Colin). The border-opening policy, started in the middle of the 1980s, then the national programme (*Quanguo xingbian fumin xingdong guiha gangyao*), whose aim is to "build prosperous border regions", and the "development of the West" policy of the 2000s thus favour directly the inclusion of Yunnan and Guangxi in the GMS. It should be noted that this connection is not only a one-way process, i.e., with partner States following the directives of the ADB, but is a two-way process thanks to concerted dialogues. Thus, the Thai government did not wait for the ADB to express explicitly its border town development policy to commit itself to this process: Elsa Lainé shows that Thai strategy for developing border towns was included in the "7th National Plan established for 1992–1996" and that Mukdahan was identified as a "border trade point" by the Thai government before being thus designated by the ADB. The only change for Mukdahan since the creation of the GMS corridors is terminological: the city is no longer simply a "border trade point" but has become an "economic gateway". The partner governments of the GMS are therefore able to make proposals that the ADB can then choose whether or not to support. Lao PDR is the country that has the most closely took into account the propositions of the ADB, partly on account of its limited financial resources, but especially because three economic corridors cross its territory. The objective shared by the ADB and the Lao government is to extract Lao PDR from its enclaved position by placing it at the centre of the subregional transport network (cf. Vatthana Pholsena). Also, Vanina Bouté and Danielle Tan emphasize that the Economic Corridors is, for the Lao government, a way of including peripheral regions in national development by orienting them towards a market economy (replacement of food crop cultivation on burnt land by cash crops), to fight against poverty and to keep a tighter control over them (eradication of opium growing, settling of border populations).

However, in spite of these converging points of view between the ADB and the governments of the GMS countries, diverging opinions can also be clearly seen between ADB proposals and national planning. In his chapter, Christian Taillard has shown at least two inconsistencies; one in Vietnam, the other in Lao PDR. In planning second-decade economic corridors (2002–12), the ABD has omitted to include the coastal motorway known as the "Mandarin Road" linking Hanoi to Ho Chi Minh City, yet offers access in the North to Guangxi and in the South to Cambodia. No mention is made either of the Ho Chi Minh motorway which runs along the Lao PDR and Cambodian border. Similarly, the "Master Plan" for the Northern region, prepared in collaboration with the Chinese, emphasizes the Lao government's difficulty to make a connection between regional integration via the East-West corridors and national integration via the North-South corridor. The division of Lao PDR between North and South, reminiscent of the colonial separation between Upper and Lower Lao PDR, does not integrate the added complexity introduced by the GMS corridor grid.

As for Myanmar, its position is an exception since for the Myanmar government, regional integration is more important than national integration (Ella Vignat). To achieve greater international legitimacy and get round sanctions imposed by the West in 1997, the junta in power until November 2011 increased the numbers of bilateral and multilateral cooperation agreements: legalization of border trade with neighbouring countries, inclusion in the GMS in 1992 and in ASEAN in 1997. The projects implemented since the country became open in 1988 have placed Myanmar at the centre of the redesigning of transport systems between Southern Asia, Southeast Asia and China and of regional energy stakes. However, these projects were only considered in second place as a tool for the unity and development of a national territory still dominated by armed conflicts in outlying areas. The North-Western corridor (Kunming-Mandalay-Rangoon) is more significant in the context of China's development policy than that of Myanmar: it enables China to secure its access to the Indian Ocean and vary its energy supply routes, and only the Mandalay-Rangoon section, which is the backbone of the territory controlled by the Myanmar authorities, has any real meaning for Myanmar territorial organization. This strategy of the Myanmar military regime, the SPDC, can only be short-term since it is difficult to conceive that foreign financial partners will continue investing in politically unstable areas.

An Unfinished Decentralization Process Unfavourable to the Autonomy of Border Regions

In theory, most countries in East Asia have, since the 1990s, engaged in a decentralization policy aimed at setting up the form of "good governance" recommended by most cooperation and development agencies. The declared objective is to involve local authorities in decision making and transfer responsibilities of drawing up and managing strategies to the centres of power closest to the local population so that they can take part in decision-making and play a full role as citizens. The other objective is to rebalance economic growth at national territorial level. In the context of the development policies of "national level border economic cooperation zones", in the 1990s the Chinese authorities opted for the decentralization of economic decision-making powers to local authorities in the provinces henceforth able to attract foreign investments, set up economic zones and manage border trade (cf. Sébastien Colin). Many Southeast Asian countries have also engaged in specific development policies for border regions, more dependent than others on changes in national policy. Since the end of the 1980s, the Thai government, supported by the ADB, has committed itself simultaneously to decentralization, development of border regions and economic cooperation with its neighbouring countries (cf. Elsa Lainé). In Vietnam, the first decentralization measures were taken in 1995, in the context of the Program of Administrative Reform (PAR), and were strengthened in the 2001 plan. In her analysis of the border town of Lào Cai, Marie Mellac describes the numerous steps taken by the Vietnamese government so that the local government could take advantage of the national border opening policy: increasing the city limits, redefining its administrative status so it has become a provincial municipality, authorizing the creation of the Lào Cai Border Gate Economic Zone and the Centre for Promotion of Trade, Investment and Tourism to attract Chinese investments, or financing out of national funds the bridge between Lào Cai and Hekou. Indonesia also set out on the route towards decentralization at the beginning of the 2000s, and from a legal point of view, the State is no longer responsible for development at a local level (cf. Muriel Charras).

In practice, whether in former Communist countries, in strongly centralizing states (Indonesia, Thailand), or even in federal states (Malaysia), the central State remains an essential element and ongoing processes of transnational integration, rather than weakening it, emphasize its undeniable weight in all political decision-making. In spite of an official government line supporting greater autonomy for provinces, the decentralization process is still incomplete in most Southeast Asian countries. In Vietnam, underlines Marie Mellac, "like all the others, the border management agenda in Lào Cai province is coherent with that at national level" and the planning system is still strongly centralized and of a "top-down" type;

although bilateral agreements between Lào Cai province and Hong Ha prefecture have been signed, these took place on each occasion after the agreement of the respective national governments. Whatever infranational administrative ranks are involved, they are never able to sign direct agreements with their counterparts on the other side of the border. Finally, most border regions are hampered by insufficient financial resources or because, in point of fact, the State is often the only body able to finance the most extensive infrastructure works (chapters by Muriel Charras, Marie Mellac, Ella Vignat and Manuelle Franck). In fact, even though the Lào Cai government's financial resources have increased with the income derived from the reattribution of land, it is still unable to decide and finance trans-border projects directly: "On the Vietnamese side, the bridge was financed by national funds, like most projects concerning the border" (Marie Mellac). It is true that, in Indonesia, Law No. 25/1999 on tax distribution favoured the return to outlying regions of income derived from the exploitation of natural resources, but the basic unit chosen for decentralization, the district rather than the province, has favoured extremely localized projects (new administrative centres, bus stations) but not structural installations for the province as a whole, whose execution still depends on the decisions of the central power (Manuelle Franck).

This opposition between a publicly displayed openness policy and the small amount of power delegated to border provinces is particularly noticeable in Vietnam and Indonesia. It is surely not a mere coincidence that both countries have experienced a considerable increase in cross-border projects, which are signs not of a real openness policy but rather a strategy of the "divide and conquer" type or a dilution of the effects of openness. The transition from 9 "Border Gate Economic Zones" (BGEZ) in 2000 to 30 projected in 2020 is likely to increase rivalry between border provinces (cf. Marie Mellac). In the same way, the extension of the IMT-GT cooperation zone to the whole of Sumatra has damaged all the specific policies relating to coastal border zones (cf. Muriel Charras). Also, this hiatus between the theory and reality of decentralization is far from being uniform even within each national territory, and it varies, on the contrary, according to the border considered or rather the neighbour involved in the cross-border cooperation process. The Vietnamese regions bordering Lao PDR and Cambodia, strongly supported by the central government, have a greater capacity for manoeuvre than those bordering China, whose cross-border development, on the contrary, is hampered (cf. Marie Mellac).

National and Local Powers; Frequently Different Views of Transnational Integration

After this study, three types of relationships between national and local powers can be clearly distinguished in the management of cross-border areas. The *first type*, the most frequent, is the opposition between the national government, which slows down the cross-border or transnational integration process despite numerous agreements signed, and local governments which, on the contrary, want to strengthen these dynamics. The central government's reticence can be explained either by the development of centrifugal forces in outlying areas of the country, as in the case of Indonesia (cf. Muriel Charras) or the South of Thailand (Abdul Rahim et al.) or by unresolved conflicts between neighbours, as between China and Vietnam (Marie Mellac). This opposition often reflects national and local authorities' very different ideas of the border. Whereas the local government in Lào Cai sees the rise of economic relations with Yunnan as a great opportunity for the province's economic development, the Vietnamese government sees it as a potential threat and a risk for internal security. Although the former wants to increase possible contacts, the latter seeks to enclose these relations in a separate enclave kept well apart from the rest of the city and even create a buffer zone obliging the population to move (cf. Marie Mellac). As for the Indonesian government, even though it is in charge of an archipelago State, its lack of consideration of trans-straits relations is an evident reflection of its incapacity to take into account the specificity of a maritime border. In its planning programmes, the east coast of Sumatra is never considered as a border area and the only part seen as a border region: "an almost terrestrial frontier" in the words of Muriel Charras, is the small island of Batam just a couple of miles from Singapore and the South of Malaysia.

The *second type*, apparently the least problematic, is the convergence of the points of view of national and local governments. This convergence can even favour joint investments: the Chinese section of the North-East corridor is financed by a multipartite agreement between the ADB, the Chinese government and the Yunnan and Guangxi provincial governments (cf. Chritian Taillard) even though in this case the initiative for cross-border development stems from the national government. In Indonesia, the development of the islands of the Riau archipelago, the only part of Indonesia to benefit from a real national trans-border development strategy, is entirely directed, planned and managed by the central power in Jakarta without the intervention of local authorities (Muriel Charras' chapter; Fau 2004). It is officially in the hands of two public actors: the BIDA (Batam

Industrial Development Authority) and the city of Batam. The BIDA is an organization created by the central government and depends entirely on it, whereas the city is dependent on the local provincial government. Decisions are all taken from above via the BIDA and not by the city.

The *third type*, the most original or the least often included in studies on borders, is the opposition between a central power encouraging regional integration and local authorities opposed to it. Only one chapter deals with this subject, that of Ella Vignat who discusses the impact of transnational projects on the development of Shan State in Myanmar. Even though it is difficult to identify accurately the local power of Shan State since "governance in Shan State is characterized by the 'mosaics of power'", this chapter underlines the sometimes violent opposition of the population and ethnic armed groups to the main projects implemented by the Myanmar government in the context of its regional integration policy. Threatening foreign investments is a way for armed groups to put pressure on the Naypyitaw government but also to contest a national policy that is implemented without consulting the local population most directly concerned in the realization of these projects. The simultaneous development of transnational projects in Shan State and increased presence of the Myanmar army, the Tatmadaw, have also fuelled tensions.

3. TRANSNATIONAL DYNAMICS AND LOCAL COMMUNITIES

A Question that is Raised in Different Terms in Mainland and Maritime Areas

The question of the impact of these transnational dynamics on local communities is raised in different terms according to the areas of Southeast Asia studied, either mainland or maritime.

In mainland areas, the chief aspect is mainly linked to infrastructure building (roads and bridges). The impact of economic corridors on local communities is nevertheless difficult to estimate since it is hard to differentiate between the results of the disenclosure process of a previously isolated area and the inclusion of a region in a transnational space. A road undeniably provides the possibility for local populations to integrate a market economy, although this possibility does not always become a reality. Vanina Bouté and Vatthana Pholsena show the multiple factors to be taken into account when analysing the impact of road construction on local society since "there is more to roads than social engineering projects, economic growth or security control" (Vatthana Pholsena). In the second part of her study of the social differences brought about by the construction of National Road 3 in Northern Lao PDR, Vanina Bouté shows that not all populations are able to benefit from the road since pressure on land is increasing and a large sum is needed for initial investment. Only traders and the local urban elite, already well established, have been able to buy rubber plantations and thus benefit from the impact of the North-South economic corridor crossing the province of Luang Namtha, whereas new migrants, farmers from the highlands, have become impoverished. The perception of the road by the local population is also an important, but often neglected factor for understanding these individual strategies. It leads to the migration of peasants from the highlands to the lowlands against all economic reason since the road is a symbol of modernity and development (Vanina Bouté). The perception of the road may also be completely out of phase with that of ADB and government planners, for whom the road is always associated with the idea of progress. For the population of the Sepon district in Savannakhet province in Southern Lao PDR, Vatthana Pholsena shows that Road 9, damaged by years of conflict, is above all associated with memories of past barbarity and atrocities. Its rehabilitation is seen by the local population as a means of reconstruction and reconnection with civilization, and the road has become the symbol of a possible rebirth.

In the maritime area, roads do not need to be built and crossings between the two sides of the Malacca Straits do not follow a single route. What counts is not so much the building of infrastructures as changes in national legislation. Straits are a transversal crossing place that can link, or on the contrary, isolate territories situated on opposite sides: this gateway between two land masses can remain open, closed, or even selectively open. The intensity of migrations between the two shores of the Malacca Straits is thus directly linked to legislative changes made by the Indonesian and Malaysian governments. Since the 1980s, the east coast of Sumatra has existed to the rhythm of the Malaysian government's immigration policy: sometimes it favours immigration, sometimes it halts it, forbids it and condemns it according to the economic situation and the internal labour market in Malaysia. Indonesians, who are sought after in times of full employment and labour shortages, become undesirable at times of economic crisis, and illegal Indonesian immigrants, or those whose contracts have come to an end, are expelled to the east coast of Sumatra (Fau 2000). The Indonesian government has also wavered in regard to free travel of Sumatrans to the other side of the straits: each time they leave the country, Indonesians have to pay a tax equivalent to about 100 euros, except for Sumatran residents who, thanks to the growth triangles, are supposed to be exempt. However, this exemption is very inconsistent, and the government sometimes allows it and sometimes abolishes it without any justification (cf. Muriel Charras). This legislative indecisiveness harms the flow of sea crossings between the two shores. It is surprising to note that although the Indonesian government has encouraged barter trade since 2004 in the Malacca Straits, in 2008 it also decided to decrease the number of ports allowed to import manufactured goods and foodstuffs, leaving only the port of Belawan on the island of Sumatra (cf. Muriel Charras and Nathalie Fau).

Legal Actors One Day, Illegal the Next

The border zones studied, either in mainland or insular Southeast Asia, are almost always "interloper" areas, the word originally designating "small craft which slipped between trading ships in the ports to buy and sell part of their goods in secret" (Brunet 1992, p. 259). According to Ruth Banomyong, who has studied data from Customs and Commerce Agencies, the total amount of informal trade in the GMS region is two or three times greater than that of formal trade. Legal and illegal trade flows cross and mingle and involve some of the same actors, small traders, entrepreneurs, customs officials or fishermen who opt for legality or illegality according to changes in the law, political circumstances or even administrative red tape. This choice may even result from the effect of the size of the firm: Danielle Tan shows that the small Chinese traders of Northern Lao PDR are often illegal, but if their firms grow larger, they prefer to legalize them by cooperating with the Lao PDR authorities. It may also be a direct consequence of national policy. In Shan State, the State monopoly in most economic sectors drives local armed groups to turn to the informal sector for funding: the black market, illegal extraction of natural resources, drug and people trafficking (cf. Ella Vignat). In the Malacca Straits, activities may be declared legal one day, then qualified by the government as illegal the next, according to changes in national legislation. For the smugglers of Riau, involved well before the independence of both countries in trade relations with Singapore, their activities are in no way illegal since they are simply pursuing their role of intermediaries, reflecting the continuity of trade in a changing political situation (cf. Nathalie Fau).

These legal and illegal flows can coexist within the same area, or form networks with different centres and meshes. The "Golden Boten City" casino and the Special Economic Zones in Bokeo in Northern Lao PDR, on the Sino-Thai border, are not only tourist destinations but also important money-laundering, prostitution and drug trafficking centres (cf. Danielle Tan's chapter). On the east coast of Sumatra, Medan attracts legal trade, whereas illegal trade is concentrated in Tanjung Balai and the islands of the Riau archipelago (cf. Nathalie Fau's chapter 1). In the Lao border market of Densavanh, legal goods take the road while illegal goods go by river (cf. Vatthana Pholsena).

Local Populations and the Expansion of the Private Sector

Private actors have not been a driving force in the construction of these transnational areas or in the financing of infrastructures either in the Malacca Straits or in the GMS. Contrary to general belief, the SIJORI growth triangle was not created at the instigation of private investment, but was entirely dependent in its initial stages on a deliberate policy of neighbouring governments (Fau 2010). The initiative for creating SIJORI in 1989 belongs to Goh Chok Tong, the then Singaporean deputy Prime Minister. Also, in the context of the economic development of the Riau islands, it may be noted that the Singaporean and Indonesian governments are at the origin of building and managing infrastructures. The development of Batam island is also entirely directed, planned and managed by the central power in Jakarta without any intervention by the local authorities. On the Singaporean side, the project was designed by the Economic Development Board (ADB), Singapore's economic development agency, and two Singaporean State companies, the Singapore Technologies Industrial Corporation (STC) and Jurong Environmental Engineering (JEE) are responsible for developing and funding all the industrial infrastructures. They apply the same techniques as those used in the development of Singapore's industrial zones. In the context of the GMS, infrastructures are either financed by the ADB, by the States concerned, or by bilateral aid, and most often by a combination of all three (cf. Christian Taillard). Even if one of the clearly announced objectives of the ADB is to create transnational areas primarily motivated by the private sector, this is not yet the case.

The slight involvement of private actors in the institutional development of these areas does not, however, mean their absence. Once the roads and infrastructures are built, they develop their own network whose logic may differ from that promoted by the ADB and national governments. In this way, private actors do not actually use the whole of the East-West corridor, but only certain sections between Bangkok, Hanoi and Ho Chi Minh City (cf. Ruth Banomyong). In the Malacca Straits, the port pair cities (Belawan-Penang or Dumai-Melaka) envisaged by the ADB do not take account of the multiplicity of sea routes taken.

This lack of cooperation between institutional and private actors partly explains the difficulty of assessing the impact of transnational projects on local communities. Many texts in this book describe the eviction of local inhabitants by private companies. In Shan State, multinational firms have taken over the country's natural resources. In Northern Lao PDR, the opening of the border favoured the installation of foreign concessions, mostly Chinese, investing mainly in the agricultural sector (rubber, maize and tea). These plantations have not only replaced dry agriculture or slash-and-burn cultivation but may lead to the change of the region's socio-economic environment. Lao farmers provide land and labour, and Chinese firms capital, technology and market access. Danielle Tan, however, shows that this system of "contract farming" is changing to that of a "concession model": since Lao labour is not sufficient, it begins to be replaced by workers from China. By thus conceding the right to use their land, or by being dispossessed by foreign investors, Laotian farmers have risk to be excluded from any participation in the rural development policy of Northern Lao PDR; they may continue to own the rights to their land, but they may no longer take decisions regarding its use. Following the massive influx of Chinese migrants, Lao traders have also lost their role of middlemen: they have been evicted from the sugar cane trading network and also goods transport networks, and Chinese markets are beginning to supplant Lao ones (Danielle Tan). The North-South economic corridor has certainly favoured the agricultural development of Northern Lao PDR, but this is taking place, at best, with small benefice for the local population, and at worst, to their detriment. The extreme image of this destitution is that of the foreign enclave on Lao PDR territory: Golden Boten is a project designed by Chinese businessmen from Fujian settled in Hong Kong, to attract Chinese tourists, and employing a mainly Chinese workforce (Danielle Tan and Sebastien Colin).

On the East coast of Sumatra, the expansion of oil-palm plantations financed by Malaysian and Singaporean capital has led to a similar expropriation of local farmers (Muriel Charras and Manuelle Franck). Since the middle of the 1990s, Malaysian palm oil producers have decreased land investments in the Malaysian peninsula and are looking for new land. Faced with rising land and labour costs in Malaysia, to remain competitive, planters must find new production areas and specialize in new processing branches derived from the palm oil industry, especially biodiesel. Investors are attracted to Sumatra by the lowest production costs in the world. The development of trans-border cooperation zones enables Malaysia to climb the technological scale and specialize in the oleo-chemical industry and also research and development while Indonesia is confined to the production of raw materials (Fau 2004). This movement towards land appropriation was exacerbated by the 1997 economic crisis and since then has been dominated by Malaysian (Guthrie) and Singaporean conglomerates, Wilmar (Hai 2013). The Wilmar group owns seventeen plantations on the island of Sumatra. Also, massive investment by the Singaporean government has transformed the Riau islands, not according to the needs of local communities but according to the demands of multinational companies based in Singapore. A genuine functional specialization of the Riau archipelago has occurred; the industrial zones of Batam house multinationals relocated from Singapore, the North coast of Bintan has become a seafront covered with luxurious resorts designed to suit the needs of holiday-making businessmen, and the island of Karinum should in due course complete Singaporean shipyards and petrochemical complexes.

Faced with the multiplication of these transnational projects on their own territory, local populations have a greater or lesser capacity to negotiate their future. The two extremes are those presented by Ella Vignat and Vatthana Pholsena. In Shan State, the participation of local populations in the central government's decision-making is completely non-existent, and is translated by numerous human rights violations near transnational development projects: militarization of areas, confiscation of land, removal of the population or even forced labour. Vatthana Pholsena shows, on the contrary, that negotiations are possible between the government and local communities and that they are not always carried out asymmetrically. In the Lao sub-district of Lakô, only a few kilometres from the Vietnamese border, the population may decide to move on the other side of the border if their living conditions have not been improved. This simple alternative, presented by an influential local authority, enabled the population to win their case.

4. TRANSNATIONAL DYNAMICS AND MOBILITY

By concentrating projects along main arteries or in urban centres and by creating new opportunities, the creation of transnational areas has had a direct impact on the reshaping of regional mobility.

Internal Migrations Linked to the Opening of Corridors and Growth Triangles

On a national scale, corridors and growth triangles are new poles of attraction for migrants. Vanina Bouté describes how villagers have moved closer to roads while enclaved villages are gradually disappearing. This strong rural mobility, brought about both by coercive government measures and by the attractiveness of new habitats situated along the roads, has profoundly changed the social composition of plains villages where highland populations are now numerous. Similarly, after its opening in 1991, Hekou's sphere of attraction has grown considerably: although the first Chinese entrepreneurs were from nearby towns in Yunnan and Guangxi, the development of the Hanoi-Kunming corridor attracted traders from Zheijiang, Guangdong, Sichuan and even provinces much further North (Caroline Grillot). In the growth triangle in the South of the Malacca Straits, Batam has become a new centre of attraction for migrants. This island is a unique case, since the Singaporean government has decided to relocate industries not to an already densely populated region, but to an island which in 1980 numbered only 38,663 inhabitants. The success of this project therefore relied from the start on a major transfer of labour and not on the exploitation of a pre-existing labour (Fau 2004). In the space of a few years, Batam has become a very attractive centre. The population of Batam has experienced major growth, and numbers have grown from 38,700 inhabitants in 1980 to 130,000 in 1992, and finally, 439,000 in 2000. This internal migration is sometimes the first step towards crossing the border, as shown by Leonore Lyons and Michele Ford (2007). These authors emphasize this link between internal and international migration using the study on Batam: this island, which has become Indonesia's new Eldorado, attracts numerous migrants, but if they are rejected by the official system due to lack of skills, they either end up in the informal sector or try their luck on the other side of the border.

Concentration of Mobility in Border Zones

Border zones concentrate very different types of mobility which can be distinguished according to duration, rhythm, frequency and reasons for crossing. The border is the limit of a State's sovereignty but also a line of economic, political and social discontinuity encouraging people to cross it.

The opening of borders in Southeast Asia favours short journeys (one or two days). So that Thais can take advantage of the Savannakhet casino in Lao PDR, there are shuttle buses from Mukdahan in Thailand (cf. Elsa Lainé). Every month, thousands of Malaysians cross the border at the weekend to enjoy the lively night-life of Danok in Thailand (cf. Abdul Rahim et al.). Chinese traders coming to Hekou for business are invited by their trading partners to cross the border to take advantage of the restaurants, karaoke bars and casino in Lào Cai (cf. Caroline Grillot).

During the working week, commuters replace trippers and tourists. The migratory flow is generally reversed: the Thai workforce from Sadao district crosses the border every day to work in the Bukit Hitam industrial zone and the Kubang Pasu rubber plantations (cf. Abdul Rahim et al.). The simplification of entry procedures to China has also encouraged a to-and-fro movement of Vietnamese workers to Hekou; they work in the transport and tourism sectors, in the small shops concentrated in the city's Vietnamese market, or as prostitutes. Some decide to move permanently and mix with the new Chinese migrant population, giving Hekou the image of a cosmopolitan city marked by the very rapid renewal of its population. Others, especially Vietnamese women of Hoa origin, take advantage of their cultural proximity to the population on the other side of the border, and play the role of intermediaries between Chinese and Vietnamese traders (cf. Caroline Grillot).

These cross-border migrations are part of a long-term movement, with processes involving both continuity and breaks. In the Malacca Straits, Minangkabau and Melayu populations experienced migratory patterns that became differentiated at the time when the Indonesian and Malaysian Nation States were created (cf. Muriel Charras). In both cases, the straits were crossed without interruption or difficulty until independence. For the Minangkabau, originally from West Sumatra, crossing the straits to Malaysia was simply continuing a movement from the highlands of Sumatra's interior to the North-East coast; it then continued from the West coast of the Malaysian peninsula to the highlands of Negeri Sembilan (Fau 2000). For the Melayus, the move from one shore to the other was favoured by their belonging to a widespread cultural area — the Malay world — where the seas unified populations rather than separating them. The geographical context of the Malay world was both mobile and multiple. The sultanates, on the other hand, came into conflict, united, separated and negotiated new alliances. Consequently the population was extremely mobile and changes of loyalty were frequent: "With its multitude of Rajas, shifting capitals, and migrating subjects, the Malay world undoubtedly presented to the European observer an image of turbulence" (Milner 1982, p. 9). This mobility nevertheless recentred itself on the national territory at the time of independence since, as Muriel Charras observes, the growing difference in economic levels between the two shores did not incite Sumatran Melayus to go more often to Malaysia, but on the contrary, to limit their journeys since the disparities in standards of living became an embarrassment to them. Conversely, the Minangkabau have remained transnational actors and contribute directly to the development of economic networks in the Malacca Straits. Danielle Tan also describes Chinese migration cycles in the North of Thailand, with breaks caused by political changes. She also qualifies the current wave of Chinese migrants to Northern Lao PDR, brought about by the opening of the North-South Economic Corridor, not as a novelty, but as "the rebirth of Chinese communities in Lao PDR".

From Cross-border to Transnational

The chapters presented in this book enable us to analyse the connection between local, national and transnational processes. The socio-economic development of the borders includes these spaces in more global processes, linked to regionalization and globalization. Territorial dynamics around and across borders cover both trans-border and transnational realities; they go outside the context of the border region to become part of much wider areas.

The increase in Sino-Vietnamese marriages analysed by Caroline Grillot in the Chinese city of Hekou enables us to grasp this connection. At the outset, these marriages were linked to the mobility of the local population which developed trans-border trading activities. They are, however, increasingly connected with the social and family networks of the Chinese traders of Hekou which extend as far as Northern China. When they visit their trading partners in Hekou, Chinese traders form relationships, of all kinds, with Vietnamese women. A result of these mixed marriages, which, to use the words of Caroline Grillot "are less the consequences of romance than the pragmatic and conjectural alliance between individuals in the need to develop their business and to improve their material life". These marriages favour the development and extension of transnational networks by associating the family, trade or administrative networks of both partners and "a simple deal between a local Hekou trader and his Vietnamese partner" becomes "a long chain of associates ranging, for instance, from a factory owner in faraway Zheijiang, to a Chinese bamboo bed-mat shop in Hanoi, transiting through a wholesaler in Hekou". More than a simple border crossing place, Hekou also plays the role of connection and bridgehead in the development of economic collaboration between China and Vietnam. The crossborder and transnational networks can thus be juxtaposed and interwoven, not only in a situation of continuity with each other. In the North of Lao PDR, whether in the agricultural sector (rubber) or in trade, Danielle Tan describes the parallel emergence of networks of small Chinese traders who have decided to cross the border, and networks of large entrepreneurs from Yunnan, but also from more distant provinces (Sichuan, Hunan and even coastal China).

In the Malacca Straits, a migratory chain has come into existence, connecting internal and trans-border or transnational migrations. It is particularly visible in Johor State (Fau 2000). This State is traditionally a labour pool for Singapore. Although, in the 1970s, Malaysian migrants were almost always unskilled, this has no longer been the case since the 1980s: attracted by the possibility of jobs that did not exist yet in Malaysia or by high salaries, managers from Johor are migrating in ever-increasing numbers to Singapore. These migrations accentuate the penury of labour in Johor State, especially in the conurbations of Johor Bahru and Pasir Gudang, where Singaporean industries have relocated. Internal migrations from rural areas to the cities have made up for some of these lacks. The lack of manpower thus shifted from the cities to the country. This demographic vacuum, in its turn, has been filled by Indonesian migrants ready to accept difficult, discredited jobs, for a salary double that earned in Indonesia. In the 1980s, this flow of migrants was concentrated on the big oil palm and rubber plantations. In the 1990s, Indonesian migrants became more numerous, and went to the cities to do the "3D" jobs — dirty, difficult and dangerous — abandoned by better-trained Malaysians, in the building and services sector (domestic workers, street vendors).

CONCLUSION: WHAT IS AT STAKE IN TRANSNATIONAL INTEGRATION STRATEGIES?

Whether in mainland or maritime areas, the regional integration process in Southeast Asia is characterized by the State's preponderant role. Despite increased numbers of actors, whether at supra- or infra-national level, the State retains the initiative in building these subregional zones (Greater Mekong Subregion and Malacca Straits Region). The ADB does not make demands, but proposals, so as not to ruffle national susceptibilities, and its directives are far from being systematically followed in national plans; as for local governments, although they have officially acquired greater prerogatives since the passing of decentralization laws, they are often incapable of implementing their own transnational integration strategy.

These subregional zones have even become a new international stake in Southeast Asia, where inter-State relations swing between cooperation and rivalry. China and Japan are both trying to impose their hegemony on the Greater Mekong Subregion, Thailand and Vietnam are using it to assert themselves in relation to their GMS neighbours, and Singapore and Malaysia also use growth triangles to capture for their own benefit, and often to their neighbour's detriment, the advantages of their location on the Malacca Straits. The construction of these transnational zones comes up against the same implementation difficulties as those faced in the building of ASEAN, but they have the advantage of being more flexible and favouring the development of bilateral and multilateral relations in spite of political tensions.

The State's major weight in the building of these transnational areas is seen, not so much in the absence of private and local dynamics as in the lack of connection of economic and spatial strategies between the different actors. Illegal actors, of course, but also the private sector and migrants, build their own transnational spaces with different boundaries and nodes from those officially recognized, since for their development they rely on family, ethnic or company networks, often much older than the official structures. De jure and de facto regional integration are both at work in Southeast Asia but often have difficulty meeting, probably because no appropriate forms of management of these subregional areas have been set up. In Southeast Asia, government rather than governance is the order of the day.

Transnational integration processes in mainland and insular Southeast Asia are, however, not all similar since the stakes are very different. The creation of the Greater Mekong Subregion is a response to almost exclusively regional stakes: its aim is to re-mesh a territory fragmented by colonization followed by decades of war. Also, the dominant actors are Asian, whether States included in the zone, neighbouring States (China and Japan) or supranational actors (ADB). Even though flows and relations in the Malacca Straits region have never been interrupted — they have sometimes changed type, from legality to illegality or vice versa — the stakes are international rather than regional: the straits are seen as an international transport route rather than an internal sea, which is a serious handicap to the construction of this transnational space. The States on its shores, especially Indonesia and Malaysia, are especially afraid of being prevented from exercising their national sovereignty over the Malacca Straits: certainly, the international users of the straits are quick to criticize their capacity to fight piracy or eradicate terrorism in the region. To prove them wrong, these States and international institutions favour investments improving security in the straits and facilitating exchanges between the two shores.

In one way, the creation of alternative routes avoiding the Malaca Straits (the Isthmus of Kra, the land bridge between Kyaukpyu in Myanmar and Kunming in China, or the Arctic sea route) could be a piece of luck for the development of this transnational space; having partly rid themselves of international constraints, States could then concentrate on stakes concerning the straits in their entirety: the protection of mangrove swamps and coasts, the management and regulation of fish stocks or the fight against pollution from land areas.

For these different reasons, the pathways and rhythms of integration of these two transnational areas are drawing further apart. The Malacca Straits, whose regional integration is historically in advance of that of the Greater Mekong Subregion, is not really making any further progress: the absence of institutionalization in this region, the scant involvement of the ADB and the lack of political will of the States concerned, centred more on their national territory than on regional integration, partly explain this slackening in growth. In the mainland area, the building of a transnational space seems, on the contrary, to be speeding up: infrastructures are financed, national legislation is modified and national plans now increasingly take into account this dimension of regional integration. The risk for insular Southeast Asian countries might be a shift of gravity within ASEAN: Indonesia, Malaysia and Singapore are founder members of ASEAN but currently the most dynamic and concrete regional integration processes can be found more in the mainland areas and insular countries could find themselves marginalized.

Notes

1 "Mekong Governments Delay the Xayaburi Dam Pending Further Study", International Rivers, 8 December 2011.

2 The Malacca Straits is a vital route for the Japanese economy, both for its supplies of raw materials and for its exports of manufactured goods to Europe (Fau 2004).

References

Brunet, R., R. Ferras, H. et Théry. "Les mots de la géographie, dictionnaire critique" [Words of geography, critical dictionary]. Montpellier/Paris/Reclus/La Documentation Française, 1992.

Fau, Nathalie. "D'une rive à l'autre: les migrations entre l'Indonésie et la Malaisie dans le détroit de Malacca" [From One Side to the Other: The Migration Between

the Indonesia and Malaysia in the Strait of Malacca]. *Revue Européenne des Migrations Internationales* 16, no. 1 (2000): 151–83.
———. "Enjeux et organisations spatiales du détroit de Malacca" [Issues and Spatial Organizations in the Straits of Malacca]. In *Intégrations régionales en Asie Orientale* [Regional Integration in Oriental Asia], edited by Christian Taillard. Paris: Les Indes Savantes, 2004.
———. "Singapore's Strategy of Regionalization". In *Making the Global City: Singapore from Srivijaya to the Twenty First Century*, edited by J.L. Margolin, Karl Hack and Karine Delaye. Singapore: National University of Singapore Publishing, 2010.
Frost, Ellen L. *Asia's New Regionalism*, 2008. Singapore Press, 2008.
Hai Teoh Cheng. "Malaysian Corporations as Strategic players in Southeast Asia's Palm Oil Industry". In *The Palm Oil Controversy in Southeast Asia, A Transnational Perspective*, edited by Oliver Pye and Jayati Bhattacharta. Singapore: Institute of Southeast Asian Studies, 2012.
Ho, Joshua. "The International Maritime Organisation-Littoral State Meetings on Enhancing the Safety, Security and Environmental Protection of the Straits of Malacca and Singapore". *Maritime Studies*, no. 152 (January/February 2007): 16–21.
Lawrence, S. and W. Prabhakar. "Maritime Security Triangulation of ASEAN-Australia-India: An Indian Perspective". In *ASEAN-India-Australia: Towards Closer Engagement in a New Asia*, edited by William Tow and Chin Kin Wah. Singapore: Institute of Southeast Studies Press, 2009.
Lyons, Leonore T. and Michele Ford. "Where Internal and International Migration Interact: Mobility and the Formation of Multi-ethnic Communities in the Riau Islands Transit Zone". *International Journal on Multicultural Societies* 9, no. 2 (2007): 236–63.
Milner A.C. *Kerajaan: Malay Political Culture on the Eve of Colonial Rule*. Tucson: The University of Arizona Press, 1982.
Ranjit, B. Rai. "India's Perspectives on the Straits of Malacca". In *Proceedings of the 6th MIMA International Conference on the Straits of Malacca: Charting the Future*, edited by Nazerih Khalid. Kuala Lumpur, 23–24 June 2009, 2010.
Robequain, Charles. *Le Monde malais* [The Malay World]. Paris, Payot, 1946.

INDEX

L

www.ingramcontent.com/pod-product-compliance
Lightning Source LLC
LaVergne TN
LVHW052210200926
848481LV00001B/70
9789814517898